TREE ROOT SYSTEMS AND THEIR MYCORRHIZAS

Developments in Plant and Soil Sciences

Volume 7

Also in this series:

1. J. Monteith and C. Webb, eds.,
   Soil Water and Nitrogen in Mediterranean-type Environments.
   1981. ISBN 90-247-2406-6

2. J. C. Brogan, ed.,
   Nitrogen Losses and Surface Run-off from Landspreading of Manures.
   1981. ISBN 90-247-2471-6

3. J. D. Bewley, ed.,
   Nitrogen and Carbon Metabolism.
   1981. ISBN 90-247-2472-4

4. R. Brouwer, I. Gašparíková, J. Kolek and B. C. Loughman, eds.,
   Structure and Function of Plant Roots.
   1981. ISBN 90-247-2510-0

5. Y. R. Dommergues and H. G. Diem, eds.,
   Microbiology and Tropical Soils and Plant Productivity.
   1982. ISBN 90-247-2624-7

6. G. P. Robertson et al., eds.,
   Nitrogen Cycling in Ecosystems of Latin America and the Caribbean.
   1982. ISBN 90-247-2719-7

7. M. R. Sarić and B. C. Loughman, eds.,
   Genetic Aspects of Plant Nutrition.
   1983. ISBN 90-247-2822-3

8. J. R. Freney and J. R. Simpson, eds.,
   Gaseous Losses of Nitrogen from Plant-Soil Systems.
   1983. ISBN 90-247-2820-7

Tree Root Systems and Their Mycorrhizas

Edited by

D. ATKINSON
East Malling Research Station

K. K. S. BHAT
East Malling Research Station
M. P. COUTTS
Forestry Commission

P. A. MASON
Institute of Terrestrial Ecology
D. J. READ
University of Sheffield

First published as *Plant and Soil*, Volume 71 (1983)

1983

MARTINUS NIJHOFF / DR W. JUNK PUBLISHERS
THE HAGUE / BOSTON / LONDON

Proceedings of a Meeting of the International Union of Forestry Research Organisations Working Party on Root Physiology and Symbiosis

Distributors

for the United States and Canada: Kluwer Boston, Inc., 190 Old Derby Street, Hingham, MA 02043, USA
for all other countries: Kluwer Academic Publishers Group, Distribution Center, P.O.Box 322, 3300 AH Dordrecht, The Netherlands

Library of Congress Cataloging in Publication Data

Main entry under title:

Tree root systems and their mycorrhizas.

(Developments in plant and soil sciences ; v. 7)
Includes index.
1. Roots (Botany)--Congresses. 2. Mycorrhiza--Congresses. 3. Trees--Congresses. I. Atkinson, D. II. International Union of Forestry Research Organizations. Working Party on Root Physiology and Symbiosis. III. Series.
QK644.T73 1983 582.16'00428 83-4008
ISBN 90-247-2821-5

ISBN 90-247-2821-5 (this volume)

PRINTED IN THE NETHERLANDS

Contents

D. Atkinson, Tree root systems and their mycorrhizas – Prelude VII

Acknowledgements IX

E. Hacskaylo, Researching the potential of forest tree mycorrhizae 1
F. T. Last, P. A. Mason, J. Wilson and J. W. Deacon, Fine roots and sheathing mycorrhizas: their formation, function and dynamics 9
D. Atkinson, The growth, activity and distribution of the fruit tree root system 23
E. K. S. Nambiar, Root development and configuration in intensively managed radiata pine plantations 37
I. J. Alexander and R. I. Fairley, Effects of N fertilisation on populations of fine roots and mycorrhizas in spruce humus 49
C. G. Langlois, L. Godbout and J. A. Fortin, Seasonal variation of growth and development of the roots of five second-year conifer species in the nursery 55
R. L. Perry, S. D. Lyda and H. H. Bowen, Root distribution of four *Vitis* cultivars 63
R. Fogel, Root turnover and productivity of coniferous forests 75
H. Å. Persson, The distribution and productivity of fine roots in boreal forests 87
A. N. Burdett, D. G. Simpson and C. F. Thompson, Root development and plantation establishment success 103
R. F. Sutton, Root growth capacity: relationship with field root growth and performance in outplanted jack pine and black spruce 111
S. A. Abod and S. Sandi, Effect of restricted watering and its combination with root pruning on root growth capacity, water status and foot reserves of *Pinus caribaea* var. *hondurensis* seedlings 123
A. Riedacker and M. Belgrand, Morphogenesis of root systems of seedlings and cuttings of *Quercus robur* L. 131
K. A. D. MacKenzie, Some aspects of the development of the endodermis and cortex of *Tilia cordata* and *Picea sitchensis* 147
M.P. Coutts and G. J. Lewis, When is the structural root system determined in Sitka spruce? 155
D. C. F. Fayle, Differences between stem and root thickening at their junction in red pine 161
E. R. C. Reynolds, The development of root systems analysed by growth rings 167
M. P. Coutts, Root architecture and tree stability 171
J. D. Deans and E. D. Ford, Modelling root structure and stability 189
V. Gianinazzi-Pearson and S. Gianinazzi, The physiology of vesicular-arbuscular mycorrhizal roots 197
S. Gianinazzi, J. Dexheimer, V. Gianinazzi-Pearson and C. Marx, Role of the host-arbuscule interface in the VA mycorrhizal symbiosis: ultracytological studies of processes involved in phosphate and carbohydrate exchange 211
F. Pons, V. Gianinazzi-Pearson, S. Gianinazzi and J. C. Navetel, Studies of VA mycorrhizae *in vitro*: mycorrhizal synthesis of axenically propagated wild cherry (*Prunus avium* L.) plants 217
F. E. Sanders and N. A. Sheikh, The development of vesicular-arbuscular mycorrhizal infection in plant root systems 223
P. A. Mason, J. Wilson, F. T. Last and C. Walker, The concept of succession in relation to the spread of sheathing mycorrhizal fungi on inoculated tree seedlings growing in unsterile soils 247
J. W. Deacon, S. J. Donaldson and F. T. Last, Sequences and interactions of mycorrhizal fungi on birch 257
L. V. Fleming, Succession of mycorrhizal fungi on birch: infection of seedlings planted around mature trees 263
F. M. Fox, Role of basidiospores as inocula of mycorrhizal fungi of birch 269
J. A. Fortin, Y. Piché and C. Godbout, Methods for synthesizing ectomycorrhizas and their effect on mycorrhizal development 275
G. Berta and P. Bonfante-Fasolo, Apical meristems in mycorrhizal and uninfected roots of *Calluna vulgaris* (L.) Hull 285
H. E. Wilcox, C. S. Yang and K. F. Lo-Buglio, Responses of pine roots to E-strain ectendomycorrhizal fungi 293
D. M. Sylvia, Role of *Laccaria laccata* in protecting primary roots of Douglas-fir from root rot 299
J. Garbaye, First results of research on the competitivity of ectomycorrhizal fungi 303
F. M. Rothwell, E. Hacskaylo and D. Fisher, Ecto- and endomycorrhizal fungus associations with *Quercus imbricaria* L. 309
J. M. Holden, G. W. Thomas and R. M. Jackson, Effect of mycorrhizal inocula on the growth of Sitka spruce seedlings in different soils 313
G. W. Thomas, D. Rogers and R. M. Jackson, Changes in the mycorrhizal status of Sitka spruce following outplanting 319
K. J. Lee and C. D. Koo, Inoculation of pines in a nursery with *Pisolithus tinctorius* and *Thelephora terrestris* in Korea 325
M. J. Daft, The influence of mixed inocula on endomycorrhizal development 331
M. H. Ivory and F. M. Munga, Growth and survival of container-grown *Pinus caribaea* infected with various ectomycorrhizal fungi 339
R. Perrin and J. Garbaye, Influence of ectomycorrhizae on infectivity of *Pythium*-infested soils and substrates 345
P. Cudlín, V. Mejstřík and J. Skoupý, Effect of pesticides on ectomycorrhizae of *Pinus sylvestris* seedlings 353
V. Mejstřík and P. Cudlín, Mycorrhiza in some desert plant species in Algeria 363
H. Peredo, M. Oliva and A. Huber, Environmental factors determining the distribution of *Suillus luteus* fructifications in a *Pinus radiata* grazing-forest plantations 367
K. K. S. Bhat, Nutrient inflows into apple roots 371
R. Habib, Phosphorus absorption (^{32}P) by apple trees under drip irrigation as influenced by the physical properties of the soil 381
D. D. Hook, D. S. DeBell, W. H. McKee Jr. and J. L. Askew, Responses of loblolly pine (mesophyte) and swamp tupelo (hydrophyte) seedlings to soil flooding and phosporus 387
J. Kummerow, Root surface/leaf area ratios in arctic dwarf shrubs 395
J. Tromp, Nutrient reserves in roots of fruit trees, in particular carbohydrates and nitrogen 401
C. P. P. Reid, F. A. Kidd and S. A. Ekwebelam, Nitrogen nutrition, photosynthesis and carbon allocation in ectomycorrhizal pine 415
C. Brownlee, J. A. Duddridge, A. Malibari and D. J. Read, The structure and function of mycelial systems of ectomycorrhizal roots with special reference to their role in forming inter-plant connections and providing pathways for assimilate and water transport 433

C. S. Bledsoe and R. J. Zasoski, Effects of ammonium and nitrate on growth and nitrogen uptake by mycorrhizal Douglas-fir seedlings 445
J. Dighton, Phosphatase production by mycorrhizal fungi 455
J. G. Buwalda, D. P. Stribley and P. B. Tinker, Increased uptake of anions by plants with vesicular-arbuscular mycorrhizas 463
F. Martin, D. Canet, D. Rollin, J. P. Marchal and F. Larker, Phosphorus-31 nuclear magnetic resonance study of polyphosphate metabolism in intact extomycorrhizal fungi 469
G. Cuenca, J. Aranguren and R. Herrera, Root growth and litter decomposition in a coffee plantation under shade trees 477
T. V. St. John, D. C. Coleman and C. P. P. Reid, Growth and spatial distribution of nutrient-absorbing organs: selective exploitation of soil heterogeneity 487
R. Van Den Driessche, Rooting of Sitka spruce cuttings from hedges, and after chilling 495
A. David, M. Faye and M. Rancillac, Influence of auxin and mycorrhizal fungi on the *in vitro* formation and growth of *Pinus pinaster* roots 501
F. E. Sanders, J. G. Buwalda and P. B. Tinker, A note on modelling methods for studies of ectomycorrhizal systems 507
J. E. Jackson, An overview of the meeting: Synthesis – progress and challenges in work on tree root systems and their mycorrhizas 513

Subject index 517

Tree root systems and their mycorrhizas – Prelude

This volume contains the proceedings of a meeting of the International Union of Forestry Research Organisations (IUFRO), working party-Root physiology and symbiosis, on tree root systems and their mycorrhizas held in September 1982.

During recent years there has been a large amount of research on both root physiology and on mycorrhizas but frequently this work is discussed at different meetings. It is often discussed without regard to the fact that most plants are infected with either ecto or endomycorrhizas, that infection may have a major influence on plant nutrition or that mycorrhizas require root growth and so are dependant upon cycles of root growth. This meeting succeeded in bringing together root physiologists and 'mycorrhizasts' to discuss work done on the functioning of tree roots and mycorrhizas under both laboratory and field conditions. Research in these subject areas is active in both forest and fruit trees but again is infrequently the subject of joint discussion. Papers on the whole range of tree types were presented and discussed.

The papers in this volume are a selection from those given as both platform and poster presentations. The selection begins with papers on the development of fruit and forest tree root systems, their distribution through the soil and estimates of underground productivity. Papers go on to consider the growth of the perennial parts of the root system and the role of roots in anchoring trees in the soil both areas of great commercial importance but which have been unfashionable, save at the sylvicultural level, in recent years. The section also contains papers discussing the importance of different types of root systems to tree survival, establishment and early growth.

The second group of papers deals with the physiology, structure and epidemiology of infection of roots with vesicular arbuscular mycorrhizas and studies of sequences of infection of forest trees with ectomycorrhizas and the effects of artificial inoculation on tree growth. Under field conditions mycorrhizas have to infect and function in these presence of other fungi, the effect of these on infection and of mycorrhizas on the effects of pathogenic fungi are discussed.

The growth of both roots and mycorrhizas involves an investment of photosynthate and this and the effects of infection on rates of photosynthesis are detailed as is the physiology of both uninfected and infected roots with respect to nutrient uptake. The role of mycorrhizas in the uptake of substances other than phosphates is discussed.

During the course of the meeting it became clear, that especially for ectomycorrhizas, there was a substantial gap between the information coming from field and laboratory experiments and that needed by those who construct models to simulate the complexities of infection and functioning. To help bridge this gulf a paper was specially prepared for inclusion in this volume, for the benefit of experimenters, on information needed to test and validate mathematical models of both infection and nutrient uptake. It is hoped that this will help experimenters to modify their trials so as to produce information helpful to the needs of modelers. The meeting and the volume conclude with a paper entitled Synthesis which sums up the main thrusts and conclusions of the meeting and which may act as a basis for a future meeting.

With any volume of proceedings inevitably total coverage is uneven; areas the subject of active research are well represented while other areas are sparely covered or absent. However, taken together the volume presents good total coverage of most aspects of current importance with respect to tree root systems and their mycorrhizas. Some papers present slightly, conflicting views of the same subject. No attempt has been made to resolve these differences wich represent real areas of uncertainty and so stress the need for future work. It is hoped that this collection of papers will act as a state of the art position for the next few years.

East Malling Research Station
Maidstone, UK
December, 1982

D. Atkinson

Acknowledgements

A successful conference is the result of much hard work by a large number of people. The obvious organisers are assisted by others who tend not to get their fair share of the credit. It is, therefore, pleasant to have the opportunity to thank all of those who assisted with the IUFRO Meeting on Tree root systems and their mycorrhizas. Prominent among these are Mrs. J M James who supervised all general arrangements and the detailed arrangements at Canterbury and without whom the meeting would not have occurred, Mrs. C Bertram and Ms. A Stratt both of the Forestry Commission who did respectively a substantial amount of essential typing and printed material for the Conference, Mr. D Solway, chief forester Glentress Forest, who arranged the Edinburgh forest visit, forestry staff at Alice Holt and Bedgebury Pinetum and Mr. G C White who assisted greatly with the Kent Visit and Mr. B F Self who arranged the East Malling visit.

In addition to the above – mentioned people we received enormous assistance from so many other staff at East Malling Research Station, Forestry Commission Northern Research Station, Institute of Terrestrial Ecology (Edinburgh Centre) and the University of Sheffield (Botany Department and Printing Unit).

We would also like to thank the very many people who chaired formal sessions or discussions sessions and the University of Edinburgh and Christ Church College, Canterbury who were hosts for the two parts of the Meeting.

The Meeting was run under the auspices of the IUFRO working group. S2-01-13 and we thank the current chairman Dr. A Riedacker and the immediate past chairman Professor E Hacskaylo for their encouragement and advice.

December, 1982 The Editors

Plant and Soil 71, 1–8 (1983). A-01

Researching the potential of forest tree mycorrhizae

E. HACSKAYLO*
USDA-Forest Physiology Laboratory, Beltsville, MD 20705, USA

Key words Forest tree Mycorrhiza

Summary During the last century important ecological and physiological principles in mycorrhizal associations were discovered. Effects of litter extracts on the microflora, nutrient uptake and translocation; hormonal relations, the M factor, and physiogical requirements of ectomycorrhizal fungi and metabolite exchanges are among those of particular significance. Reconsideration of this era of mycorrhizal research is worthy of current consideration since progress particularly in physiology has been slow. To derive the potentially vast benefits from mycorrhizal associations in field applications and to achieve genetic alteration to improve physiological benefits derived from mycorrhizas, fundamental characteristics and biochemical mechanisms in the fungus-root complex must be better understood.

Introduction

Professor Harley[18] in his opening address to the conference on endomycorrhizae at Leeds, noted that 'mycorrhizasts' are thought to spend too much time reviewing their subject. Some of us have done this several times. Today I want to emphasize the current status of some areas of mycorrhizal research with important implications, but which are under-investigated. Many early observations and interpretations based primarily on structure and morphology of ectomycorrhizae were amazingly accurate. Frank[9] and Stahl[37], for example, suggested that nitrogen absorption and translocation to roots by the associated fungi was greater than in non-mycorrhizal roots. Later research[19,25], has verified these concepts. By 1930 many fine descriptions were published encompassing essential morphological and anatomical features of most mycorrhizae. Back in 1917 however, Melin, rather than interpret possible physiological interactions based upon observation, began well-controlled laboratory experiments on ectomycorrhizal fungi and their hosts. For over a half century he and his students established high standards for those working on mycorrhizal physiology. Harley and his students have also had a major impact.

Research on mycorrhizal associations has increased dramatically during the past decade, particularly in applied areas. However, certain important phyological and biochemical problems are not receiving the attention they merit.

* Present address: USDA-Mycology Laboratory, BARC-West, Beltsville, MD 20705, USA

The boundaries of mycorrhizae

An ectomycorrhiza is not limited to the small lateral root with its mantle and Hartig net. A clear example of hyphal extensions into the substrate can be seen in the colonization of anthracite coal wastes[31]. Hyphal strands were often attached to the bases of the ectomycorrhical fungus mantle, and wound along the roots and beyond into the substrate for many centimeters. Other studies[32,33] on the distances mycorrhizal hyphae extend into the soil showed that absorbed phosphate could be translocated 12 cm through soil by hyphal strands of *Rhizopogon roseolus* to the roots of *Pinus radiata*. Concepts of the physiological parameters of mycorrhizal fungi, in association with host roots, should include extramatrical hyphae in mycorrhizal complexes. Until accurate methods are devised to measure the extent of physiologically active mycorrhizal mycelia beyond the mantle, nutrient cycling studies will continue to have serious deficiencies. Harley[17] estimated that 10% of host carbohydrate was utilized by ectomycorrhizal fungi, but this may be too conservative. Mycorrhizal fungi however seem to more than compensate the carbohydrate drain by extensive networks of hyphae that assist in absorption and recycling of mineral nutrients. Ecological studies on nutrient cycling are difficult but promising and some investigators[7,40] are attempting to accurately measure mycorrhizal production and turnover in forests. Such studies should ultimately include the impact of extramatrical hyphae in belowground ecosystems, as more accurate measuring techniques develop.

The establishment of mycorrhizae

Observations of the development of ectomycorrhizae of pines in monoxenic culture suggested that short roots were quickly enveloped by mycelia, as soon as the root apex emerged from its mother root. From a pseudoparenchymatous mantle, hyphae invade the middle lamellae by secreting enzymes that permit invasion of the intercellular regions of the cortex. A compact layer of intercullar hyphae becomes the Hartig net. Recently Nylund[29] suggested that the Hartig net is derived from hyphae that morphologically change into a labyrinthic mode of growth between the cortex cells and that this process *precedes* mantle development. He found little evidence of fungal enzymatic action during cortical penetration. Others[20,41] have suggested enzymatic activity at hyphal tips. Nylund[29] observed plasmodermata in the invaded cortex.

The careful analysis of ectomycorrhizal development by Nylund[29] differs from previously concepts, but suggests the need for continued studies on host/fungus syntheses and the ontogeny of mycorrhizae.

Nitrogen/carbohydrate interactions

Increased uptake of nitrogen by mycorrhizae was recognized early in mycorrhizal studies[9,23,37]. Nitrogen isotopes have been used to demonstrate the uptake and translocation of nitrogen, from several sources[25,26]. Ectomycorrhizal fungi that can utilize both ammonium and nitrate nitrogen have been reported many times but only recently the movement of nitrogen from the soil into host plants via mycorrhizal fungi has been followed systematically. Where the movement of ions is slow, the rate of transfer of ions from the soil into the host plant can be very limited unless the root geometry is extended by mycorrhizal hyphae[3,4]. Again the question of the physical limits of a mycorrhiza are being addressed to quantify the uptake activity of any particular mycorrhizal association.

The importance of the studies on nutrient uptake and carbon supply cannot be over-emphasized. The establishment of mycorrhizae is dependent upon the availability carbon in and around the host roots. If carbon compounds exuded from roots of host plants support metabolic activities of mycorrhizal fungi on the young lateral roots, then colonization will be directly related to the internal carbohydrate status of the young lateral roots[2,16]. This means that light intensity, photoperiod, temperature, availability of nutrients and translocation rates, and other factors affecting internal concentrations and the composition of carbon compounds within the root will have a direct influence upon establishment and maintenance of mycorrhizae. France and Reid have presented a conceptual model for carbon and nitrogen interactions in ectomycorrhizae which indicates the areas that are still unresolved, particularly those related to nitrogen combination and translocation.

Growth factors

The 'M-factor'[24] is an unidentified active principle produced by roots. The stimulus that attracts mycorrhizal fungi to roots and the subsequent orderly invasion of the root cortex has been assumed, at least in part attributable to root exudates. Among these is a substance, possibly the 'M-factor', that stimulates spore germination and mycelial growth in certain ectomycorrhizal fungi[1,10,24].

The ectomycorrhizal fungus *Thelephora terrestris* requires association with host roots to complete its life cycle[15]. Attempts to germinate spores of *T. terrestris* axenically were never successful although spontaneous mycorrhizal infection of pine seedlings in the greenhouse from spores is so common that pines or spruces free of the fungus are rare. Recently Birraux and Fries[1] reported that *T. terrestris* spores germinated in the presence of axenically grown roots of *Pinus sylvestris* and so *in vitro* culture of the mycelia was attained from spores. Attention is refocused on Melin's[24] studies and critical, biochemical research on the exudate identity and activity is needed. If it is the first step in the establishment of and perhaps maintenance of the symbiotic association a stimulus, perhaps the 'M factor' must have a great impact on the survival of mycorrhizal associates.

Aside from the 'M-factor' and its role in establishing and maintaining the mycorrhizal association, there are other compounds, *e.g.* auxins, and cytokinins produced by the fungi, that are physiologically significant. However, again few fundamental facts have been elicited about fungal auxins and cytokinins.

Auxin and cytokinins were detected as products of ectomycorrhizal fungi many years ago[27,34,35], although their roles in establishing and maintaining the symbiotic relationships are unclear. Slankis[36] in a review of hormonal relationships in mycorrhizal development concluded that some hormones must be provided by the fungal associate to maintain a symbiotic relationship with roots. He noted that increased nitrogen concentrations reversed the mycorrhizal state with root apices resuming rapid elongation if the symbiotic relationship was terminated. Slankis attempted to relate this phenomenon to the distribution of fungal auxin at high nitrogen levels, rather than to levels of soluble sugars in roots previously suggested by Bjorkman[2]. There have been other[28,38,39] more limited studies on auxin production by mycorrhizal fungi. Do all ectomycorrhizal fungi produce cytotoxins and auxins? Our observations with soybean callus assays indicate this is difficult to determine (unpublished data). We were unable to demonstrate cytokinin and auxin activity in most of our isolates from ectomycorrhizal fungi. It is possible that compounds are produced at levels too low to detect by these methods or that synthesis occurs only when the fungus is in association with roots. It seems likely however that auxin and cytokinin are produced universally, at least in the ectomycorrhizal fungus-root complex and are essential for establishment and maintainance. They probably are produced in small quantities as secondary metabolites, synthesized from root-originated substrates and present at fairly constant, low concentrations. Characteristics of ectomycorrhizae which might be attributed to these compounds are: (1) reduction of the rate of meristematic activity in the root tip, (2) delay of maturation of the cortex cell walls to permit enzymatic and/or mechanical intrusion into the middle lamella; hence Hartig net formation, (3) delay in maturation of the primary walls; hence elasticity resulting in the isodiametric shape of cell walls, (4) changes in membrane permeability so regulating translocation rates of ions and compounds, (5) increases in lateral root initiation; hence increased absorption area, and (6) delayed senescence of root tissues. These are not proven effects of fungul hormones, but are possible manifestations. We need to understand factors essential for successful establishment and maintenance of ectomycorrhizae as attempts to manipulate the genetic constituents of mycorrhizal fungi will certainly become a reality. Elements required for the symbiotic state will have to be maintained in the altered organism, but only intensive research will clarify which should be protected.

Genetic manipulation

It is somewhat surprising that only limited efforts have been made to modify the genetic constituion of mycorrhizal fungi. Giles and Whitehead[11,12,13,14] reported that spheroplasts of *Azotobacter vinelandii* were successfully fused with those of the ectomycorrhizal fungus *Rhizopogon roseolus*. The bacterium is capable of fixing atmospheric nitrogen and to the objective was to incorporate N-fixing ability into hyphae of *R. roseolus*. Reassociated hyphae seemed able to fix atmospheric nitrogen, as evidenced by acetylene reduction. However there was some tendency for a strain of the fungus to become pathogenic toward *Pinus radiata* when attempts were made to synthesize ectomycorrhizas. I am aware of only two preliminary experiments now in progress in the United States directed toward incorporating N-fixing capabilities into ecto- and endomycorrhizal fungi. Others may be in progress elsewhere. This would be an elegant approach to incorporating NIF gene potential into root systems.

If NIF genes could be incorporated into the roots of agricultural crops, particularly grains, Nitrogen fertilizers could be conserved but yields of agricultural crops, or of trees, on soils of low fertility, increased. Nitrogen fixation occurs in legumes and actinorhizal plants. Separate lateral roots, have mycorrhizas, on the same individual host plants. Greater growth and yields of peanuts, clovers, black locust resulted from endomycorrhizal and N-fixing bacterial associations together rather than from either relationship separately[5,30]. Similar relationships have been noted on actinorhizal plants in combination with either ectomycorrhizal or endomycorrhizal fungi and *Frankia* sp.[6]. Although there are formidable problems involved the already established natural compatibility of mycorrhizal fungi with root tissues of higher green plants should give priority, to modifying mycorrhizal fungi to fix nitrogen, over attempts to modify the genetic constituents of root cells. In addition to root tissuefungus compatibility the transfer of NIF genes or incorporation of spheroplasts or organelles or DNA from bacteria or actinomycetes into true fungi might be more easily attainable than attempting to incorporate the N-fixing capability from those micro-organisms into root tissues. The evolutionary distance is far less among the lower groups. The preference may ultimately be with Frankia DNA incorporated into either ecto- or endomycorrhizal fungi or both since the anaerobic conditions for N-fixation are less stringent compared to nodule bacteria and fungi and actinomycetes are evolutionarily more closely allied. The potential for use of endomycorrhizal fungi appears to be great, particularly in agricultural and forest crops. The wide host range of endomycorrhizal fungi could be a distinct advantage with one species or strain as a mycorrhizal associate for a wide variety of hosts. The impact of these possibilities is self-evident in phosphorous-deficient soils where endo mycorrhizal fungi greatly increase nutrient uptake.

In the future, the knowledge and techniques will exist to alter the genetic constituents of the fungi, not only by incorporating N-fixation, but genetic materials or configurations to provide the best of many attributes. Tolerance to various ranges in temperature, soil pH and moisture levels, among other environmental factors; high efficienc in the uptake of nutrients, particularly N and P; wider ranges of host/fungus compatibility, and even selectivity toward an ectomycorrhizal fungi with edible sporocarps are possible goals. Whether any or all of these objectives can be accomplished by incorporating spheroplasts, organelles, DNA molecules or biochemical alterations of genetic constituents can only be solved by long-range planning and by intensive efforts by multidisiplinary teams.

Conclusion

Many interesting areas of physiological/biochemical research on mycorrhizae remain fertile areas for exploration. To derive the potentially vast benefits from mycorrhizal associations in field applications and to achieve genetic alterations so as to improve physiological benefits fundamental characteristics and biochemical mechanisms need research to a much higher level.

References

1 Birraux D and N Fries 1981 Germination of *Thelephora terrestris* basidiospores. Can. J. Bot. 59, 2062–2069.

2 Bjorkman E 1942 Über die Bedingungen der Mykorrhizabildung bei Kiefer und Fichte. Symb. Bot. Ups. 6, 1–190.

3 Bowen G D 1973 Mineral nutrition of ectomycorrhizae. *In* Ectomycorrhizae: their Ecology and Physiology. Eds. G C Marks and T T Kozlowski. pp 151–205. Academic Press. New York.

4 Bowen G D and S E Smith 1981 The effects of mycorrhizas on nitrogen uptake by plants. *In* Terrestrial Nitrogen Cycles. Eds. F E Clark and T Rosswall. pp 237–274. Ecol. Bull. Stockholm. Vol. 33.

5 Daft M J and A A El-Giahmi 1976 Studies on modulated and mycorrhizal peanuts. Ann. Appl. Biol. 83, 273–276.

6 Daft M J and E Hacskaylo 1976 Arbuscular mycorrhizas in the anthracite and bituminous coal wastes of Pennsylvania. J. Appl. Ecol. 13, 523–531.

7 Fogel R 1980 Mycorrhizae and nutrient cycling in natural forest ecosystems. New Phytol. 86, 199–212.

8 France R C and C P P Reid 1982 Interactions of nitrogen and carbon in the physiology of ectomycorrhizae. Proc. 5th North American Conf. on Mycorrhizae. (*In press*)

9 Frank A B 1894 Die Bedeutung der Mykorrhizapilze für die gemeine Kiefer. Forstwiss. Centralbl. 38, 185–190.

10 Fries N and D Birraux 1980 Spore germination in Hebeloma stimulated by living plant roots. Experimenta 36, 1056–1057.

11 Giles K L and H C M Whitehead 1975 The transfer of nitrogen fixing ability to a eukaryote cell. Cytobios 14, 49–61.

12 Giles K L and H Whitehead 1976 Uptake and continued metabolic activity of Azotobacter within fungal protoplasts. Science 193, 1125–1126.

13 Giles K L and H Whitehead 1977 The localization of introduced Azotobacter cells within the mycelium of a modified mycorrhiza (Rhizopogon) capable of nitrogen fixation. Plant Sci. Lett. 10, 367–372.

14 Giles K L and H C M Whitehead 1977 Reassociation of a modified mycorrhiza with the host plant roots (*Pinus radiata*) and the transfer of acetylene reduction activity. Plant and Soil 48, 431–452.

15 Hacskaylo E 1965 *Thelephora terrestris* and mycorrhizae of Virginia pine. For. Sci. 11, 401–404.

16 Hacskaylo E 1973 Carbohydrate Physiology of Ectomycorrhizae. *In* Ectomycorrhizae: Their Ecology and Physiology. Eds. G C Marks and T T Kozlowski. Ch. 6. pp 207–230. Academic Press, New York.

17 Harley J L 1973 Symbiosis in the Ecosystem. J. Nat. Sci. Council Sri Landa 1, 31–48.

18 Harley J L 1975 Problems of Mycotrophy. *In* Endomycorrhizas Eds. F E Sanders, B Mosse and P B Tinker, pp 1–24. Academic Press, Inc. (London)Ltd.

19 Hatch A B 1937 The physical basis of mycotrophy in Pinus. Black Rock Forest Bull. No. 6.

20 Lindeberg Gosta and Marga Lindeberg 1977 Pectinolytic ability of some mycorrhizal and saprophytic hymenomycetes. Arch Microbiol. 115, 9–12.

21 Melin E 1917 The Mycorrhiza of Coniferous Trees and their Significance to Tree Development on drained Beat Bogs. Uppsala.

22 Melin E 1925 Untersuchungen über die Bedeutung der Baummykorrhiza. Verlag Gustav Fischer, June. 152 p.

23 Melin E 1927 Studien über die Entwicklung der Nadelbaumpflanze in Rahhunus. II. Medd. Statens Skogsforskningsinst. Swed. 23, 433–494.

24 Melin E 1954 Growth factor requirements of mycorrhizal fungi of forest trees. Sven. Bot. Tidskr. 48, 1, 86–94.

25 Melin E and H Nilsson 1952 Transport of labelled nitrogen from an ammonium source to pine seedlings through mycorrhizal mycelium. Sven. Bot. Tidskr. 46, 3–4, 281–285.

26 Melin E and H Nilsson 1953 Transfer of labelled nitrogen from glutamic acid to pine seedlings through the mycelium of *Boletus variegatus* (Sw.) Fr. Nature London 171, 134.

27 Miller Carlos O. 1967 Zeatin and zeatin riboside from a mycorrhizal fungus. Science 157, 1055–1057.

28 Moser M 1959 Beiträge zur Kenntnis der Wuchsstoffbeziehungen im Bereich ectotropher Mycorrhizen I. Arch. Mikrobiol. 34, 251–269.

29 Nylund J 1981 The formation of ectomycorrhiza in conifers: Structural and physiological studies with special reference to the mycobiont, *Piloderme croceum* Erikss & Hjorst. Acta Univ. Ups. 1–34.

30 Powell C L 1976 Mycorrhizal fungi stimulate clover growth in New Zealand hill country soils. Nature London 264, 436–438.

31 Schramm J R 1966 Plant Colonization Studies on Black Wastes from Anthracite Mining in Pennsylvania. Torreya 461 p.

32 Skinner M F and Bowen G D 1974 The uptake and translocation of phosphate by mycelial strands of pine mycorrhizas. Soil Biol. Biochem 6, 53–56.

33 Skinner M F and Bowen G D 1974 The penetration of soil by mycelial strands of ectomycorrhizal Fungi. Soil Biol. Biochem. 6, 57–61.

34 Slankis V 1949 Wirkung von β-Indolylessigsäure auf die dichtomische Verzweigung isolierter Wurzeln von *Pinus sylvestris*. Sven. Bot. Tidskr. 43, 603.

35 Slankis V 1951 Über den Einfluss von β-Indolylessigsäure und andren Wuchstoffen auf das Wachstum von Kiefernwurzeln. I. Symb. Bot. Ups. 11, 1–63.

36 Slankis V 1973 Hormonal relationships in mycorrhizal development. *In* Ectomycorrhizae: Their Ecology and Physiology Eds. G C Marks and T T Kozlowski. pp 231–298. Academic Press. New York.

37 Stahl E 1900 Der Sinn der Mycorrhizenbildung. Jahrb. Wiss. Bot. 34, 539–668.

38 Strzelczyk E, Sitek J M and S Kowalski 1977 Synthesis of auxins from trypotophan and trypotophanprecursors by fungi isolated from mycorrhizae of pine (*Pinus silvestris* L.). Acta Microbiol. Pol. 26, 255–264.

39 Ulrich J M 1960 Effect of mycorrhizal fungi and auxins on root development of sugar pine seedlings (*Pinus lambertiana.* Dougl.) Physiol. Plant. 13, 494–504.

40 Vogt K A, Edmonds R I and Grier C C 1981 Seasonal changes in biomass and vertical distribution of mycorrhizal and fibrous-textured conifer fine roots in 23- and 180-year-old subalpine *Abies amabilis* stands. Can J. For. Res. 11, 223–229.

41 Worthington, S J, Black H D and Coons L B 1981 Entry of *Pisolithus tinctorius* hyphae into *Pinus taeda* roots. Can. J. Bot. 59, 2135–2139.

Plant and Soil 71, 9–21 (1983). A-02

Fine roots and sheathing mycorrhizas: their formation, function and dynamics

F. T. LAST, P. A. MASON, J. WILSON
Institute of Terrestrial Ecology, Bush Estate, Penicuik, Midlothian, EH26 OQB, Scotland

and J. W. DEACON
Microbiology Department, School of Agriculture, West Mains Road, Edinburgh, EH9 3JG, Scotland

Key words Fine roots Forest trees Host-dependence of mycorrhizal fungi Mycorrhizal successions Provenance effects Root biomass Root/shoot ratios Soil moisture Spatial and temporal distribution of roots Structural roots

Summary Many factors affect root development. While considering seasonal and other abiotic influences, the major, genetically controlled differences, attributable to provenance (site of origin), should not be ignored.

Site preparation, involving the development of drainage channels and the inversion of some soil horizons, leads to an increasingly complex milieu for root development with a changed availability of nutrients and the restriction of root growth to largely linear configurations.

Fruitbodies of sheathing mycorrhizal fungi do not occur at random. Their distribution, in time and space, is host-dependent: there is strong evidence of a succession. Together, fruitbody observations and controlled inoculations suggest that there are functional differences between fungi occurring in the early stages of the mycorrhizal succession *e.g.* species of *Hebeloma* and *Laccaria*, and those occurring at a later stage *e.g.* species of *Amanita*. While both groups of fungi readily form sheathing mycorrhizas with tree seedlings in axenic conditions, only early-stage fungi form mycorrhizas with seedlings growing in unsterile soils. The difference seems to be related to the occurrence of other soil microbes; it highlights the need to adopt an epidemiological approach to a three membered biological complex of host, mycorrhizal fungi and other soil microbes.

For the future the functional relationship between fine roots and tree growth, and the effects of changes in the soil environment, should be extended to include mycorrhizas and their external wefts of strands and hyphae.

Introduction

When concentrating on roots and their environments, it is nonetheless advisable to be aware of the inter-relations between root and shoot growth. In the past, interest in sheathing (ecto-) mycorrhizas has centred on the amelioration of P-uptake[5] and its consequent effects on tree survival. Surprisingly, the important set of observations made by Daft and Okusanya[7] have been largely ignored. They found that endomycorrhizal associations with *Glomus macrocarpus* var. *geosporus* had profound effects on the stem structure of tomatoes and petunias. Similarly the stem structure of birch (*Betula* spp.) seedlings was altered when roots were inoculated with the sheathing mycorrhizal

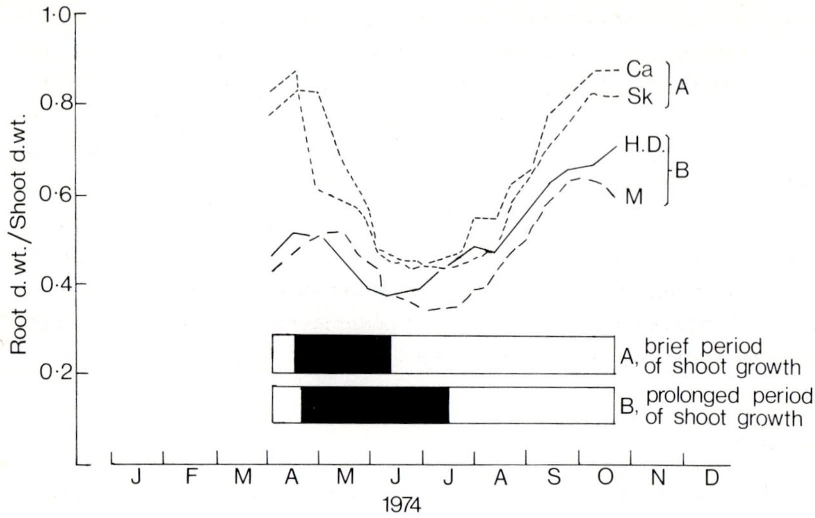

Fig. 1. Changing root : shoot ratios (dry wt.) in the third year after sowing 4 provenances of *Pinus contorta*[2]. (Provenances – Ca, Cascadia; Sk, Skagway; H. D, Hauser Dunes and M, Masset).

fungus *Amanita muscaria* (L. ex Fr.) Pers. ex Hooker[21]. Stem diameters were appreciably increased, the initiation and development of bark was hastened, cortex was composed of a greater proportion of more rounded cells and, additionally, xylem thickening seemed to be more uniform and intense. Our latest observations on Sitka spruce inoculated with an isolate of *Laccaria*

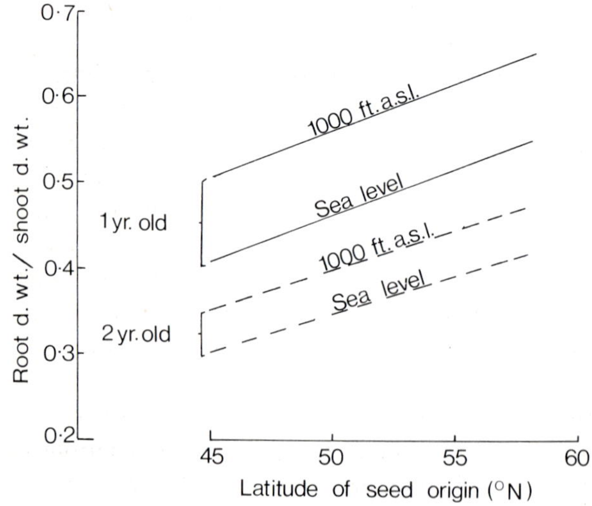

Fig. 2. Association of root : shoot ratios (dry wt.) with latitudes and elevations at the origins of 16 provenances of *Pinus contorta*. Data for first and second seasons after planting[2].

'*laccata*' (accession number 0003) confirm that mycorrhizal fungi can greatly affect stem development although in this instance bark formation was not accelerated by the development of an epidermal phellogen but instead by the premature development of a subepidermal layer of cells containing large concentrations of lipids. Whether or not these changes simply reflect the hastening of normal maturity, they are likely to benefit the early stages of establishment after putting transplants into exposed field conditions. Accelerated bark formation could conceivably protect stems against drying atmospheres whereas the presence of intercellular spaces in the cortex may aid gaseous exchanges so decreasing the effects of damaging gases commonly found in anaerobic waterlogged conditions typical of upland forestry.

Root growth

While it is readily accepted that root growth varies seasonally, too little regard is usually paid to inherent factors. As shown by Cannell and Willett[2], provenance (or site of origin) can have a large effect on the root/shoot ratios of lodgepole pine (*Pinus contorta* Dougl.). When seedlings derived from four different locations in N.W. America were grown side-by-side near Edinburgh, root/shoot ratios were smaller in mid season than either at the beginning or end of the growing season. The extent of the mid-season trough, when root/shoot ratios were minimal, was directly related to periods of active shoot extension (Fig. 1). Thus, for seedlings which had brief periods of shoot growth like those derived from seed collections made at Cascadia and Skagway, the trough was shortlived compared with that of seedlings with prolonged periods of shoot growth, like those from collections made at Hauser Deans and Masset. These data, like those in the horticultural literature concerned with apple rootstocks, show that root/shoot ratios can vary appreciably within a species, presumably reflecting the different environmental selection pressures operating at different locations within the natural range of a species. These differences have been systematically characterised for a range of forest trees but rarely, if ever, for species of 'top-fruit'. There are many reports in the forestry literature showing that the performance of a range of provenances follows a predictably systematic, clinal, pattern. The root/shoot ratios (dry wt.) of northern provenances (55°–60°N) of lodgepole pine were larger than those of southern (45°–50°N) provenances; seedlings derived from collections made about 300 m (1000 ft) above sea level had larger root/shoot ratios, overwinter, than those grown from seed collected at sea level (Fig. 2).

Horticulturists concerned with top-fruit are usually privileged to work with fertile, well drained and uniform sites. In contrast foresters are often confronted with the challenge of extreme conditions, either too dry (with no prospect of irrigation) or too wet. They recognize, as was shown with Monterey pine, *Pinus radiata* D. Don, by Moir and Bachelard[23] that most fine roots grow in the top

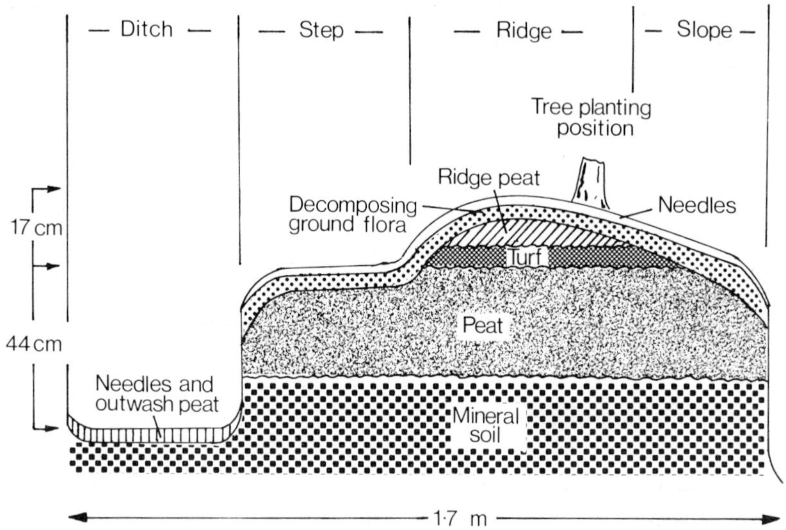

Fig. 3. Arrangement of soils after ploughing a peaty-gley site in southern Scotland for planting with *Picea sitchensis*[11].

15 cm of soil, but until recently the impact of site preparation had not been investigated in detail. In cutting a ditch (by ploughing) when preparing a peaty gley site in the Southern Uplands of Scotland, Ford and Deans[11] identified the formation of 'step', 'ridge' and 'slope'. Within each of these zones there are different soil profiles. In the step region the mineral soil is overlain by peat, decomposing ground vegetation and a needle layer; in the ridge zone the mineral soil is overlain by (i) peat, (ii) decomposing ground vegetation and peat inverted from the ditch, (iii) decomposing ground vegetation and (iv) a needle layer (Fig. 3). The precise details are unimportant for the purpose of this paper; instead it should be recognised that a great many man-made forests are growing in highly structured root environments in which the importance of different strata may change as plantations get older. Sites are commonly prepared to (i) improve soil aeration by decreasing waterlogging and (ii) minimize competition by weeds during establishment. In achieving these objectives the natural concentric growth of roots will be replaced by linear configurations constrained by parallel ditches. Additionally, by inverting and burying the sod when ploughing, the normal pattern of nutrient availability from decomposing vegetation will have been altered with two layers of decomposing vegetation spaced 10 cm apart (vertically) in the ridge zone. When estimating the weights and lengths of fine roots of Sitka spruce, *Picea sitchensis* (Bong.) Carr., Deans[8] found significantly more roots per unit volume in the decomposing layers of vegetation (9.7–19.1 cm ml^{-1}; 2.9–6.9 mg ml^{-1}) than in the ridge peat (9.3 cm ml^{-1}; 2.6 mg ml^{-1}) or needle layer (5.5 cm ml^{-1}; 2.0 mg ml^{-1}) (Table 1). Interestingly, the length per

Table 1. Amounts of fine roots in the different soil zones at a ploughed peaty gley site planted with *Picea sitchensis*[8]

	Zones after ploughing				Undisturbed peat	
	Needles	Decom-position	Ridge peat	Turf	Top 10 cm	Deep peat
Root length (cm ml^{-1})	5.5	19.1	9.3	9.7	4.8	0.7
Root weight (mg ml^{-1})	2.0	6.9	2.6	2.9	1.6	0.3

unit weight of root ranged from 2.33 in deep peat to 3.58 in ridge peat. Based on these data and the known volumes of the different soil horizons, Ford and Deans[11] estimated that 'their' plantation of Sitka spruce had about 68,000 km of fine roots per hectare 11 years after planting, with 290–1500 km ha^{-1} in the mineral soil and needle horizons; 3700–9000 km ha^{-1} in the layers of decomposing vegetation and 14000 km ha^{-1} in the ridge peat (Table 2).

But, it should be remembered, fine roots (< 2 mm diameter) represent only one of three categories of roots, the others measuring 2–5 mm and > 5 mm (structural roots). When sampled 16 years after planting, the fine roots of Sitka spruce accounted for 14% of the total biomass compared with 80% contributed by structural roots. In contrast most (62%) of the annual below ground production was directed to the formation of fine roots, 1.38 of 2.22 kg tree^{-1}, compared with only 37% to structural roots (Table 3). Further, because the annual production of fine roots exceeded the contribution of fine roots to the total biomass, 1.38 compared with 0.93 kg tree^{-1}, the data serve to illustrate that

Table 2. Spatial distribution of fine roots (< 0.5 mm diam.) in the different horizons of a ploughed site planted 11 years previously with *Picea sitchensis*[11]

Soil horizon	Position across the planting unit (see Fig. 3)			
	Ditch	Step	Ridge	Slope
Needles	–	1010	1480	410
Decomposition	3710	5360	9020	3810
Ridge peat	–	–	14380	–
Turf	–	–	7480	–
Undisturbed peat	–	5770	7860	4800
Mineral	1100	580	490	290

Total length of roots = 67500 km ha^{-1}

Lengths of root km ha^{-1}

Table 3. Biomass and annual production of roots of different sizes in a stand of *Picea sitchensis*, 16 years-old[9]

Root diameter (cm)	Biomass		Annual production	
	Kg/ha	Kg/tree	Kg/ha	Kg/tree
0–0.2 cm, fine roots	3530	0.93	5240	1.38
0.2–0.5 cm	1370	0.36	38	0.01
>0.5 cm, structural roots	20100	5.29	3150	0.83
Total	25000	6.58	8428	2.22

there is a continuing process of make-and-break which might profoundly affect the development of mycorrhizas if the fungi involved, also their rhizoplane competitors, respond differently to seasonally fluctuating environmental conditions.

Douglass[10] outlined the seasonally changing distribution of soil moisture associated with a stand, 16 years-old, of loblolly pine (*Pinus taeda* L.) planted with individual trees 6.1 m (20 ft) apart. Early in the year amounts of soil moisture between trees were uniformly distributed but as the growing season advanced amounts near trees (within 1.2 m (4 ft)) were depleted sooner than those at distances of 1.8–3.0 m (6–10 ft) (Fig. 4). Deans found that the roots of

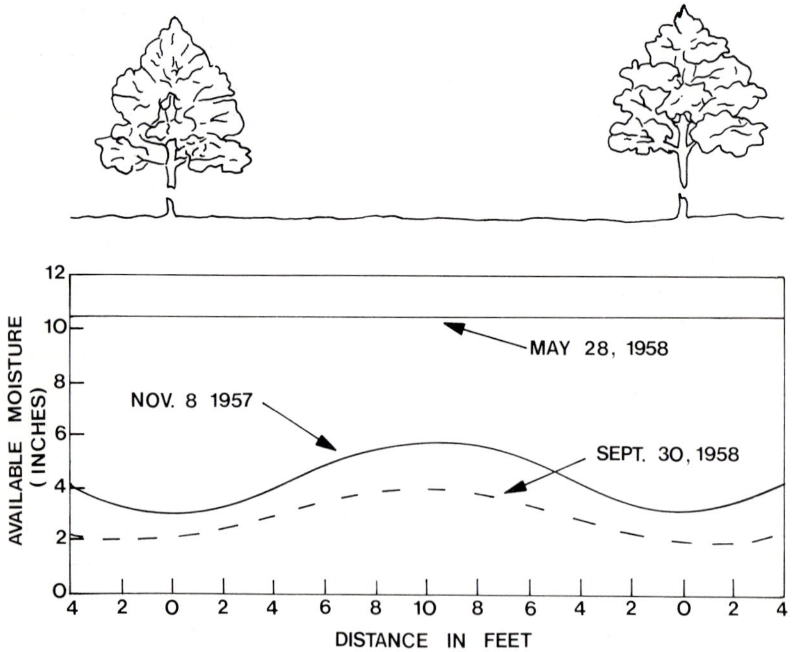

Fig. 4. Distribution of soil moisture at different distances from individual trees in a stand of *Pinus taeda*, 16 years-old[10]. (Available moisture refers to top 4 ft of soil profile).

Sitka spruce were very sensitive to soil moisture deficits, with root growth being detectably decreased by deficits of only −0.2 bars. Although not doubting the accuracy of Dean's observations it might be argued that the functional relationship between soil moisture, fine roots and tree growth should be extended to include assessments of the 'external' wefts of fungal strands and hyphae permeating soil from mycorrhizas, the strands and hyphae often transporting water and nutrients over relatively large distances. Fungi are usually thought to be inactive in dry soils[28] but there is evidence to suggest that some mycorrhizal fungi are better equipped than others to grow on substrates with appreciable moisture deficits[22]. These fungi could conceivably give their hosts a competitive advantage.

Nowadays there is an increasing awareness of the ways in which the chemistry of incident rain is altered as it comes into contact with the foliage of trees. During the International Biological Programme (IBP) many measurements were made of throughfall (including leaf drip) and stemflow. Those series of measurements, mainly concerned with the flow of nutrients, showed that the inputs to soil in an oak (*Quercus petraea* (Mattuschka) Liebl.) woodland were patterned, with concentrations of potassium, calcium and magnesium nutrients being larger, and those of nitrogen and phosphorus being smaller, in stemflow than in throughfall[3]. With the present concern for acid rain the range of assessments has been increased to include SO_4^{2-}, NO_3^- and pH, so reflecting the involvement of atmospheric pollutants, particulate SO_4^{2-} being derived from the oxidation of gaseous SO_2 and particulate NO_3^- from gaseous oxides of nitrogen[15]. Without going into the many details, it is clear that rain, which is probably acidic even in remote areas of industrialised countries, will become increasingly acidic as it courses its way through the foliage of conifers – virtually nothing is known of events in stands of broadleaved trees. Throughfall, collected beneath a stand of Scots pine (*Pinus sylvestris* L.), had a pH of 3.7 compared with pH 4.2 for incident rain, that is an increase of × 3; stemflow had an average (over two years) pH of 3.3, an increase of × 10[24]. How does this acidified rain affect biological activities in soil? Are the effects likely to change, at least in intensity, as stands of trees get older? Evidence from the pollution literature suggests that we should be concerned primarily, but not exclusively, with conifers growing on already acid soils where the incidence of acid rain, and modified acid rain, is unlikely to be countered by widespread and repeated applications of lime. In these circumstances, where soils have aluminium/strong acid buffering systems, acidic inputs will solubilize aluminium, manganese and other elements which may reach phytotoxic concentrations. In Western Scotland the general yield class (GYC) of Sitka spruce was negatively correlated with foliar concentrations of aluminium which in turn were directly related to numbers of non-mycorrhizal 'beaded' roots[14]. Because numbers of 'beaded' roots were inversely related to numbers of functional mycorrhizal roots it seems that large concentrations of aluminium may, as already suggested by Ulrich and his colleagues in the Federal

Republic of Germany[27], deleteriously affect the establishment of mycorrhizas. Accepting that the effects of aluminium, and probably other elements, in soil solution are strongly dependent upon concomitant concentrations of calcium, is it conceivable that acid rain may affect the epidemiology of mycorrhizas even though fruitbody evidence suggests that aluminium, and the associated manganese, are not accumulated in members of the Agaricaceae and Clavariaceae[6,26]? Could it be, that the damaging effects of drought on root growth may, at least in part, be explained in terms of phytotoxic concentrations of aluminium etc. in soil solution? But, a word of caution is needed: beaded roots have also been attributed to phosphorus deficiencies[1] which, however, can be induced by aluminium[4] – a complex situation that needs further investigation.

Mycorrhizal associations

To many microbiologists the idea of working with root colonising fungi is somewhat daunting – you can't see what is happening. But is this really so? A few minutes spent with some of the more ecologically oriented books, used for identifying fungi, will enable the compilation of a considerable list of macro-fungi associated with broadleaved trees[25]. Pegler has identified at least 23 toadstool-producing genera (Table 4): his list could be augmented by including the elf-cups of *Peziza badia*, the earth fans of *Thelephora terrestris* and the false truffles of *Hymenogaster tener*.

Some of these fungi are weak parasites, others decompose litter while a third group form, or are thought to form, sheathing (ecto-) mycorrhizas enabling a two-way interchange, the mycorrhizal fungus aiding tree growth whereas the

Table 4. Genera of fungi that commonly produce mushrooms and toadstools in association with broad-leaved trees[25]

	Agaricus	*	Lactarius
*	Amanita		Lepiota
*	Cantharellus		Lepista
	Clitocybe		Marasmius
	Collybia		Mycena
	Coprinus		Pleurotus
	Entoloma		Pluteus
	Flammulina		Psathyrella
*	Hebeloma		Pseudoclitocybe
	Hygrophorus	*	Russula
*	Inocybe	*	Tricholoma
*	Laccaria		

* Genera with species proven or suspected of being able to form sheathing mycorrhizas with birch (*Betula* spp.)

tree, when in leaf, enables mycorrhizal fungi to produce expanded fruitbodies[18]. By analogy with the succession of microbes colonizing leaves it would be surprising if the roots of young and old trees were equally suitable substrates for all mycorrhizal fungi. In southern India it has been found that numbers of *Amanita muscaria* toadstools are directly related to the age of their *Pinus patula* Schl. and Cham. hosts, the monthly production increasing from 3 per thousand trees in plantations 5 years after planting to 310 associated with trees 16 years-old[16]. There therefore seems to be an element of age-dependency which has subsequently been corroborated by Mason *et al.* who annually recorded the occurrence of fruitbodies developing in association with *Betula pendula* Roth. and *B. pubescens* Ehrh. grown on a brown earth[20]. They identified a sequence in time starting with toadstools of *Hebeloma crustuliniforme* and *Laccaria* sp. in the second year after planting followed by the earthfans of *Thelephora terrestris* and toadstools of *Hebeloma sacchariolens*, *Inocybe lanuginella*, *Lactarius pubescens*

Table 5. Succession of fruitbodies of proven, or suspected, sheathing mycorrhizal fungi appearing in a stand of birches (*Betula* spp.) planted at Bush Estate, near Edinburgh

Years after planting	Fungi
1	Nil
2	*Hebeloma crustuliniforme* (Bull. ex St. Amans) Quélet
	Laccaria sp.
3	*Thelephora terrestris* Ehrenb. ex Fr.
4	*Hebeloma fragilipes* Romagnési
	H. sacchariolens Quélet
	H. mesophaeum (Pers. ex Fr.) Quélet
	Inocybe lanuginella (Schroet.) Konrad and Maublanc
	Lactarius pubescens (Fr. ex Krombh.) Fr.
6	*Cortinarius* sp.
	Hebeloma leucosarx P. D. Orton
	Hymenogaster tener Berkeley and Broom
	Inocybe petiginosa (Fr. ex Fr.) Gillet
	Leccinum roseofracta Watling
	L. scabrum
	L. versipelle (Fries and Hök) Snell
	Peziza badia Persoon ex Mérat
	Ramaria sp.
7	Other *Cortinarius* spp.
	Other *Hebeloma* spp.
	Lactarius glyciosmus (Fr. ex Fr.) Fr.
	Leccinum subleucophaeum Dick and Snell
10	*Hebeloma vaccinum* Romagnési
	Russula betularum Hora
	R. grisea (Pers. ex Secr.) Fr.
	R. versicolor J. Schaeff.

with species of *Cortinarius* and *Leccinum* in year 6 and of *Russula* in year 10 (Table 5). As yet species of *Amanita* (notably *A. fulva* and *A. muscaria*) and *Tricholoma*, which are considered to be characteristic of birchwoods, have not appeared.

Toadstools of mycorrhizal fungi do not occur at random, a conclusion that is as applicable to their spatial distribution as it is to their occurrence in time[12,20]. With some exceptions, notably *Laccaria* spp., fruitbodies of mycorrhizal fungi tend to occur in rings, whose radii, like those of fairy ring fungi in grasslands, increase from year to year. Thus there seem to be successive waves of fungi with the first to appear, *Hebeloma crustuliniforme*, progressively moving outwards and being substituted near the tree by *Inocybe lanuginella*, which in turn moves out to be replaced by *Lactarius pubescens*, to be followed at a later stage by *Leccinum* spp. and so on. Thus around a tree 6–8 years-old it is often possible to identify the elements of 3 or 4 rings with the first species in the temporal

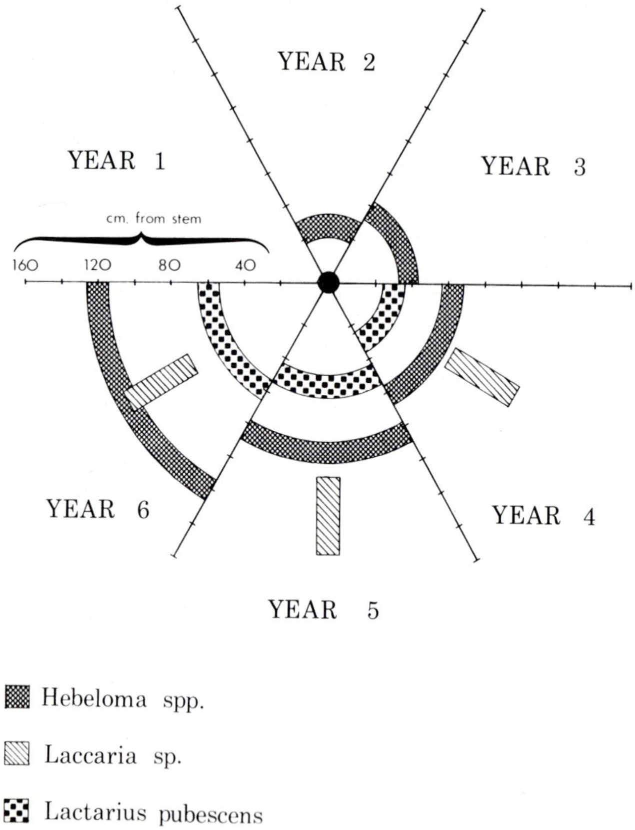

Fig. 5. Mean spatial distribution around the bases (O) of silver birch trees (*Betula pendula*) of fruitbodies of sheathing mycorrhizal fungi appearing during the first six years after planting.

succession being furthest from the stem and the most recent species in the temporal succession being closest to the stem (Fig. 5).

What does all this imply? As Harper and Webster have warned, the temporal sequence of fungal fruitbodies appearing on different substrates may have nothing to do with a phased sequence of colonisation but may instead reflect the period of time necessary to accumulate the resources to enable fruiting[13]. Caution is therefore required. However because Hebeloma-type mycorrhizas are found below *Hebeloma crustuliniforme* toadstools, Lactarius-type mycorrhizas beneath *Lactarius pubescens* toadstools and Leccinum-type mycorrhizas near fruitbodies of *Leccinum versipelle* and *L. roseofracta*, it seems reasonable to suggest, even though other types of mycorrhizas may be present, that there is a mycorrhizal succession. Early-stage fungi such as *Hebeloma crustuliniforme* and *Inocybe lanuginella* colonize the roots of young trees and the peripheral roots of older trees whereas the appearance of late- or later-stage fungi, such as species of *Leccinum* and *Russula*, is delayed and confined to the space bounded externally by the rings of early-stage fungi. But as is already known early- (*Hebeloma* and *Laccaria*) and late- (*Amanita* and *Leccinum*) stage fungi seem to form mycorrhizas with equal facility on tree seedlings grown in axenic conditions[19]. But what happens when seedlings with mycorrhizas established in axenic conditions are transplanted into unsterile field soils?

Seedlings of *Betula pendula*, from surface sterilised seeds, were grown in a

Table 6. Percentages of root fragments with sheathing mycorrhizas after inoculating seedlings of *Betula pendula* during propagation with either *Hebeloma sacchariolens* or *Amanita muscaria*[17]

	% of root fragments with mycorrhizas attributable to	
	Inocula used during propagation	Fungi other than those used during propagation
A. Observed at end of propagation in sterile conditions and immediately before transplanting to unsterile soils		
Inoculated with:		
Hebeloma sacchariolens	100%	Nil
Amanita muscaria	100%	Nil
B. Observed 121 days after transplanting to unsterile soils		
Inoculated with:		
Hebeloma sacchariolens	100%	Nil
Amanita muscaria	Nil	100%

sterilised vermiculite/peat substrate inoculated with either *Hebeloma sacchariolens* (an early-stage fungus) or *Amanita muscaria* (a late-stage fungus). After 65 days propagation some of the replicate seedlings were destructively sampled whereas the remainder were transplanted into pots with a variety of unsterilised soils and harvested 121 days later. As was expected, all fungi formed mycorrhizas with equal facility in the axenic conditions prevailing during propagation (Table 6). In contrast none of the mycorrhizas found at the end of the experiment on seedlings initially propagated with *Amanita muscaria* could be ascribed to that fungus whereas all of those found on seedlings inoculated during propagation with *Hebeloma sacchariolens* were attributable to that fungus. It seems, therefore, that there may be a functional difference between early- and late-stage mycorrhizal fungi, the latter unlike the former being unable to overcome the 'barriers' imposed by rhizophere and rhizoplane microbes. These observations and deductions highlight the need for an epidemiological approach to the study of mycorrhizas if the concept of *succession* is to be established beyond doubt[17]. It is desirable to consider, as with destructive root pathogens, a three membered biological complex of host, mycorrhizal fungi and other microbes – a system in which the effects of the abiotic environment on (i) the establishment of mycorrhizas and (ii) the permeation of soil by mycorrhizal hyphae and strands, should be investigated.

References

1 Beslowd T, Hackskaylo E and Melhuish J H 1970 Effect of environment on beaded root development. Bull. Torrey Bot. Club 97, 248–252.

2 Cannell M G R and Willet S C 1976 Shoot growth phenology, dry matter distribution and root : shoot ratios of provenances of *Populus trichocarpa*, *Picea sitchensis* and *Pinus contorta* growing in Scotland. Silvae Genet. 25, 49–59.

3 Carlisle A, Brown A H F and White E J 1967 The nutrient content of tree stem flow and ground flora litter and leachates in a sessile oak (*Quercus petraea*) woodland. J. Ecol. 55, 615–627.

4 Clarkson D T 1969 Metabolic aspects of aluminium toxicity and some possible mechanisms for resistance. *In* Ecological Aspects of Mineral Nutrition of Plants. Ed. I H Rorison. Blackwell Scientific Press, Oxford pp 381–397.

5 Clarkson D T and Hanson J B 1980 The mineral nutrition of higher plants. Annu. Rev. Plant Physiol. 31, 239–98.

6 Cromack K, Todd R L and Monk C D 1975 Patterns of basidiomycete nutrient accumulation in conifer and deciduous forest litter. Soil Biol. Biochem. 7, 265–268.

7 Daft M J and Okusanya B O 1973 Effect of *Endogone* mycorrhiza on plant growth VI Influence of infection on the anatomy and reproductive development in four hosts. New Phytol. 72, 1333–1339.

8 Deans J D 1979 Fluctuations of the soil environment and fine root growth in a young Sitka spruce plantation. Plant and Soil 52, 195–205.

9 Deans J D 1981 Dynamics of coarse root production in a young plantation of *Picea sitchensis*. Forestry 54, 139–155.

10 Douglass J E 1960 Soil moisture distribution between trees in a thinned loblolly pine plantation. J. For. 58, 221–222.

11 Ford E D and Deans J D 1977 Growth of a Sitka spruce plantation: spatial distribution and seasonal fluctuations of lengths, weights and carbohydrate concentrations of fine roots. Plant and Soil 47, 463–485.

12 Ford E D, Mason P A and Pelham J 1980 Spatial patterns of sporophore distribution around a young birch tree in three successive years. Trans. Br. Mycol. Soc. 75, 287–296.

13 Harper J E and Webster J 1964 An experimental analysis of the coprophilous fungus succession. Trans. Br. Mycol. Soc. 47, 511–530.

14 James H, Court M N, Macleod D A and Parsons J W 1978 Relationships between growth of Sitka spruce (*Picea sitchensis*), soil factors and mycorrhizal activity on basaltic soils in Western Scotland. Forestry 51, 105–119.

15 Last F T 1982 Effects of atmospheric sulphur compounds on natural and man-made terrestrial and aquatic ecosystems. Agric. Environ. 7, 299–381.

16 Last F T, Mason P A, Smith R, Pelham J, Bhoja Shetty K A and Mahmood Hussain A M 1981 Factors affecting the production of fruitbodies of *Amanita muscaria* in plantations of *Pinus patula*. Proc. Indian Acad. Sci. (Plant Science) 90, 91–98.

17 Last F T, Mason P A, Wilson J and Deacon J 1983 Succession: its relevance when selecting sheathing mycorrhizal fungi for inoculating young trees. Nature London, *in litt.*

18 Last F T, Pelham J, Mason P A and Ingleby K 1979 Influence of leaves on sporophore production by fungi forming sheathing mycorrhizas with *Betula* spp. Nature London 280, 168–169.

19 Mason P A 1975 The genetics of mycorrhizal associations between *Amanita muscaria* and *Betula verrucosa*. *In* The Development and Function of Roots. Eds. J G Torrey and D T Clarkson. Academic Press, London pp 567–574.

20 Mason P A, Last F T, Pelham J and Ingleby K 1982 Ecology of some fungi associated with an ageing stand of birches (*Betula pendula* and *B. pubescens*). For. Ecol. Manag. 4, 19–39.

21 Mason P A, Pelham J and Last F T 1977 Stem anatomy and sheathing mycorrhizas in the *Betula verrucosa* – *Amanita muscaria* relationship. Nature London 265, 334–335.

22 Mexal J and Reid C P P 1973 The growth of selected mycorrhizal fungi in response to induced water stress. Can. J. Bot. 51, 1579–1588.

23 Moir W H and Bachelard E P 1969 Distribution of fine roots in three *Pinus radiata* plantations near Canberra, Australia. Ecology 50, 658–662.

24 Nicholson I A, Cape N, Fowler D, Kinnaird J W and Paterson I S 1980 Effects of a Scots pine (*Pinus sylvestris* L.) canopy on the chemical composition and deposition pattern of precipitation. *In* Ecological impact of acid precipitation. Eds. D Drabløs and A Tollan. Proceedings of International Conference, Sandefjord. SNSF Project, NISK, 1432 Ås-NLH, Norway pp 148–149.

25 Pegler D N 1981 The Mitchell Beazley Pocket Guide to Mushrooms and Toadstools. Mitchell Beazley Publishers, London 168 p.

26 Tyler G 1982 Accumulation and exclusion of metals in *Collybia peronata* and *Amanita rubescens*. Trans. Br. Mycol. Soc. 79, 239–245.

27 Ulrich B 1980 Die Wälder in Mitteleuropa: Messergebnisse ihrer Umweltbelastung, Theorie ihrer Gefährdung, Prognose ihrer Entwicklung. Allg. Forstz. 44, 1–5.

28 Waid J S 1960 The growth of fungi in soil. *In* The Ecology of Soil Fungi. Eds. D Parkinson and J S Waid. University Press, Liverpool pp 55–75.

Plant and Soil 71, 23–35 (1983). A-03

The growth, activity and distribution of the fruit tree root system

D. ATKINSON
Pomology Department, East Malling Research Station, Maidstone, Kent ME19 6BJ, UK

Key words Apple Mycorrhizas Phosphorus Root activity Root growth Water

Summary The paper reviews information, much of it obtained from studies using the East Malling root observation laboratories, on the growth and development of the fruit tree root system. The production of new white root varies from year-to-year, generally being highest in the early years. As trees age, woody roots constitute an increasing fraction of total root length although the contribution made by new root growth to the total root length of established trees is also affected by soil management, being higher for trees under grass than under herbicide. Soil management also affects the balance of short (lateral) to long (extension) roots; under grass there are more lateral roots.

Calculation of the rate of water uptake per unit root length needed at various times in the year to meet transpirational demand, suggests that woody roots, which recent experimental work has shown to be capable of absorbing water, must be responsible for much of total water supply.

Measurements of VA mycorrhizal infection in field-grown trees indicated, for part of the season, higher per cent infection in trees grown under irrigated grass than under herbicide management. It is suggested that this, which is associated with raised leaf phosphorus levels, may be due at least partly to higher numbers of lateral roots, the root type which becomes infected. The growth and functioning of the root system under field conditions depend upon the production and integration of a range of root types.

Introduction

The root systems of trees, including those of fruit trees, differ from annual plants by their perenniality. Trees develop a framework of roots ramified through the soil volume. As a result later-formed roots can be initiated throughout the soil; the volume being exploited without the necessity for further growth from the plant base. The root system, at almost any time, consists of a mixture of roots of different types and ages[2]. The balance within the total length between these types varies with tree species, tree age, between seasons and with soil and management treatments. All of these factors can also affect the development of the root system and the need for new growth. New root growth can be most easily followed using *in situ* observation methods. Data on changes in new root production, its significance and its fate, are discussed here.

Recent experimental work has shown that roots of all ages can absorb both water and mineral nutrients[9,22]. Uptake is not restricted to white newly-produced roots. To ascertain under what circumstances older roots will, under field conditions, function in absorption, information on the relative dynamics of white and brown root production and available lengths is needed. Calculations relevant to this are presented.

Under field conditions most tree root systems are infected with mycorrhizas;

usually VA mycorrhizas. Infection depends upon both the tree root system and the fungus. Less is known of the role of the former. Within fruit tree root systems only thin lateral roots with limited growth-potential seen to become infected. Their production is influenced by growing conditions, *e.g.* soil management. Root characteristics and levels of infection found under field conditions are discussed. The above aspects are discussed together in relation to the functioning of the integrated fruit tree root system.

Materials and methods

The paper reviews data from a number of experimental plantings for which relevant experimental details and references to published descriptions are given, together with the appropriate results.

Lengths of both white and brown roots were assessed using an intersection method[13] to record root lengths visible in the observation panels of a root laboratory[18]. Lengths of woody roots, in field plots, were calculated from profile wall distributions[5], assuming that the density of roots revealed at the wall represented that throughout the soil volume to the stated depth.

Mycorrhizal infection was scored as the per cent of 2 mm lengths of root infected, usually containing either hyphae or vesicles, in a sample of roots stained with cotton blue[17].

Results and discussion

The growth and development of the root system

Structural root production In a root laboratory experiment[3], where trees of Cox and Golden Delicious/M.9 were planted in 1970 at a range of densities with $0.09\,m^2$, $0.36\,m^2$, $1.44\,m^2$ and $5.76\,m^2$ soil surface per tree (orchard 1), the length of white root was recorded at weekly intervals and its survival as either brown, suberized or woody root annually[4]. Data for trees at the $0.36\,m^2$ spacing are shown in Figure 1. All root appears initially as white root which, after 3–4 weeks during most of the growing season, turns brown. Some brown roots disappear rapidly, as a result of animal activity[5] while others remain in the soil until the next season and a proportion develop secondary tissues to become woody permanent members of the root system. At the beginning of this trial all the root present against the observation windows was white. As a result of new growth and of survival, by the end of 1970, there was about 2 m per $1.2\,m^2$ observation window of older root, brown + woody <2 mm diameter. During 1971 some of this root, about 20%, increased in diameter to make up the >2 mm diameter fraction while others either increased less in diameter, therefore remaining <2 mm, or died. The <2 mm fraction was increased as a result of the new growth which occurred in 1971. New growth was less in 1972 and 1973 than previously, because the trees began to crop. The length of brown, <2 mm diameter, root decreased although that >2 mm continued to increase. The period during the year when the maximum length of root was present was always while the trees were in full leaf although white root production was usually minimal during the period of active shoot growth. Although root length varied during the season and between years, the maximum root length was usually about 12 cm cm^{-2} soil surface. The significance of this is discussed later in relation to root activity. New root production can, as in 1971, have a large impact on the total length available for absorption.

Annual variation in root production Variation in new root production from year-to-year was studied in a trial involving trees of Worcester/MM.104 planted, in 1961, adjacent to a root laboratory at a spacing of 3.6 × 5 m (orchard 2). Data for two trees, both grown with a sown S50 *Phleum nodosum* sward, are shown in Table 1 for the period 1961–1971. New growth was at a maximum either in the year of planting or early in the life of the tree. As for orchard 1 (Fig. 1), growth was reduced after the trees began to crop, from 1969 onwards. With tree 2 the best year subsequent to planting (1962) produced 54% of initial growth. For tree 1, growth was 110% in 1962 and 95% in 1966 relative to initial growth. Averaged over the period 1961–71 growth was similar for the two trees, 79 and 73 cm $window^{-1}$ respectively, although in 5 of the 10 years their recorded growth was very different. The minimum length of white root produced during the period when the trees were in leaf was, for both trees, similar and low. In many years there were periods when no new growth occurred during the time when the trees were in leaf.

Table 1. The maximum and minimum lengths of white root (cm per window) produced by trees of Worcester/MM.104 while the trees were in leaf (May–October) during the period 1961–1971. Both trees were grown under grass and defruited until 1969

Tree		1961	1962	1963	1964	1965	1966	1967	1968	1970	1971
1	max	132	146	57	40	97	126	29	88	20	60
	min	7	0	1	2	4	2	4	0	0	3
2	max	236	127	54	95	42	38	54	49	10	26
	min	17	0	2	4	3	1	11	0	0	1

The length and type of new growth The importance of new growth to the length of root present is affected by orchard soil management. Trees of Cox/M.26 planted at 5.76 m^2 were grown adjacent to a root observation laboratory with the following treatments[2]: a) soil kept completely free of vegetation with herbicides, b) a complete cover of S50 *Phleum nodosum*, c) as in b) but with irrigation to maintain soil water potentials above -0.03 MPa (orchard 3). Estimates of the length of root m m^{-2} soil surface were calculated from profile wall measurements for brown and woody root and from the observation windows for white root, assuming that the density visible at the glass represented a layer of soil 3 mm thick. Although a single 1975 estimate of woody root is used and both of these measurements represent variable spatial estimates, involve substantial extrapolation to get from a plane surface to a volume measurement, and do not include samples from close to the tree trunk where root density is highest[9], they give an acceptable estimate of the relative importance of brown and white root length. That for herbicide managed trees is comparable to Fig. 1 in the relative amounts of white and woody root on cropping trees. The relative amounts of woody root shown in Table 2, 1.00 : 0.52 : 1.02 for herbicide, grass and grass + water treatments compare well with weight data obtained by excavation, ratios of 1.00 : 0.73 : 1.22 respectively. In all cases, and in both years (Table 2), new growth increased the length of root in the soil appreciably, up to 50% for herbicide-treated plots, 226% for grassed plots and 176% for trees in the grass + water treatment. Increases were somewhat larger in 1974 than 1973.

Table 2. Variation in the estimated length, m m^{-2} soil surface, of root at 0–30 cm depth and at different times during the year for trees of Cox/M.26 under a number of soil management treatments. Winter values estimated from soil trenches, May–August from root laboratory + trenches

Treatment	Winter	May	June	July	August
1973					
Herbicide	44	46	46	48	63
Grass	23	35	45	52	31
Grass + water	47	55	90	87	86
1974					
Herbicide	44	45	54	66	46
Grass	23	45	50	75	101
Grass + water	47	53	113	130	108

Soil management, in addition to effects on root production, also affects the balance of long (extension) roots, which appear to have unlimited growth potential and short (lateral) roots, which rarely exceed 2 cm in length. These roots differ in their mean diameter. Measurements on samples of 100 typical extension and lateral roots gave a mean diameter of 1.75 mm (95% confidence limits 0.9–2.7 mm) for long roots and 0.62 mm (95% confidence limits 0.3–1.1 mm) for short roots. Information on the relative amounts produced is important: long roots survive to become woody root members and are responsible for the radial and vertical spread of the root system through the soil volume. Short roots can be infected with VA mycorrhizas and contribute much of the root length within a given volume of soil. For the trees from orchard 1 with 0.09 m^2 soil surface per tree, roots <0.5 mm contributed 40% of the total root system length and roots 0.5–2 mm, a further 40%. Comparable values for trees with 5.76 m^2 per tree were 15% and 70%[7]. Data on the relative numbers of lateral and extension roots present in 1973 and 1974 are shown in Table 3 for orchard 3. In May, although the total numbers of roots were small, most were

Table 3. The numbers of lateral (L) and extension (E) roots (no. per 2 observation windows) visible at the observation windows of a root laboratory for trees of Cox/M.26 growing with a number of soil management treatments

Treatment	May		June		July		August	
	E	L	E	L	E	L	E	L
1973								
Herbicide	1	0	0	1	1	2	4	7
Grass	4	1	4	4	6	7	2	3
Grass + water	3	1	11	11	11	6	10	10
1974								
Herbicide	1	0	3	6	5	16	1	1
Grass	3	5	5	17	18	23	13	36
Grass + water	1	2	10	28	18	48	11	24

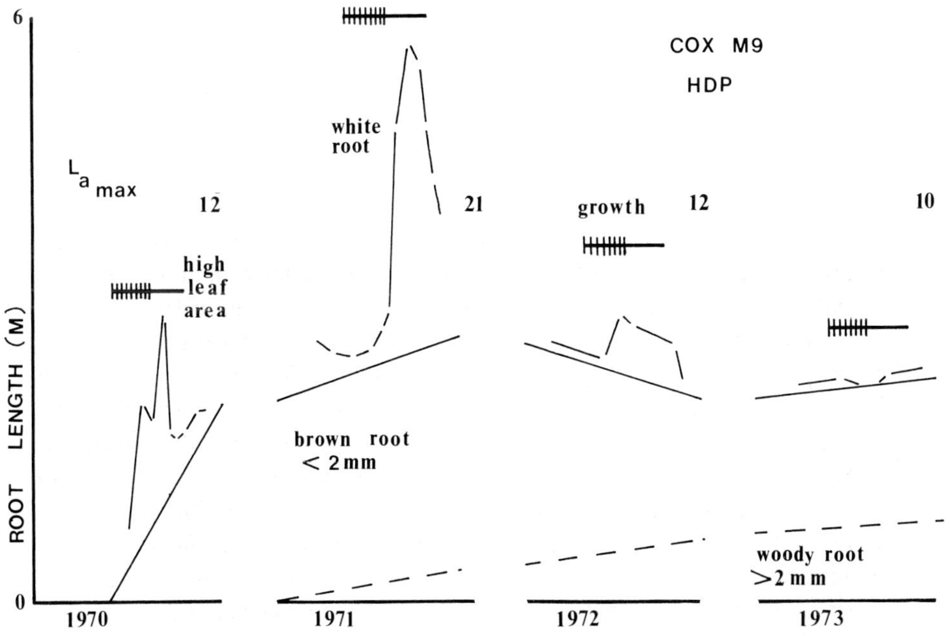

Fig. 1. Variation in the length of white, brown and woody roots per 1.2 m² observation window for trees of Cox/M.9 planted in 1970 and grown at 0.6 × 0.6 m under total herbicide management. The timing of leaf presence —— and active shoot growth ////// are shown as horizontal bars. Maximum values of LA cm cm^{-2} are shown.

extension roots. With the herbicide treatment substantial numbers of lateral roots did not appear until relatively late in the season. Numbers were always much lower than in the grass and grass + water treatments. The ratio of lateral to extension roots can be affected by soil treatment (Table 3) and varies during the year (Fig. 2). Data for orchard 1 showed that the ratio was highest in July, as total root numbers were beginning to increase rapidly and in the autumn when they were decreasing. Trees with 0.09 and 0.36 m² soil surface per tree showed similar patterns. Soil water potential also affects the ratio of lateral to extension roots. For trees of Cox/M.9, grown at spacings of 2.4 × 0.6 m, 1.2 × 1.2 m and 0.6 × 0.6 m with and without irrigation, the ratio, averaged over the whole of the year of planting, changed from 2.4 to 1.7, 5.0 to 1.1 and 3.0 to 1.3 respectively, in the absence of irrigation even under herbicide management.

Variations in the ratio of lateral to extension roots as well as variation in the proportion of extension roots which survive as woody roots, the extent to which they develop woody tissues and inherent limitations of the observation method[4], account for the poor relationship between root growth as seen through the root observation laboratory window and final weight at grubbing (Fig. 3). For orchard 2, although the final weight of the root system was usually higher for trees with higher mean root lengths as seen in the observation windows, scatter

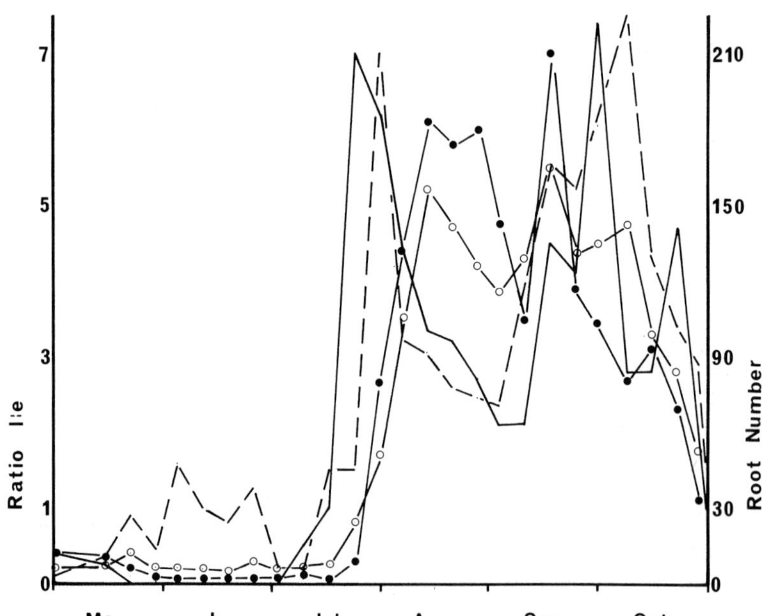

Fig. 2. Variation in the ratio of lateral to extension roots and in the total number of roots per observation window for trees with 0.09 and 0.36 m² soil surface per tree in orchard 1 in 1971. —— ratio, 0.09; — — ratio 0.36; ● — ● number 0.09; ○ — ○ number 0.36.

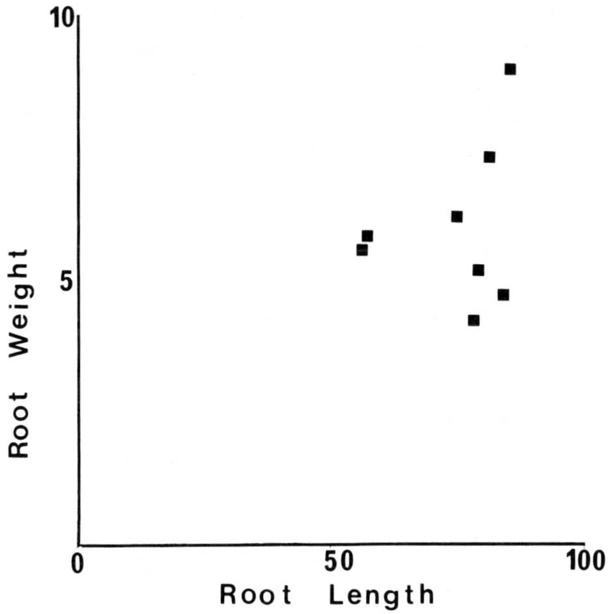

Fig. 3. The relationship between the mean length of white root cm per observation window produced by trees from orchard 2 over the period 1961 to 1972 and the weight of the tree root systems at harvest, kg.

Table 4. The effect of tree type on root weight at harvest, g tree^{-1}, and white root production, m per observation window. Root production was calculated as total white root length present between May and October in 2 years/4 (the number of weeks roots remain white)

Rootstock		Wt of root		White root	Ratio Weight/length
		>2 mm	<2 mm		
Plum	Pixy	62	9.7	12.3	5.9
	St Julien A	166	7.2	14.8	11.1
Cherry	Colt	317	62	21.3	16.7
	F12/1	383	36	18.9	20.0

was appreciable (Fig. 3). Relationships between root length, weight and other parameters vary between fruit species and rootstocks (Table 4). Root production and the ratio root weight/root length were greater for cherry than for plum trees. Within both fruit species the ratio varied with rootstock, as did the ratio of >2 mm/<2 mm diameter roots. With Colt 16% of total root weight was <2 mm diameter, for F12/1 the comparable figure was 8%.

Table 5. The relative lengths of white root, cm per 2 observation windows, produced by apple trees, Cox/M.26 and an S50 *Phleum nodosum* sward in 1973

Treatment	Species	May	June	July	August
Grass	Apple	15	27	35	10
	Grass	5433	3485	1789	899
Grass + water	Apple	10	52	48	47
	Grass	2609	2364	790	1363

The functioning of the apple root system

Root density The relative length of apple root under a given area of soil surface is low compared with that found for a range of other crops. Typical values for apple[5,8], 2–24 cm cm^{-2}, are much lower than those reported for other species[16], *e.g.* 200–4000 for cereals. The relative lengths of white root produced by the trees and grass in orchard 3 differed by a factor varying from 362 to 16 depending upon time during the year and treatment (Table 5). Except for August, when irrigated trees and grass both produced more root, grass root length was usually greater in the unirrigated treatment.

The low root densities found in apple have important implications for the absorption of water. The consequences of a low root density will be most apparent when high rates of water inflow are needed to maintain transpiration, *i.e.* limitations to water supply will be most obvious when evaporative demand is substantial and the ratio of leaf area to root length is high. At high rates of inflow, 20–30 μl mm^{-1} d^{-1}, the soil resistance to water movement to the root surface is likely to become limiting, even at high soil water potentials[20]. In apple this appears to occur relatively often[6].

Water absorption in apple Depletion of soil water at a rate of 3 mm d^{-1} occurs commonly. For this rate of depletion rates of water inflow have been calculated

Table 6. Ratios of water uptake by apple roots needed to sustain a transpiration rate of 3 mm d^{-1} at different times in the year assuming white root length as for orchard 1 in 1971 (μl mm^{-1} d^{-1})

Functioning root system	Rate needed	
	Early July	Mid-August
White root tips	100	9
All white root	60	4
Whole root system	4	2
Root system below 20 cm depth	7	4

Table 7. The absorption of water (μl mm^{-1} d^{-1}) by roots of different crop species

Species	Rate	Reference
Cherry	1.6–10	22
Wheat	0.5– 1.0	14, 15, 21
Broad-leaved crops	0.2– 4.0	12, 16
Maize	2.4	19
Cotton	0.7	19

for the 0.36 m^2 trees of orchard 1 in 1971 (Fig. 1), assuming water uptake to occur through varying proportions of the root system (Table 6). The leaf area index of these trees was >2 and for most of the season an almost complete leaf area able to transpire at a high rate was present. Calculations suggest that in July if only new root tips, often suggested as the major site of water uptake, absorb water then very high rates of inflow, greater than those reported for a wide range of species (Table 7), would be required. Only if the whole root system functions in uptake are rates within the normally reported range. In mid-August, when new root production is at its maximum (Fig. 1), much lower rates of inflow are needed. In August, however, soil water potentials are often low and so resistances to flow within the soil will be increased necessitating relatively higher root densities to maintain the same total flow. Recent experimental work[8,9,22] has shown that all apple roots, white, brown and woody are potentially able to absorb water and mineral nutrients. The above calculations of rates of water uptake needed to supply transpirational needs indicate that the whole of the root system, including woody roots, must function in absorption.

Farré[11] investigated the effect of soil management on root density (LA, cm cm^{-2}). He found that for trees under grass the total LA was 9.1, for herbicide-managed trees LA was 7.6. For grassed orchards 81% of root lenght was at 0–30 cm depth and only 12%, an LA of 1.1, below 50 cm depth. Under herbicide management comparable values were 90% and 4% with an LA of 0.3 below 50 cm depth. Under grass, by early July, only soil below 30 cm depth was at a soil water potential >1.2 MPa and by early August only that below 50 cm was at a potential >0.5 MPa. Thus from mid-summer, July-August, only roots below 50 cm were in soil which could easily supply water to the tree. With the effective root density available, LA = 1.1, high rates of water flow would be needed to supply normal transpiration. In this situation new root growth (Table 2) would be important for high water extraction. Studies of ^{32}P uptake from the soil[1] have shown a good relationship for young trees between new root production and absorption. Although white root production is not absolutely essential to tree root system functioning in most situations it is important.

Table 8. The effect of soil management on the phosphorus levels, % DW, found in 1979 in the leaves of six-year-old trees of Cox/MM.106 grown under a number of soil management treatments

Treatment	27 Jul	16 Aug	22 Aug	5 Sep	21 Sep
Herbicide	0.230	0.213	0.210	0.218	0.220
Grass	0.245	0.232	0.228	0.250	0.260
Grass + water	0.270	0.248	0.252	0.285	0.287

The role of mycorrhizas

Phosphorus uptake Measurements of the concentration of phosphorus in the leaves of the trees of Cox/MM.106 (orchard 4) grown with the same treatments as in orchard 3 showed enhanced levels in trees in grass, especially where the grass was irrigated (Table 8). Levels of soil phosphorus, extractable in sodium bicarbonate were higher in the herbicide-treated than in the grassed plots[10]. The enhanced phosphorus uptake could therefore result from 1) increased root activity, although[25] the increased root growth (Table 2) is often associated with a shoot system increased even more and so gives a reduced root to shoot ratio, 2) a difference in infection with VA mycorrhizal fungi. Under grass, with its high relative root density (Table 5) the potential for infection as a result of contact, with hyphae, infected roots or spores associated with roots, is much higher than under herbicide-managed soil, where there are relatively few roots (Table 2). Measurements of the number of *Glomus mosseae* spores in soil under these treatments indicated similar numbers for orchards of varying ages (Sewell, personal communication).

Mycorrhizal infection Roots were extracted from soil cores taken on a number of occasions in 1980 from the herbicide and grass + water treatments of orchard 4. The % of root infected with VA mycorrhizas (Table 9) indicated relatively similar levels of infection in May, probably resulting from infections in 1979, higher levels under grass + water in July and similar levels in October. Infection

Table 9. The per cent of root infected with VA mycorrhizas from two depths (cm) on plants receiving different soil management treatments

Soil management	Depth	Per cent infection		
		May	July	October
Herbicide	0–15	9.8	5.6	17.2
	15–30	8.7	0.6	10.9
Irrigated Grass	0–15	3.7	25.9	11.1
	15–30	11.4	7.7	2.0

with VA mycorrhizal fungi requires both the presence of the fungus and of suitable root tissue, *i.e.* lateral roots capable of being infected. Under the grass treatments the length of lateral root available for infection is higher than under herbicide (Table 3). The low soil water potentials found under unirrigated grass would make infection difficult and reduce the effectiveness of both mycorrhizal and other roots in absorption. Trees in grass showed elevated phosphorus levels in early and late season when soil water potentials would be highest.

Unworked apple rootstocks were planted in soil obtained either from grassed or herbicide-treated areas, within an established orchard. The plants were grown in a greenhouse for 6 months. Root samples were scored for infection with VA mycorrhizas. A further set of plants was grown in a sand loam mix and at harvest the ratio of small to large roots determined on samples. Although differences were small, with the exception of M.9, % infection was higher with the grassed soil. M.9 rootstocks, which have relatively little thin root, were less infected than the others which were all infected to a similar extent (Table 10).

Table 10. The effect of apple rootstock on infection (%) with VA mycorrhizas in previously grassed or herbicide-treated soil and on the ratio of thin (<1 mm diameter) to thick roots (>1 mm)

Infection	Rootstock				
	M.27	M.9	M.26	MM.106	MM.111
Grass	65	49	73	73	71
Herbicide	61	58	66	67	69
Ratio thin/thick roots	–	0.04	0.07	–	0.11

Conclusions

The development of the fruit tree root system into the large and complex structure which can be revealed by excavation[5] involves many complex processes. The root system contains a number of root types which differ in morphology, anatomy and age. During both the life of the tree and a single season the relative amounts of different root types vary. With age, the woody part of the root system contributes an increasing fraction of root length although, even in established trees, new growth has a major impact upon total root length. Soil management treatments can alter both the amount of new growth and the relative amounts of short lateral roots and long extension roots being produced. This seems to have significant implications for the ease of infection of the root system with VA mycorrhizal fungi and for root survival.

Studies of changes in root length indicate that to sustain normal rates of transpiration the whole root system, white and woody roots, must function in absorption. This is particularly important for trees in grass, where soil water potentials in much of the available soil volume for part, at least, of the year are low. The supply of large amounts of phosphorus, as in many other crops, seems dependent upon infection with VA mycorrhizas.

The integrated functioning of the whole range of root types is needed for the maintenance and activity of the total root system.

Acknowledgement I thank Mrs. C. M. Crisp for assistance with the studies of mycorrhizas.

References

1 Atkinson D 1974 Some observations on the distribution of root activity in apple trees. Plant and Soil 40, 333–342.

2 Atkinson D 1977 Some observations on the root growth of young apple trees and their uptake of nutrients when grown in herbicide strips in grassed orchards. Plant and Soil 49, 459–471.

3 Atkinson D 1978 The use of soil resources in high density planting systems. Acta Hortic. 65, 79–89.

4 Atkinson D 1980 The growth and activity of fruit tree root systems under simulated orchard conditions. *In* Environment and Root Behaviour pp. 171–185. Ed. D N Sen Geobios, Jodhpur, India.

5 Atkinson D 1980 The distribution and effectiveness of the roots of tree crops. Hortic. Rev. 2, 424–490.

6 Atkinson D 1980 Water use and the control of water stress in high density plantings. Acta Hortic. 114, 45–56.

7 Atkinson D, Naylor D and Coldrick GA 1976 The effect of tree spacing on the apple root system. Hortic. Res. 16, 89–105.

8 Atkinson D and Wilson S A 1979 The root soil interface and its significance for fruit tree roots of different ages. *In* The Soil Root Interface. Eds. J L Harley and R Scott Russell. pp 259–271. Academic Press, London.

9 Atkinson D and Wilson S A 1980 The growth and distribution of fruit tree roots: some consequences for nutrient uptake. *In* Mineral Nutrition of Fruit Trees. Eds. D Atkinson, J E Jackson, R O Sharples and W M Waller. pp 137–150. Butterworth, London.

10 Atkinson D, Swain R W and Fricker D 1980 The effect on soil condition of herbicide use in fruit tree plantations. *In* Agrochemicals in Soils. Eds. A Banin and U Kafkafi. pp 179–189. Pergamon, New York.

11 Farré J M 1979 Water use and productivity of fruit trees: effects of soil management and irrigation. Ph. D. Thesis, University of London.

12 Graham J, Clarkson D T and Sanderson J 1974 Water uptake by the roots of marrow and barley plants. ARC Letcombe Laboratory Annu. Rept. for 1973 , 9–11.

13 Head G C 1966 Estimating seasonal changes in the quantity of white unsuberized root on fruit trees. J. Hortic. Sci. 41, 197–206.

14 Herkelrath W N 1973 Water uptake by plant roots. Ph. D. Thesis, University of Wisconsin.

15 Lawlor D W 1969 Resistance to water flow in soil and plant. I. Soil resistance in relation to amounts of root: theoretical estimates. J. Appl. Ecol. 6, 1–12.

17 Phillips J M and Hayman D S 1970 Improved procedures for clearing roots and staining parasitic and vesicular-arbuscular mycorrhizal fungi for rapid assessment of infection. Trans. Br. Mycol. Soc. 55 158–

18 Rogers W S and Head G C 1962 Studies of growing roots of fruit plants in a new underground root observation laboratory. Proc. 16th Intern. Hortic. Congr. 311–318.

19 Taylor H M and Klepper B 1978 The role of rooting characteristics in the supply of water to plants. Adv. Agron. 30, 99–128.

20 Tinker P B 1976 Roots and water: transport of water to plant roots in soil. Phil. Trans. R. Soc. Lond. B. 273, 445–461.

21 Williams J 1974 Root density and water potential gradients near the plant root. J. Exp. Bot. 25, 669–674.

22 Wilson S A 1981 Nutrient and water absorption by fruit tree roots of different ages. Ph. D. Thesis, University of London.

Plant and Soil 71, 37–47 (1983). A-04

Root development and configuration in intensively managed radiata pine plantations

E. K. SADANANDAN NAMBIAR
CSIRO, Division of Forest Research, Mount Gambier, South Australia 5290

Key words Ecological aspects Nutrient uptake Pine *Pinus radiata* D. Don Physiological aspects Root development Water absorption

Summary In south-east Australia, where radiata pine (*Pinus radiata* D. Don) is grown on sandy soils low in nutrients and short of water, early establishment, and rapid growth to canopy closure lead to increased productivity. At this stage demands for nutrients and water are high, and trees respond vigorously to silvicultural inputs.

For several months after transplanting in winter roots are confined within a narrow planting wedge, low temperature restricts new root growth and slows recovery from water stress in plants. From spring, depending upon the configuration and vigour of the roots transplanted, lateral roots extend radially throughout the soil.

Although there were small decreases in concentration of roots radially from the stems of very young trees, such spatial differences disappeared between ages 2 and 3, so that rooting density was independent of distance from the stem. The pattern of vertical distribution of lateral roots was not influenced by age and 80–90% of the lateral roots were within the top 30 cm soil. Roots developed rapidly as the trees grew towards canopy closure, but in general the rooting densities of these pines are among the lowest reported for plants. In rapidly growing trees approaching canopy closure, the secondary thickening of the lateral roots was sufficient to double the weight of roots without altering root length.

Knowledge about root growth and root configuration during the early phase of plantation development will assist management decisions where intensive silviculture is practiced, and hence ensure the most efficient use of nutrients and water.

Introduction

In south-east Australia, in order to sustain and increase the productivity of radiata pine (*Pinus radiata* D. Don) plantations growing on generally infertile podzolized sandy soils, contrasting silvicultural practices are being advocated: intensive silviculture (cultivation, weed control and heavy fertilization) to achieve canopy closure by age 4 to 5[20], or retention and management of slash, thereby influencing both water (mulching effect) and the nutrients available to trees[3,17]. The growth of trees during the early phase of a plantation (period of 'get away' or 'site capture') when young trees respond vigorously to silvicultural manipulations is recognised as critical. During this period the roots confined initially as a mat in a narrow planting wedge, ramify throughout the soil.

In this paper some aspects of the development of roots in radiata pine plantations from transplanting to canopy closure are examined. The work shows how a detailed understanding of the ecological and physiological aspects of root development and function can aid management decisions and optimise silvicultural practices.

Transplanting stresses and the growth of new roots

The environmental and physiological factors that influence the ability of seedlings to produce new roots when transplanted and practical implications of the root growth potential for survival and early growth of outplanted seedlings have been emphasised[1,2,6,13,19]. Two important factors influencing the new root growth and early establishment of pines are the temperature during the planting season[1,6,7], and the root configuration of the planting stock[7].

Effect of soil temperature on root regeneration

Fig. 1 shows the effect of soil temperature on the growth of non-mycorrhizal new roots, and on the needle water potential in transplanted radiata pine. During the experiments, pots to which seedlings were transplanted were maintained at constant soil temperatures, but shoots were exposed to non-constant air temperatures ranging from a daily minimum of 6–10°C to a maximum of 14–20°C. While the numbers, weight and lengths of root responded markedly to soil temperature, the shape of the response curves differed between root characters. Number of root apices increased linearly as soil temperature

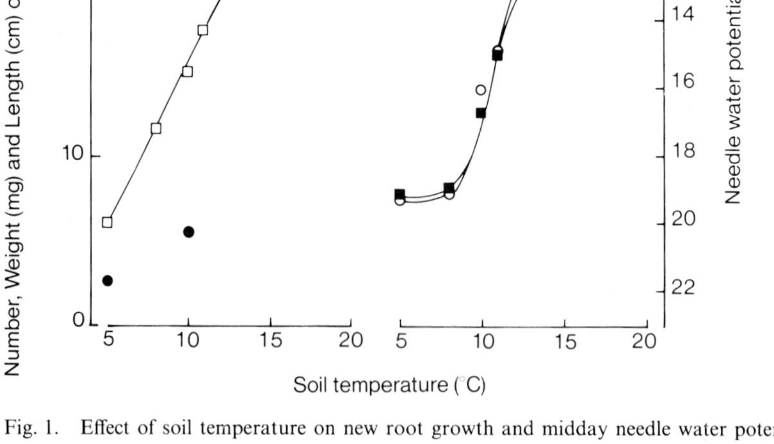

Fig. 1. Effect of soil temperature on new root growth and midday needle water potential in 8-month-old radiata pine seedlings, 32 days after transplanting to a sandy forest soil. (Data from 6, 10)

increased from 5 to 15°C, with little further increase in root growth from 15 to 20°C. In contrast, length and weight were little affected by a rise in temperature from 5 to 8°C, but a further rise to 14–15°C increased the weight and length considerably. Temperatures greater than 15°C increased the length more than weight. Soil temperatures of less than 10°C restricted root elongation and dry weight accumulation more than root initiation. Between 10 and 15°C response curves of all root characters were parallel, and the maximum growth of roots occurred at 20°C. There is a close relationship between growth of new roots and needle water potential (Fig. 1)[6]. The suppressive effect of low soil temperature (5–10°C) on root growth is primarily responsible for the commonly observed slow recovery from water stress in transplants despite the adequate supply of soil water during the planting season[6,9].

In South Australia seedlings transplanted to the field in July initiate few new roots until late September (mean soil temperature 11°C) and midday needle water potentials from −16 to −19 bar are common in winter. More favourable needle water potentials of −10 to −11 bar are found in October–November when soil temperatures rise to 12–16°C, even though mean rainfall in October is, in general, about one-third of that in July (Table 1)[8,9]. The soil temperature below which new root growth is restricted, and the optimum temperature for maximum new root growth vary between species[1,6,18,19] and between families[10]. In a comparison of the root regenerating capacity of 8 radiata pine families, it was found that the differences in the number of new root apices produced between the first and last ranking families were ninefold at 8°C, threefold at 11°C and sevenfold at 14°C; and also that root regeneration characters were highly heritable[10]. The root system also adopts strategies to minimise the effect of nutritional and environmental stress on root growth and configuration[7]. It is also important to recognise that seedlings are able to remobilize nutrients internally to support new root growth, when factors such as low nutrient supply, sub-optimal temperature, loss and damage to roots, and water stress restrict nutrient uptake. For example, the net amount of nitrogen and phosphorus mobilized from needle, stem and old roots during the period of active new root growth may account for 36 to 46% of nitrogen and phosphorus present in these parts, large enough to meet the nutrient needs of new roots (unpublished data).

Stresses due to loss or damage to roots

To ensure the development of a root system adequate for early recovery from transplanting stress, and for survival and growth, the configuration of the transplanted root system, particularly the number and vigour of the first order lateral roots (roots which originate directly from the tap root), is critical. These parameters provide the basic framework for a regenerating root system[6]. There is a close relationship between the number of first order lateral roots retained at transplanting, and the ability of plants to cope with subsequent environmental

Table 1. Midday needle water potential, nitrogen content, survival and stem diameter of young radiata pine transplanted into a sandy soil at Mount Gambier, in relation to the number of first order lateral roots (1°L) retained on the planting stock

Number of 1°L on planting stock	Weeks after transplanting					
	1 (winter)	11 (spring)	26 (summer)	20 Needle nitrogen (%)	52 Survival (%)	52 Stem diameter (mm)
	Needle water potential (–bar)					
All (14–20)	16.4	10.9	14.6	1.84	100	12.2
	(0.3)	(0.2)	(0.6)	(0.17)		(0.2)
6	17.2	12.6	15.9	1.60	100	8.8
	(0.5)	(0.3)	(0.4)	(0.32)		(0.1)
2	18.5	14.4	19.7	1.26	78	6.8
	(0.3)	(0.3)	(2.6)	(0.12)		(0.1)
1	19.8	17.1	*	*	*	*
	(0.4)	(1.8)				

* Not measured due to low and variable survival (20–60%).
Figures in parenthesis are standard errors.

stresses (Table 1). The loss of first order lateral roots during transplanting would increase the water stress in trees throughout the growing season and reduce nutrient concentration, growth and survival (Table 1).

Root configuration in rapidly developing plantations

Detailed information on the spatial variation in lateral root distribution, vertical distribution of lateral roots, and the effects of the age of young trees on root configuration, are necessary to understand the efficiency of silvicultural practices promoting rapid early growth.

Root distributions of 15, 34 and 46 months old trees growing on deep podzolized sandy soils low in organic matter are described. Trees were grown from half-sib families of seed-orchard seed, raised in the same nursery, and were of similar quality. They were planted 2–2.2m apart within the row and 2.4 m between rows and subsequently managed following local practice, 2–3 ploughings, herbicide applications at planting and about a year later, and application of complete fertilizers in 5–7 instalments before canopy closure. At the time of root sampling trees were 1.4, 3.1 and 5.4 m tall, and stem diameters were 3.6, 6.8 and 12.4 cm (at 15 cm from the ground) for 15, 34 and 46 months old trees respectively. The tallest trees were approaching the closed canopy stage. Root samples were collected from several positions between and within rows using soil cores (surface area 225 cm^2, depth 50 cm). Roots were collected by

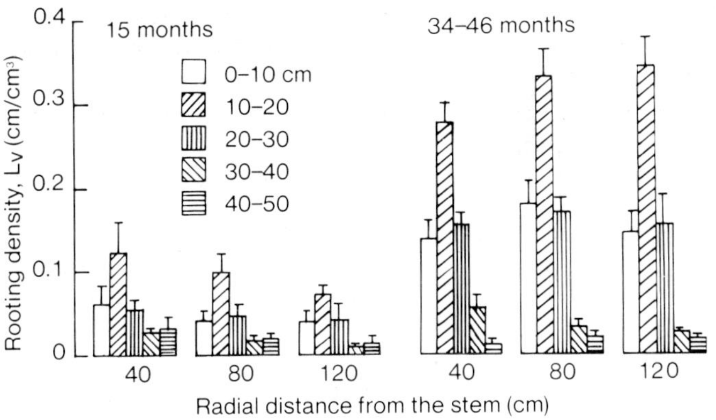

Fig. 2. Rooting density in relation to radial distance from the stem in young radiata pine growing on Mount Burr Sand. (Vertical bars are SE, n = 8)

gently washing the cores over a one mm sieve. Depending upon the sample size and root diameter, root lengths were measured directly or by intercept counting. In addition the number of root intercepts greater than 1 mm in diameter in a 30 × 10 cm vertical plane were counted at 16 positions within and between rows, within each age. There were no differences in root distribution attributable to direction within and between rows.

The distribution patterns of lateral roots are shown in Fig. 2. At 15 months the rooting density decreased from the tree to the mid-point of the rows, the pattern being consistent to 50 cm depth. However, roots had ramified throughout the plantation. By 34 months the rooting density had more than doubled, but was largely independent of the radial distance from the tree. In bigger trees rooting density at 10–20 cm depth, where maximum roots were found, tended to increase towards the mid-point of the rows, suggesting that the roots of trees from adjacent rows may be intermingling to give higher concentrations of roots in the middle of the row. However, when there is soil mechanical impedence to root growth, practices like deep ripping will influence spatial distribution of roots without affecting root growth[15].

The vertical distribution of lateral roots over time is shown in Fig. 3. From 15 to 34 months the length and weight of roots increased 2– to 3–fold to a depth of 30 cm. From 34 to 46 months there was little increase in root length except for a small increase in the surface horizon. In contrast, there was a large increase in root weight at all depths, largely due to roots greater than 1 mm in diameter. Most lateral roots (80% at 15 months and 91% at 34–46 months) were within the top 30 cm soil profile. Similarly, roots less than 1 mm in diameter contributed 87.3 and 85.5% of the total root length at 34 and 46 months respectively, but only 38.9 and 22.2% of the weight.

The contribution of roots greater than 1 mm to total length was very small, but they contributed substantially to the weight and their distribution (number) in

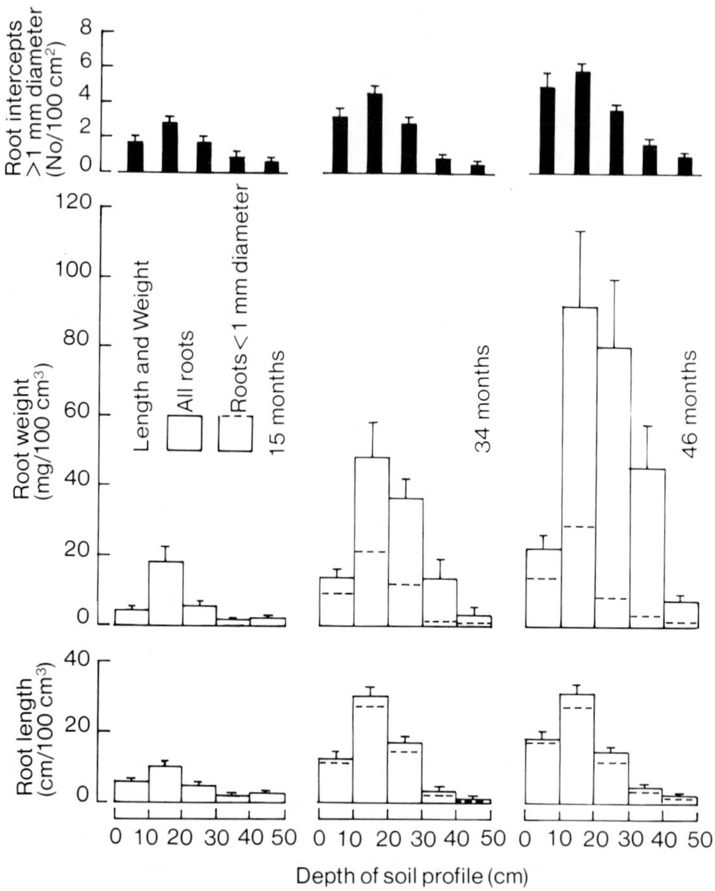

Fig. 3. Distribution of lateral roots in intensively managed radiata pine plantations from establishment to canopy closure. (Horizontal bars are SE for all roots, n = 20)

the profile generally followed the same pattern as length and weight. The contrast in the length and weight of roots found between 0–10 and 10–20 cm depth was less apparent in the number of root (greater than 1 mm) intercepts.

Similarly, roots greater than 2 mm in diameter were few, their diameter ranged from 2.0 to 11.0 mm (mean 2.5 mm) at 34 months, and 2.0 to 18.8 mm (mean 3.4 mm) at 46 months). Of the 20 cores examined within each age, 0, 4 and 13 of them had roots greater than 5 mm in diameter at 15, 34 and 46 months respectively.

Relationship between weight and length of lateral roots

The increased occurrence of thicker roots as seedlings grew to juvenile trees appears to have had a strong influence on the weight of roots, and weight-length relationship. This was examined by regressing root length (y) on weight (x). The

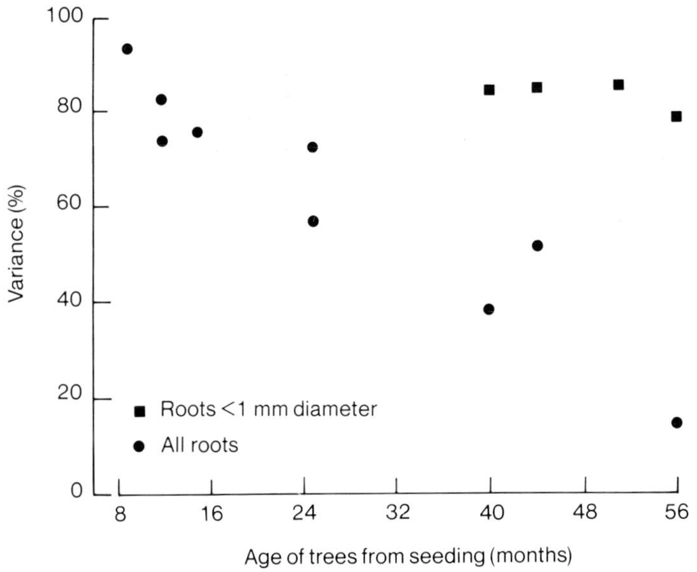

Fig. 4. Variance in lateral root length explained by the regression between length and weight of roots in relation to age.

linear regression included data on lateral roots from several experiments concernėd with seedlings, post-transplanting, and trees from establishment to canopy closure (Fig. 4). Except in the case of 9-month-old seedlings when only 14 data sets were available ($n = 14$) the regressions were based on data sets with n ranging from 57 to 95. The proportion of the variance in length of all roots explained by root weight decreased progressively with age, and after about 3 years the weight of all roots was poorly related to length. Weight-length relationships were not significantly improved by non-linear regressions (analysis not included). There was, however, a close relationship between weight and length in roots less than 1 mm in diameter (which contributed most to root length) regardless of age. The doubling of the weight of lateral roots without increasing the root length between 34 to 46 months (Fig. 3) as the tree height increased from 3.5 to 5.5 m may indicate the need for relatively thick lateral roots with secondary thickening for anchorage. These demands may be greater when rapid increase in biomass production is encouraged through intensive management.

General discussion

In environments where tree seedlings are transplanted in winter, low soil temperature restricts new root growth which in turn slows recovery from water stress in plants[6,9]. Apart from the effect of soil temperature on new root growth, the lower hydraulic conductance of old roots compared to new roots would also influence the water relations of transplants. For example, in 8-month-old loblolly pine seedlings (*Pinus taeda* L.) the average hydraulic conductance of old (brown, suberized) roots was 7.55×10^{-7} cm sec^{-1} bar^{-1} compared to 1.95×10^{-6} cm sec^{-1} bar^{-1} of new (white, newly formed) roots[14]. The extent of suberization in newly initiated roots is influenced by temperature[6].

The critical importance of the first order lateral roots as a framework for new root growth[6,7], and the strong influence of these roots on subsequent water relations and growth of young trees (Table 1) emphasises the need to minimise the loss of, and damage to, these roots while lifting, transporting and planting. Even under very favourable conditions 30–40% of the first order lateral roots do not produce new apices after transplanting[6,7,10]. There seems little opportunity to manipulate the number of first order lateral roots in radiata pine by nursery practices[6,7] or genenetically[10]. Root wrenching produces a 'fibrous root system' by increasing the growth of second and third order laterals (unpublished data). However, the large genetic differences and the high estimated heritabilities in new root characters suggest that more research is needed to exploit genetic effects on root growth for greater survival and better early establishment[10].

The root systems described here did not have many mycorrhizal roots. Mycorrhizae do not appear to be abundant on roots of trees which are grown with intensive weed control and fertilization. Mycorrhizal apices on radiata pine seedling roots were reduced 10-fold by fertilization in sandy soils[7]. There are other reports of inhibition of mycorrhizal development by herbicides and fertilizers[5,11]. In another study, fertilization increased the proportion of mycorrhizas on radiata pine roots at a poor site but significantly inhibited mycorrhizas at a better site[16]. Further investigations are needed on this important question.

When seedlings are transplanted in winter, roots are confined to a narrow planting wedge. When favourable conditions follow in spring, and if the ensuing competition for water by weeds is minimised, roots ramify laterally throughout the soil. Although there was a decrease in rooting density radially from the tree in 15-month-old trees, there was no significant difference in spatial distribution in 3-year-old trees. In older plantations the distribution of fine roots is known to be parallel to the pattern of throughfall[4], and the greater stemflow may lead to higher concentrations of fine roots near the trunk[12].

Rooting densities in young pine plantations (0.05 to 0.32 cm cm^{-3}) are very low compared to those in older plantations (0.8 to 5.0 cm cm^{-3})[9,16] or agricultural plants (*e.g.* 5 to 25 cm cm^{-3} in wheat, 30 to 50 cm cm^{-3} in pastures)[9]. Such low root concentrations present considerable problems for efficient use of fertilizers by young trees, particularly in sandy soils prone to leaching (unpublished data). Some field practices, such as planting and fertilizing at the same time to save on operational costs, and placement of soluble fertilizers in a clump or band near young trees, will adversely affect uptake efficiency by

Plate 1. Root distribution of radiata pine on a podzolized sandy soil about 8 months after transplanting one-year-old seedling.
Arrow points to the remains of a fertilizer mixture (superphosphate, ammonium sulphate, potassium sulphate and trace elements) placed in a slot about 10 cm away and parallel to the planting slot 3–4 weeks after planting, according to the local field practice[20].

minimising root-nutrient contact (see Plate 1). Application of fertilizers within a wide weed-free strip along the tree row would enhance nutrient uptake efficiency even in young trees.

During summer the water content of the surface soil in the plantations where the present studies were carried out was usually between 0.02 and 0.04 g g^{-1} (suction often exceeding 15 bars), depending upon the weed cover. Summer water stress, largely induced by weeds, has a critical effect on the growth of young trees[8,17]. Because the fine roots are predominantly in the surface soil (63 to 70% of the roots in 0–20 cm; Fig. 2 and 3)[16], even a relatively small increase in surface soil water through the mulching and weed control effect of slash and litter retention can benefit tree growth[3,17]. This may be even more important when nutrients are concentrated in the surface soil, as is usually the case.

The doubling of the weight of the lateral roots without increasing the root length between 34 and 46 months, as the tree height increased from 3.5 to 5.5 m, may indicate the need for relatively thick (5 to 18.6 mm diameter) lateral roots with secondary thickening for anchorage. Another reason for secondary thickening could be to increase the cross sectional area to maintain non-limiting axial resistance for the continually ramifying root system below. These demands may be even greater in intensively managed radiata pine plantations where rapid increase in biomass production is the aim.

Despite the fact that lateral roots are largely confined to the top 30 cm of the profile, even 16-month-old trees withdraw water from at least 2 m depth, maintaining midday needle water potentials of −11 to −13 bar in summer when surface soil was very dry[9] (Sands and Nambiar, unpublished). Thus there is a need to critically examine the factors that influence the distribution and function of tap or sinker roots in relation to soil profile characteristics and water uptake.

Acknowledgements I am very grateful to Messrs P. S. Keeley and M. Howard for their enthusiastic technical support and to Ms C. M. Skinner for assistance in preparing the manuscript. I also thank Dr R. Sands (CSIRO, Division of Soils) and Messrs B. A. Mitchell and R. Cromer (CSIRO, Division of Forest Research) for comments on the manuscript.

References

1 Abod S A, Shepherd K R and Bachelard E P 1979 Effect of light intensity, air and soil temperatures on root regeneration potential of *P. caribaea* var. *hondurensis* and *P. kesiya* seedlings. Aust. For. Res. 9, 173–184.

2 Burdett A N 1979 New methods for measuring root growth capacity: their value in assessing lodgepole pine stock quality. Can J. For. Res. 9, 63–67.

3 Farrell P W, Flinn D W, Squire R O and Craig F G 1981 On the maintenance of productivity of radiata pine monocultures on sandy soils in south-east Australia. Proceedings Division 1 XVII IUFRO World Congress, Japan, pp. 117–128.

4 Ford E D and Deans J D 1978 The effects of canopy structure on stemflow, throughfall and interception loss in a young sitka spruce plantation. J. Appl. Ecol. 15, 905–917.

5 Iloba C 1975 The effect of some herbicides on the development of ectotrophic mycorrhiza of *Pinus sylvestris*. Eur. J. For. Pathol. 5, 339–343.

6 Nambiar E K S, Bowen G D and Sands R 1979 Root regeneration and plant water status of *Pinus radiata* D. Don seedlings transplanted to different soil temperature. J. Expt. Bot. 30, 1119–1131.

7 Nambiar E K S 1980 Root configuration and root regeneration in *Pinus radiata* seedlings. N.Z. J. For. Sci. 10, 249–263.

8 Nambiar E K S and Zed P G 1980 Influence of weeds on the water potential, nutrient content and growth of young radiata pine. Aust. For. Res. 10, 279–288.

9 Nambiar E K S 1981 Ecological and physiological aspects of the development of roots: from nursery to forest. Proc. Aust. For. Nutr. Workshop: Productivity in Perpetuity, Canberra, Australia, August, pp. 117–129.

10 Nambiar E K S, Cotterill P P and Bowen G D 1982 Genetic differences in the root regeneration of radiata pine. J. Exp. Bot. 33, 170–177.

11 Powell C L 1980 Effect of phosphate fertilizers on the production of mycorrhizal inoculum in soil. N.Z. J. Agric. Res. 23, 219–223.

12 Reynolds E R C 1970 Root distribution and the cause of its spatial variability in *Pseudotsuga taxifolia* (Poir) Britt. Plant and Soil 32, 501–517.

13 Ritchie G A and Dunlap J R 1980 Root growth potential: its development and expression in forest tree seedlings. N.Z. J. For. Sci. 10, 218–248.

14 Sands R, Fiscus E L and Reid C P P 1982 Hydraulic properties of pine and bean roots with varying degrees of suberization, vascular differentiation and mycorrhizal infection. Aust. J. Plant Physiol. 9, 559–569.

15 Somerville A 1979 Root anchorage and root morphology of *Pinus radiata* on a range of ripping treatments. N.Z. J. For. Sci. 9, 294–315.

16 Squire R O, Marks G C and Craig F G 1978 Root developments in a *Pinus radiata* D. Don plantation in relation to site index, fertilizing and soil bulk density. Aust. For. Res. 8, 103–114.

17 Squire R O, Flinn D W and Farrell P W 1979 Productivity of first and second rotation stands of radiata pine on sandy soils. I. Site factors affecting early growth. Aust. For. 42, 226–235.

18 Stone E C and Schubert G H 1959 Root regeneration by ponderosa seedlings at different times of the year. For. Sci. 5, 322–332.

19 Stone E C, Jenkinson J L and Krugman S L 1962 Root regeneration potential of Douglas-fir seedlings lifted at different times of the year. For. Sci. 8, 288–297.

20 Woods R V 1976 Early silviculture for upgrading productivity on marginal *Pinus radiata* sites in the south-eastern region of South Australia. Bulletin 24, Woods and Forests Department, South Australia.

Plant and Soil 71, 49–53 (1983). A-05

Effects of N fertilisation on populations of fine roots and mycorrhizas in spruce humus

I. J. ALEXANDER and R. I. FAIRLEY
Department of Botany, The University, Aberdeen, Scotland

Key words Fertiliser Fine roots Mycorrhizas Nitrogen *Picea sitchensis*

Summary Fertilisation with N alters the numbers, biomass and seasonal pattern of activity of fine roots and mycorrhizas in spruce humus. Production and mortality decrease, and longevity increases. Fertilisation reduces mycorrhizal infection, alters the relative proportions of mycorrhizal types and permits the appearance of new types.

Introduction

Boreal and north temperate forest ecosystems are often nitrogen stressed and fertilisation is an option for forest management. We wished to obtain information on the effect of N fertilisation on numbers, biomass and seasonal variation of fine roots and mycorrhizas, on their production, longevity and death, on degree of mycorrhizal infection and on frequency of mycorrhizal types. This paper combines data from a number of different investigations to be reported in detail elsewhere.

Materials and methods

Field experiment

A randomised block experiment was set up in a 35y old Sitka spruce (*Picea sitchensis*) plantation. Three levels of nitrogen (75,150 and 300 kg ha^{-1} N as $(NH_4)_2SO_4$ granules) were applied in March 1979. Each treatment was replicated 3 times. Only data from the control and highest level of application are presented here. Each month for 2 years fifteen 4 cm diam cores were taken at random from the top 10 cm of the soil. All living and dead root material was extracted manually and counts made of living mycorrhizal and non-mycorrhizal root tips. *Cenococcum geophilum* mycorrhizas were recorded separately. Material was then divided into i) living roots and mycorrhizas < 1 mm in diam, ii) living roots 1–5 mm in diam and iii) dead root fragments, for oven dry wt determination. The contribution of mycorrhizas to (i) was calculated from the counts on the basis of sample weighings.

The biomass data in these 4 compartments (mycorrhizas, roots < 1 mm, roots 1–5 mm, dead roots) were used in a simple population model where by balancing the fluxes into and out of each compartment to account for monthly changes in the size of each it is possible to arrive at estimates of annual production and mortality for each category of root. This model assumes that transfers between live categories are small compared to mortality and production of new tissue.

In September and December of each year a random sample of c.40 root tips from each block were fixed, embedded, sectioned and stained for anatomical investigation.

Pot experiment

Fourteen months after fertiliser application organic material from AoF and AoH horizons, including fine roots, of control and fertilised plots was sieved, mixed 2 : 1 with coarse sand and used to pot up 5 week old non-mycorrhizal Sitka spruce seedlings. After 20 weeks in the greenhouse the root systems were harvested and all root tips on randomly selected segments amounting to 10% of total root length were excised. A sub-sample of c.60 tips per plant were assessed microscopically for mycorrhizal infection and the mycorrhizas assigned to types on the basis of structure and extent of sheath and net, hyphal characteristics and staining properties.

Results and discussion

Numbers, biomass and seasonal variation (Fig. 1)

In the summer following fertilisation (1979) root numbers and biomass in fertilised plots did not decline and the differential between fertilised and control plots was maintained throughout the autumn peak of root activity. This response was most evident in the roots <1 mm diam. In 1980 seasonal

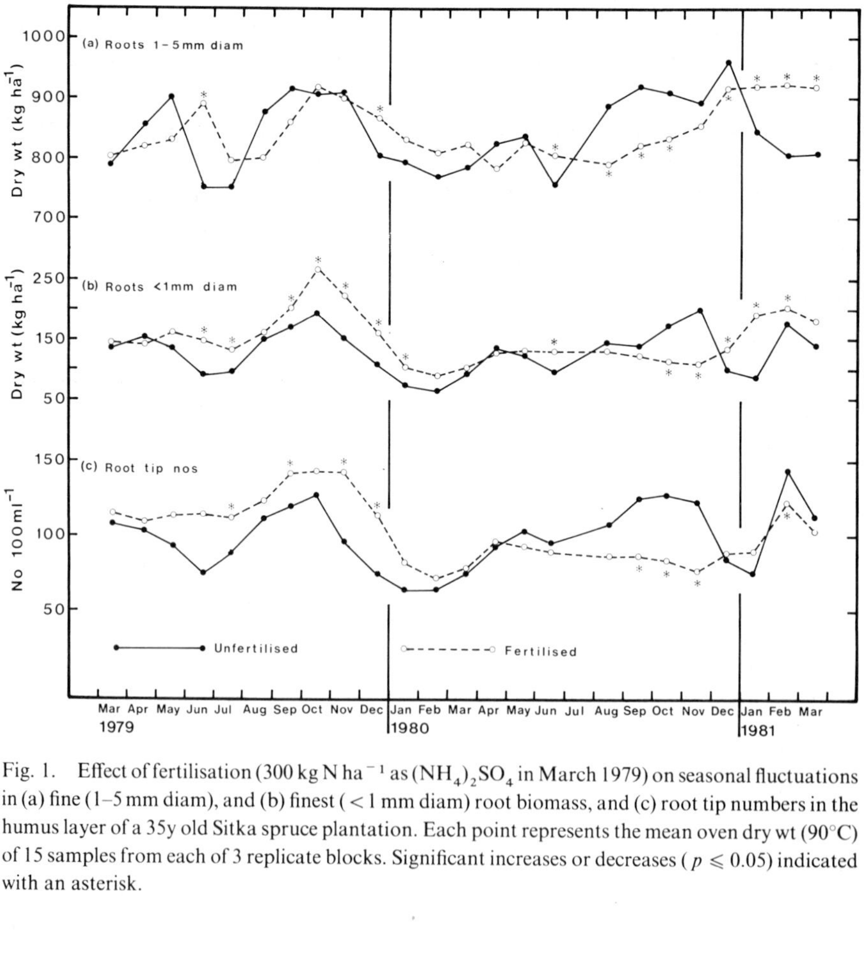

Fig. 1. Effect of fertilisation (300 kg N ha^{-1} as $(NH_4)_2SO_4$ in March 1979) on seasonal fluctuations in (a) fine (1–5 mm diam), and (b) finest (< 1 mm diam) root biomass, and (c) root tip numbers in the humus layer of a 35y old Sitka spruce plantation. Each point represents the mean oven dry wt (90°C) of 15 samples from each of 3 replicate blocks. Significant increases or decreases ($p \leqslant 0.05$) indicated with an asterisk.

fluctuation in root activity was reduced in fertilised plots and for much of the year numbers and biomass were less than in control plots. Towards the end of the year biomass, particularly of roots 1–5 mm diam, started to rise and this continued into the spring of 1981 so that when the experiment terminated the standing biomass in fertilised plots was considerably higher, though root tip numbers remained lower, than in control plots. We interpret these changes as follows. In the months following fertilisation the fine roots remain an effective sink for carbon by virtue of their increased capacity to supply N and are not shed. In the next year there is a readjustment of the functional balance between root and shoot, in the light of increased N availability, leading to reduced root growth. Finally the increased net assimilation resulting from fertilisation permits an overall increase in growth and by spring 1981 this is reflected in increased fine root biomass, particularly in the larger, more structural, 1–5 mm diam size class. At the same time a reduced density of root tips indicates less intensive exploitation of the substrate.

Production, mortality and longevity (Table 1)

Over the 2y period fertilisation decreased the production of mycorrhizas, finest and fine roots by 15%, 22% and 31% respectively. At the same time mortality was reduced by 30%, 30% and 36%. This differential between production and mortality leads to increased mean biomass in each category and indicates that fertilisation increases the longevity of the elements of the fine root system. A measure of this is given by the increased turnover times (Mean biomass ÷ Mortality) which indicate that following fertilisation mycorrhizas and the finest roots that bear them live on average for about 16 weeks, while the

Table 1. Effect of N-fertilisation on the mean biomass and dynamics of spruce fine roots and mycorrhizas in humus estimated from 24 monthly samples. Weights are oven dry (90°C)

	Biomass ± se kg ha^{-1}	Production kg ha^{-1} a^{-1}	Mortality kg ha^{-1} a^{-1}	Turnover time y
Unfertilised				
Mycorrhizas	87 ± 4	377	413	0.2
Finest roots < 1 mm diam	42 ± 3	213	233	0.2
Fine roots 1–5 mm diam	845 ± 11	289	216	3.9
Total	974 ± 5	879	862	
Fertilised				
Mycorrhizas	99 ± 4	323	289	0.3
Finest roots < 1 mm diam	51 ± 2	166	163	0.3
Fine roots 1–5 mm diam	850 ± 6	199	138	6.2
Total	1000 ± 10	686	590	

1–5 mm diam roots live for over 6 years. An important consequence of fertilisation may be a reduction in turnover of the fine root system allowing reallocation of those resources elsewhere in the tree. Fertilisation does not alter the position of the mycorrhizas as the most dynamic part of the fine root system: they account for only 10% of the biomass but up to 50% of the turnover.

Mycorrhizal infection

From 90%–97% of root tips from the field were judged to be mycorrhizal on the basis of external morphology. Infection was lower in the fertilised plots particularly during the winter months but not significantly so. Anatomical investigation revealed that these percentages were overestimates and that there was indeed a real effect of fertilisation and an even more pronounced effect of sampling time (Table 2). The disparity between morphological and anatomical classification is mainly caused by root tips which lack root hairs, possess distinct peripheral layers of tanned cells and have varying amounts of intracellular and surface hyphae but no net or sheath. Infection of potted plants was lower than of the trees in the field and greatly reduced in fertilised humus (Table 3).

Table 2. Effect of N-fertilisation on the mean % mycorrhizal infection of root tips in spruce humus sampled in September and December of successive years

	1979		1980	
	September	December	September	December
Unfertilised	90.7	76.7	95.3	83.7
Fertilised	83.5	62.5	90.7	64.8

Analysis of Arcsin transformed data shows significant main effect of month of sampling ($p \leq 0.001$) and fertilisation ($p \leq 0.05$). Year of sampling and all interactions are not significant.

Table 3. Mycorrhizal infection and distribution of mycorrhizal types on spruce seedlings grown for 20 weeks in N-fertilised and unfertilised humus. n = 12

	Mean % mycorrhizal tips	Total no. of types	Mean % mycorrhizas of type					
			1	2	9	10	11	Others
Unfertilised	71	8	24	19	0	0	0	57
Fertilised	34**	11	2**	10**	13	15	6	54

** Mean significantly different ($p \leq 0.01$). Arcsin transformation.

Mycorrhizal types

Fertilisation significantly reduced the incidence of *C. geophilum* mycorrhizas on field material. No other types could be distinguished. In fertilised humus the number of mycorrhizal types on the potted plants increased and their relative proportions were altered (Table 3). The most widespread type (1) in unfertilised conditions was virtually absent from fertilised pots where three new types (9, 10, 11) made up over 30% of the mycorrhizas.

Plant and Soil 71, 55–62 (1983). A-06

Seasonal variation of growth and development of the roots of five second year conifer species in the nursery

C. G. LANGLOIS,
Nurseries Division, MER, 200 B, Chemin Ste-Foy, Quebec, Canada, GIR 4X7

L. GODBOUT
Provincial Nursery manager

and J. A. FORTIN
Faculty of Forestry, Univ. Laval, Ste-Foy, Quebec, Canada, GIK 7P4

Key words *Abies balsamea* Coniferous species *Picea glauca* *Picea mariana* *Pinus banksiana* *Pinus resinosa* Root growth

Abstract Every other week over their second growing season, stem height, collar diameter, shoot and root dry masses, number of lateral roots and length of the tap root were measured on nursery grown seedlings of *Abies balsamea* L. Mill., *Pinus banksiana* Lamb., *Pinus resinosa* Ait., *Picea mariana* Mill. BSP and *Picea glauca* Moench Voss. Root elongation, branching and mycorrhizal development were also recorded.

Given species showed distinct seasonal growth patterns. The rate and timing of maximum root growth (mg/dry weight/week) differed markedly between species.

Except for the increase in height of *Pinus banksiana*, root and shoot growth were not negatively correlated.

The results are discussed in relation to the performance of tree seedlings in the nursery.

Introduction

The main objective, of forest nurserymen, is to produce seedlings which survive and grow well after outplanting[2,8,13,22,25]. Seedlings need to be of high quality[24], preferably with efficient mycorrhizae[16,17,18].

The growth and development of tree seedlings vary between species and with environmental conditions[15,21,23]. Lack of information on the growth and development patterns of a number of important species makes optimum scheduling of nursery operations difficult.

To obtain more precise knowledge of the interrelationships between shoot and root and their growth, *Abies balsamea* L. Mill., *Pinus banksiana* Lamb., *Pinus resinosa* Ait., *Picea mariana* Mill. BSP and *Picea glauca* (Moench) Voss., were studied from seed to shipping age.

Material and method

Seeds of two provenances of Black spruce (*Picea mariana*) and of four other species were sown in an experimental plot. All seed came from the 45th parallel except for Black spruce, provenance II, which

Table 1. The increase of the root dry weight for different coniferous species during their second year in the nursery

Date of measurement	*P. banksiana*		*P. resinosa*		*P. glauca*		*A. balsamea*		*P. mariana* (1)		*P. mariana* (2)	
	Root weight	± E(%)*	Root weight	± E(%)	Root weight	± E(%)	Root weight	± E(%)	Root weight	± E(%)	Root weight	± E(%)
04.30.81	191,3	14,3	117,5	10,4	27,5	14,1	28,3	9,7	13,4	9,5	18,3	8,4
05.11.81	281,9	8,6	156,6	7,1	37,9	9,4	31,6	11,0	23,7	8,5	19,7	9,7
05.26.81	279,5	5,9	156,9	6,2	36,6	9,1	28,1	8,0	21,7	9,2	20,6	8,5
06.08.81	245,9	10,4	181,9	6,5	38,3	10,8	29,6	7,0	23,1	8,0	21,7	15,6
06.22.81	246,8	8,5	157,6	8,3	47,1	9,2	35,4	7,6	25,8	8,3	25,3	11,0
07.08.81	335,4	8,0	170,4	7,9	63,5	7,4	38,6	4,8	33,3	4,7	35,9	9,3
07.20.81	413,3	10,5	225,9	7,5	75,7	10,4	54,6	8,3	32,5	6,2	35,8	8,5
08.03.81	485,6	5,8	285,0	6,2	106,5	6,6	73,7	13,9	51,4	13,1	54,9	13,3
08.17.81	671,7	13,5	379,6	4,5	146,5	11,7	100,1	8,7	82,8	5,7	81,2	14,8
09.01.81	978,4	31,2	469,2	14,5	171,1	5,0	119,3	16,0	117,7	24,7	114,1	10,6
09.25.81	1142,0	10,3	576,9	5,6	190,7	9,2	127,2	9,3	127,3	12,4	140,4	15,2
10.16.81	1308,6	13,4	710,3	5,9	204,6	8,3	152,2	9,0	141,9	10,6	139,7	9,2

* $\pm E\% = \pm \frac{E \times 100}{\text{Mean}}; E = \frac{\text{SED} \times t}{\sqrt{n}}$

came from the 49th parallel. The experimental plot consisted of five seven-row nursery beds, 150 m long. Each bed was divided into 25 m plots, to which 5 replicates of each species were randomly allocated. Every other week, 11 samples/replicate were brought to the laboratory, washed, and a minimum of 3 undamaged seedlings per sample selected for detailed study. A minimum of 165 seedlings per species were analysed on each occasion. During the second season, 11 samplings were conducted between April 30th and October 16th.

Shoot height, stem diameter at collar level, length of the tap root, and number of first order laterals were measured on each seedling. The number of ramifications of the shoot and the mean length of the first 4 lateral roots were measured on 30 seedlings out of 165. Phenological state, root elongation and proliferation, as well as abundance of ectomycorrhizae were observed. Samples from each replicate were bulked up and roots and shoots were dried and weighed in groups of ≅ 10. Variation was greatest for root weight although even here the errors on the means were under ± 15%.

Results

Root weight increase

The final weight of the different species showed large differences in root weight (Table 1): *P. banksiana* was 1.8 and 9.3 times respectively larger than *P. resinosa* and *P. mariana*.

Relative changes in root weight (multiples of initial value) (Fig. 1) show the largest increases occuring in the second half of the season. The size of changes varied between species, and ranged from a 5.3 times increase for *A. balsamea* to 10.6 for *P. mariana* I. Even at the final harvest, root weight was still increasing, except for the northern provenance of *P. mariana* II.

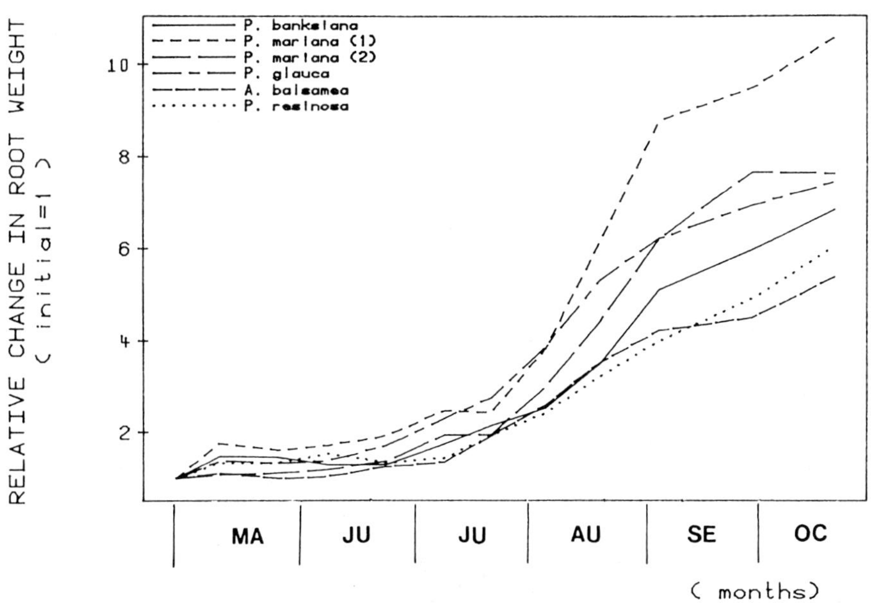

Fig. 1. The change in root weight during the second season.

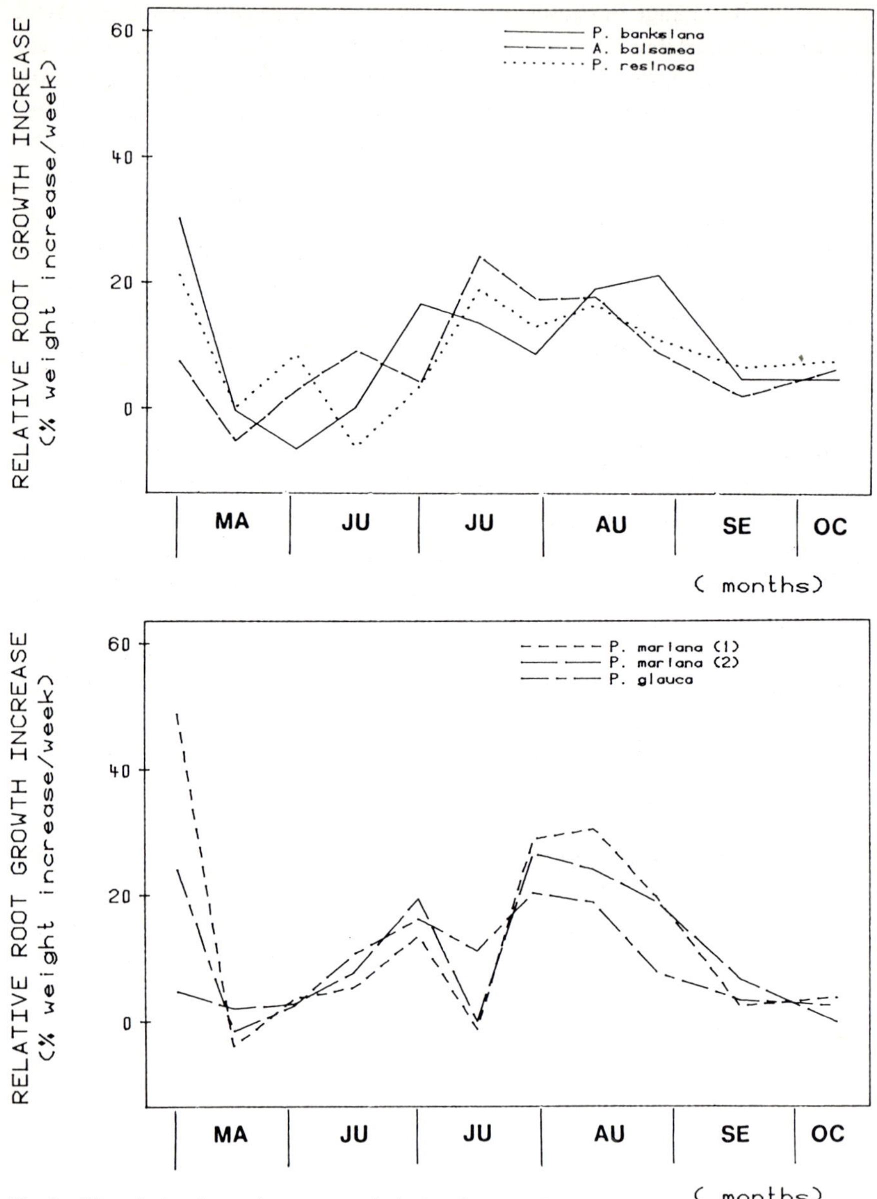

Fig. 2. The relative change in root growth during the second season.
Legend: The root weight increase of each interval between collection dates divided by the corresponding nomber of weeks was expressed as a percentage of the weight at the beginning of that interval.

% weight increase/week

$$\frac{\frac{W_2 - W_1}{W_1} \times 100}{\frac{T_2 - T_1}{7}}; \quad \frac{\frac{W_3 - W_2}{W_2} \times 100}{\frac{T_3 - T_2}{7}}; \textit{ etc} \ldots$$

The relative change in root growth (% increase/week) showed active growth occuring earlier in Balsam fir and Red pine than in Jack pine (Fig. 2a). At the very beginning of the season, Black spruce I (southern) and White spruce were much more active than Black spruce II (Fig. 2b). Growth activity increased in June, decreased in early July and increased again from mid-July to mid-August when it was at a maximum. After August, relative activity decreased faster in White spruce than in Black spruce.

The rate of root growth as percentage of maximum growth rate (mg/week) was reached earliest with *A. balsamea* and *P. resinosa*. Growth patterns for the two Black spruce provenances were similar, until the end of the season. White spruce reached a maximum value before Black spruce.

Shoot growth

An increase in tree height began in May and for most species was complete by the end of July. The timing of the onset of shoot extension and the duration and rate of extension varied between species; shoot elongation began earlier with *P. banksiana* than *P. mariana* while *P. resinosa* had the shortest elongation period.

Shoot weight increased over a longer period than did height (Fig. 3). Weight increased from June to late August and some species again in October. During this period, the timing and magnitude of the increase varied between species.

The change in shoot weight was similar in White and Black spruce I. Only Black spruce II showed no weight increase during October.

Like shoot weight, the change in stem diameters, at collar level, reached a

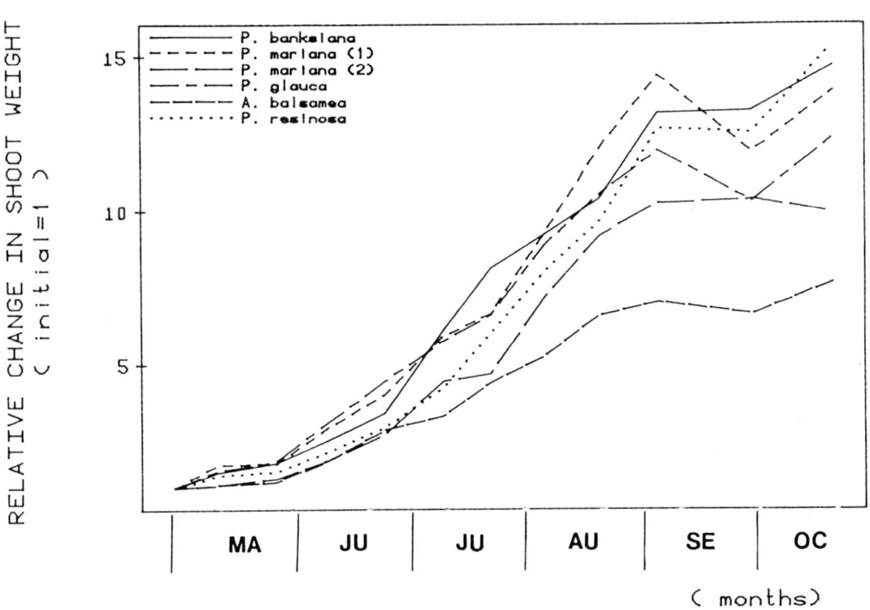

Fig. 3. The change in shoot weight during the second season.

maximum in August. During the season, there were a number of periods, which varied between species, showing an increase.

Relationships between root and shoot growth

Correlation coefficients were obtained for the relationships between root and shoot growth parameters during the season. In *P. banksiana*, the rate of increase in height and the rate of increase in root weight were negatively related ($R^2 = 0.65$). In all other cases, correlations were either positive or inconclusive. In all species, root weight was positively correlated with shoot weight, although their growth rates and relative growth increases were not. The same was true for correlations between root masses and stem diameters. No positive correlation was observed between root weight and shoot height, or their respective growth rates or relative growth increases.

Variation in type of root growth

Qualitative observations on root elongation and proliferation indicate that the root systems of all the different species were active in the period between mid-June and late August.

The general root development pattern observed consists of the growth of the tap root and first order laterals, then the proliferation of rootlets on slow growing laterals which more or less coincided with the onset of the emergence of higher order lateral roots and an increase in root diameter. The most active elongation phase occurred when relative root weight was rising while the proliferation phase and root diameter increase phase were coincident with the largest changes in root dry weight. In *A. balsamea*, an elongation phase occurred in July and August while lateral root diameters increased simultaneously. The emergence of second order roots was localised between mid-June and mid-July and between mid-August and mid-September.

In *P. mariana*, the number of short rootlets significantly increased between mid-July and mid-August. Their number was proportional to root length over the whole season in *P. banksiana*, *P. resinosa*, *A. balsamea* and *P. glauca*.

Natural mycorrhization

Mycorrhizal colonization was recorded as the presence of a fungal mantle on small rootlets and the proportion of the total number of rootlets infected. In *Pinus* spp., the presence of dichotomy and corralloid branching was also considered. The ectomycorrhizal colonization of all species was poor in the second season, although an infection phase seemed to occur in late August early September.

Discussion

The root growth of the various species followed distinct developmental patterns during the second growing season. The major phases were in the summer months[15,23], July and August, rather than before shoot elongation[22].

All seedlings from the 45th parallel showed a short phase of intense activity in early spring[6,19]. Root weights measured at the beginning of the third year indicated that autumn growth persisted weakly after the last sampling in year 2.

Root growth activity can potentially occur during most of the vegetative season[1,10,11,12,15] if cultural conditions, especially soil moisture[7,23], are favorable. The most active phases of root growth occurred after active shoot elongation[12], and at about the time of shoot mass and stem diameter increase. Root weight increased with increase in day length ($R^2 = \geqq .78$) and was also correlated with relative weekly increase in degree days ($R^2 = \leqq .70$).

Previous comparisons between shoot elongation and root elongation have indicated antagonism for some species[20,21] between aerial and underground growth and not for others[4,15,,27]. Here there was a negative correlation only in *P. banksiana*. In other species, no such correlation were found although root growth took place after the main flush of shoot elongation, presumably because the bulk of sugars are for a time diverted to shoot elongation[5,9,14].

The usefulness of root weight as a growth parameter is often doubted (3). However, we feel that the root weight growth rates, as well as relative growth increases, are useful to locate the active phases during the season. Root elongation phases coincided with rises in relative growth increases and root proliferation activity coincided with maximal root mass growth rates.

It is important for nurserymen to locate the root mass growth phases to allow better timing of cultural operations.

References

1 Armson K A 1960 White spruce seedlings: The growth and seasonal absorption of nitrogen, phosphorus and potassium. For. Bull. No. 6 Univ. Toronto, 37 p.

2 Armson K A and Sadreika V 1979 Forest tree nursery soil management and related practices Ont. Min. Nat. Res. Can., 179 p.

3 Böhm W 1979 Methods of studying root systems. Ecological studies: Analysis and synthesis 33 Springer-Verlag, Berlin, Heidelberg, New York, 188 p.

4 Büsgen M 1901 The shape and manner of growth of tree roots. Allg. Forst-u. Jagdztg. 77, 273–278; 303–309.

5 Carlson L W The effect of defoliation on conifer seedling root initiation. Can. For. Serv. Bi-monthly Res. Notes 33.1.

6 Crossley D I 1956 The possibility of continuous planting of white spruce throughout the frost free period. Tech. Note For. Br. Can. No. 32, 31 p.

7 Day R J 1980 Basic nursery soil physical properties in: Proc. North Amer. For. Tree Nurs. Soils Works. Syracuse, N. Y., 52–71.

8 Day R J, Stupendick J T and Butler J M 1976 Root periodicity and root regeneration potentiel

are keys to successful plantation establishment. *In* Proc. Ont. Min. Nat. Res. and Can. For. Serv. Symp. on 'Plantation Establishment' Kirkland Lake, Ont. 16 p.

9 Glerum C 1980 Food sinks and food reserves to trees in temperate climates. N. Z. J. For. Sci. 10, 176–185.

10 Kramer P J and Kozlowski T T 1960 (Eds.) Physiology of Trees. McGraw Hill, N. Y., 642 p.

11 Ladeloged K 1939 Investigations on the periodicity of eruption and length of roots of some of our commonest tree species. Department of Forestry, Ottawa. Translocation from Forstl. Forsøgsv. Danm. 16, 1–256.

13 Leaf A L, Rathakette P and Solan F M 1978 Nursery seedling quality in relation to plantation performance *In* van Eerden and Kinghorn (Eds.) pp 45–52.

14 Loach K and Little C H A 1973 Production storage and use of photosynthate during shoot elongation in balsam fir (*Abies balsamea*). Can. J. Bot. 51, 1161–1168.

15 Lyr H and Hoffmann G 1967 Growth rates and growth periodicity of tree roots. Int. Rev. For. Res., N. Y., 181–236.

16 Marx D H 1977 The role of mycorrhizae in forest production. Tappi Conf. Pap., Annu. Meet., Atlanta, Ga., 151–161.

17 Mexal J G 1980 Aspects of mycorrhizal inoculation in relation to reforestation. N. Z. J. For. Sci. 10, 208–217.

18 Mikola P 1970 Mycorrhizal inoculation in afforestation. Int. Rev. Forest Res. 3, 123–196.

19 Mullin R E 1963 Growth of white spruce in the nursery. For. Sci. 9, 68–72.

20 Resa F 1878 The phenology of root formation especially of woody plants. Forstl. Bl. 15, 321–331.

21 Riedacker A 1976 Rythmes de croissance et de regénération des racines des végétaux ligneux. Ann. Sci. Forst. 33, 109–138.

22 Ritchie G A and Dunlap J R 1980 Root growth potential: Its development and expression in forest tree seedlings. N.Z. J. For. Sci. 10, 218–248.

23 Sutton R F 1969 Form and development of conifer root systems. Commonw. For. Bur. Tech. Comm. No. 7, Oxford, England, 131 p.

24 Sutton R F 1979 Planting stock quality and grading. Forest Ecol. Manage. 2, 123–132.

25 Van den Driessche R 1980 Health vigour and quality of conifer seedlings in relation to nursery soil fertility in: Proc. North Amer. For. Tree Nurs. Soils Works. Syracuse, N.Y., 100–120.

26 Van Eerden E V, Kinghorn J M 1978 (Eds.) Proceedings of the root form of planted trees symposium. B.C.M.F./C.F.S. Joint Report No. 8, 357 p.

27 Wierler A 1894 Periodicity in the root formation of plants. Forstwiss Cbl. 16, 333–349.

Plant and Soil 71, 63–74 (1983). A-07

Root distribution of four *Vitis* cultivars

R. L. PERRY, S. D. LYDA* and H. H. BOWEN
Horticulture Department, Michigan State University, East Lansing, MI 48824, USA

Key words Calcium nitrate Distribution Roots Vitis

Summary Root distribution of four *Vitis* cultivars at one location in Texas was asessed using trench profile and soil core sampling methods. Greater differences were obtained among cultivars utilizing the latter method when comparing root dry weights. Surface area of roots extracted from soil samples was estimated using a calcium nitrate slurry dip. This technique was found to be unreliable. *Vitis champini* cv. Dogridge maintained the greatest above-ground canopy and the most extensive root system. In measuring root numbers and dry weight, *V. rotundifolia* cv. Noble ranked at or near the bottom with *V. vinifera* cv. Barbera and *V. labruscana* cv. Concord being intermediate. Overall schematic root system patterns for each cultivar indicate that Noble is shallow rooted, having more than 35% of the total roots in the 0–15 cm depth. In contrast, ten percent of the total Dogridge roots were found at the 90–105 cm depth.

Introduction

Wide differences in distribution of *Vitis* species and cultivar adaptation are probably the result of variation in soil and climatic conditions in the USA. The primary limitations to grape production in the southwestern U.S., are high soil pH, high salt accumulations, drought stress, mineral deficiencies, and Phymatotrichum root rot[15]. All of these factors are edaphic, providing obstacles to grape root systems.

Differences in root distribution and tolerance to mineral deficiencies, water stress and diseases are influenced by the local environment and plant genetic make-up[3]. Plants with deep root systems are considered more tolerant of drought stress than those with shallow systems[13]. Root distribution studies are difficult and little information is available on the root patterns of *Vitis* species, although growers are warned to avoid deep cultivation of muscadine (*Vitis rotundifolia*) and so root injury because of its shallow rooting habit[2,5].

The objective of this study was to characterize the root distribution pattern, by profile wall and auger methods, of four Vitis cultivars, each representing a different species. An estimate of root surface area was made of roots collected from augered soil samples.

Materials and methods

Establishment and maintenance

One-year-old vines of *V. vinifera* L. cv. Barbera, *V. labruscana* Bailey cv. Concord. *V. rotundifolia* Mich. × cv. Noble and *V. champini* Planch cv. Dogridge were planted on April 25, 1974 in a field

* Plant Sciences Dept., Texas A&M Univ., College Station, TX 22843, USA.
Michigan Agricultural Experiment Station Journal Article No 10517.

Table 1. Comparison of several methods for estimating total root volume of four Vitis cultivars, 4-years-old

Cultivar	Measuring techniques				
	Root count[w]			Dry wt.[x] (g)	Surface adsorption[y] (g)
	1 mm	1 mm	Total		
Barbera	230 a[z]	67 a	296 a	7.2 b	31.2 ab
Concord	228 a	58 a	287 a	5.3 b	25.3 b
Dogridge	201 a	69 a	270 a	10.6 a	32.9 a
Noble	186 a	45 a	231 a	5.8 b	25.2 b

[w] Diameter size ranges counted on vertical trench face.
[x] Root samples measured per 2651.4 cu cm soil.
[y] Measured gravimetrically with $Ca(NO_3)_2$ slurry.
[z] Mean separation within columns by Duncan's multiple range test, 5% level.

consisting of Selden fine sand at the Texas Agricultural Experiment Station Fruit Testing Laboratory, Montague, Texas. Plants were arranged in single plant plots, in a randomized complete block design and replicated 4 times. Vines were spaced 2 m by 6 m and trained to a 122 cm supporting stake. Plants were set through a 1 meter wide, 0.102 mm thick black plastic mulch to prevent weed growth and conserve moisture during the first year. Weed growth was controlled mechanically to a depth of 7.5 cm between rows and by spot treatment with a contact herbicide applied twice during the season down the trellis row.

One shoot was selected to form a trunk during the following growing season and was trained into a head system at the top wire of a 2 wire trellis by the end of 1975. In february 1976, each plant was pruned, allowing two, 2-bud spurs. At this time, 50 g of a pelleted fertilizer (13-13-13) was incorporated in the soil around each vine. Training for the next 3 growing seasons was limited to deshooting trunks. Four, 2-bud spurs were allowed each vine during pruning in March of 1977 and four, 4-bud spurs per plant in winter of 1978. Fruit clusters were first allowed in the fourth growing season when 50% were retained at 20 days post bloom. Water was supplied as needed by sprinkler irrigation. A caliper measurement of each trunk was taken at 5 cm above the ground level in February 1976 and again just prior to root mapping and sampling in June of 1978.

Data collection

Root counts and sizes in profile sections were obtained by the trench profile method[4]. Dry weight and surface area estimates were made with roots extracted from soil core samples taken with the auger method[4].

Trench profile method A trench 90 cm wide and 120 cm deep was dug parallel to and on the north side of the vine row on June 26, 1978 (Fig. 1). The inside wall of the trench was 38 cm from the vine trunk row. Soil (approximately 8 cm) surrounding roots on each wall was carefully removed. A rectangular metal frame 105 cm × 240 cm was divided into a grid with 15 × 30 cm sections and centered at each vine trunk against the 2 walls. Roots were mapped and counted within the 15 × 30 cm. Roots were categorized into a range of 0–1 mm, 1–2 mm, 2–5 mm, 5–10 mm, 10–20 mm and 20–30 mm diameter.

Auger method Soil cores were taken on July 10, 1978 on the south side of the vine rows at 30, 60, 90 and 120 cm distances from the trunks. Nine core samples, pooled for 15 cm depths, at each distance per vine were collected from points 10 cm apart, with a steel frame, hydraulic cylinder and a 90 × 5 cm sampling tube. Collection points were parallel to the vine row (Fig. 1). Soil samples, 75 cm or more in length (depth), 5 cm in diameter, were collected and divided into 6 segments (15 cm); with a volume of 2651.4 ml for each depth and lateral distance. Samples were stored at 7°C until evaluation.

A. Root recovery Soil samples were soaked 24 h in sodium hexametaphosphate solution (28 g in 8 l of water) to deflocculate and soften core samples prior to washing[18]. Soil core segments were mounted on a reciprocating shaker. A continuous flow of water sprayed the samples in order to wash the soil through the sieves, and the remaining particles consisting of roots, leaves, small stones and shells were brought to the laboratory where fleshy and woody grape roots were teased and separated. Samples were stored in water at 6°C prior to measurement.

B. Root surface area Each sample of roots was air dried, placed in a 6 mm mesh wire basket and dipped in a 6:1 solution of calcium nitrate[7]. Root adsorptive area was measured as a gravimetric difference beeen the amount of freshly made $Ca(NO_3)_2$ slurry absorbed on root samples and the basket. The basket with roots was dipped for 10 sec. then drained for 30 sec. Without roots, the basket was dipped twice every 10 samples to monitor basket slurry absorption. Calculation for root adsorption of slurry were based on differences from empty basket weights.

Fresh slurry was added to the weighing beaker to correct volume when a maximum of 20 ml had been lost. A greater deficit decreased the area of basket in contact with slurry which increased variation.

The accuracy of this method was compared with wooden dowels treated similarly. One, two and three pieces of wooden dowels with the dimensions: 3.0 mm × 52 mm ($504.29\ mm^2$), 4.0 mm × 52 mm ($678,67\ mm^2$), 5.2 mm × 52 mm ($865.94\ mm^2$) and 8.8 × 52 mm ($1476.5\ mm^2$) were dipped.

C. Dry Weight After roots were dipped, they were washed immediately, dried at 100°C for 2 h and weighed.

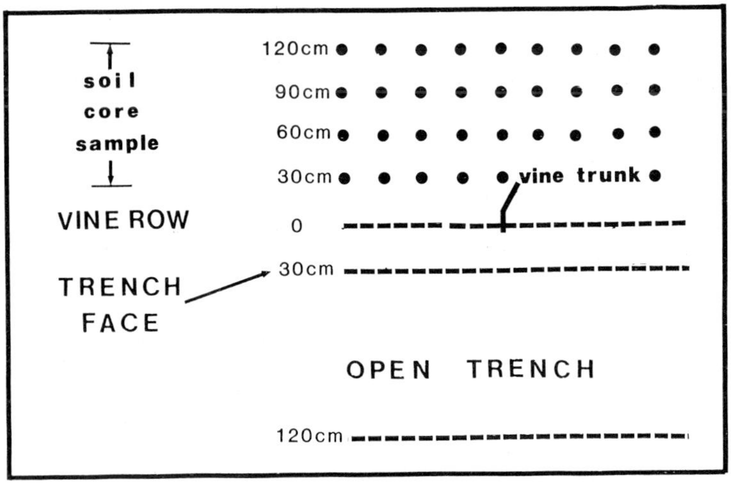

Fig. 1. Soil core sampling (●) and trench placement with respect to each vine. Soil Cores (z): 5 × 75 cm.

Table 2. Mean quantity of roots of 4 Vitis cultivars sampled from a vertical profile wall and soil core samples to a 75 cm depth and 120 cm lateral distance

DEPTH (cm)	Cultivar	Measuring techniques: Count <1 mm	Count >1 mm	Dry wt. (mg)	Surface area (g)	Count <1 mm	Count >1 mm	Dry wt. (mg)	Surface adsorption (g)
0	Dogridge	66	21	3539 a[z]	8.8	44	13	1801	5.9
	Barbera	75	32	2551 ab	8.0	32	8	1120	6.2
	Concord	57	24	1474 b	6.2	33	11	1396	6.2
	Noble	71	18	1926 b	7.9	26	11	911	5.4
30									
	Dogridge	52	23	3413 a	9.9 b	39	13 a	1835 a	8.0 a
	Barbera	92	20	2436 b	11.6 a	30	7 b	1033 b	5.4 b
	Concord	85	16	1662 b	8.3 bc	52	9 b	991 b	4.6 b
75	Noble	63	13	1493 b	6.8 c	26	3 b	803 b	5.1 b
					60				120

LATERAL DISTANCE (cm)

[z] Mean separation within columns and quadrants by Duncan's multiple range test, 5% level.

Table 3. Correlation coefficients between roots counted, dry weight, and root surface area from roots of 4 Vitis cultivars sampled in 15 × 30 cm profile sections

	Root count	Dry weight (mg)
Root count	–	–
Dry weight (mg)	0.27*	–
Root surface (g)	0.30*	0.58*

* Significant at the 5% level.

Results

Only the root count data for the trench face nearest the vine row is presented since there were few roots on the outside wall. Due to lack of roots >2 mm in the 15 × 30 cm sections consolidating the size ranges were amalgamated to <1 mm and >1 mm.

Cultivar comparison

Root mass 'Dogridge' was the most vigorous of the cultivars in this study as indicated by the increase in trunk cross sectional area. It also had the greatest, mass of roots number of roots >1 mm and total potential surface area (Table

Table 4. Mean gravimetric measurements (g) and coefficient of variability (cv) for wooden dowels dipped in $Ca(NO_3)_2$ slurry as 1 piece, 2 pieces and 3 pieces for each of 4 sizes: 504 mm^2, 679 mm^2, 886 mm^2 and 1476 mm^2

	Dowel size							
	504 mm^2	CV	679 mm^2	CV	866 mm^2	CV	1476 mm^2	CV
1 dowel	0.11	28.7	0.22	19.2	0.34	15.2	0.49	14.7
2 dowels	0.20	33.3	0.32	13.3	0.48	13.2	0.67	15.8
3 dowels	0.31	10.2	0.53	21.8	0.65	13.1	0.84	11.5

1). 'Barbera' had the greatest number of roots < 1 mm and was second to 'Dogridge' in root surface area. 'Noble' ranked at or near the bottom, with 'Barbera' while 'Concord' was intermediate for dry weight, count (>1 and <1 mm size) and surface area.

Profile data, pooled into 4 vertical quadrants, showed 'Dogridge' with the

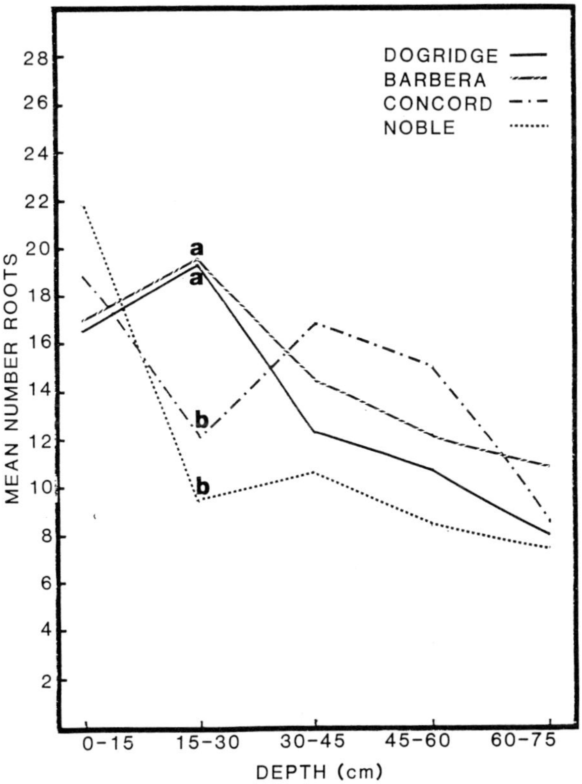

Fig. 2. Number of roots on trench face for each 15 cm depth of 4 Vitis cultivars. Means pooled for 4 distances from vine trunk: 30, 60, 90 and 120 cm. Mean separation within Duncan's multiple range test, 5% level.

greatest dry weight in 3 of the 4 quadrants (Table 2). 'Dogridge' had the greatest root volume in the lower quadrants, 'Noble' was lowest throughout. There were no significant differences among cultivars in root counts throughout the profile except in the outer, lower quadrant where 'Dogridge' had more roots >1 mm. No significant difference among cultivars in root surface area was found in the upper 30 cm of the profile. 'Barbera' was significantly greater in the lower quadrant nearest the trunk while 'Dogridge' dominated the outer, lower quadrant. 'Noble' was lowest overall. In all categories except small root count, 'Dogridge' had the highest root volume in the lower outer quadrant.

'Barbera' and 'Dogridge' were significantly higher in root quantity measured by all methods at 15–45 cm except at 30–45 cm where difference in root counts was not significant (Figs. 2–7). Although not statistically different, 'Noble' and 'Concord' had the highest root quantity for all 3 parameters measured at 0–5 cm depth.

Rooting pattern The percentage of roots by depths to 105 cm, 120 cm from vine, for each cultivar show 'Noble' as shallow rooted with more than 34% at

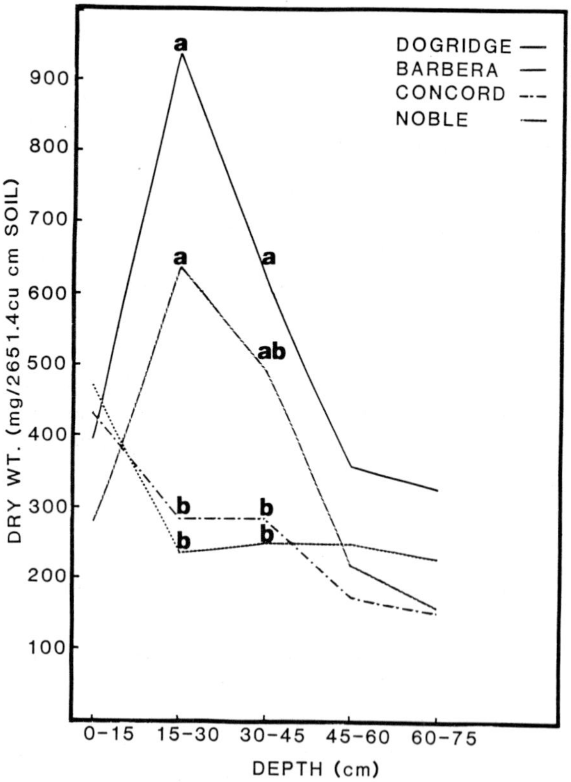

Fig. 3. Dry weight of roots for 4 Vitis cultivars extracted from soil cores at 5 depths. Means pooled for 4 distances from vine trunks; 30, 60, 90 and 120 cm. Mean separation within depths by Duncan's multiple range test, 5% level.

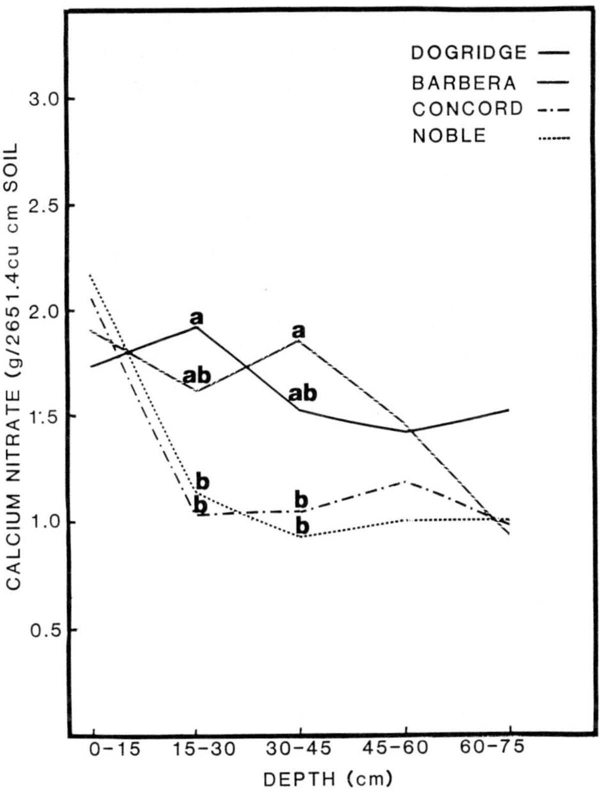

Fig. 4. Gravimetric measurements of the adsorptive surface area of roots extracted from soil cores using Ca $(NO_3)_2$ slurry dip for 4 Vitis cultivars at 5 depths. Means pooled for 4 distances for 4 distances from vine trunks; 30, 60, 90 and 120 cm. Mean separation within depths by Duncan's multiple range test, 5% level.

0–15 cm depth. 'Dogridge' possessed a relatively deep root system with root counts above 10% at the 90–105 cm depth. 'Barbera' and 'Concord' demonstrated a progressive decline of root volume through the profile with 'Barbera' having a more concentrated system.

Methods comparison

The results from the different methods used to assess root distribution were significantly correlated (Table 3). However, lower values indicate weak relationships. Correlation coefficients from a multivariate analysis of root counts > 1 mm and < 1 mm in diameter and dry weight for each pooled quadrant of the profile were not significant. The lack of statistical significance suggests either insufficient replicates or an influence of other factors with a major influence on root distribution, *e.g.* soil bulk density, porosity and vine age.

We question reliability in estimating surface area of extracted roots by gravimetric measurement of the amount of $Ca(NO_3)_2$ slurry absorbed. When

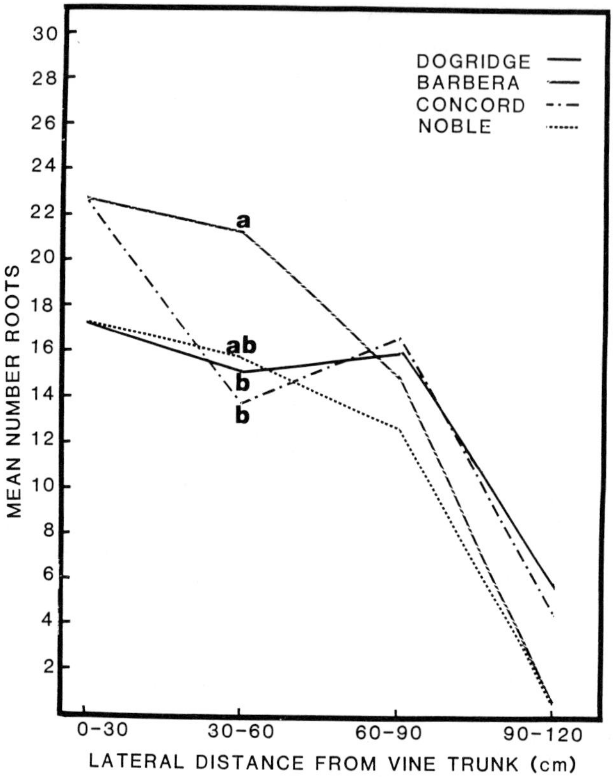

Fig. 5. Numbers of roots on trench face 75 cm deep for each 30 cm distance from vine trunk of 4 Vitis cultivars. Mean separation within distances by Duncan's multiple range test, 5% level.

wooden dowels of known surface area were assessed, the coefficient of variability was high and inconsistent (Table 4). Measurements may have been distorted when more than one piece were dipped together[7]. Roots in solution attracted to one another due to tension causing adjacent surfaces to be uncovered by slurry.

Discussion

Cultivar comparison

'Dogridge' had the most above-ground vigor and the deepest, most extensive root system laterally among the cultivars. These results compare similarly with citrus, where a direct relationship was found between tree height and rooting depth with sweet orange scion on several rootstocks[8]. Apple, in contrast, when grown on dwarfing[9] rootstock possesses a deep concentrated (vertical) rooting pattern, while the same scion variety on more vigorous M1 and M16 rootstocks have a greater percentage of roots in the top 30 cm of the soil profile[17].

A universally compatible scion cultivar would have been desirable but is not available for all 4 Vitis cultivars. In apple, scion cultivar does not affect the

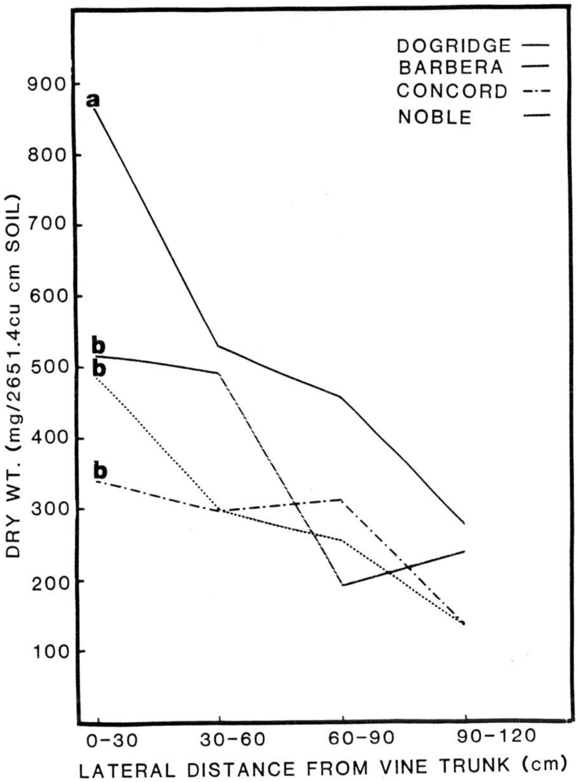

Fig. 6. Dry weight of roots for 4 Vitis cultivars extracted from soil cores 75 cm deep at 4 distances from vine trunks. Mean separation within distances by Duncan's multiple range test, 5% level.

character and morphology of the rootstock root system[16,19] and in orange scion cultivar did not affect root volume and depth of Cleopatra mandarin rootstock[11].

A deep, extensive root system, as possessed by 'Dogridge', may play a major role in avoiding moisture stress. 'Dogridge' is considered a vigorous grape rootstock and it is recommended for sandy, droughty soils in the San Joaguin Valley of California[12]. Tree fruit non-structural root volume is correlated with soil moisture depletion[6,9,10] and in 4 stocks studied, rough Lemon citrus had the largest daily water loss, root weight most extensive root system [9] and highest leaf water potential.

Depth of rooting and avoidance of soil moisture stress may play an important role in geographical adaptation of Vitis. For example, the shallow rooted *V. rotundifolia* is indigenous to the high humidity frequent rainfall area of southeastern U.S.[2] *V. champini* is native to the low rainfall arid west central area of Texas[12].

While 'Dogridge' demonstrated a deep and extensive root system 'Noble' confirmed previous reports of a shallow root system. However, most roots were

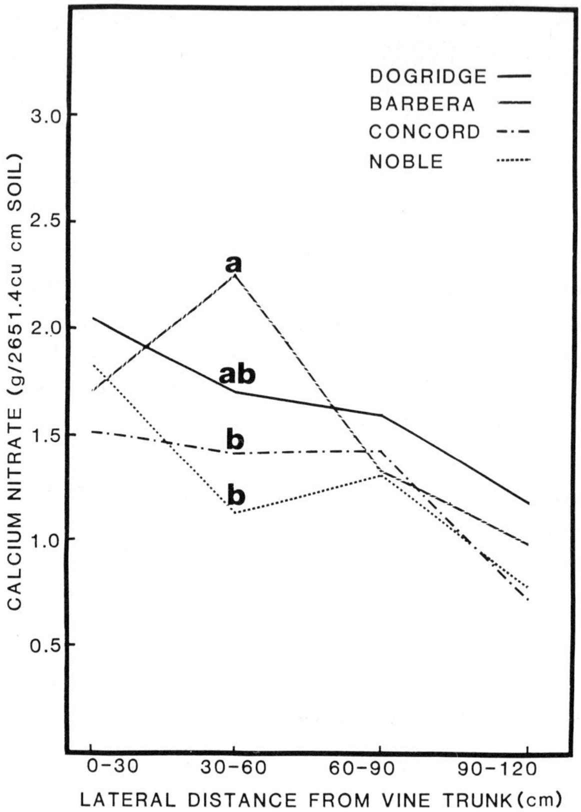

Fig. 7. Gravimetric measurements of the adsorptive surface area of roots extracted from soil cores using $Ca(NO_3)_2$ slurry dip for 4 Vitis cultivars 75 cm deep at 4 distances from vine trunks. Mean separation within distances by Duncan's multiple range test, 5% level (no letters: not significant).

not in the upper 30 cm, previously suggested [5], but over 52% were below 30 cm. Tree root distribution is highly influenced by soil profile composition, particularly bulk density, porosity and available oxygen supplies [20]. The soil in this study was uniform and not considered a factor in root distribution among vines although the percent clay in the B horizon increased with depth. Roots of all cultivars were found at the 105 cm depth and over 52% of the roots of 'Noble', the most shallow rooted cultivar, were in the B horizon. Root distribution can interact with soil conditions, although cherry, apple and peach roots all possessed similar deep rooting patterns in well drained soils and shallow patterns in poorly drained soils [14]. Citrus responded similarly [1] and plum root distribution [14] was unaffected by changes in soil type. Apple rootstocks grown in sand, loam and clay soils, while modified, retained distinctive rooting patterns [17].

Differences in plant age prevents making more detailed root distribution comparisons of vines in this study with older tree fruits in previous studies. A greater volume of roots is found shallower and more concentrated in younger plants[4]. Rogers and Vyvyan[17] found that root systems of mature apple trees maintained distinctive characters of different rootstocks observed when they were young.

Methods comparison

The trench profile and auger methods have shortcomings and benefits. Qualitative analysis of root distribution related to the soil matrix and roots of adjacent vines is easier with the trench profile. The sampling of roots from one geometric plane limits accurate quantitative analysis, *e.g.* roots were often missing deep and in the outer extremities of the profile for one side of a vine but not for the other. The destructive nature of the process prevents retain sampling.

The auger method can give a more accurate quantitative analysis and allow periodic non-destructive sampling. However, for vines of this age, fewer roots are found in the profile so increasing the chances of negative or inaccurate sampling. The lack of high correlation and statistical significance among methods, cultivars and profile sections, suggests the need to increase plant replications and perhaps diameter of core samples.

References

1 Adriance G W and Hampton H E 1949 Root distribution in citrus as influenced by environment. J. Am. Soc. Hortic. Sci. 53, 103–108.

2 Anonymous 1971 Muscadine grapes, a fruit for the south. USDA Farmers Bull. #2157, 14 p.

3 Bilan M V 1969 Some aspects of tree root distribution. pp 69–80. *In* Mycorrhizae. Ed. E Hacsckaylo. Proc. 1st. N. Am. Conf. Mycorrhizae, Misc. Publ. #1189 USDA.

4 Böhm W 1979 Methods of studying Root Systems. Springer-Verlag, NY. 188 p.

5 Brooks J F 1972 Muscadine grapes production guide for North Carolina. N. C. State Coop. ext. Service Bull. 32 p.

6 Cahoon G A and Stolzy L H 1959 Estimating root density and distribution in citrus orchards by the neutron moderation methods. J. Am. Soc. Hortic. Sci. 74, 322–327.

7 Carley H E and Watson R D 1966 A new gravimetric method for estimating root surface areas. Soil Sci. 102, 289–291.

8 Castle W S and Krezdorn A H 1974 Effect of citrus rootstocks on root distribution and leaf mineral content of Orlando Tangelo trees. J. Am. Soc. Hortic. Sci. 100, 1–4.

9 Castle W S and Krezdorn A H 1977 Soil water use and apparent root efficiencies of citrus trees on four rootstocks. J. Am. Soc. Hortic. Sci. 102, 403–406.

10 Conrad J P and Veihmeyer F J 1929 Root development and soil moisture. Hilgardia 4, 113–134.

11 Ford H W 1959 Growth and root distribution of orange trees on two different rootstocks as influenced by depth to subsoil clay. J. Am. Soc. Hortic. Sci. 74, 313–321.

12 Kasimatis A N and Lider L 1975 Grape rootstock varieties. Univ. Calif. Coop. Ext. Leafl. #2780, 16 p.

13 Kramer P J 1969 Plant and Soil Water Relationships: a modern Synthesis, McGraw-Hill, N.Y.

14 Oskamp J 1932 The rooting habits of deciduous fruits on different soils. J. Am. Soc. Hortic. Sci. 29, 213–219.

15 Perry R L and Bowen H H 1974 A feasibility study for grape production in Texas. Tex. Agric. Exp. Sta. Tech. Rep. 74–43.

16 Rogers W S and Beakbane A G 1956 Stock and scion relations. Annu. Rev. Plant. Physiol. 8, 217–236.

17 Rogers W S and Vyvyan M C 1934 Rootstock and soil effect on apple root systems. J. Pomol. Hortic. Sci. 12, 110–150.

18 Schuurman J J and Goedewaagen M A J 1971 Methods for the Examination of Root Systems and Roots. Agric. Publ. and Documentation, Wageningen, Netherlands.

19 Tukey H B and Brase K D 1933 Influence of the scion and of an intermediate stem-piece upon the character and development of roots of young apple trees. N.Y. Agric. Exp. Sta. Tech. Bull. 281.

20 Wiersum L K 1980 The effect of soil physical conditions on roots and uptake. pp 11–121. *In* Mineral Nutrition of Fruit Trees. Eds. D Atkinson, J E Jackson, R O Sharples and W M Walter. Butterworths, London.

Plant and Soil 71, 75–85 (1983). A-08

Root turnover and productivity of coniferous forests

R. FOGEL
Herbarium, University of Michigan, Ann Arbor, MI, USA

Key words Ecosystem Mycorrhiza Root biomass

Summary Fine roots and mycorrhizae have recently been shown to produce a major portion of the organic matter entering decomposition. Roots and mycorrhizae constitute 63 to 70% of total net primary production in Douglas-fir and Pacific silver fir stands. The importance of roots in primary production makes the method of root extraction from the soil important. Wet-sieving with small mesh screens is more effective than hand-sorting for fine roots and mycorrhizae. Screen size, the efficiency of recovery, the physiological status of the roots and conversion factors to derive biomass from the numbers of root tips should be stated. Published data is enhanced if the phenological status of the stand, its age, tree density, and soil texture are quoted.

Given the large fluxes in fine root and mycorrhiza populations, single biomass estimates are not useful in studies of ecosystem structure and function. A better understanding needs accurate methods to distinguish live and dead roots, data on the production and turnover of large roots, and data on the transfer of nutrients accompanying the large input of roots to decomposition.

Introduction

The role of roots and mycorrhizae in the primary production and mineral nutrition of coniferous forests is attracting increased attention. The exclusion of mycorrhizae, until recently, reflects the problems in separating them from soil and studying them *in situ*. Data on the belowground ecosystem is needed to extrapolate to field conditions information on ion uptake and translocation, and respiration.

This review examines the contribution of roots (mycorrhizal and non-mycorrhizal) to the productivity of coniferous forests, provides information on the efficiency of two methods for estimating root biomass, and compares three methods of calculating root turnover.

Root classification

Several different systems have been employed to classify roots. One common system divides roots on the basis of the order they arise from a primary root, *e.g.*, first order lateral producing a second order lateral, *etc*[44]. Roots providing structural support for the tree may be termed long lateral roots (or any of 20 or more other names) while the roots sometimes suggested as functioning in the absorption of water and nutrients may be called short roots and are very often mycorrhizal[44]. Among the conifers, short roots are found throughout genera in the Pinaceae but not in the Cupressaceae[44]. Lateral roots may be classified into large and fine roots on the basis of diameter. There is no established convention defining the diameter size range of fine roots, although many biomass studies use 5 mm or 2 mm as the upper limit in diameter[43]. Some researchers further subdivide fine roots into a number of size classes: 2–5 mm in diam, less than 2 mm in diam, less than 1 mm in diam, *etc*.

Mycorrhizae should be treated as a separate class of roots because they result from a mutualistic symbiotic association between a host and a fungus and have a profound effect on fleshiness, amount and type of branching, rate of elongation, root hair formation, and suberization. Beyond the morphological and anatomical differences, mycorrhizae play a major role in the mineral nutrition of trees and may confer disease or drought resistance[45]. The association between host and fungus in mycorrhizae is so integrated that complete separation of the host and fungus is virtually impossible. In ectomycorrhizae, the most common type in the Pinaceae, short roots are covered with a mantle of hyphae of variable thickness from which hyphae extend out into the soil and intercellularly into the primary cortex to form a Hartig net. An individual hypha may extend more than 2 m from a mycorrhiza and form more than 120 lateral branches; 200 to over 2000 individual hyphae have been counted emerging from a single mycorrhiza[45]. From the standpoint of function, mycorrhizae are extensions of the plant root system greatly extending the region of nutrient absorption beyond the 1–2 mm zone generally ascribed to roots[38].

Endomycorrhizae are found in the Cupressaceae and Taxodiaceae, and have relatively few external hyphae and root hairs near the apex of infected roots[45]. Estimates of endomycorrhizal biomass in coniferous forests have not been reported, although there are values for fine root biomass (diam <2 mm) for Cryptomeria (Taxodiaceae) and Chamacyparis (Cupressaceae)[22].

Structure of root systems

The form and development of each root system is unique. Initial development appears to be under genetic control, although modified by soil (bulk density, moisture, temperature, texture, *etc*.) and plant factors (*e.g*., root competition, photosynthate availabilty, tree density)[19,44]. On level terrain, Douglas-fir first order lateral roots extend more or less equidistantly around the tree but asymmetrically on steeply sloping ground[28]. First order laterals in Douglas-fir may extend 2.5 m from the stem[28]. In large leaf pine, lateral roots may spread 14.3 m[20], 25 m for pines in sandy soils[19].

Rooting density (root diam >2 mm) may reach 19 roots/0.25m^2 in Douglas-fir stands because of the spreading nature of the rooting systems, the extensive production of higher order laterals, and the intermingling of roots from different trees[28]. Roots extend considerably beyond the width of the crown although rooting density decreases with distance from the stem[19,33]. Spatial variation in fine root (diam <2 mm) density has been reported independent of the proximity of stems[31,33,43], although in a young Scots pine stand the largest amount of fine roots was 0.5 to 1.0 m from the stem[33]. Other studies indicate fine root biomass decreases with distance from the stem[11,39].

Numerous studies have shown that most roots are found in the upper 50 cm of soil, and most root activity and mycorrhizae in the top 20 cm depending on soil aeration and fertility[9,19]. Roots have been found at depths of 6 m in sandy soils[19] and ectomycorrhizae at depths of 1 to 3 m[9]. Localized concentration of roots may occur in decaying roots, channels formed by decaying roots, decaying logs, and thick litter layers[15,28]. Localized concentrations in more than one soil horizon may also occur due to fossil soil horizons[19] or obstructions such as fragipan, bed rock, or cobbles and boulders[24]. Completion of the vertical expansion of the structural root system has been reported to require 30 years in a number of species, but may require 70 years in Scots pine[19].

The problems inherent in the excavation of root systems have favored indirect methods for estimating root biomass. One commonly used method for large roots (diam >10 mm) is based on an allometric equation that relates root biomass to an easily measured tree parameter, *e.g.*, diameter at breast height[19,43]. The applicability of such equations is probably restricted to similar conditions. Fine roots are usually lost during excavation of the trees used to develop the model. Soil cores have been utilized extensively to sample fine roots and circumvent the problems caused by excavation methods[4]. The usefulness of the coring method may be reduced if small diameter cores are used in soils with low rooting densities unless replication is increased.

Estimates of fine root and mycorrhiza biomass resulting from soil-coring methods are difficult to compare given the number of different methods employed to separate roots from cores. The main methods involve hand sorting, dry sieving, wet sieving, plus various permutations of these methods combined with magnification, and air or water separation of organic matter[10,12,23,27,33,43]. A number of procedural and reporting problems hinder comparisons of the results of these different methods. Root tip numbers have been reported without an associated biomass conversion factor. Differences exist in the depth of soil sampled, phenological status of the stand (often not stated), stand age, tree density (often not stated), size and functional classification of roots, and uncertainty about the efficiency of different sampling methods. Consider the efficiency of different methods for recovering mycorrhizae. Many ectomycorrhizae range in size from 0.5 to 3.0 mm long and from 0.15 to 0.6 mm in diam[9]. Consequently, researchers using screens with openings larger than 0.5 mm may not recover all of the mycorrhizae, a common problem[15,29,30,46]. Some researchers have deliberately discarded mycorrhizae before weighing fine roots[31,41], and other have been ambiguous about the method used[34] or fail to state mesh size[13]. An additional problem arises when only 'active' mycorrhiza biomass is reported without the proportion of total mycorrhiza biomass this fraction represents being stated[15,16,46] or total mycorrhiza biomass alone presented[9,10].

Table 1. Increase in estimated root biomass after soil samples previously hand-sorted at 7x magnification were wet sieved. Cores extracted 23 May 1982 in a 35 year-old *Pinus strobus* stand near Ann Arbor, MI

Method	Core	Kg roots per ha (% total)	
		2–5 mm dia.	< 2 mm. dia.[b]
Hand-sort	1	5,986 (100)	7,378 (69)
	2	2,389 (97)	5,970 (60)
Wet sieve[a]	1	0 (0)	3,351 (31)
	2	74 (3)	4,053 (40)

[a] method of Fogel and Hunt[10]

[b] includes mycorrhizae

Is wet sieving more efficient than hand sorting for estimating root biomass? Cores were removed from the top 30 cm of soil in a white pine plantation. After the soil was screened (2 mm), roots were hand sorted at 7 × magnification then wet sieved[10] to recover any roots missed during hand sorting. Results (Table 1) clearly indicate that hand sorting is adequate for roots 2 to 5 mm in diam, but that wet sieving produces a 30 to 40% increase in the biomass estimate for roots <2 mm diam. Higher values of non-mycorrhizal fine roots (diam 0.4–3 mm) were also produced by wet sieving than by excavation in a radiata pine stand[31]. The effect of mesh size on recovery was also shown in tests where a third more roots were retained by a 0.53 mm screen than by a 0.91 screen[36]. Clearly the method of root extraction from soil should be given careful consideration and the method selected carefully described in subsequent publication.

Reported root biomass estimates

The estimates available[43] indicate that root biomass generally comprises 15 to 25% of total tree biomass in coniferous forests[2,14,40], although individual estimates range from 9 to 44 percent. Root biomass estimates for conifer stands less than 200 years old range from 3 to 85 Mg/ha; an estimated 209 Mg/ha has been reported for an old growth Douglas-fir stand[43]. Reported fine root biomass estimates [43] range from 1 to 12.6 Mg/ha with a mean of 5 Mg. It is unclear but highly likely that mycorrhizae have not been included in these estimates. Mycorrhiza account for an additional 8% of total tree biomass in young Douglas-fir stands[10]. Given the large input of fine roots and mycorrhizae to the decomposition process[8,10,13], it is crucial to determine mycorrhiza biomass. Unfortunately, the few published estimates available (Table 2) illustrate the difficulties in comparing different studies. The biomass of Douglas-fir (*Pseudotsuga menziesii*) mycorrhizae appears to be much larger than that of pine

Table 2. Estimates of mycorrhiza biomass in coniferous forests

Species and location	Stand age	Sample depth (cm)	Stems ha^{-1}	Standing Crop ($kg\ ha^{-1}$)	Physiological class	Separation method	Reference
Abies amabilis USA	23	25	110,500	2,000°	active	1 mm screen-wet	46
Abies amabilis USA	180	25	510	2,700°	active	1 mm screen-wet	46
Picea glauca-Abies lasiocarpa Canada	?	100	329	1,870	total	hand-sort	23
Pinus radiata Australia	42	15	296	1,433	total	0.5 mm screen-wet	27
Pinus resinosa USA	38	50		1,450–2,380**	total	1.5 mm screen-wet	5
Pinus rigida-Quercus spp. USA	?	21	?	$4.5–9.9 \times 10^{-3}$*	active	1 mm screen-wet	6
Pinus sylvestris France	10	32.5	26,875	989	active?	2 mm screen	29
Pinus sylvestris Sweden	20	30	1,100	50–380	active	2 mm screen	32
Pinus sylvestris-Fagus sylvatica France	65	26	2,200	1,380	active?	2 mm screen	29
Pseudotsuga menziesii UK	36	107	774	13,700+	active	0.9 mm screen-wet	36
Pseudotsuga menziesii USA	50	100	1,626	9,762–25,032	total	0.5 mm screen-wet	10
Pseudotsuga menziesii-Larix USA	250	38	4,631	18,377++	active	2 mm sieve-hand	15

++ assuming each tip weighs 19.7×10^{-5} g dwt, recalculated from data presented in reference

* assuming each tip weighs 12.5×10^{-5} g dwt.

\+ mostly for roots 0.5 mm.

°estimated from fig. 3 in reference cited

** root diam. 1 mm

Table 3. Comparison of fine root biomass in stands of different species composition near Ann Arbor, Michigan

Species	Site	Stand age	Kg roots per ha in top 30 cm of soil		
			<2 mm dia. (% total)	mycorrhizae (% total)	total fine roots
Pinus strobus	Michigan[a]	35	11,750 (94)	736 (6)	12,486
			15,815 (95)	900 (5)	16,715
Quercus/Carya	Michigan[a]	60	11,259 (88)	1,526 (12)	12,785
Pseudotsuga menziesii	Michigan[a]	30	1,661 (22)	5,978 (78)	7,639
			2,760 (33)	5,491 (67)	8,251
	Oregon	50			
	maximum[b]		11,916 (49)	12,199 (51)	24,115
	minimum[b]		1,948 (29)	4,752 (71)	6,700

[a] same soil type, cores extracted 23 May–8 June 1982
[b] recalculated from Fogel and Hunt[10]

or spruce. The biomass of Abies mycorrhizae may be similar to Douglas-fir if the proportion of active to total mycorrhizae is the same (mean = 8%) as measured for Douglas-fir (Fogel unpubl). The resulting standing crops in the 23 and 180 year-old Abies stands (25.2 and 34.0 Mg/ha) compare very well to the 25.0 Mg/ha reported for Douglas-fir[10]. Mycorrhiza biomass was compared in stands of mixed hardwood (Quercus and Carya dominates), white pine, and Douglas-fir growing on the same sandy glacial kame. Cores were extracted and wet sieved as described earlier[10]. The results of this study (Table 3) indicate that among the Michigan stands the Douglas-fir plantation has the smallest fine root biomass, but has the largest biomass and proportion of mycorrhizae. The estimates for the Michigan Douglas-fir stand approximate the minimum standing crop reported for a similarly aged Douglas-fir stand in Oregon[10].

Root production and turnover

Direct measurement of mycorrhiza and fine root production in established stands are extremely difficult *in situ*[39] and most studies have estimated production and turnover on the basis of changes in the biomass of various fractions over a given period. No significant seasonal differences in the biomass of large conifer roots have apparently been detected[28,32,43]. Several researchers have reported that 30–86% of the fine roots turnover annually[10,32,42], similar to the values reported for deciduous trees[9]. The large turnover in fine roots and mycorrhizae represents a considerable 'cost' to trees and constitutes a rather large sink of carbon, from 63 to 77% of total net primary production in coniferous forests[1,14]. These results indicate that roots are more important than leaf and branch litter in returning organic material to the soil, 4 to 5 times more material is returned by roots than by litter[10,32]. Studies in which root turnover has been assumed to contribute half as much as litter to the soil should be reexamined[32,47].

The large turnover of mycorrhizae and fine roots results from grazing and/or senescence accompanied by decomposition. Estimates of total grazing do not exist although partial estimates indicate that soil fauna consume 0.3 to 10% of the estimated production of roots in temperate forests[14,26]. Senescence of fine roots has been attributed to stresses imposed on the root system by drought, periodic high water tables, extreme soil temperatures, defoliation, parasitism, over-fertilization, and in some instances by internal hormonal control[11,17,25,35,37,48].

The high turnover of fine roots has been explained as an 'adaptive strategy'. Root sloughing, with subsequent microbial immobilization, could help retain nutrients in the upper soil horizons when the tree is under stress[14]. Reynolds[37] has calculated that during drought, roots consume energy reserves in one week equivalent to their own dry weight. This high 'cost' for maintenance respiration would make it 'advantageous' for a tree to maintain a small mass of roots during periods of stress. Withdrawal of essential nutrients just prior to root senescence, analogous to deciduous leaves, is an attractive hypothesis that deserves study. The rate of withdrawal, however, is important because this mechanism requires the tree to be stress tolerant thereby reducing its capacity for stress avoidance.

A second explanation is based on the assumption that most absorption of water and nutrients occurs immediately behind the root apex[37] and that localized regions of water and nutrient depletion occur in soil. If this were the case shedding roots in depleted zones would avoid the high costs of respiration allowing new roots to exploit untapped soil zones for water and nutrients[7,37]. This explanation fails to take into account the role of mycorrhizal fungi in extending the absorption zone of roots beyond the 1–2 mm zone generally ascribed to roots[38].

Given the importance of roots in primary production and the difficulties in estimating biomass, it is important to examine the different assumptions made in deriving production and turnover estimates. A direct comparison of the three methods employed was attempted by utilizing monthly estimates of mycorrhiza biomass and decaying roots (generally included in biomass by most researchers) for a Douglas-fir ecosystem (Fogel unpubl, Fogel and Hunt[10]). The first method assumes an acceptable method of separating live (active) from dead (inactive) roots does not exist. Root production (growth) is calculated by summing the differences between significantly different peaks and throughs in monthly or seasonal biomass estimates[10,13]. Positive increments are used to estimate production and decrements the return to soil (throughput fide Fogel and Hunt[10]). Calculated production in the Douglas-fir ecosystem totals 15.3 Mg/ha and throughput 14.6 Mg/ha; mortality is not estimated by this approach.

Method 2 for estimating production and turnover assumes that differences in standing crops are real even if they are not significantly different statistically[32]. No attempt is made in this approach to reconcile simultaneous changes in live and dead standing crops. Production and mortality are estimated as the sum of all positive increments, minus a correction factor, in the standing crops of live and dead root fractions respectively. A minimum estimate of production by this method is 2.4 Mg/ha, mortality 37.0 kg/ha, and throughput 36.4 Mg/ha.

Method 3 is similar to method 2, but adds the assumption that production and mortality occur simultaneously. Production, mortality, and throughput estimates are obtained by balancing changes in the standing crop of each root fraction during each sampling interval[42]. For example, if live root standing crop decreases 10 kg/ha and dead root standing crop decreases 5 kg, production would be 0 kg, mortality would be 10 kg, and throughput would be 15 kg/ha. Estimated production in the Douglas-fir ecosystem is 2.4 Mg/ha, mortality 45.1 Mg/ha, and throughput 46.9 Mg/ha.

Methods 1 and 3 accurately estimate the measured total standing crop of mycorrhizae on the last sample date when throughout is subtracted from the sum of the initial standing crop, production, and mortality. Method 2 overestimates the final measured standing crop by 3 Mg. The production estimated by methods 2 and 3 is identical. Method 3 produces estimates 22% larger for mortality and 29% larger for throughput than method 2. Santantonio[42] reported differences in the estimates of method 3 compared to method 2 of 30–60% in production, 50–70% in mortality, and 50–65% in throughput. Method 1 yields what are generally regarded as conservative estimates because steady state conditions are assumed and mortality is not taken into account. Throughput estimated by the latter method is lower than those estimated by methods 2 and 3, 40% and 31% respectively, but production is 530% larger than calculated by either of the other methods. The much larger production estimated by method 1 is due to the inclusion of mortality in the production estimate and perhaps an underestimation of the live biomass in methods 2 and 3. Method 3 produces estimates very similar to those of method 1 if only the maximum and minima preceding and following the maximum are used in the calculation rather than all 12 monthly standing crops. All three methods of estimating production and turnover indicate large fluxes in mycorrhiza production are matched by almost equally large fluxes in throughput.

Resolution of the question of which of the three methods yields the best estimate of production will depend on obtaining accurate estimates of the quantities of live and dead mycorrhizae. In practice, it has been very difficult to distinguish live and dead mycorrhizae or for that matter roots and the possibility for individual errors is great[4]. The problem of classifying the physiological status of ectomycorrhizae is compounded by their small size and the possibility of host tissues continuing to be active after death of the fungal symbiont. Common criteria for characterizing live mycorrhizae include: turgidity, color, integrity of root apex, association with mycelia, hyphal strands, or rhizomorphs, and when sectioned, color of host tissues[15]. Generally no attempt is made to determine the proportion of live to dead mycorrhizae. Underestimates of live mycorrhiza biomass may result from partial separation from soil, errors in classifying the physiological status of individual mycorrhizae, and from improper handling. For example, exposure to dry air for only a few minutes is sufficient to make small roots shrivel and die[24].

Vital stains have been used by several investigators in an attempt to avoid subjective use of morphological criteria in distinguishing live and dead roots. The stains employed have included 2, 3, 5-triphenyltetrazolium chloride, 2,3,4-triphenyltetrazolium bromide, and congo red[4].

Unfortunately, roots of various plant species react quite differently to staining. Thus, dye adsorption is apparently related to the physical properties of the roots as well as to root cation exchange capacity.

References

1 Ågren G I, Axelsson B, Flower-Ellis J G, Linder S, Persson H, Staaf H and Troeng E 1980 Annual carbon budget for a young Scots pine. Ecol. Bull. Stockholm 32, 307–313.

2 Assmann E 1970 The Principles of Forest Yield Study. 1st English edition. Pergamon Press, New York, NY, 506 p.

3 Berndt H W and Gibbons R D 1958 Root distribution of some native trees and understorey plants growing on three sites within ponderosa pine watersheds in Colorado. U.S. Dep. Agr., For. Serv., Rocky Mountain For. Range Exp. Sta. Pap. No. 37.

4 Bohm W 1979 Methods of Studying Root Systems. Springer-Verlag, New York, NY. 188 p.

5 Braekke F H and Kozlowski T T 1977 Distribution and growth of roots in *Pinus resinosa* and *Betula papyrifera* stands. Medd. Nor. Inst. Skogforsoeksves. 33, 437–451.

6 Bucholz K and Motto H 1981 Abundances and vertical distributions of mycorrhizae in plains and barrens forest soils from the New Jersey, USA, pine barrens. Bull. Torrey Bot. Club 108, 268–271.

7 Caldwell M M 1979 Root structure: The considerable cost of below-ground function. *In* Topics in Plant Population Biology. Eds. O T Solbrig, S Jain, G B Johnson and P H Raven. Columbia Univ. Press, New York, NY. pp 408–430.

8 Edwards N T and Harris W F 1977 Carbon cycling in a mixed deciduous forest floor. Ecology 58, 431–437.

9 Fogel R 1980 Mycorrhizae and nutrient cycling in natural forest ecosystems. New Phytol. 86, 199–212.

10 Fogel R and Hunt G 1979 Fungal and arboreal biomass in a western Oregon Douglas-fir ecosystem: Distribution patterns and turnover. Can. J. For. Res. 9, 245–256.

11 Ford E D and Deans J D 1977 Growth of Sitka spruce plantation: Spatial distribution and seasonal fluctuations of lengths, weights and carbohydrate concentrations of fine roots. Plant and Soil 47, 463–485.

12 Grier C G, Vogt K A, Keyes M R and Edmonds R L 1981 Biomass distribution and above- and belowground production in young and mature *Abies amabilis* zone ecosystems of the Washington Cascades. Can. J. For. Res. 11, 155–167.

13 Harris, W F, Kinerson R S Jr and Edwards N T 1977 Comparison of belowground biomass of natural deciduous forest and loblolly pine plantations. Pedobiolobia 17, 369–381.

14 Harris W F, Santantonio D and McGinty D 1980 The dynamic belowground ecosystem. *In* Forests: Fresh Perspectives from Ecosystem Analysis. Ed. R H Waring Oregon St. Univ. Press, Corvallis. pp 119–129.

15 Harvey A E, Larsen M J and Jurgensen M F 1976 Distribution of ectomycorrhizae in a mature Douglas-fir/larch forest soil in western Montana. For. Sci. 22, 393–398.

16 Harvey A E, Larsen M J and Jurgensen M F 1980 Partial cut harvesting and ectomycorrhizae: Early effects in Douglas-fir-larch forests of western Montana. Can. J. For. Res. 10, 436–440.

17 Head G C 1973 Shedding of roots. In Shedding of Plant Parts. Ed. T T Kozlowski. Academic Press, New York, NY. pp 237–293.

18 Heikurainen L 1957 Ueber Veraenderungen in den Wurzelverhaeltnissen der Kiefernbestaende auf Moorboeden im Laufe des Jahres. Acta For. Fenn. 65, 1–54.

19 Hermann R K 1977 Growth and production of tree roots: A review. *In* The Belowground Ecosystem: A Synthesis of Plant-Associated Processes. Ed. J K Marshall. Range Sci. Dep. Sci. Ser. No. 26, Colo. St. Univ., Ft. Collins, CO, pp 7–28.

20 Hodgkins E J and Nichols N G 1977 Extent of lateral roots in natural longleaf pine as related to position and age of trees. For. Sci. 23, 161–166.

21 Kalela E K 1954 Mäntysiemenpuiden ja-kuusikoiden juurisuhteista. Acta For. Fenn. 61, 1–16.

22 Karizumi N 1977 Root biomass. *In* Primary Productivity of Japanese Forests. Eds. T Shidei and T Kira. Univ. Tokyo Press. Vol. 16, 45–52.

23 Kimmins J P and Hawkes B C 1978 Distribution and chemistry of fine roots in a white spruce – subalpine fir stand in British Columbia: Implications for management. Can. J. For. Res. 8, 265–279.

24 Lyford W H 1975 Rhizography of non-woody roots of trees in the forest floor. *In* The Development and Function of Roots. Eds. J G Torrey and D T Clarkson. Academic Press, New York, NY. pp 179–196.

25 Lyr H and Hoffman G 1967 Growth rates and growth periodicity of tree roots. *In* International Review of Forestry Research. Eds. J A Romberger and P Mikola. Academic Press, New York, NY. Vol. 2, pp 181–236.

26 Magnusson C and Schlenius B 1980 Root consumption in a 15–20 year old Scots pine stand with special regard to phytophagous nematodes. Ecol. Bull., Stockholm 32, 261–268.

27 Marks G C, Ditchburne N and Foster R C 1968 Quantitative estimate of mycorrhiza populations in radiata pine forests. Aust. For. 32, 26–38.

28 McMinn R G 1963 Characterization of Douglas-fir root systems. Can. J. Bot. 41, 105–122.

29 McQueen D R 1968 The quantitative distribution of absorbing roots of *Pinus silvestris* and *Fagus sylvatica* in a forest succession. Oecol. Plant. 3, 83–99.

30 McQueen D R 1973 Changes in understory vegetation and fine root quantity following thinning of 30-year *Pinus radiata* in central North Island, New Zealand. J. Appl. Ecol. 10, 13–21.

31 Moir W H and Bachelard E P 1969 Distribution of fine roots in three *Pinus radiata* stands near Canberra. Ecology 50, 658–662.

32 Persson H 1978 Root dynamics in a young Scots pine stand in central Sweden. Oikos 30, 508–519.

33 Persson H 1980 Spatial distribution of fine-root growth, mortality and decomposition in a young Scots pine stand in central Sweden. Oikos 34, 77–87.

34 Quesnel H J and Lavkulich L M 1981 Comparison of the chemical properties of forest floors, decaying wood, and fine roots in three ecosystems on Vancouver Island. Can. J. For. Res. 11, 215–217.

35 Redmond D R 1955 Mortality of rootlets in balsam fir defoliated by the spruce budworm. For. Sci. 5, 64–69.

36 Reynolds E R C 1970 Root distribution and the cause of its spatial variability in *Pseudotsuga taxifolia* (Poir.) Britt. Plant and Soil 32, 501–517.

37 Reynolds E R C 1975 Tree rootlets and their distribution. *In* The Development and Function of Roots. Eds. J G Torrey and D T Clarkson. Academic Press, London. pp. 163–177.

38 Rhodes L H and Gerdemann J W 1975 Phosphate uptake zones of mycorrhizal and non-mycorrhizal onions. New Phytol. 75, 555–561.

39 Roberts J 1976 A study of the root distribution and growth in a *Pinus silvestris* L. (Scots pine) plantation in East Anglia. Plant and Soil 44, 607–621.

40 Rodin L E and Bazilevich N I 1967 Production and Mineral Cycling in Terrestrial Vegetation. Oliver and Boyd, London, 288 p.

41 Safford L O and Bell S 1972 Biomass of fine roots in a white spruce plantation. Can. J. For. Res. 2, 169–172.

42 Santantonio D 1979 Seasonal dynamics of fine roots in mature stands of Douglas-fir of different water regimes: A preliminary report. *In* Root Physiology and Symbiosis. Eds. A Riedacker and J Gagnaire-Michard. Proc. IUFRO Symp. Root Physiol. Symbiosis, Nancy, France, Sept. 1978. CNRF, Champenoux, France. Vol. 6.

43 Santantonio D, Hermann R K and Overton W S 1977 Root biomass studies in forest ecosystems. Pedobiologia 17, 1–31.

44 Sutton R F 1980 Root system morphogenesis. N.Z. J. For. Sci. 10, 264–292.

45 Trappe J M and Fogel R 1977 Ecosystematic functions of mycorrhizae. *In* The Belowground Ecosystem: A Synthesis of Plant-Associated Processes. Eds. J K Marshall. Range Sci. Dep. Sci. Ser. No. 26, Colo. St. Univ., Fort Collins. pp. 205–214.

46 Vogt K A, Edmonds R L, Grier C C and Piper S R 1980 Seasonal changes in mycorrhizal and fibrous-textured root biomass in 23- and 180-year-old Pacific silver fir stands in western Washington. Can. J. For. Res. 10, 523–529.

47 Whittaker R H and Marks P L 1975 Methods of assessing terrestrial productivity. *In* Primary Productivity of the Biosphere. Eds. H Leith and R H Whittaker. Ecol. Stud. 14, 55–118.

48 Zak B 1964 Role of mycorrhizae in root disease. Ann. Rev. Phytopathol. 2, 377–392.

Plant and Soil 71, 87–101 (1983). A-09

The distribution and productivity of fine roots in boreal forests

HANS Å. PERSSON
Department of Ecology and Environmental Research, Swedish University of Agricultural Sciences, S-750 07 Uppsala, Sweden

Key words *Calluna vulgaris* Core sampling Fine roots Fine-root biomass Fine-root dynamics Fine-root necromass Fine-root production Forest ecosystem Ingrowth cores *Pinus sylvestris* *Vaccinium vitis-idaea*

Summary The important role of fine roots in the energy expenditure of boreal forests has only recently been illuminated in ecosystem research. As demonstrated in recent research, fine roots are in constant flux, with death and replacement taking place simultaneously. Closely related to the energy costs of development and maintenance of the different roots systems are their functional effectiveness as organs for the absorption of water and nutrients. In forests the dynamics of fine roots need to be considered further in order to make reliable comparisons of plant productivity on an ecosystemic level.

Data on the temporal variations of fine-root death and replacement are given from two Scots pine stands (*Pinus sylvestris*) of different ages in Central Sweden. Two methods were adopted in order to investigate the fine-root dynamics: (i) using data obtained by sequential core sampling and (ii) measuring the ingrowth of new roots into root-free cores removed at regular intervals. Only a few vascular species, *viz. P. sylvestris*, *Calluna vulgaris* and *Vaccinium vitis-idaea*, were to be found in this fairly simple type of ecosystem and all of them had distinctive morphological rooting features. The ratio between the production estimates from either method (i) or (ii) and the average standing crop (biomass + necromass) of the fine roots (< 2mm in diameter) varied between 2.0 and 1.5; 0.6 and 0.5; and 1.5 and 0.4 for *P. sylvestris*, *C. vulgaris* and *V. vitis-idaea* respectively in a young Scots pine stand and between 0.9 and 0.7; 0.6 and 0.7; and 0.1 and 0.1 in a mature Scots pine stand. The greatest death and replacement of fine roots of the tree component was estimated in the mineral soil, whereas that of the superficially distributed dwarf shrubs was found in the F/H layer.

Introduction

Plant roots are the connecting link between the plant and the soil. In many ways, the growth of the root system is decisive both for the development of visible above-ground structures and for soil formation. Up until about twenty years ago quantitative root studies in forests at an ecosystemic level were usually of the gross root systems of individual trees. The role of small-diameter roots and mycorrhiza in total forest-biomass dynamics were neglected[7,29,51,62,63]. In many previous studies rough estimates of below-ground turnovers were accepted as an alternative to having no estimates at all[22,36,41,42]. Many workers assumed that the growth dynamics of fine roots were similar to those of shoots. There is, however, no evidence for this assumption as the ratio of above-ground to below-ground production may differ considerably, even in the same species[43,45]. In the absence of hard data, unrealistic speculations may be introduced.

There is a general shortage of information concerning the rate of death and renewal of fine roots in forest ecosystems in relation to changes in abiotic and biotic environmental factors[29,63,65]. This paucity of information on below-ground dynamics may be illustrated by some questions, still poorly answered questions formulated over a hundred years ago by Resa[57], *i.e*:

1 Do roots have an annual growth periodicity?

2 Is the growth periodicity, if any, different from that of above-ground growth?

3 Does there exist an annual fine-root mortality?

What controls the death and replacement of fine roots of forest trees is a matter of conjecture[7,29,52,53,65]. However, during recent years, many authors have come to the conclusion that fine-root growth occurs relatively independently of shoot growth and that the temporal variation is largely determined by changes in environmental conditions. Most authors[29,39,62,63] share the opinion that the factors determining root growth are complex and involve environmental conditions of both above below-ground plant parts, nutritional status, carbohydrate supply and allocation in the plant, respiratory costs, growth regulatory substances (root hormones), and genetic influences[9,11,12,17,18,29,31,32,50,53,55,56,60,61,63,65,67,69]. Furthermore, in a forest stand and in forest trees competition between individuals of the same species and interspecific competition between different plant populations influence the process of root growth. Methodological difficulties and lack of uniform investigatory techniques make it practically impossible, at present, to draw general conclusions from available data in literature.

Most works on the temporal pattern of fine root growth in boreal forests indicate considerable variations in fine roots during the growing season[9,10,13,20–24,28,30,33,44,46,47,49,50,51,53,58]. Fine roots are in a state of constant flux, with death and replacement taking place simultaneously and although there is a growing evidence of large fluctuations occurring in fine root length, biomass, there are few estimates of their rates of production and turnover. Noteworthy exceptions are:[10,21,22,23,24,39,44,46,47,48,49,50,51].

Fine roots exert a significant influence in soil profile development and, upon dying, they contribute substantially to the organic matter content of soil[8,27,40,46,53]. Thus, much fine-root biomass is constantly being transformed into necromass (dead fine-root tissue), frequently not estimated or accounted for turnover calculations. Fine-root necromass may vary considerably during the growing season[27,40,44,46,47,49,50,51,53,55,61]. Together with litter from the aerial parts of plants, shedded roods provide the main bulk of material for the complex decomposition cycles that occur in the soil. These include bacteria, fungi and soil animals[54].

Methods used in the study of fine-root production

The interpretation of available root data in literature and comparisons at an ecosystemic level (or between species) are often impossible due differences in methods of measuring root growth differently. Reviews of recent methods are to be found in[6,19,26,29,34,41,42,59]. At present there are no revolutionary techniques in sight so future ecological root investigations wil probably involve mechanized sampling, then washing and sorting of samples.

The simplest way of studying the periodic changes in fine roots is from volumetric soil samples which reflect the standing crop (biomass and necromass) throughout the year. This method is however best on sandy forest soils where the soil is easily penetrated by sampling. The excavation of large soil blocks (monoliths) is currently the only feasible method on till so quantitative data on fine-root dynamics of forest ecosystem on till are therefore presently conspicuous by their absence.

Two methods were used within the Swedish Coniferous Forest Ecosystem Project SWECON to investigate fine root dynamics: (1) using data obtained by sequential core sampling and (2) measuring the ingrowth of new roots into root-free cores (ingrowth cores) removed at regular intervals[11,16,44,46,47,49,50,51,53]. The ingrowth method consists of removing a core and refilling the hole with sifted soil containing no roots. Thus, the core initially consists of cylindrical 'net stockings' filled with sifted sand. The core is then resampled after a period of time and the roots that have grown into it are measured.

Most root studies hitherto conducted in forest ecosystems, generally fail to characterize where and when the root growth takes place. Existing published data has the following limitations: (1) sampling carried out at a few intervals during the growing season, (2) the fine-root biomass includes an unknown amount of dead roots, (3) the dead fine-root fraction (necromass) was not estimated, (4) the number of replicate samples was insufficient to detect significant differences between samplings, (5) the estimated quantitative variables (*e.g.* the number of root tips, root length *etc.*) are not directly convertable into changes in root weight (6) the root sampling was not carried out on an area basis and (7) the root fragments were not classified into species or diameter-fractions.

Similar criteria, in order to distinguish live fine roots from dead ones, have been elaborated by different workers[39,44,46,47,49–51,53,61,62,63]. Living rootlets are far more resilient than dead ones and are not so easily broken if twisted or bent. Actively growing root tips are light-coloured, succulent and unsuberized. Dead rootlets are often wrinkled and dark in colour, fine ramifications are often shed and, if resistant, easily broken off. When it is difficult to see whether a root-fragment is living or dead, it can be determined by the degree of cohesion between the cortex and periderm. The rootlets are then cut lengthwise with a sharp dissecting knife and the section surface is examined under a stereo microscope (× 60)[44,47].

A number of different research workers have estimated fine root production by adding increments in the fine-root biomass only for an annual period without estimating the dynamic changes in the dead roots[23,24,49,50]. This assumes that death and replacement of fine roots do not take place simultaneously since root mortality is ignored (*e.g.* the accumulation of fine root necromass in the soil). Only a few workers[39,44,46,47,49–51,53,61,62,63] have so far incorporated the dynamics of dead fine roots into their calculations. A correction for overestimation is necessary in all these culculations, if only increments (or decrements) are added[47]. At present, all net primary production estimates of fine roots are fairly conservative, *e.g.* they indicate the lowest possible value. The ingrowth method (see above) can be used to estimate annual fine root production, from the amount of new roots which have grown into the cores during a period of one year. The estimate is then calculated as the difference between the highest and lowest estimate during an annual period. Both living and dead root ramifications can be included in the calculation.

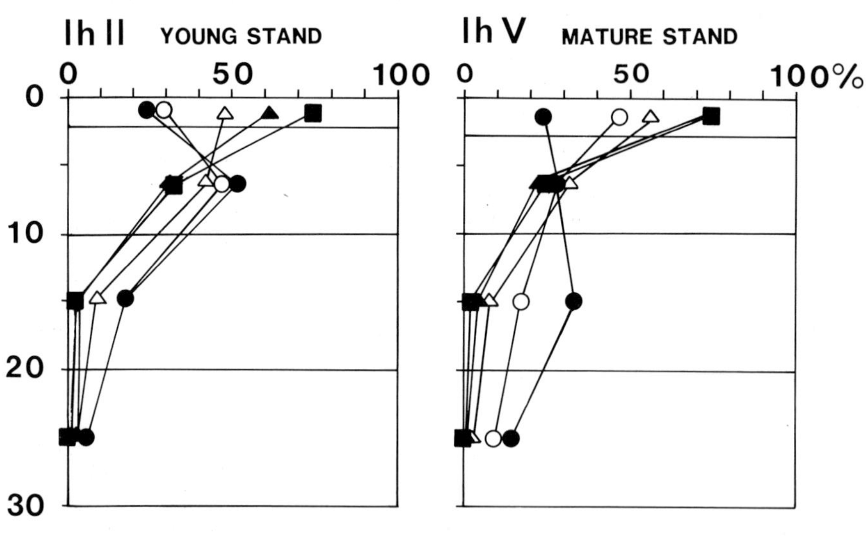

Fig. 1. Vertical distribution of root biomass (< 10 mm in diameter) of *Pinus sylvestris* (●; ○ = fine roots < 2 mm), *Calluna vulgaris* (▲; △ = fine roots < 2 mm) and *Vaccinium vitis-idaea* (■; all roots < 2 mm). The diagrams show determinations in the different soil layers (F/H, mineral soil down to 10, 10–20 and 20–30 cm) expressed as percentage of the total root amount in the whole core. The thickness of the F/H layer was 23 ± 1 mm ($\bar{x}$ ± s.e; n = 412) in Ih II and 26 ± 1 mm (n = 188) in Ih V.

Recent data on root dynamics from SWECON

During recent years, new information on the growth dynamics of fine roots has been obtained by the root studies portion of the SWECON Project[1,11,16,44,45,46,49,50,51,53,54]. A general summary of these studies is given below.

The bulk of the fine roots of all species (*cf* Fig. 1) was concentrated to the upper 10 cm of the soil profile (*e.g.* in the F/H layer and in the upper layer of the mineral soil). While the dwarf shrubs (*viz. Calluna vulgaris* and *Vaccinium vitis-idaea*) were fairly superficially distributed with most root weight in the F/H layer, the roots of Scotch pine (*Pinus sylvestris*) were concentrated immediately below the upper part of the mineral soil (*cf* also Table 1). As regards the total-root biomass an almost identical vertical distribution was observed, except in the case of *P.*

Table 1. The distribution of fine-root biomass (FRB), fine-root necromass (FRN), total root biomass (TRB) and total root necromass (TRN) in the F/H layer, the mineral soil down to 30 cm and total sampling depth (0–30 cm) in a young Scots pine stand (Ih II) and in a mature Scots pine stand (Ih V). The total number of samples in Ih II was 416 (distributed in two strata I and II) and in Ih V was 210 during the 1974–75 samplings. Estimates (g dw m^{-2}) are pooled averages from 13 samplings in Ih II and 8 samplings in Ih V. FRB and FRN = <2 mm in diameter. TRB and TRN = <10 mm in diameter

Area/Species	Soil layer	FRB	FRN	TRB	TRN
		g DW m^{-2}			
Ih II (I–II)					
Pinus sylvestris	F/H	7.8 ± 0.6	8.4 ± 0.7	12.3 ± 1.7	16.7 ± 2.1
	m.s.	18.3 ± 1.1	59.3 ± 1.9	37.5 ± 2.8	148.0 ± 7.1
	total	26.1 ± 1.2	67.6 ± 2.2	49.8 ± 2.3	164.7 ± 7.4
Calluna vulgaris	F/H	23.9 ± 1.4	9.4 ± 0.7	63.5 ± 7.8	38.1 ± 2.7
	m.s.	26.3 ± 1.0	8.3 + 0.6	39.5 + 2.8	16.4 ± 1.6
	total	50.2 ± 1.8	17.6 ± 1.3	102.9 ± 8.6	54.5 ± 5.1
Vaccinium vitis-idaea	F/H	31.7 ± 1.8	30.5 ± 2.2	31.7 ± 1.8	30.5 ± 2.2
	m.s.	17.0 ± 1.2	9.5 ± 0.8	17.0 ± 1.2	9.5 ± 0.8
	total	48.7 ± 2.2	40.0 ± 2.4	48.7 ± 2.2	40.0 ± 2.4
Ih V					
Pinus sylvestris	F/H	58.5 ± 2.5	8.5 ± 0.9	78.8 ± 5.3	10.3 ± 2.4
	m.s.	64.0 ± 2.8	55.7 ± 2.8	234.3 ± 13.6	106.3 ± 7.8
	total	122.5 ± 3.7	64.2 ± 3.1	313.1 ± 21.6	116.6 ± 8.2
Calluna vulgaris	F/H	36.0 ± 3.3	9.2 ± 1.5	117.8 ± 10.5	41.3 ± 5.3
	m.s.	25.8 ± 1.3	5.8 ± 0.8	40.2 ± 3.3	16.8 ± 3.0
	total	61.8 ± 3.7	15.0 ± 1.7	158.0 ± 11.1	58.2 ± 6.1
Vaccinium vitis-idaea	F/H	117.8 ± 5.3	25.0 ± 2.3	117.8 ± 5.3	25.0 ± 2.3
	m.s.	40.1 ± 3.0	7.0 ± 0.9	40.1 ± 3.0	7.0 ± 0.9
	total	157.9 ± 5.9	32.0 ± 2.6	157.9 ± 5.9	32.0 ± 2.6

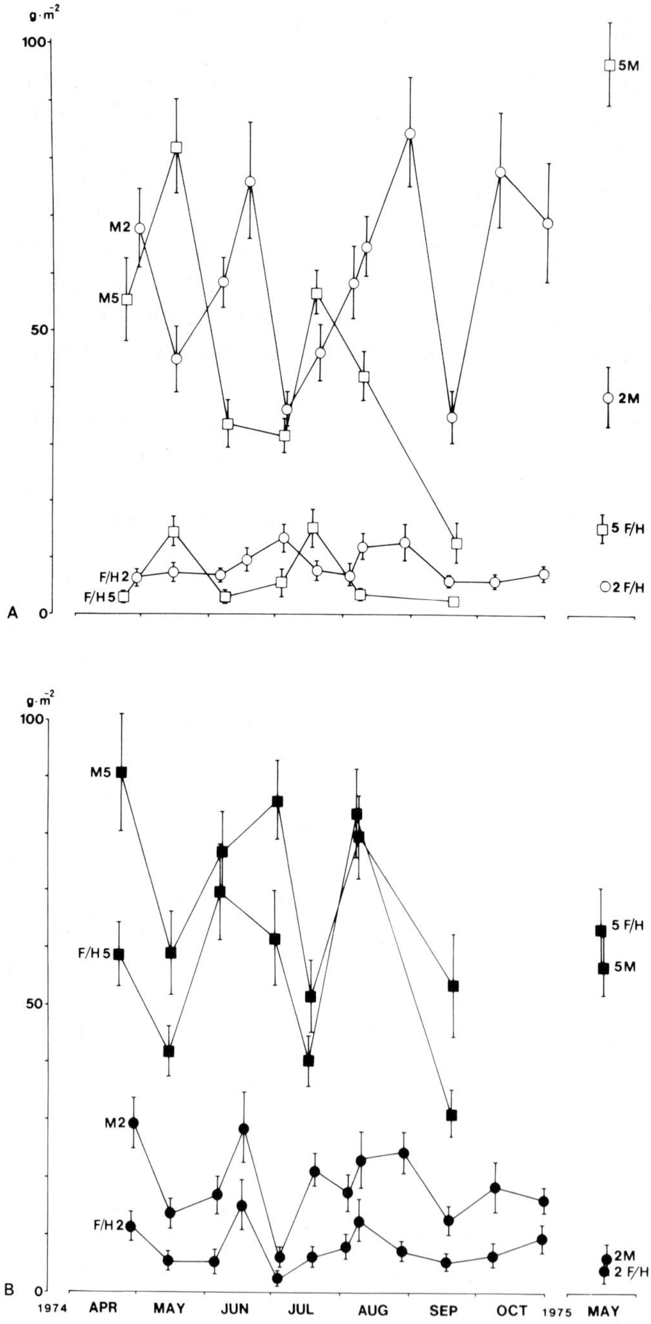

Fig. 2. Seasonal fluctuations in below-ground fine-root biomass (filled symbols) and necromass (empty symbols) of *Pinus sylvestris* < 2 mm in diameter in a young Scots pine stand (2 = Ih II) and in a mature Scots pine stand (5 = Ih V). Seasonal variation is indicated for both the F/H layer (F/H) and the mineral soil down to 30 cm (M).

sylvestris, where a more substantial part of the root biomass was recorded in the mineral soil.

The fine-root biomass (FRB) represents a varying proportion of the total accumulated organic matter of the forest stand. As demonstrated by the present results (Table 1, Fig. 2), the ratio of live to dead fine roots differed distinctively between a young Scots 20-year-old pine stand (Ih II) and a mature 120-year-old stand (Ih V). The young stand had proportionally more dead fine roots (FRN) of *P. sylvestris* than the mature stand (FRB/FRN = 0.39 in Ih II and 1.91 in Ih V). Similarly *C. vulgaris* and *V. vitis-idaea*, the amount of live fine roots was higher in Ih V than in Ih II (FRB/FRN = 2.81 and 1.22 in Ih II and 4.12 and 4.93 in Ih V for *C. vulgaris* and *V. vitis-idaea*). Large variations in FRB and FRN were found in both stands.

Estimates of FRB and FRN are average scores for the 1974–75 growing season (Table 1) as are average total root biomass (TRB) and necromass (TRN) (< 10 mm in diameter), the weights of which were determined to a large extent from the proportion of FRB and FRN (< 2 mm in diameter). FRB was 52, 49 and 100% of TRB, as for *P. sylvestris*, *C. vulgaris* and *V. vitis-idaea* in Ih II and 55, 26 and 100% in Ih V. The related estimates of FRN were 41, 32 and 100% in Ih II and 55, 26 and 100% in Ih V. All below-ground plant parts of *V. vitis-idaea* had diameters < 2 mm (viz. FRB and FRN = TRB and TRN). Except for *C. vulgaris*, where substantial adventitious root growth took place in autumn, from the stem parts in moist soil, no distinct seasonal pattern was found in the changes.

The overall level of fluctuations in FRB for *P. sylvestris* was considerably lower in Ih II than in Ih V – the pooled average of FRB in Ih II was only 0.2 times that in Ih V (Table 1). The related FRN fluctuated, however, at nearly the same level in both stands probably indicating larger annual supply of dead fine roots to the soil in Ih II (Fig. 2). The overall level of FRB and FRN as regards *C. vulgaris* did not differ greatly between the two stands. As regards *V. vitis-idaea*, the FRB was considerably lower in Ih II than in Ih V, *i.e.* only 0.3 times the latter estimate. However, the related estimate of FRN did not differ very much, indicating, also in the case of *V. vitis-idaea*, a relative larger annual supply of dead fine roots to the soil in Ih II.

The overall levels of FRB and FRN of *P. sylvestris* are relatively low compared with other structural tree components[2,3,4,5,14,15,16], FRB is a low portion of total tree-root biomass including the coarse roots and stumps as much root biomass accumulates in the latter structures. Diameter analyses, however, show that the major part of the length and surface area of the tree root systems is made up of fine roots[44]. Structural data on above-ground plant parts of the field layer species are to be found in[45,48,52].

In the young stand, the tree root systems are still expanding, both horizontally and vertically and new soil volumes are successively occupied. Because of the varying ratios of live to dead fine roots, it was judged as more correct to relate the

Table 2. Below-ground production and cumulative necromass totals (mean ± standard error) of fine roots (< 2 mm in diameter) during the 1974-75 samplings in a young Scots pine stand (Ih II) and in a mature Scots pine stand (Ih V). The number of replicates in each of the 13 samplings in Ih II was 16 in both of the strata I and II (416 core samples in total). The number of replicates in each of the 8 samplings in Ih V were, in progressive order 25, 30, 25, 30, 35, 20 and 30 (210 core samples in total). The estimates in parenthesis did not consist of significant (5% level of significance) increments. The test of significance was a two-tailed test of the distribution of F, in this case used to test whether the variance of the changes of positive sign differed from the variances of the samples. Estimates are calculated for fine-root biomass (b), fine-root biomass + necromass (b + n) and fine-root necromass (n) from the first to the last sampling (j = 13 in Ih II; = 8 in Ih V)

	Sum of increments (g DW m^{-2} yr^{-1}) for:		
	Pinus sylvestris	*Calluna vulgaris*	*Vaccinium vitis-idaea*
Ih II (I–II)			
F/H layer			
b_j	21 ± 7	33 ± 14	(54 ± 16)
$(b + n)_j$	27 ± 9	41 ± 18	121 ± 17
n_j	10 ± 5	(24 ± 12)	87 ± 11
Mineral soil			
b_j	44 ± 11	(12 ± 8)	(24 ± 11)
$(b + n)_j$	160 ± 25	(16 ± 11)	(44 ± 16)
n_j	123 ± 19	19 ± 5	25 ± 9
Total			
b_j	61 ± 14	38 ± 16	59 ± 22
$(b + n)_j$	183 ± 26	(30 ± 22)	(135 ± 37)
n_j	122 ± 20	(26 ± 13)	104 ± 25
Ih V			
F/H layer			
b_j	77 ± 16	(33 ± 19)	(47 ± 30)
$(b + n)_j$	65 ± 18	(21 ± 20)	(28 ± 36)
n_j	37 ± 5	(7 ± 10)	(21 ± 13)
Mineral soil			
b_j	68 ± 6	29 ± 9	(0 ± 18)
$(b + n)_j$	115 ± 17	18 ± 11	(15 ± 16)
n_j	145 ± 15	20 ± 5	(9 ± 6)
Total			
b_j	139 ± 21	(43 ± 14)	(14 ± 35)
$(b + n)_j$	159 ± 27	(44 ± 23)	(0 ± 40)
n_j	179 ± 16	24 ± 11	(28 ± 15)

Table 3. Pooled average standing crop of the fine roots (FRB + FRN) in Ih II and Ih V during the 1974–75 samplings. Also given are the below-ground production of fine roots calculated either from the sum of increments for FRB of FRB + FRN. R = the ratio between the fine-root production and the average standing crop. Estimates are given for different soil layers (*cf* Table 1). The estimates in parenthesis did not consist of significant (5% level of significance) increments (*cf* Table 2). FRB and FRN = <2 mm in diameter

Area/Species fraction	Soil layer	FRB + FRN (g DW m^{-2})	Fine-root production (g DW m^{-2} yr^{-1})	R
Ih II (I–II)				
Pinus sylvestris	F/H	16 ± 1	27 ± 9	1.7
	m.s	78 ± 3	160 ± 25	2.1
	total	94 ± 3	183 ± 16	2.0
Calluna vulgaris	F/H	33 ± 2	41 ± 14	1.2
	m.s	35 ± 1	(16 ± 11)	0.4
	total	68 ± 2	38 ± 16	0.6
Vaccinium vitis-idaea	F/H	62 ± 3	121 ± 17	2.0
	m.s.	26 ± 2	(44 ± 16)	1.7
	total	89 ± 4	135 ± 37	1.5
Ih V				
Pinus sylvestris	F/H	67 ± 3	77 ± 16	1.2
	m.s.	120 ± 4	115 ± 17	1.1
	total	187 ± 5	159 ± 27	0.9
Calluna vulgaris	F/H	45 ± 4	33 ± 19	0.7
	m.s.	32 ± 2	29 ± 9	0.9
	total	77 ± 4	(44 ± 23)	0.6
Vaccinium vitis-idaea	F/H	143 ± 6	(47 ± 30)	0.3
	m.s.	47 ± 3	(15 ± 16)	0.3
	total	190 ± 7	(14 ± 35)	0.1

Table 4. The ingrowth of new roots into root-free cores (ingrowth cores) removed on successive sampling occasions. The estimates are the differences between the highest and lowest estimate of the actual root amount in the cores during the 1975–76 and 1976–77 growing periods. Also given is the ratio (R) between the fine root production estimated in this way and the average standing crop of the fine roots in Ih II and Ih V during the 1974–75 samplings (*cf* data in Table 3)

Area/Species fraction	Ingrowth 1975–76 (g m^{-2} yr^{-1})	R	Ingrowth 1976–77 (g m^{-2} yr^{-1})	R
Ih II (I–II)				
Pinus sylvestris	139 ± 80	1.5	22 ± 49	0.2
Calluna vulgaris	26 ± 10	0.4	11 ± 7	0.2
Vaccinium vitis-idaea	26 ± 16	0.3	16 ± 14	0.2
Ih V				
Pinus sylvestris	128 ± 38	0.7	202 ± 69	1.1
Calluna vulgaris	36 ± 9	0.5	53 ± 18	0.7
Vaccinium vitis-idaea	5 ± 4	0.0	27 ± 6	0.1

estimates of the fine-root production (Table 2) to the average fine root biomass + necromass (FRB + FRN; *cf* Table 3 and 4) *viz*. the standing crop of the fine roots. The fine-root production estimates in the different soil environments are, probably, closely related to the degree of exploitation by the different root systems.

The whole profile is well exploited by the root systems. The greatest fine-root production of *P. sylvestris* in Ih II took place in the mineral soil horizon, whereas that of *C. vulgaris* and *V. vitis-idaea*, which are both shallow-rooted, was in the F/H layer. In the mature stand Ih V, no decisive differences were obtained between the F/H layer and the mineral soil. The estimates calculated from increments in FRB + FRN for *P. sylvestris* were considerably higher than those calculated only from increments in biomass (Table 2). The latter indicates that a substantial amount of FRB is constantly transformed into FRN.

The magnitude of the fine-root production estimates, calculated from the sum of increments in FRB or in FRB+FRN (see methods of calculation in[44,47]) were confirmed by data obtained from the ingrowth experiment. Thus, since the death of fine roots was proportionally low during the 1975–76 growing season and the ingrowth of new roots took place fairly rapidly, the accumulated standing crop in the cores must be considered as an indication of the annual below-ground production. Note, however that the sequential core sampling series were obtained during the growing season 1974–75. The continual ingrowth of *P. sylvestris* fine roots in Ih V compared with that in Ih II during the 1976–77 growing season must relate to the larger average level of fine-root standing crop in Ih V (Table 4). The two sets of estimates, from the sequential core sampling series (Table 2) are not strictly comparable, since the number of samplings differed (13 in Ih II and 8 in Ih V, during the growing season 1974–75).

The values for the below-ground production (Table 2) should be regarded as minimum estimates of the absolute magnitude. With an increasing number of samplings there is an increased chance of measuring extreme figures, – *i.e.* 'local' minima and maxima which would affect the productivity estimates. The sampling frequency employed (every 2 weeks in Ih II, every month in Ih V during the growing season) seems to have covered major fluctuations but probably not all the changes which occurred. Growth which took place during the winter months (November 1974 – April 1975) was not estimated, since sampling could not be undertaken in frozen ground.

Supporting evidence from other root studies – Conclusions

Fine roots constitute a dynamic component of the boreal forest and recent root studies on an ecosystemic level indicate that fine roots play a much greater part in carbon cycling than has been conventionally accepted[1,5,7,8,11,37,38,54,64]. Although there is much supporting evidence to suggest considerable changes in below-ground organs, there are comparatively few estimates of fine-root production *sensu stricto*. Closely related to the energy costs in development and maintenance of the different root systems are their functional effectiveness as organs for absorption and transport of water and nutrients[7,10,17,24,25,32,50,56,59,60,61,62,68,69].

In forests and other ecosystems, the dynamics of fine roots need to be considered to interprete descriptions and comparisons of plant productivity and other related structural and functional attributes of the ecosystems[16]. In most existing data root weight and distribution in relation to changes in environmental conditions has been obtained by sampling on a few specific occasions rather than over substantial periods of time. There is a general lack of information concerning the rate of death and renewal in relation to changes in abiotic and biotic environmental factors[29,50,59,62,65]. One conclusion, from one of the SWECON experiments is that the relative allocation of carbon for root growth decreased drastically with fertilization and especially with irrigation plus fertilization[1,5,52].

Roots are the least measured part of plants. The fine root ramifications are inherently difficult to study *in situ* so within SWECON considerable efforts were devoted to developing methods for fine root production[14,51]. Similar methods have been used by other investigators for boreal forests[10,31,39,61,62].

Although, most existing studies of fine roots are 'preliminary', some general working hypotheses may be summerized:

i. Carbohydrate supply does not normally limit root growth; most forest plants can invest substantial energy, during the year, into fine root growth (and mycorrhizas) from a relative carbohydrate surplus. The occurence of mycorrhiza on forest tree roots should probably be regarded as a result of a carbohydrate surplus.

ii. The ion and water uptake in dry forest soil, when it is partially wetted is confined to a zone of the new roots immediately behind the apex of the root tips. The latter zone is generally short-lived in its effectiveness, since suberization occurs rather soon and local depletion of nutrients takes place around the root tips, therefore they must continually be renewed.

iii. The whole soil profile is most efficiently exploited by the different root systems. Growth is then promoted in soil horizons where they experience the best water and nutrient conditions. If the quality of the soil is poor (*e.g.* in the mineral soil), there is a tendency for the tree root system to develope long roots which penetrate large soil volumes rapidly. On the other hand, if the soil conditions are better (*e.g.* in the F/H layer) the development of ramifications with numerous short root (*e.g.* mycorrhiza) is promoted, leading to an exploitation of available moisture and nutrients.

iv. The root dynamics differ along a moisture gradient from one forest stand to another. A wet forest stand has more living than dead roots (average live/average dead > 1) and the opposite holds for a dry stand. The major influence of increased soil moisture is to improve the decomposition and mineralization of dead roots so that the flux of critical nutrients in the soil is increased.

v. Fine-root production may vary substantially from one year to the next. The latter differences may be even more substantial than the differences between sites. There appears to be as much variation in fine-root dynamics within as between species.

At present, we have to admit that relatively little is known of the mechanism controlling root growth and the range of fine root production in forest ecosystems. However, root studies during the last decade have created the first basis for reliable comparisons of plant productivity in forests. Such detailed data are indispensable to an intelligent simplification of how the ecosystem functions[16]. The need for further research in this particular field of root biology in the next generation of ecosystem studies is evident.

References

1 Ågren G, Axelsson B, Flower-Ellis J G K, Linder S, Persson H, Staaf H and Troeng E 1980 Annual carbon budget for a young Scots pine. *In* Structure and Function of Northern Coniferous Forests – An Ecosystem Study. Ed. T Persson. Ecol. Bull. (Stockholm) 32, 307–313.

2 Albrektson A 1980 Biomass of Scots pine (*Pinus sylvestris* L.) amount – development – methods of mensuration. University of Agricultural Sciences, Dept. of Silviculture Rep. 2, 189 p.

3 Albrektson A 1980 Total tree production as compared to conventional forestry production. *In* Structure and Function of Northern Coniferous Forests – An Ecosystem Study. Ed. T Persson. Ecol. Bull. (Stockholm) 32, 315–327.

4 Albrektson A and Andersson B 1978 Biomassa i en gammal barrblandskog i Dalarna. Swed. Conif. For. Proj., Int. Rep. 79, 40 p.

5 Axelsson B 1981 Site differences in yield-differences in biological production or in redistribution of carbon within trees. Swed. Univ. of Agr. Sciences, Dept. Ecology and Environmental Res. Rep. 9, 1–11.

6 Böhm W 1979 Methods of studying root systems. Ecological Studies 33, 188 p., Springer-Verlag, Berlin, Heidelberg, New York.

7 Caldwell M M 1979 Root structure: the considerable costs of belowground function. *In* Topics in Plant Population Biology. Eds. O T Solbrig, S Jain, G B Johnson and P H Raven. pp. 408–427. Macmillan Press LTD, London, Basingstoke.

8 Coleman D C 1976 A review of root production processes and their influence on soil biota in terrestrial ecosystems. *In* The Role of Terrestrial and Aquatic Organisms in Decomposition Processes. Eds. J M Anderson and Macfadyen. Blackwell Scientific Publ., Oxford, pp 417–434.

9 Deans J D 1979 Fluctuations of the soil environment and fine root growth in a young Sitka spruce plantation. Plant and Soil 52, 195–208.

10 Deans J D 1981 Dynamics of coarse root production in a young plantation of *Picea sitchensis*. Forestry 54, 139–155.

11 Ericsson A and Persson H 1980 Seasonal changes in starch reserves and growth of fine roots of 20-year-old Scots pines. *In* Structure and Function of Northern Coniferous Forests – An Ecosystem Study. Ed. T Persson. Ecol. Bull. (Stockholm) 32, 239–250.

12 Farrel E P and Leaf A L 1974 Effects of fertilization and irrigation on root numbers in a red pine plantation. Can. J. For. Res. 4, 366–371.

13 Ford E D and Deans J D 1977 Growth of Sitka spruce plantation: spatiel distribution and seasonal fluctuations of lengths, weights and carbohydrate concentrations of fine roots. Plant and Soil 47, 463–485.

14 Flower-Ellis J G K, Albrektson A and Olsson L 1976 Structure and growth of some Scots pine stands: (1) dimensional and numerical relationships. Swed. Conif. For. Proj., Tech. Rep. 3, 98 pp.

15 Flower-Ellis J G K and Olsson L 1978 Litterfall in an age series of Scots pine stands and its variation by components during the years 1973–76. Swed. Conif. For. Proj., Tech. Rep. 16, 62 p.

16 Flower-Ellis J G K and Persson H 1980 Investigations of structural properties and dynamics of Scots pine stands. *In* Structure and Function of Northern Coniferous Forests – An Ecosystem Study. Ed. T Persson. Ecol. Bull. (Stockholm) 32, 125–138.

17 Fogel R 1980 Mycorrhizae and nutrient cycling in natural forest ecosystem. New Phytol. 86, 199–212.

18 Fogel R and Hunt G 1979 Fungal and arboreal biomass in western Oregon Douglas-fir ecosystem: distribution patterns and turnover. Can. J. For. Res. 9, 245–256.

19 Ghilarov M S, Kovda V A, Novichkova-Ivanova L N, Rodin L E and Sveshnikova M V (Eds.) 1968 Methods of Productivity Studies in Root Systems and Rhizosphere Organisms. Nauka Leningrad, USSR, 240 p.

20 Göttsche D 1972 Verteilung von Feinwurzeln und Mykorrhizen im Bodenprofil eines Buchen- und Fichtenbestandes im Solling. Hamburg: Kommissionsverlag, Buchhandlung Max Wiedebusch, 103 p.

21 Grier C C, Vogt K A, Keyes M R and Edmonds R L 1981 Biomass distribution and above- and belowgrund production in young and mature *Abies amabilis* zone ecosystems of the Washington Cascades. Can. J. For. Res. 11, 155–167.

22 Harris W F, Goldstein R A and Henderson G S 1973 Analyses of forest biomass pools, annual primary production and turnover of biomass for a mixed deciduous forest watershed. *In* IUFRO Biomass Studies. Ed. H E Young. Univ. Maine at Orono, pp 43–64.

23 Harris W F, Kinerson R S and Edwards N T 1977 Comparison of belowground biomass natural deciduous forests and loblolly pine plantations. Pedobiologia 17, 369–381.

24 Harris W F, Santantonio D and McGinty D 1980 The dynamic belowground ecosystem. *In* Forests: Fresh Perspectives from Ecosystem Analysis, Ed. R H Waring. pp 119–129. Oregon State University Press, Corvallis.

25 Head G C 1967 Effects of seasonal changes in shoot growth on the amount of unsuberized root on apple and plum trees. J. Hortic. Sci. 42, 169–180.

26 Head G C 1970 Methods for the study of production in root systems. *In* Methods of Study in Soil Ecology. Ed. J Philipson. Proc. Paris Symp. pp 151–157. UNESCO, Paris.

27 Head G C 1973 Shedding of Roots. *In* Shedding of Plant Parts. Ed. T T Kozlowski. pp 237–293. Academic Press, New York, London.

28 Heikurainen L 1955 Über Veränderungen in den Wurzelverhältnissen der Kiefernbestände auf Moorböden im Laufe des Jahres. Acta Forest. Fenn. 65, 70 p.

29 Hermann R K 1977 Growth and production of tree roots: a review. *In* The Belowground Ecosystem: A Synthesis of Plant-Associated Processes. Ed. J K Marshall. pp 7–28. Fort Collins: Range Science Dept. Sci. Ser. 26. Colorado State Univ.

30 Kalela E K 1955 Über Veränderungen in den Wurzelverhältnissen der Kiefernbestände im Laufe der Vegetationsperiode. Acta Forest. Fenn. 65, 42 p.

31 Keyes M R and Grier C C 1981 Above- and belowground net production in 40-year-old Douglas-fir stands on low and high productivity sites. Can. J. For. Res. 11, 599–605.

32 Kimmins J P and Hawkes B C 1978 Distribution and chemistry of fine roots in British Columbia: implication for management. Can. J. For. Res. 8, 265–279.

33 Kohman K 1972 Root-ecological investigations on pine (*Pinus silvestris*). I. Problems of methodology and general root relationships. Det Norske Skogsforsøksvesen 30, 325–357.

34 Köstler I N, Brückner E and Bibelreither H 1968 Die Wurzeln der Waldbäume. Verlag Paul Parey, Hamburg, Berlin 284 p.

35 Kummerow J, Krause D and Jow W 1977 Root systems of chaparral shrubs. Oecologia Berlin 29, 163–177.

36 Lieth H 1968 The determination of plant dry-matter production with special emphasis on the underground part. *In* Functioning of Terrestrial Ecosystems at the Primary Production Level. Ed. F E Eckart. pp 179–186. Proceedings of the Copenhagen Symposium, UNESCO.

37 Linder S and Troeng E 1981 The seasonal variation in stem and coarse root respiration of 20-year-old Scots pine (*Pinus sylvestris* L.). *In* Radial Growth in Trees. Ed. W Tranquillini. Mitt. Forst. Bundesversuchsants., Wien 142, 125–139.

38 Linder S and Axelsson B 1982 Changes in carbon and allocation patterns as a result of irrigation and fertilization in a young Scots pine stand. Manuscript.

39 Lyr H and Hoffman G 1967 Growth rates and growth periodicity of tree roots. Int. Rev. Forest Res. 2, 181–236.

40 McClaugherthy C A, Aber J D and Melillo J M 1982 The role of fine roots in the organic matter and nitrogen budgets of two forested ecosystems. Ecology 63, 1481–1490.

41 Newbould P J 1967 Methods for estimating the Primary Production of Forests. IBP Handbook No. 2. Oxford: Blackwell Scientific Publications. 62 p.

42 Newbould P J 1968 Methods of estimating root production. *In* Functioning of Terrestrial Ecosystems at the Primary Production Level. Ed. R E Eckart. Proc. Copenhagen Symp. pp 187–190. UNESCO.

43 Persson H 1975 Deciduous woodland at Andersby, Eastern Sweden: Field-layer and below-ground production. Acta Phytogeogr. Suec. 62, 1–72.

44 Persson H 1978 Root dynamics in a young Scots pine stand in Central Sweden. Oikos 30, 508–519.

45 Persson H 1979 The possible outcomes and limitations of measuring quantitative changes in plant cover on permanent plots. *In* The Use of Ecological Variables in Environmental Monitoring. Ed. H Hyttenborn. The National Swedish Environment Protection Board, Rep. PM 1151, 81–87.

46 Persson H 1979 Fine-root production, mortality and decomposition in forest ecosystem. Vegetatio 41, 101–109.

47 Persson H 1980 Spatial distribution of fine-root growth, mortality and decomposition in a young Scots pine stand in Central Sweden. Oikos 34, 77–87.

48 Persson H 1980 Structural properties of the field and bottom layers at Ivantjärnsheden. *In* Structure and Function of Northern Coniferous Forests – An Ecosystem Study. Ed. T Persson. Ecol. Bull. (Stockholm) 32, 153–163.

49 Persson H 1980 Death and replacement of fine roots in a mature Scots pine stand. Ibid 32, 251–260.

50 Persson H 1980 Dynamic aspects of fine-root growth in a Scots pine stand, with and without near optimum nutrient and water regimes. Acta Phytogeogr. Suec. 68, 101–110.

51 Persson H 1981 Death replacement of fine roots in forest ecosystem – a neglected area of research. Swed. Conif. For. Proj., Tech. Rep. 29, 1–25.

52 Persson H 1981 The effect of fertilization and irrigation on the vegetation dynamics of a pine-health ecosystem. Vegetatio 46, 181–192.

53 Persson H 1982 Changes in the tree and dwarf shrub fine-roots after clear-cutting in a mature Scots pine stand. Swed. Conif. For. Proj., Tech. Rep. 31, 1–19.

54 Persson T, Bååth E, Clarholm M, Lundkvist H, Söderström B E and Sohlenius B 1980 Trophic structure, biomas dynamics and carbon metabolism of soil organisms in a Scots pine forest. *In* Structure and Function of Northern Coniferous Forests – an Ecosystem Study. Ed. T Persson. Ecol. Bull. (Stockholm) 32, 419–459.

55 Reynolds E R C 1970 Root distribution and the cause of its spatial variability in *Pseudotsuga taxifolia* (Poir.) Britt. Plant and Soil 32, 501–517.

56 Reynolds E R C 1974 The distribution pattern of fine roots of trees. *In* Ecology and Physiology of Root Growth. II. Ed. G Hoffmann. Int. Symp., pp 101–112. Berlin: Akademie-Verlag.

57 Resa F 1877 Ueber die Periode der Wurzelbildung. Acad. Diss., Bonn, 39 p.

58 Roberts J 1976 A study of root distribution and growth in a *Pinus silvestris* L (Scots pine) plantation in East Anglia. Plant and Soil 44, 607–621.

59 Russel S R 1977 Plant root systems: their function and interaction with the soil. London, McGraw-Hill Comp., 298 p.

60 Safford L O 1974 Effect of fertilization on biomass and nutrient content of fine roots in a beech-birch-maple stand. Plant and Soil 40, 349–363.

61 Santantonio D 1979 Seasonal dynamics of fine root in mature stands of Douglas-fir of different water regimes – a preliminary report. *In* Physiologie des Racines et Symbioses. Eds. A Riedacker and J Gagnaire-Michard. C.R. des Reunions du Groupe d'Etude des Racines, pp 190–203. Nancy, France.

62 Santantonio D 1982 Production and turnover of fine roots of mature Douglas-fir in relation to site. Doctor thesis, Oregon State University, School of Forestry, Corvallis, Oregon, 93 p.

63 Santantonio D, Herman R K and Overton W S 1977 Root biomass studies in forest ecosystems. Pedobiologia 17, 1–31.

64 Schulze E D, Fuchs M and Fuchs M I 1977 Spacial distribution of photosynthetic capacity and performance in a mountain spruce forest of Northern Germany. III. The significance of the evergreen habit. Oecologia Berlin 30, 239–248.

65 Sutton R F 1969 Form and development of conifer root systems. Commonwealth Agr. Bureaux, Techn. Commun. 7, 1–131.

66 Sutton R F 1980 Root system morphogenesis. N. Z. J. For. Sci. 10, 264–292.

67 Torrey J G 1976 Root hormones and plant growth. Annu. Rev. Plant Physiol. 27, 435–459.

68 Trappe J M and Fogel R D 1977 Ecosystematic function of mycorrhizae. *In* The Belowground Ecosystem: A Synthesis of Plant Associated Processes, Ed. J K Marshall. pp 205–214.

69 Vogt K A, Edmonds R L, Grier C C and Piper S R 1980 Seasonal changes in mycorrhizal and fibrous-textured root biomass in 23- and 180-year-old Pacific silver fir stands in western Washington. Can. J. For. Res. 10, 523–529.

Plant and Soil 71, 103–110 (1983). A-10

Root development and plantation establishment success

A. N. BURDETT, D. G. SIMPSON and C. F. THOMPSON
British Columbia Ministry of Forests, 1450 Government Street, Victoria, Canada, V8W 3E7

Keywords Chemical root pruning Picea Pinus Root growth capacity Root morphogenesis Shoot growth

Summary Batches of 2 + 0, bare-root lodgepole pine (*Pinus contorta* Dougl.) and white spruce (*Picea glauca* (Moench) Voss) seedings differing in provenance, lifting date, or nursery of origin varied greatly both in root growth capacity (RGC) as measured by a laboratory assay and in early survival and growth after field planting. Most of the variation in field performance was accounted for by the variation in RGC.

The lateral roots of lodgepole pine seedlings were pruned chemically by raising them in containers coated on the inside with cupric carbonate in latex paint. When field planted, the chemically root-pruned stock grew appreciably faster than control stock raised in unpainted containers. The difference in growth rate was associated with a difference in root form 4 years after planting.

Introduction

A newly planted tree has a root system which, however great in mass, provides access to only a limited volume of soil. This makes the tree prone to both desiccation and nutrient deficiency. It is presumably for this reason that the ability of planted trees to become established often depends on the speed with which they put out new roots[5,10,12,13,16,18,19,21].

The form of root development also affects plantation performance. This it may do through effects on either mechanical stability, as in young plantations of certain pines[4,20], or on shoot growth[6,8,11,20].

Both the rate and the form of root development are dependent on environmental factors[2]. In planted trees, however, early root development is also dependent on the morphological and physiological characteristics of the trees when planted. Some of these characteristics vary greatly according to the cultural and handling methods used in the production of the seedlings[3,6,10,14,15,17,18,21]. Thus there is scope for improving plantation establishment succes through measures to control the root growth- and the root morphogenetic-potential of forest planting stock. This is demonstrated by the experiments reported here.

Materials and methods

To investigate the relationship between the root growth capacity of 2 + 0, bare root lodgepole pine (*Pinus contorta* Dougl.) and white spruce (*Picea glauca* (Moench) Voss) seedlings and their performance when planted, batches of stock of one or several provenances (Table 1) of each species

Table 1. Provenance and nursery of origin of stock used. Both nurseries are located in the interior of British Columbia; Red Rock at 53°40′N 122°40′W; Skimikin at 50°40′N 119°15′W.

Seedlot number	Nursery of origin	Species	Location of parent stand	Elevation (m)
1743	Red Rock	*Pinus Contorta*	55°10′N 127°50′W	610
2201	Red Rock	*Pinus contorta*	55°40°N 121°40′W	1006
2233	Skimikin	*Pinus contorta*	49°40′N 116°10′W	1372
2484	Skimikin	*Pinus contorta*	51°10′N 116°10′W	1372
1868	Red Rock	*Picea glauca*	56°40′N 122°10′W	823
2737	Red Rock	*Picea glauca*	53°40′N 123°10′W	915
2750	Red Rock	*Picea glauca*	53°10′N 120°10′W	1281
2502	Skimikin	*Picea glauca*	50°50′N 118°40′W	1402

were lifted at two nurseries (Table 1) on 5 to 7 dates between September 12 and April 17. During May the Stock, which had been cold-stored, was tested for root growth capacity (RGC)[5]. Twenty-four trees of each batch of stock were potted and placed in a controlled environment chamber providing the following conditions: 30°C day- and 25°C night-temperature; 16 hours daily photoperiod (15,000 Lux); 75% relative humidity. After one week the trees were scored for the number of roots elongated during the test as follows: 0 = no newly elongated roots; 1 = some newly elongated roots, none more than 1 cm; 2 = 1 to 3 roots elongated by more than 1 cm; 3 = 4 to 10 roots elongated by more than 1 cm; 4 = 11 to 30 roots elongated by more than 1 cm; 5 = 31 to 100 roots elongated by more than 1 cm.

Within two days of the start of the root growth test, 192 trees of each batch of stock were planted on clear-cut forest sites. The pine was planted on a site previously occupied by pine at 49°30′N 115°4′W, elevation 1500 m; and the spruce on a site previously occupied by spruce at 53°30′N 122°20′W, elevation 1000 m. The trees were planted in 6 blocks, each containing one row of each batch of stock. The order of rows within the blocks was random. At the end of the first growing season survival and height growth of the planted trees was recorded.

To examine the effect of chemical root pruning on the field performance of container-grown stock, three provenances of lodgepole pine were raised in both copper-painted and unpainted containers[3]. The containers were 2 cm in diameter and 11 cm deep. They were painted by dipping them in exterior latex paint containing 0.1 kg/l basic cupric carbonate powder[3,7]. The containers were filled with a 3 : 1, peat : vermiculite mix containing dolomitic lime (12 mesh and finer). The containers were seeded in April. The trees were raised according to standard procedures at a British Columbia government forest nursery[3]. In November of the year of sowing the trees were extracted from the containers and placed in cold storage for planting the following spring. The trees were planted in rows of 25 in two blocks. In each block there was one row of each provenance of root-pruned stock and two rows of each provenance of control stock. Height growth of the planted trees was recorded for 4 years. Trees were excavated after 4 years to assess the effect of the root pruning treatment on root morphology.

Results

The relationship between RGC and first season survival of the 2 + 0, bare-root lodgepole pine is illustrated in Fig. 1. In stock without any detectable RGC mortality was 100%. Survival increased with RGC approaching 100% as RGC exceeded the mid-range value of 2.5.

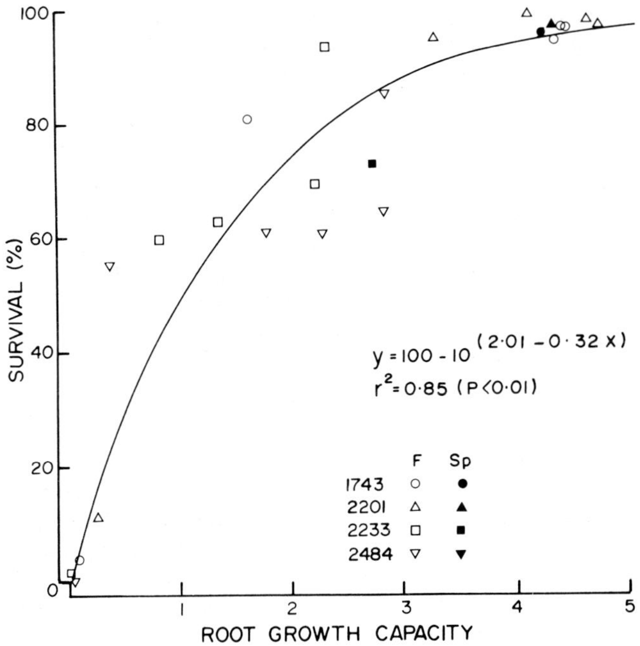

Fig. 1. The relationship between root growth capacity and first season survival in spring-planted lodgepole pine. F and S are fall-and spring-lifted samples, respectively.

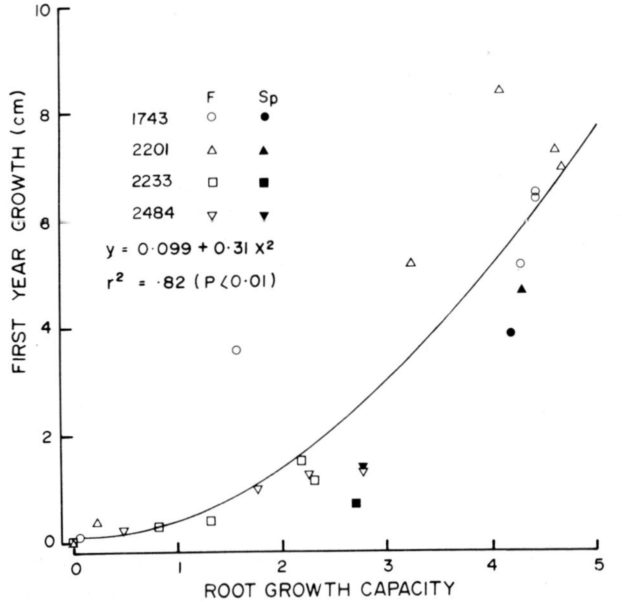

Fig. 2. The relationship between root growth capacity and first season shoot extension in spring-planted lodgepole pine. F and S are fall- and spring-lifted samples, respectively.

First season shoot extension in the pine also increased with RGC (Fig. 2). But, whereas survival increased asymptotically with increasing RGC, shoot growth (mean for all trees including those dead at the time of measurement) increased exponentially. This indicates that the effect of RGC on shoot extension was at least partly independent of its effect on surval.

In the spruce survival increased asymptotically from near zero to almost 100% with increasing RGC (Fig. 3). Shoot extension in the spruce also increased with RGC (Fig. 4). In this case the relationship was linear. This means that shoot extension increased with RGC even when RGC exceeded the level necessary for almost 100% survival. Thus in the spruce, as in the pine, RGC had an effect on shoot extension independent of that on survival.

The effect of the chemical root pruning treatment on root morphogenesis in container-grown, Lodgepole pine is illustrated by fig. 5. The unpruned seedlings extended few roots except from the tip of the root plug. These were first- and higher-order laterals which had grown down the walls of the seedling container prior to planting. The main lateral roots of the chemically root-pruned trees, however, emerged from the sides of the root plug; especially the uppermost half. These were laterals that were prevented from growing down, or around the

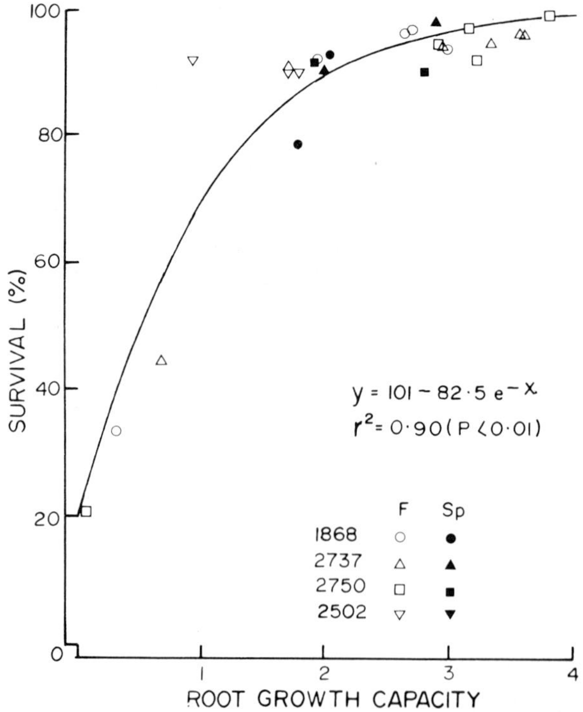

Fig. 3. The relationship between root growth capacity and first season survival in spring-planted white spruce. F and S are fall- and spring-lifted samples, respectively.

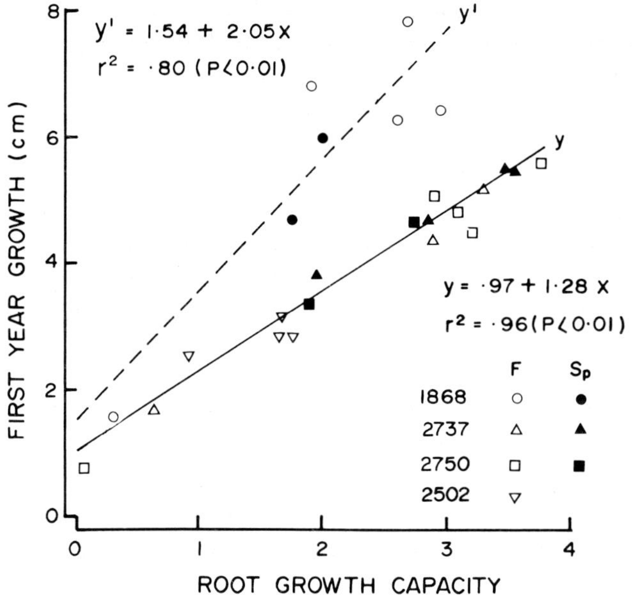

Fig. 4. The relationship between root growth capacity and first season shoot extension in spring-planted white spruce. F and S are fall- and spring-lifted samples respectively.

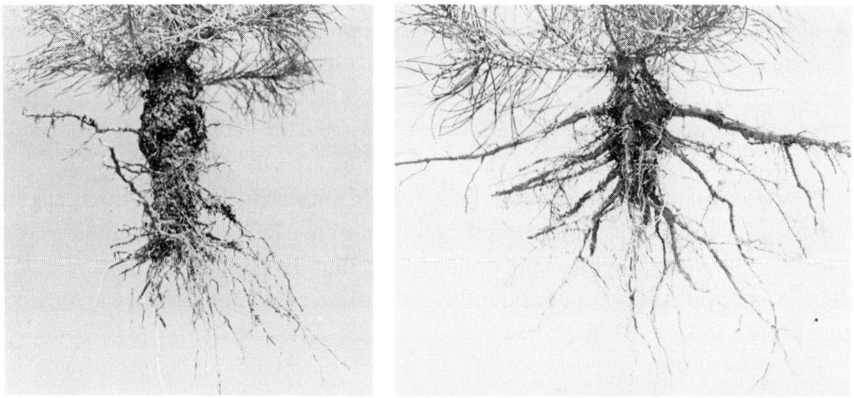

Fig. 5. The roots of an untreated (left), and a chemically root-pruned lodgepole pine seedling 4 growing seasons after field planting. The depth of the root plug is approximately 11 cm.

container wall by contact with the copper paint, but which resumed elongation when the trees were planted.

Another effect of the chemical root pruning treatment was to increase height growth in the third and fourth years after planting (Table 2). In the fourth year the effect was highly significant.

Table 2. Height growth in chemically root-pruned and unpruned lodgepole pine seedlings*

Year	Mean increase in height (cm)**		
	Pruned	Unpruned	Pruned as % of unpruned
1	3.7	3.8	97
2	9.3	9.0	103
3	23.6	20.6	115
4	29.3	25.6	115

* Taken with permission from reference[6].
** Mean value for all provenances.

Discussion

A relationship between the RGC of forest tree seedlings and their survival when planted has been reported in a number of conifers including lodgepole pine[5,10,16,18,21], although not previously in white spruce. The present study is the first, however, to provide unequivocal evidence that root growth capacity can affect shoot growth independently of its effect on survival.

How RGC measured under standard conditions in the laboratory predicts the performance of stock planted in the field remains to be demonstrated. The assumption, however, underlying the development of root growth tests for grading forest planting stock, is that relative RGC under standard conditions is correlated with relative RGC in the field[12]. This assumption is supported by the correlations reported here and elsewhere between the RGC of tree seedlings measured in the laboratory and their survival or growth in the field.

The practical implication of the effect of RGC on survival is clear. Every effort should be made to ensure that the trees have a high RGC at the time they are planted. This study confirms previous reports that, in the choice of lifting date, there is an opportunity for considerable control over the root growth capacity of cold-stored stock[10,14,15,18,21].

The silvicultural importance of the effect of RGC on shoot growth is uncertain. Most of the variation among batches of stock in first season shoot extension was attributable to variation in RGC capacity. In relative terms the variation was large, but in absolute terms, it amounted to no more than a few centimetres. The direct effect of such differences on plantation productivity will be insignificant. Often, however, trees that grow poorly during the first year after planting remain checked in their growth for a number of years afterwards[1,9]. Sometimes, therefore, the effect of RGC on shoot extension may persist for a number of years, thereby measureably affecting productivity.

Even if the direct effect of RGC on shoot extension is only of short duration it will affect the ability of planted trees to dominate competing vegetation. Under some circumstances this effect will be a key factor determining plantation establishment success.

The chemical root pruning treatment tested with lodgepole pine was intended to improve mechanical stability by promoting the development of lateral roots growing straight out from the tap root close to the soil surface[3]. In this the treatment was highly effective. It also promoted height growth.

Increased height growth was probably a consequence of the effect on root form. The upper soil layers into which the lateral roots of the root-pruned stock developed must have differed in a number of respects from the deeper horizon into which the first order laterals of the control stock grew. It seems likely that these differences, for example in fertility, temperature, moisture content, and density, were adequate to account for the difference in shoot growth between the root-pruned, and the control stock.

The persistence of the effect of the chemical root pruning treatment on shoot growth remains to be seen. Effects of nursery treatments and planting on tree root form diminish with time[2], however, and the same can be expected of chemical root pruning. But even if quite transient, the effect on growth could influence plantation performance through an effect on the ability of trees to contend with competing vegetation.

The experiments reported here demonstrate great variation in the root growth- and the root morphogenetic-potential of forest planting stock. Furthermore, they confirm that such variation can have a major impact on plantation establishment success. Control of the form and physiological condition of forest planting stock to promote satisfactory root development after planting thus provides an effective means of improving plantation establishment success and productivity.

References

1 Armson K A 1958 The effect of two planting methods on the survival and growth of white spruce (*Picea glauca* (Moench) Voss) in eastern Ontario. For. Chron. 34, 376–379.

2 Armson K A 1978 Roots of the new forest. Proc. Root Form of Planted Trees Symposium Victoria, Canada May 16–19, 1978, Eds. E Van Eerden and J M Kinghorn. pp. 325–328.

3 Burdett A N 1978 Control of root morphogenesis for improved mechanical stability in container-grown lodgepole pine. Can. J. For. Res. 8, 483–486.

4 Burdett A N 1979 Juvenile instability in planted pines. Ir. For. 36, 36–47.

5 Burdett A N 1979 New methods for measuring root growth capacity: their value in assessing lodgepole pine stock quality. Can. J. For. Res. 9, 63–67.

6 Burdett A N 1981 Box-pruning the roots of container grown tree seedlings. Proc. Canadian Containerized Tree Seedling Symposium, Toronto, Ontario, Sept. 14–16, Ed. J. Scarratt. pp 203–206.

7 Burdett A N and Martin P A F 1982 Chemical root pruning of coniferous seedlings. Hortic. Science 17, 622–624.

8 Hay R L and Woods F 1974 Roots deformation correlated with sapling size for loblolly pine. J. For. 72, 143–145.

9 Mullin R E 1963 Planting check in spruce. For. Chron. 39, 252–259.

10 Ritchie G A and Dunlap J R 1980 Root growth potential: its development and expression in forest tree seedlings. N.Z. J. For Sci. 10, 218–248.

11 Rudolf P O 1939 Why forest plantations fail. J. For. 37, 377–383.

12 Stone E C 1955 Poor survival and the physiological condition of planting stock. For. Sci. 1, 89–94.

13 Stone E C and Jenkinson J L 1971 Physiological grading of ponderosa pine nursery stock. J. For. 69, 31–33.

14 Stone E C, Jenkinson J L and Krugman S L 1962 Root-regenerating potential of Douglas fir seedlings lifted at different times of year. For. Sci. 8, 288–297.

15 Stone E C and Schubert G H 1959 Root regeneration of ponderosa pine seedlings lifted at different times of year. For. Sci. 5, 322–332.

16 Sutton R F 1980 Planting stock quality, root growth capacity and field performance of three boreal conifers. N.Z. J. For. Sci. 10, 54–71.

17 Sutton R F 1980 Root system morphogenesis. N.Z. J. For. Sci. 10, 264–292.

18 Todd G 1964 Do Douglas fir seedlings have roots? J. For. 62, 561–563.

19 Wakeley P C 1948 Physiological grades of southern pine nursery stock. Proc. 49th Annual Meet. Soc. Am. For. pp. 311–322.

20 Wibeck E 1923 Om missbildning av tallens rotsystem vid spettplantering. Meddelanden fran statens Skogsforsoksanstalt 20, 261–303.

21 Winjum J K 1963 Effects of lifting date and storage on 2 + 0 Douglas fir and Noble fir. J. For. 61, 648–654.

Plant and Soil 71, 111–122 (1983). A-11

Root growth capacity: relationship with field root growth and performance in outplanted jack pine and black spruce

R. F. SUTTON
Great Lakes Forest Research Centre, Canadian Forestry Service, Sault Ste. Marie, Ontario, Canada, P6A 5M7

Key words Boreal forest Bräcke Mounding *Picea mariana* *Pinus banksiana* Planting Planting stock quality Root growth capacity Site preparation

Summary First- and second-year results from a study of mounding site preparation for regenerating boreal forest cutovers in Ontario using bareroot jack pine and black spruce are reported. At each of 4 sites 5 treatments were investigated: (a) control; (b) shoulder of standard Bräcke patch; (c) bottom of Bräcke patch; (d) a 20-L mound of mineral soil on the mineral soil on the shoulder of Bräcke patch; (e) a similar mound on the double layer of organic matter (original forest floor + inverted organic matter from Bräcke patch) immediately distal to the patch. At each site, 600 pines or spruces were planted as 4 replicates of 30-tree plots for the 5 treatments. Field root growth (FRG) was determined on 5 trees for each plot 30 days post-planting. Survival and growth were assessed after 1 and 2 growing seasons.

Root growth capacity (RGC) tests revealed large differences between lots of planting stock. Only with black spruce was a higher RGC associated with better nursery performance. Only one third of a tree's RGC could be explained in multiple regression with initial top height, stem diameter, and root area index. Field performance of outplants was highly site- and treatment-specific, but in both species on all sites, stem diameter increment and FRG were consistently high for the mound treatment. FRG, ($r > 0.50$ in 8 of 16 cases), was best correlated with first- + second-year stem diameter increment. Correlations between FRG and RGC were inconsistent and generally low.

Introduction

The forester's interest in root growth capacity (RGC) is based on his need to know the quality and growth potential of forest tree planting stock[3,4,11,14,15,17]. Until relatively recently, performance potential was estimated solely from morphological and chromatic features[16], a basis often found wanting[14,18]. Physiological attributes of planting stock must be taken into account.

This study examines the variation in field performance of outplanted trees on various sites where conditions were influenced by site preparation with the Bräcke scarifier and assesses interrelationships between RGC, nursery performance, field root growth (FRG) after 30 days, and field performance.

This provides a basis for evaluating the mounding method of site preparation, in the regeneration of cutover boreal forest sites in Ontario, using spring-lifted, bareroot jack pine (*Pinus banksiana* Lamb.) and black spruce (*Picea mariana* (Mill.) BSP).

Materials and methods

In autumn 1979, mounding[1,2,5,7,9,12,13] was carried out on clearcut sites that earlier in the year had been treated with the standard Bräcke Scarifier to create patches scalped through the organic matter of the forest floor into mineral soil. Five treatments were compared (Fig. 1): (a) control; undisturbed forest floor; (b) upper part (shoulder) of the standard Bräcke patch; (c) lower part (bottom) of the standard Bräcke patch; (d) 20 L of mineral soil heaped on mineral soil on the shoulder of the Bräcke patch; (e) a similar mound but on the double layer of organic matter, original forest floor + inverted LFH layer from the Bräcke patch immediately distal to the Bräcke patch. In accomplishing the original purpose of patch scarification *i.e.* production of such patches, the scarifier scuffs up an inverted minimound of mainly organic matter onto undisturbed floor: treatment (e) was effected by heaping mineral soil on this double layer. The mineral soil scuffed up by the Scandinavian patch scarifiers is insufficient to suppress competing vegetation[10].

On each of eight sites, located between 48° and 51°N, 80° and 91°W and 275–450 m elevation, 600 bareroot trees, either 2 + 0 jack pine seedlings or $1\frac{1}{2} + 1\frac{1}{2}$ black spruce transplants, 4 sites each (Table 1), were spring-outplanted by hand in 1980 using the operational slit method. Each of the five treatments had four replicates of 30-tree plots. Different seed lots were used to plant up some of the sites although all trees were lifted between 1 and 14 May, stored at 1–2°C and planted between 16 and 30 May. The lot 1 stock planted at Savant Lake was significantly ($P = 0.05$) larger than the lot 2 plants used elsewhere (Table 1). Trees used at Cochrane, Foleyet and White River were similar in terms of height and ground-level stem diameter.

All outplanted trees were assessed 30 days after planting. Five randomly selected trees were excavated for each plot (30 trees) and the number of unlignified root tips > 1 cm in length and their aggregate length determined. Tree condition was scored as: – 1 = tree in good condition; 2 = tree in good/moderately good condition but with leading shoot defective; 3 = tree alive but in poor condition; and 4 = tree dead or virtually so. The remaining 25 trees were assessed for survival, annual height increment, ground level stem diameter, and condition, at the end of the first and second growing seasons.

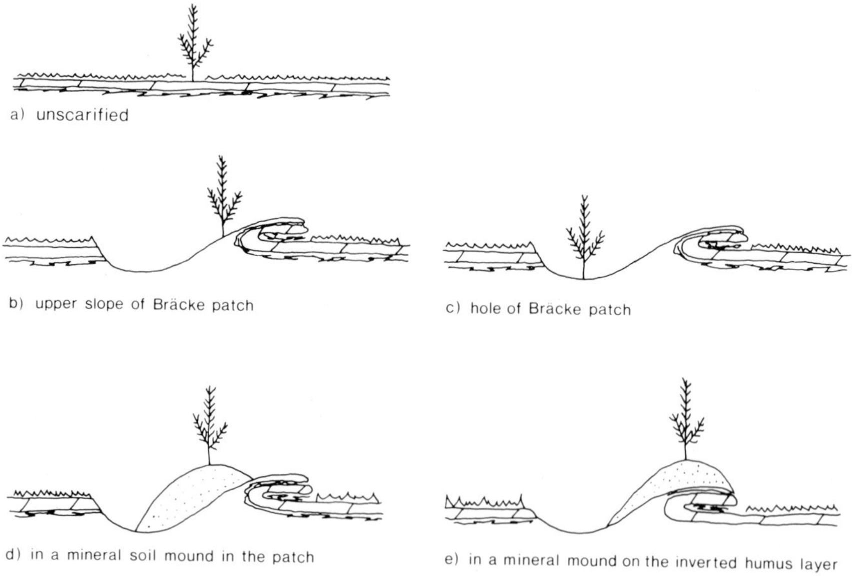

Fig. 1. Diagrammatic representation of the 5 treatments. (After Edlund[5]).

A subsample of the planting stock used on each site was planted in a nursery to determine the viability of the planting stock at planting. These trees were assessed 30 days post planting, and as above at the end of one growing season.

In addition, lot 1, Savant Lake, and the White River portion of lot 2 stocks were subsampled to give 60 trees per species per lot for RGC determinations. Top length, ground-level stem diameter, root system volume (by displacement, root area index, initial number and number after 30 days in a growth chamber of unlignified roots > 1 cm long, the initial and 30 day aggregate lengths of these roots, the initial and 30 day assessments of condition; the initial and 30 day numbers of white roots < 1 cm long were recorded.

Statistical analysis, *e.g.* analyses of variance, Tukey's Multiple Comparison test, chi-square tests and correlation analysis has been applied to data as appropriate.

Results and discussion

Nursery test planting

The nursery test planting showed survival among jack pine and black spruce at the end of the growing season as 96% and 98% respectively. Growth and condition were good with the Savant Lake trees (lot 1) significantly ($P = 0.01$), bigger than at other sites with respect to both height increment and stem diameter, (Table 2).

This may be a reflection of the high RGC exhibited by the Savant Lake stock (lot 1) compared with White River (lot 2) stock (Table 2). The nursery tests showed low correlations between the initial size and first-year performance.

Table 1. Details of jack pine and black spruce planting stock used

Site	Seed lot	Site region	Lift date	Storage	Pickup date	Plant date	Top ht cm	Stem diam mm
(*a*) *Jack pine*								
Cochrane	M77–147	3215	May 1	N–R	May 23	May 30	16.0a[1]	3.9b
Foleyet	M77–147	3215	May 1	N–R	May 23	May 27	16.5b	3.5a
Savant Lake	M77–147	3215	May 1	N	May 14[2]	May 16	16.4b	4.3c
White River	M77–147	3215	May 1	N–B	May 20[3]	May 22	17.0c	3.6a
(*b*) *Black spruce*								
Cochrane	73–182	3200B	May 5	N	May 20	May 29	18.7a	3.4a
Foleyet	73–182	3200B	May 5	N	May 20	May 26	18.4a	3.5a
Savant Lake	M76–122	3400	May 14	–	May 14[2]	May 17	20.1c	4.5b
White River	73–182	3200B	May 5	N	May 20[3]	May 23	19.4b	3.5a

[1] Within species, values not followed by the same letter differ significantly (P 0.05).

[2] RGC determined on a subsample, Lot 1.

[3] RGC determined on a subsample, Lot 2.

Table 2. Nursery test planting: details of stock and performance: figures in parentheses are overall means from field outplanted stock

Stock	Initial top ht cm	Initial stem diam. mm	1st-yr ht increment cm	1st-yr stem increment mm
(a) Jack pine				
Cochrane	14.9a[1] (16.0)	3.7a (3.9)	14.2b (6.3)	3.0b
Foleyet	15.3a (16.5)	3.5a (3.5)	12.2a (9.5)	2.5b
Savant Lake	15.9a (16.4)	4.8b (4.3)	14.7b (9.5)	1.6a
White River	16.3a (17.0)	3.8a (3.6)	14.5b (7.8)	2.9b
	NS[2]	**	**	**
All	15.6	3.9	13.9	2.5
(b) Black spruce				
Cochrane	19.4b (18.7)	3.5b (3.4)	8.6a (3.4)	1.7a
Foleyet	17.8ab (18.4)	3.0a (3.5)	8.3a (4.7)	1.8a
Savant Lake	18.5ab (20.1)	4.2c (4.5)	17.0b (5.8)	3.4b
White River	17.4a (19.4)	3.1ab (3.5)	8.0a (3.6)	1.6a
	*	**	**	**
All	18.3	3.4	10.5	2.1

[1] Within species, values followed by the same letter do not differ significantly (*P* 0.05) by t-test.
[2] NS, *, ** = not significant, significant at *P* 0.05, and significant at *P* 0.01 levels respectively by analysis of variance.

Root growth capacity (RGC)

Differences between lots The 2 lots, Savant Lake and White River, differed significantly ($P = 0.01$) in aspects of initial morphology and in RGC (Table 3). The Savant Lake jack pines (lot 1) produced 3.6 times as many roots > 1 cm long as did the White River (lot 2) stock, and their aggregate length was 5.3 times greater. The Savant Lake (lot 1) black spruce produced 8.7 times as many new roots, 12.5 times longer, than the White River (lot 2) black spruce.

The RGC values for the White River (lot 2) black spruce were low and showed a reduction in unlignified root tips during the RGC test.

RGC and tree size Tree size was a poor predictor of RGC. For jack pine, the multiple regression of top height, stem diameter, and root area index as independent variables, (significant at $P = 0.01$), explained only 31% of variation in root number and 26% in aggregate length. For black spruce, the equivalent figures were 32% and 35%.

Ground-level stem diameter was most highly correlated with both root number and length in jack pine and black spruce, (r^2's ranging from 0.26 to 0.35).

Table 3. Root growth capacity

Lot	n	Initial morphology						RGC	
		Top ht cm	Stem diam mm	Root vol cm^3	RAI[1] cm^2	Unlig root tips no.	Unlig root tips[3] cm	Unlig root tips no.	Unlig root tips[3] cm
(a) Jack pine									
1. Savant Lake	60	17.0a[2]	3.9b	3.4b	36b	0	0	59.4b	444.0b
2. White River	60	18.0a	3.2a	1.9a	24a	0	0	18.1a	91.3a
(b) Black spruce									
1. Savant Lake	60	20.8a	4.3b	3.6a	31a	12.2a	21.1a	57.5b	179.5b
2. White River	60	19.9a	3.4a	3.0a	25a	12.3a	19.9a	8.1a	17.6a

[1] RAI = root area index (Morrison and Armson 1968).
[2] Within species, values followed by the same letter do not differ significantly by t-test (P 0.01).
[3] Aggregate length of unlignified root tips > 1 cm.

RGC and nursery performance The superiority in RGC of lot 1 over lot 2 jack pine was not reflected in nursery performance. The RGC of the White River stock in a low stress situation was adequate for good survival and growth. With black spruce, however, the high-RGC lot 1 stock performed twice as well in the nursery test as did the low-RGC lot 2 stock.

The similarity in nursery performance between the lot 2 plantings (White River, Cochrane, and Foleyet) (Table 2) suggests that RGC determined on White River stock only may also apply to the other trees. RGC/nursery performance relationships are not simple, however.

Field performance

Initial survival Initial survival was good despite hot and dry weather during planting and subsequently; 30 days post-planting, survival was ≥90% in 38 of the 40 treatment/sites. After 2 growing seasons, 95% of the jack pine and 85% of the black spruce were in good condition (classes 1 + 2).

Influence of site Survival and height increment were highly site-specific in both species. At each assessment, (30 days, end of years 1 and 2) there were significant ($P = 0.01$) differences (chi-square test) in the condition of trees among sites. Second-year mortality in jack pine ranged from 0.6% at Cochrane to 8.4% at White River; comparable figures for black spruce are 6.8% at Foleyet to 23.4% at White River (Table 4).

Influence of microsite Survival differed among treatments at 2 of the jack pine sites and all black spruce sites (Table 4). The effect of a given treatment varied among sites. For instance, by the end of year 2 there was no jack pine mortality on the control (treatment a) at Cochrane, at White River there was 21% mortality.

Table 4. Cumulative mortality (%) during the first 30 days after outplanting and after the first and second growing seasons, by species, site, and microsite

Site	Assessment[1]	Treatment a	b	c	d	e	a–e (mean)
(a) Jack pine							
Cochrane	1	0	0	0	0	1	0.2
	2	0	0	0	1	1	0.4
	3	0	0	1	1	1	0.6
Foleyet	1	1	2	5	0	11	3.8**[2]
	2	5	3	5	1	11	5.2*
	3	5	3	7	2	11	5.6*
Savant Lake	1	0	1	0	3	3	1.6**
	2	3	1	0	3	4	2.2
	3	3	1	1	3	4	2.4
White River	1	14	3	2	1	6	5.4**
	2	19	6	2	4	6	7.6**
	3	21	7	3	5	6	8.4**
(b) Black spruce							
Cochrane	1	0	0	0	2	22	4.8**
	2	0	6	8	4	22	8.0**
	3	3	10	16	8	27	12.8**
Foleyet	1	2	0	0	1	6	1.8**
	2	2	0	3	2	7	2.8*
	3	7	1	11	5	9	6.6*
Savant Lake	1	4	1	0	0	9	3.0**
	2	11	5	0	3	17	7.2**
	3	14	5	16	3	18	11.2**
White River	1	0	1	3	1	0	1.0
	2	5	17	39	3	1	13.2**
	3	6	26	64	16	5	23.4**

[1] Assessment 1 after 30 days planting, assessment 2 after 1 growing season, and assessment 3 after 2 growing seasons.
[2] * and ** differences between treatments contributing to the mean significant at the $P = 0.05$ and $P = 0.01$.

Patterns of mortality also varied: 21 of the 22 jack pine that died in treatment e during the 2 seasons did so within 30 days of planting. For the control only half of mortalities occurred within the first 30 days. Similarly 63% of the black spruce that died on treatment e succumbed within the first 30 days. For the control only 27% succumbed within the first 30 days. Only at the Savant Lake and Cochrane

Table 5. Field performance: 1st- and 2nd-year height increment and second year ground-level stem diameter increment % initial stem diameter

Site		Variable	Treatment					
			a	b	c	d	e	a–e (mean)
(a) Jack pine								
Cochrane	Height	Δh1 cm	6.5a[1]	6.3a	5.8a	6.8a	6.0a	6.28
	Height	Δh2 cm	9.9ab	10.9b	8.7ab	10.9b	9.0a	10.08*[2]
	Diam.	Δd1 + 2%	57	78	82	100	100	
Foleyet	Height	Δh1 cm	8.8a	9.0a	10.3ab	10.6b	8.7a	9.46**
	Height	Δh2 cm	29.5a	28.0a	26.2a	27.2a	28.0a	27.77
	Diam.	Δd1 + 2%	186	195	167	213	218	
Savant Lake	Height	Δh1 cm	6.4a	9.8ab	11.1b	10.4ab	9.6a	9.46**
	Height	Δh2 cm	22.9a	25.4ab	28.4b	24.0ab	23.8a	24.90**
	Diam.	Δd1 + 2%	79	102	98	102	106	
White River	Height	Δh1 cm	7.7ab	9.3b	6.9a	8.4ab	6.8a	7.82**
	Height	Δh2 cm	19.6a	23.9b	19.3a	21.6ab	19.8a	20.85**
	Diam.	Δd1 + 2%	172	171	136	192	190	
(b) Black spruce								
Cochrane	Height	Δh1 cm	3.9a	3.5a	3.3a	3.7a	2.5a	3.38
	Height	Δh2 cm	9.2b	9.2b	8.2ab	9.7b	7.0a	8.67**
	Diam.	Δd1 + 2%	45	48	43	53	53	
Foleyet	Height	Δh1 cm	4.0a	5.0a	4.8a	5.2a	4.3a	4.67*
	Height	Δh2 cm	7.0ab	8.8b	6.9ab	7.4ab	6.4a	7.30*
	Diam.	Δd1 + 2%	54	74	58	94	77	
Savant Lake	Height	Δh1 cm	5.4ab	5.7ab	4.8a	6.8b	6.1ab	5.75**
	Height	Δh2 cm	7.7a	6.7a	5.6a	7.1a	6.5a	6.72
	Diam.	Δd1 + 2%	48	57	40	65	65	
White River	Height	Δh1 cm	3.0a	3.5a	3.4a	3.7a	4.4a	3.60
	Height	Δh2 cm	10.6b	6.4a	5.8a	8.2a	6.3a	7.46**
	Diam.	Δd1 + 2%	46	41	30	53	41	

[1] Within rows, values not followed by the same letter differ significantly ($P = 0.05$) by Tukey's Multiple Comparison test.

[2] * and ** differences between microsites contributing to the mean significant at $P = 0.05$ and $P = 0.01$.

jack pine sites were there no significant differences in mortality between treatments after 2 seasons.

The effect of treatments on height increment varied among sites (Table 5) *e.g.* treatment a was best and c worst for second-year height increment in jack pine at Foleyet. At Savant Lake this was reversed. However, second-year height increments were poor for treatments c and e on 3 of the 4 jack pine and all black spruce sites. Nevertheless, microsite e and the other mound treatment d usually produced the greatest increment in ground-level stem diameter on all sites and in both species.

Influence of initial tree size The size of outplant at the time of planting and the subsequent first- and second-year height increments were generally very poorly correlated, particularly in black spruce. Initial height and initial ground-level stem diameter in both species correlated quite well (r's between 0.48 and 0.79) with stem diameters at the end of both the first and second growing seasons.

Field root growth (FRG)

The trees excavated 30 days post-planting differed significantly from their non-excavated counterparts in only 3 of 40 treatment/site combinations, with respect to initial height and 2 of 40 for initial stem diameter, and 4 of 40 for condition. Although these frequencies are somewhat greater than expected by chance there seems no real difference between the FRG subsample and the non-excavated trees.

FRG varied significantly ($P = 0.01$) between species, sites within species, and treatments within sites (Table 6). For jack pine the control, treatment a, was consistently low and, except Savant Lake, the mounding treatments (d and e) consistently high. With black spruce, treatment c was poorest and on all 4 sites. The mounding treatments e and d highest. The mounding method of site preparation seems to influence the amount and rate of root growth made by newly outplanted trees, presumably due to changes in soil temperature and moisture.

Field root growth and field performance Mean FRG of the excavated trees, 30 days post-planting, was examined in relation to the mean field performance of the remaining trees.

Overall sites field performance of jack pine was almost totally independent of either the number or length of roots 30 days post-planting (Table 7). Black spruce increment and field root growth showed a maximum correlation coefficient of 0.30.

Within sites, White River had the highest correlation between FRG, number and length, and field performance. The variable most consistently correlated with FRG was ground-level diameter increment, first year + second year combined: in 8 of 16 cases, r exceeded 0.50.

Table 6. Field root growth: mean number of turgescent white roots 1 cm or more long and their mean total length determined on outplants excavated 30 days after planting, by species, site and microsite, n = 20

Site	Treat-ment	No. of roots	Stand. error	Length of roots cm	Stand. error
(a) Jack pine					
Cochrane	a	32.0	5.5	57.1	10.6
	b	53.8	5.0	120.2	15.0
	c	43.1	5.8	88.2	14.8
	d	77.2	9.5	210.2	30.6
	e	74.7	7.9	210.5	29.3
Foleyet	a	20.1	3.6	39.0	7.6
	b	31.9	4.8	63.3	9.2
	c	27.0	2.2	52.8	4.0
	d	72.4	7.8	213.9	33.6
	e	60.1	10.6	150.1	24.4
Savant Lake	a	44.0	11.6	85.6	22.1
	b	48.8	7.2	98.9	15.9
	c	65.8	9.0	145.1	18.7
	d	63.2	8.7	144.0	21.9
	e	48.2	6.3	113.6	16.4
White River	a	20.2	4.9	45.1	12.6
	b	34.0	7.0	68.0	15.2
	c	21.2	3.7	42.0	7.7
	d	48.2	5.7	121.5	19.2
	e	38.4	4.6	96.6	13.0
(b) Black spruce					
Cochrane	a	20.7	3.8	38.5	7.4
	b	19.1	3.4	32.3	6.1
	c	16.7	2.7	28.4	5.0
	d	37.2	7.4	62.8	13.8
	e	37.9	6.4	76.4	13.3
Foleyet	a	16.2	2.7	30.6	6.6
	b	13.0	2.4	23.0	4.5
	c	3.7	0.9	4.9	1.2
	d	21.4	4.2	48.0	10.9
	e	32.8	3.7	59.3	7.6
Savant Lake	a	7.6	2.2	12.8	4.3
	b	15.2	3.4	22.0	5.3
	c	8.8	1.5	11.2	2.0
	d	21.4	3.1	29.2	4.3
	e	38.0	5.7	57.0	8.6
White River	a	9.8	2.6	17.0	4.8
	b	5.8	1.9	9.7	3.5
	c	5.9	1.7	9.6	3.3
	d	14.0	1.9	26.0	4.1
	e	19.1	3.8	34.0	7.5

[1] See text.

Field root growth and root growth capacity By matching excavated FRG trees with RGC trees of the same initial height and stem diameter the relationship between the root growth of outplants, in the field 30 days post-planting, and by matched trees in the RGC test was tested. Correlations between FRG and RGC also were usually low and inconsistent. Correlations (r) between numbers of roots produced in the field and in the growth chamber ranged from 0.96 (n = 5) for

Table 7. Field root growth (determined 30 days after outplanting) and 1st- and 2nd-year field performance: correlation coefficients (r) between mean root growth of the 5 excavated trees in each plot and mean performance of the 25 associated non-excavated trees per plot

Site	n	Survival at end of growing season		Increment		
		1	2	Height 1st yr	Height 2nd yr	Ground-level stem diam 1st + 2nd yr
Basis: number of roots						
(a) Jack pine						
Cochrane	20	0.01	0.21	0.27	0.32	0.71
Foleyet	20	0.16	0.15	0.27	−0.05	0.58
Savant Lake	20	0.17	0.11	0.09	0.02	−0.05
White River	20	0.43	0.38	0.38	0.47	0.56
Overall	80	0.31	0.32	0.09	−0.09	−0.07
(b) Black spruce						
Cochrane	20	0.14	0.10	0.14	0.26	0.18
Foleyet	20	−0.30	0.03	−0.06	−0.21	0.37
Savant Lake	20	−0.57	−0.24	0.18	0.01	0.55
White River	20	0.65	0.59	0.03	0.43	0.41
Overall	80	0.05	0.15	0.04	0.19	0.30
Basis: length of roots						
(a) Jack pine						
Cochrane	20	0.05	0.18	0.19	0.19	0.61
Foleyet	20	0.16	0.17	0.36	−0.09	0.58
Savant Lake	20	0.11	0.07	0.19	0.02	0.04
White River	20	0.34	0.32	0.27	0.37	0.57
Overall	80	0.26	0.27	0.09	−0.12	0.01
(b) Black spruce						
Cochrane	20	−0.04	−0.07	0.10	0.18	0.26
Foleyet	20	−0.23	0.10	−0.10	−0.24	0.45
Savant Lake	20	−0.62	−0.27	0.14	0.04	0.54
White River	20	0.63	0.59	−0.06	0.43	0.34
Overall	80	0.05	0.15	−0.07	0.21	0.27

Savant Lake black spruce treatment a to −0.42 (n = 9) for Savant Lake jack pine treatment b. For root length, correlations ranged from 0.85 (n = 8) for White River black spruce treatment e to −0.30 (n = 6) for White River black spruce treatment d (Table 7).

Conclusions

RGC

Large differences in RGC occurred between different lots of jack pine nursery stock of the same provenance, perhaps associated with either morphological differences, of common occurrence between batches, or differences in storage or handling conditions or a combination of these.

Within a given lot of jack pine or black spruce, tree size gave little indication of RGC.

In jack pine, large differences in RGC did not affect nursery performance. This may not hold for plantings subjected to greater stress or with larger differences in RGC[7]. In black spruce differences in RGC of an order of magnitude were associated with significant differences in nursery performance. RGC and performance are not linearly related.

Field performance

The variation in field performance of Lot 2 stock at Cochrane, Foleyet, and White River contrasts with limited variation in the nursery planting. Site-site variation complicates the RGC-performance relationship, so interpretations must be site/weather specific. Both stem diameter increment, though not height increment and FRG were strongly promoted by mounding.

Acknowledgements The work described in this paper forms part of the Ontario Ministry of Natural Resources Forest Research Project 'Mineral mound and humus mound planting in the boreal forest of Ontario' which is being conducted collaboratively with the Great Lakes Forest Research Centre of the Canadian Forestry Service, with partial funding from the Canada Department of Regional Economic Expansion. Field work was carried out by KBM Inc., Forestry Consultants, of Thunder Bay, Ontario. The expert and timely support by Wendy Beilhartz, assisted by Nova Bailey, in data processing, and by Marilyn Ruprechter in typing are greatly appreciated.

References

1 Anon. 1979 Tests show survival rate of seedlings is high when new Swedish machine used. Can. Pulp Pap. Indus. 32, 13.

2 Apt K Z, Parolin R W and Lane R W 1979 Development and testing of a spot scarifier. Am. Soc. Agric. Eng. Paper 79-1612, 8 p.

3 Day R J and MacGillivray G R 1975 Root regeneration of fall-lifted white spruce nursery stock in relation to soil moisture content. For. Chron. 51, 196–199.

4 Day R J, Stupendick J T and Butler J M 1977 Root periodicity and root regeneration potential are keys to successful plantation establishment. pp 19–21. *In* Plantation Establishment Symposium. Ed. R F Sutton. Dep. Environ., Can. For. Serv., Sault Ste. Marie, Ontario, Rep. 0-P-5.

5 Edlund L 1980 Mineral mound and humus mound methods: two alternative soil scarification methods applied in forest land in northern Sweden. pp. 427–436. *In* Symposium on Stand Establishment Techniques and Technology, Proc. IUFRO Subject Group S3.02-00, Moscow.

6 Edlund L 1980 The results and findings of fifteen years large scale mechanical site-preparation experience in Sweden. Contrib. paper, Silvic. Equip. Exhib., Thunder Bay, Ontario, 26 p.

7 McMinn R G 1980 Root growth capacity and field performance of various types and sizes of white spruce stock following outplanting in the central interior of British Columbia. pp 37–41. *In* Characterization of Plant Material. Ed. H. Schmidt-Vogt. IUFRO Working Group S1.05-04 Meeting Proc., Waldbau-Inst. Univ. Freiburg.

8 Morrison I K and Armson K A 1968 The rhizometer: a device for measuring the roots of tree seedlings. For. Chron. 44, 21–23.

9 Odin H 1979 Impact of environmental factors on forest development. pp 1–21. *In* Symposium on Forest Meteorology. World Met. Org. Publ. No. 527.

10 Parolin R W, Read A and McMinn R G 1981 Operational trial of a spot scarifier. pp 240–247. *In* Forest Regeneration. Symposium on Engineering Systems for Forest Regeneration. Proc. Am. Soc. Agric. Eng., ASAE Publ. 10–81.

11 Schubert G H and Adams R S 1971 Reforestation practices for conifers in California. Calif. Res. Agency, Dep. Conserv., Div. For., Sacramento, 359 p.

12 Söderström V 1977 Problem och metoder i skogsföryngringsarbetet. IV Försök med markberedningsmetoder. Skogs-o. Lantbr.-akad. Tidskr. 116, 43–58.

13 Söderström V, Bräcke J, Byfalk R and Jonsson C 1978 Jämförelse mellan plantering i jordrabatter och efter andra markberedningsmetoder. Skogshögskolan, Instit. f. Skogsskötsel, Rapp. och uppsatser No. 11, 177 p.

14 Stone E C 1955 Poor survival and physiological condition of planting stock. For. Sci. 1, 90–94.

15 Stone E C, Jenkinson J L and Krugman S L 1962 Root-regenerating potential of Douglas-fir seedlings lifted at different times of the year. For. Sci. 8, 288–297.

16 Sutton R F 1979 Planting stock quality and grading. For. Ecol. Manage. 2, 123–132.

17 Sutton R F 1980 Planting stock quality, root growth capacity and field performance of three boreal conifers. N.Z.J. For. Sci. 10, 54–71.

18 Wakeley P C 1948 Physiological grades of southern pine nursery stock. Proc. Soc. Am. For. 43, 311–322.

Plant and Soil 71, 123–129 (1983). A-12

Effect of restricted watering and its combination with root pruning on root growth capacity, water status and food reserves of *Pinus caribaea* var. *hondurensis* seedlings

SHEIKH ALI ABOD*
Faculty of Forestry, Universiti Pertanian Malaysia, Serdang, Selangor, Malaysia

and SIMON SANDI**
Universiti Pertanian Malaysia, Serdang, Selangor, Malaysia

Key words Bare-root planting *Pinus caribaea* Plant food reserves Plant water status Restricted watering Root growth capacity Root pruning Tropical climate

Summary Six-month old *Pinus caribaea* var. *hondurensis* seedling were subjected to various regimes of restricted watering and their combinations with root pruning to determine a suitable technique of producing seedlings with optimum physiological condition for successful bare – root planting.

From the range of treatments tested, plants subjected to watering once every 9 days (D9) were found to be the 'hardest' in terms of highest root growth capacity, leaf water content, root starch content and in having a moderately low transpiration rate. *P. caribaea* seedlings subjected to this preconditioning treatment may be more able to survive bare-root planting.

Introduction

Large scale plantation establishment of *Pinus caribaea* Mor. var. *hondurensis* B & G in Peninsular Malaysia started in 1974 after investigations by UNDP/FAO had shown that the establishment of long-fibred tropical pine species for domestic pulp and paper production was biologically and economically feasible[4]. Current techniques for establishing the species attempts to minimize transplantation shocks, and so reduce losses, by using container-grown stock[13]. This technique is expensive and the cost of establishing a plantation can be reduced substantially by using open-root stock[7].

In Malaysia, the use of open-rooted *P. caribaea* seedlings has failed[9,12]. The simple techniques for producing open-root nursery stock of *P. caribaea*[7,8] and *P. radiata* D. Don[15] by repeated root wrenching, may be adequate where field planting conditions are not severe[23]. However, open-rooted stocks may be subjected to a more severe water stress when planted in Malaysia where, in addition to the hot tropical climate, droughts are also common and may last for 3 weeks or more even in the wet season.

The literature on effects of nursery practices on conifers[1,3,5,10,11,16] show that preconditioning of seedlings by restricted watering in the nursery can

* Lecturer

** Final year Forestry student (1981/82 session)

prepare them for short dry spells in the field. This study was aimed to test the effectiveness of various regimes of restricted watering and some combinations with root pruning to give a suitable technique for producing *P. caribaea* seedlings in an optimum physiological condition for successful bare-root planting in Malaysia.

Materials and methods

Three-month old *P. caribaea* var. *hondurensis* seedlings (seed source: Fiji) were planted bare-rooted in open beds for another three months to simulate an open-root nursery. The beds, filled with an equal mixture of sand and top soil, were covered with transparent plastic sheets to keep rain out. The plants were kept well supplied with water and nutrients.

Fourty seedlings were used in each treatment. The treatments were:

D1	watering once daily
D3	watering once every 3 days
D6	watering once every 6 days
D9	watering once every 9 days
D3W6	watering once every 3 days and fortnightly root pruning
D3W3	watering once every 3 days and monthly root pruning
D6W3	watering once every 6 days and monthly root pruning
C	control *i.e.* potted stocks, watering once daily and no root pruning

After 3 months, the following parameters were measured.

root growth capacity[19,21] (10 plants)

transpiration at field capacity (5 plants) and under limiting soil moisture (5 plants), watered to field capacity initially then allowed to dry over 15 days

leaf relative turgidity and xylem water potential (15 plants)

starch content in roots, stems and needles (5 plants)

Plants were removed from the bed and all white root tips were pinched off to simplify the recognition of new roots. Subsequently, they were planted under standard conditions for one month. RGC was determined from either or both of the total number or the total length of new roots per plant. The transpiration rate was measured by weighing and expressed as gram water loss per needle dry weight per day. To measure leaf relative turgidity and xylem water potential, plants were transplanted from the beds into 18.5 cm diameter × 16.0 cm height pots filled with the bed soil. The pots were first watered to field capacity and then allowed to dry over 15 days. One seedling was harvested each day for leaf relative turgidity (R.T.) and xylem water potential measurements. Relative turgidity[26] was calculated as $\text{R.T.}\,(\%) = \dfrac{\text{Fresh wt} - \text{Dry wt}}{\text{Turgid wt} - \text{Dry wt}} \times 100$. The xylem water potential was measured with a pressure 'bomb'.

All data were subjected to analysis of variance and the significance of differences between group means were tested with Duncan's new multiple range test[17].

Results and discussion

The effect of the preconditioning treatments tested, on root growth capacity (RGC) of *Pinus caribaea* var. *hondurensis* seedlings can be summarized as: D9 > D6 > D3 > D1 > C > D3W3 > D6W3 > D3W6. (Table 1). The RGC of potted seedlings and those subjected to restricted watering treatments were significantly higher than the RGC of plants subjected to combinations of

Table 1. Effects of restricted watering and its combination with root pruning on root growth capacity (RGC) and plant starch content. Bracketed means are not significently different ($P = 0.05$)

Root growth capacity				Starch content (%)					
*RGC_N		**RGC_L		Root		Stem		Needles	
D9	58	D9	384	D9	35	C	24	D9	44
D6	45	D6	265	D6W3	31	D1	21	D3W6	43
D3	40	D3	256	D3W6	29	D6W3	20	C	38
D1	37	D1	230	D3W3	28	D3W3	18	D6W3	36
C	35	C	200	D3	26	D3W6	17	D1	35
D3W3	23	D3W3	57	D6	25	D6	17	D3W3	33
D6W3	20	D6W3	49	D1	21	D9	16	D3	30
D3W6	18	D3W6	47	C	17	D3	15	D6	23

* Total number of new roots per plant

**Total length (cm) of new roots per plant.

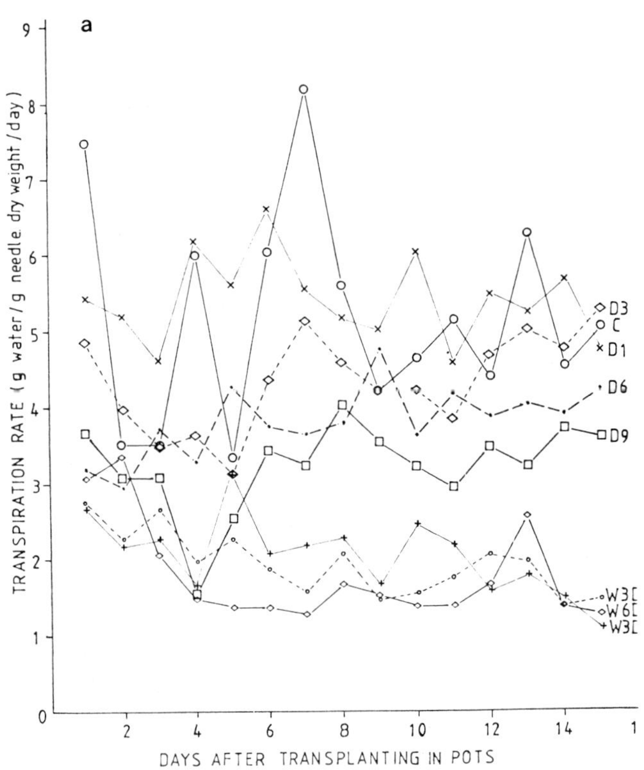

Fig. 1a. Daily transpiration rate of plants in soil kept at field capacity moisture condition over 15 days.

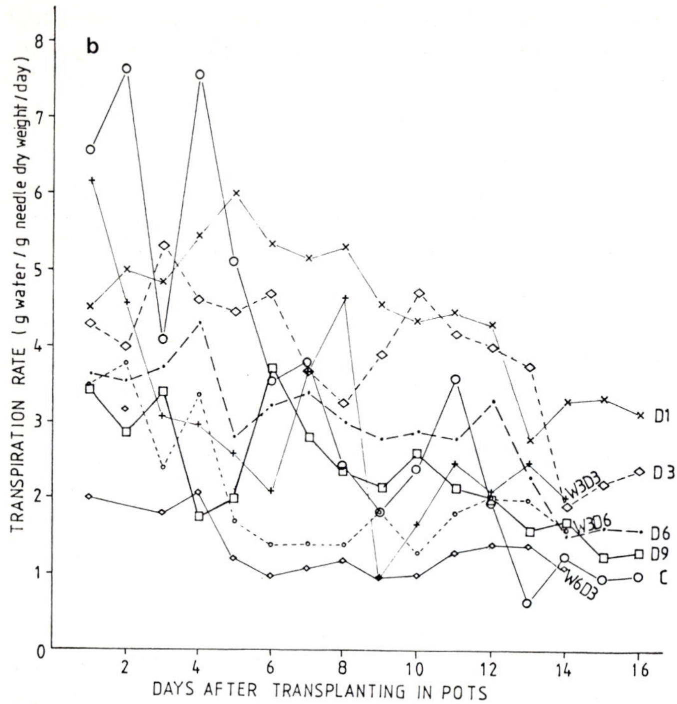

Fig. 1b. Daily transpiration rate of plants in soil watered to field capacity only once at the beginning of measurement and then allowed to dry over 15 days.

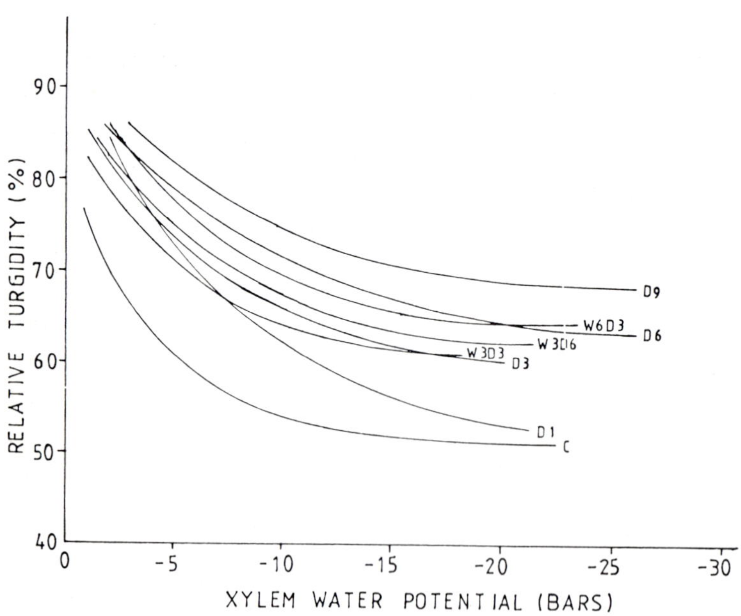

Fig. 2. Relationship between leaf relative turgidity and xylem water potential.

restricted watering and root pruning. Only plants preconditioned by watering once every 9 days for 3 month had RGC which was significantly higher than all other treatments. More severe watering regimes need to be tested for effects on RGC to determine the optimum preconditioning. The intitial survival of planted seedlings depends chiefly on the capacity of their root systems to regenerate in the first few weeks after out-planting so as to re-establish contact with the surrounding soil and tap water and nutrients[18,20,21,22]. Nevertheless, it has been suggested however that RGC values alone are an insufficient basis for predicting field survival because in many situations low RGC may depress transpiration without affecting the plant survival.

Restricted watering combined with root pruning depressed root growth (Table 1) but these plants had lower rates of transpiration than potted stocks and plants subjected to restricted watering treatments alone under both the field capacity (Fig. 1a) and water stress conditions (Fig. 1b). The general trends in the rate of transpiration of plants at the two soil moisture regimes tested can be summarized as follows: C > D1 > D3 > D6 > D9 > D3W3 > D6W3 > D3W6. The results suggest that the rate of transpiration of plants was directly related to the severity of the preconditioning treatment.

The effect of treatments on the water status of plant foliage was less consistent than effects on transpiration. The curves relating relative turgidity of needles and xylem water potential (Fig. 2) reveal that for a given xylem water potential, the amount of water present in the foliage of plants in different treatments was as follows: D9 > D6 > D3W6 > D6W3 > D3 > D3W6 > D1 > C. Nevertheless, the results clearly show that D9 plants were the most efficient in maintaining a high level of hydration in the needles under both adequate and water stress conditions (Fig. 2), while at the same time were able to maintain a relatively moderate rate of transpiration (Fig. 1a and 1b). The moderately high rate of transpiration in D9 plants would suggest that the seedlings were able to carry on active photosynthesis soon after being transplanted. A higher level of metabolic activity was suggested also by the presence of many new root tips. Rook[14] suggested that rates of transpiration per se were not direct important to plant survival. Internal water deficits largely determine whether or not seedlings survive. Several investigators, working on *P. radiata* seedlings, found an increase in water saturation deficit to decrease stomatal opening and the influx of CO_2, thereby reducing photosynthesis.

The Starch content of plants was highest in needles followed by roots and stems. There was no significant difference between treatment means for needles and stems (Table 1). Starch was however, found to be significantly higher in the roots of D9 plants (35%) compared to potted stocks (17%); the effect of treatments can be ranked as D9 > D6W3 > D3W6 > D3W3 > D3 > D6 > D1 > C indicating a general increase in starch content with increasing severity of treatment. The results do not show any direct relationship between RGC and starch content in the roots, stems or leaves. This agrees with results obtained in a study on seasonal changes in RGC and carbohydrates in *Pinus resinosa.* Ait. and *Picea glauca* (Moench.) Voss seedlings[25].

References

1 Aldhous J R 1972 Nursery practice. Brit. For. Comm. Bull. No. 43, 184 p.

2 Allen S E, Grimshaw H M, Parkinson J A and Quarmby C 1974 Chemical Analysis of Ecological Material. Blackwell Scientific Publ., Oxford, London.

3 Anonymous 1957 Tree planting practices in Tropical Asia. F.A.O. Forestry Development Paper No. 11, 171 p. Rome.

4 Anonymous 1975 Plantation management procedures for large-scale plantations in Peninsular Malaysia. FO: DP/MAL/72/009. Working paper No. 36, 205 p.

5 Awang K B 1973 The development of nursery practices for plantation forests, with particular reference to tropical and subtropical conditions. Hons. thesis submitted to Aust. Nat. Univ., 143p. Unpublished.

6 Bacon G J and Bacheland E P 1978 The influence of nursery conditioning treatments on some physiological responses of recently transplanted seedlings of *Pinus caribaea* Mor. var. *hondurensis* Barr and Golf. Aust. For. Res. 8, 173–83.

7 Bacon G J and Hawkins P J 1977 Studies on the establishment of open root Caribbean pine planting stock in Southern Queensland. Aust. For. 40, 173–91.

8 Bacon G J and Hawkins P J 1980 Establishment trials with *Pinus caribaea* var. *hondurensis* in South-East Queensland. Mal. For. 43, 24–36.

9 Briscoe C B 1962 Early lifting of pine seedlings. Inst. Trop. For., Puerto Rico. Note 10.

10 Chavasse C G R and Weston G C (Ed.) 1969 Forestry nursery and establishment practice in New Zealand. N.Z. For. Serv., F.R.I. Symp. No. 9.

11 Darby S P 1962 Intensified nursery practices – the key to high quality, custom – grown field graded forest tree seedlings. Tree Plant Notes 52, 7–11.

12 Lamb A F A 1973 *'Pinus caribaea.'* Vol.1. Fast growing timber trees of the lowland tropics. Comm. For. Inst. Oxford. No. 6.

13 Paul D K 1972 A handbook of nursery practice for *Pinus caribaea* var. *hondurensis* and other conifers in West Malaysia. FAO UNDPFO: SF/MAL 12. Working paper No. 19, 139 p.

14 Rook D A 1969 Water relations of wrenched and unwrenched *Pinus raidata* seedlings on being transplanted into conditions of water stress. N.Z.J. For. 14, 50–58.

15 Rook D A 1971 Effect of undercutting and wrenching on growth of *Pinus radiata* D.Don. Seedlings. J. Appl.Ecol.8, 477–90.

16 Rook D A 1972 Conditioning of radiata pine seedlings to transplanting, by restricted water. N. Z. J. For. Sci. 3, 54–69.

17 Steel R G D and Torrie J H 1960 Principles and Procedures of Statistics. McGraw Hill, New York.

18 Stone E C 1955 Poor survival and the physiological condition of planting stock. For. Sci. 1, 90–94.

19 Stone E C 1967 The root regenerating capacity of seedling transplants and the availability of soil moisture. Ann. Arid Zone 6, 42–57.

20 Stone E C and Benseler R W 1962 Planting ponderosa pine in the California pine region. J. For. 60, 462–66.

21 Stone E C and Jenkinson J L 1970 Influence of soil water on root growth capacity of ponderosa pine transplants. For. Sci. 16, 230–39.

22 Stone E C and Schubert G H 1959 The physiological condition of ponderosa pine (*P. ponderosa* Laws.) planting stock as it affects survival after cold storage. J. For.7, 837–41.

23 Stupendick J T and Shepherd K R 1980 Root regeneration of rootpruned *Pinus radiata* seedlings. II. Effects of root pruning on photosynthesis and translocation. N.Z.J. For. Sci. 10, 148–58.

24 Sutton R F 1980 Planting stock quality, root growth capacity and field performance of three boreal conifers. N.Z.J. For. Sci. 10, 54–71.

25 Driessche R van den 1978 Seasonal changes in root growth capacity and carbohydrates in red pine and white spruce nursery seedlings. Proc. IUFRO Symp. on root physiology and symbiosis, Nancy, France, 11–15 Sept., 1978.

26 Weatherley P E 1950 Studies in the water relations of the cotton plant. I. The field management of water deficits in leaves. New Phytol. 49, 81–97.

Plant and Soil 71, 131–146 (1983). A-13

Morphogénèse des systèmes racinaires des semis et boutures de chêne pédonculè

Morphogenesis of root systems of seedlings and cuttings of Quercus robur L.

ARTHUR RIEDACKER
INRA, Centre de Recherches Forestières d'Orléans, Ardon, F-45160 Olivet, France

and MICHELINE BELGRAND
INRA, CNRF, Champenoux, F-54280 Seichamps, France

Key words Geotropism Juvenility *Quercus robur* L. Root morphogenesis Waterlogging

Summary Primary roots are orthogeotropic on seedlings (Fig. 3) and on cuttings taken from seedlings (Fig. 23 to 25) and semi-plagiogeotropic on cuttings taken from sprouts on old stumps. (Fig. 6-8-10). And when these roots are decapitated, they regenerate generally roots with the same geotropism (Fig. 5 and 11). These properties cannot be easily changed. Several years after rooting, primary roots of cuttings from old stumps are still semi-plagiogeotropic (Fig. 14-15 and 21). Orthogeotropism can be considered as a juvenile character. Growth and geotropism of lateral roots are under control of the tip of the primary root. When longitudinal growth of the main root is blocked (Fig. 16) lateral roots become longer thicker and acquire the same geotropism as the primary root. But when growth of the main root is only reduced (Fig. 1 and 2) or when the tip of the taproot regenerates new roots continuously (Fig. 17) secundary roots may become longer but remain subhorizontal. A shallow watertable on a clay soil blocks the tip of the taproot and increases root porosity of all the roots (Fig. 19). Geotropism and growth of a given root tip seems to depend on the tip of the root of lower order, on the properties of the root tissue (cutting taken from seedlings, or from sprouts on old stumps, taproot or primary root of the cutting...) and perhaps also on the shoot. Up to now we could not obtain systematically orthogeotropic roots on cutting taken from old stumps. Some observations however suggests that this might not be impossible.

Mots clefs Ennoyage Géotropisme Juvénilité Morphogenese *Quercus robur* L. Racinaire

Résumé Les pivots des semis et racines principales de boutures issues de semis sont orthogéotropes, ce qui favorise la prospection du sol en profondeur. Les racines principales de boutures issues de rejets de vieilles souches sont en revanche semi-plagiogéotropes et restent en surface. Il en résulte des morphogénèses de systèmes racinaires très différentes, aussi bien sur un milieu bien drainé, qu'en présence d'une nappe temporaire. Diverses manipulations du système racinaire permettent de préciser les facteurs responsables de la croissance et du géotropisme des racines.

Introduction

L'architecture des systèmes racinaires peut être décrite par l'orientation, les dimensions et les ramifications des principales racines. Celle des semis de Chêne est caractérisée par la présence d'un pivot portant de fines racines secondaires subhorizontales à croissance peu durable[1,8,11,14,20]. Mais, la dimension finale d'une racine résulte à la fois de ses aptitudes initiales, de la richesse nutritive et du régime hydrique du milieu exploré[4], ainsi que des corrélations actives de croissance qui peuvent être modifiées par l'hétérogénéité du milieu. Les racines secondaires des semis de Chêne peuvent ainsi s'allonger, grossir davantage et éventuellement devenir orthogéotropes, en particulier lorsque la croissance du pivot est bloquée[12,20].

Nous examinerons l'effet de divers types de blocages du pivot, ainsi que les modifications de la morphogénèse qui se produisent lorsque le pivot rencontre une nappe d'eau superficielle.

Mais, comme la qualité du bois détermine fortement la valeur des grumes de chêne, il était intéressant de chercher à bouturer cette essence. Cornu *et al.*[2,3] y parvinrent en prenant des boutures herbacées de rejets de souches. Il convenait ensuite de vérifier si la qualité du bois des plants ainsi produits allait être comparable à celle des pieds mères, et si leur développement aérien et souterrain allait être satisfaisant?

En effet, bien que l'on sache bouturer l'Hévéa depuis longtemps[13,22], cette méthode de multiplication n'est toujours pas employée à échelle industrielle, en particulier parce que les boutures obtenues jusqu'ici sont facilement déracinées par le vent[23], leurs sytèmes racinaires étant superficiels, contrairement à ceux des semis qui sont pivotants[16].

Matériel et méthode

La morphogénèse des semis et boutures de Chêne pédonculé (*Quercus robur* L.) a été étudiée en observant d'une part la croissance des racines à travers les faces transparentes des minirhizotrons placés en serre et d'autre part en déterrant des plants élevés in situ. Les minirhizotrons (Fig. 6)[18], sont des parallélépipèdes plats (1 à 6 cm d'épaisseur) en plexiglass transparent qu'on peut remplir de tourbe ou de terre. Les racines visibles sur les faces transparentes sont dessinées sur une feuille de polyéthylène, avec des crayons marqueurs de différentes couleurs. Cette méthode permet en particulier d'étudier le géotropisme des racines de jeunes semis ou boutures, ainsi que de plants plus âgés repiqués en minirhizotron. Pour cela, il suffit de faire subir à ces minirhizotrons une rotation de 90° dans le plan vertical. Une racine qui aurait pu avoir une croissance verticale après déviation par la plaque transparente ne redeviendra dans ces conditions verticale que si elle est orthogéotrope[20]. Si elle est plagiogéotrope, elle se réorientera, mais adoptera une croissance oblique avec une pente en général inférieure ou égale à 45°. Enfin, si elle ne se réoriente que lorsqu'on dirige sa pointe vers le ciel, et non lorsqu'on la dirige dans le même sens que la pesanteur, elle sera dite semi-plagiogéotrope.

Nous avons également étudié la réaction des racines verticales rencontrant un obstacle plat. Le résultat observé est alors être la récette dernière étant d'autant plus forte que l'obstacle est plus court, sultante de réactions géotropiques et de réactions exotropiques[19]. Le blocage du pivot et des racines principales a été réalisé avec des blocs de bois ou de plâtre[12] et la dessication des pointes de pivot à l'air, grâce à des fenêtres découpées dans la face des minirhizotrons (Fig. 17).

Pour étudier l'influence d'une nappe d'eau, nous avons préféré utiliser un sol hydromorphe lorrain, en l'occurence un horizon humifère A_1, provenant d'un pseudogley typique selon la classification française des sols, caractérisé par une hydromorphie temporaire. Le niveau de la nappe d'eau était maintenu constant à −2 cm, −4 cm et −8 cm de la surface du sol. Seul ce dernier cas sera décrit. Par suite des remontées capillaires, on peut sans doute considérer que l'alimentation en oxygène sera limitante pour les racines dans la couche de sol au-dessus de la nappe, soit sur une épaisseur d'au moins 4 cm.

L'augmentation de la porosité racinaire (rapport du volume des espaces intercellulaires des racines/volume total) serait, selon de nombreux auteurs, une des adaptations racinaires à la survie dans des conditions asphyxiantes[4,6,7,15] notamment en permettant peut-être le transport interne d'oxygène des parties aériennes vers les racines[17]. Elle a été déterminée par mesure pycnométrique[9]. Après 60 jours d'ennoyage de la pointe du pivot, la nappe d'eau a été spprimée. Les schémas des figures[1,2,16,17,18,19] sont ceux de plants jugés les plus représentatifs du lot, les observations ayant porté sur 6 à 10 plants par traitement.

Le bouturage a été réalisé sous serre, dans un brouillard, à l'aide de rameaux herbacés de 10 cm de long, privés de leur bourgeon terminal, à 5 feuilles, trempés au préalable dans une poudre de talc contenant 0,5% d'AIB[3]. Ils provenaient soit de rejets de l'année sur souches âgées de plus de 100 ans, soit de pieds mères issus de ces rejets, soit de semis âgés de quelques semaines, dont la hauteur atteignait 15 à 20 cm.

Les comptages pour les boutures issues de rejets ont porté sur 335 boutures enracinées. La comparaison de l'aptitude à l'enracinement et des réactions géotropiques des boutures issues de semis ou de rejets de souche a porté sur deux lots de 60 boutures placés sous brouillard le 12 Août 1971, sevrés le 21 Septembre, puis transférés en minirhizotrons pour examen des réactions géotropiques de leurs racines au cours des deux mois suivants. Les expériences de décapitation et de blocage des racines principales ont été réalisées l'année du bouturage. Par ailleurs, un premier lot de boutures (Lot I), obtenu en 1977, a été repiqué en pépinière en 1978, puis transplanté deux ans plus tard en 1980 sur une station à sol brun lessivé à pseudogley.

Les racines de ces boutures ont été observées à leur sortie de la pépinière en 1980 et cinq boutures représentatives on été transférées en minirhizotron avec les pointes de leurs racines principales sectionnées. Durant l'hiver 1971–72, douze autres boutures de ce lot on été déterrées, observées et repiqués en minirhizotrons pour étude de la réaction géotropique de leurs racines.

Un second lot de boutures (Lot II) obtenu en 1975 et élevé en conteneur rond de 5 litres a été mis sur le terrain au printemps 1977 sur une station à sol brun lessivé à pseudogley et à nappe temporaire hivernale et printanière à une vingtaine de cm de profondeur. Dans cette même station, avaient été repiqués en même temps des plants issus de glands de chêne de même âge du commerce. Nous avons observé durant l'hiver 1971–72, dans ce deuxième lot, la morphogénèse du système racinaire de 14 boutures et de 4 plants issus de glands prélevés dans les divers carreaux de l'expérience.

Résultats

Morphogénèse des systèmes racinaires de semis et de boutures en l'absence de contraintes

Semis Les semis dont la croissance du pivot n'est pas freinée ou bloquée par des obstacles ou des conditions de milieux défavorables présentent un pivot orthogéotrope à forte croissance radiale et axiale (Fig. 1 et 3). Lorsqu'il rencontre un obstacle, il le longe et redevient vertical dès qu'il n'est plus gêné (Fig. 4). Après décapitation, il y a néoformation sur son extrémité distale de 2 à 7 grosses racines généralement strictement orthogéotropes (Fig. 5), dont la croissance radiale est comparable au pivot et que nous appelons pivots de remplacements.

Boutures issues de rejets de souches Les boutures issues de rejets de souches présentent 1 à 7 racines principales, dont la pente varie entre 0 et 80° (Fig. 6 - 7 - 12 - 13). Dans certains cas, on peut même obtenir des racines verticales. Néanmoins, 80% des boutures n'ont pas plus de 3 racines. Certaines

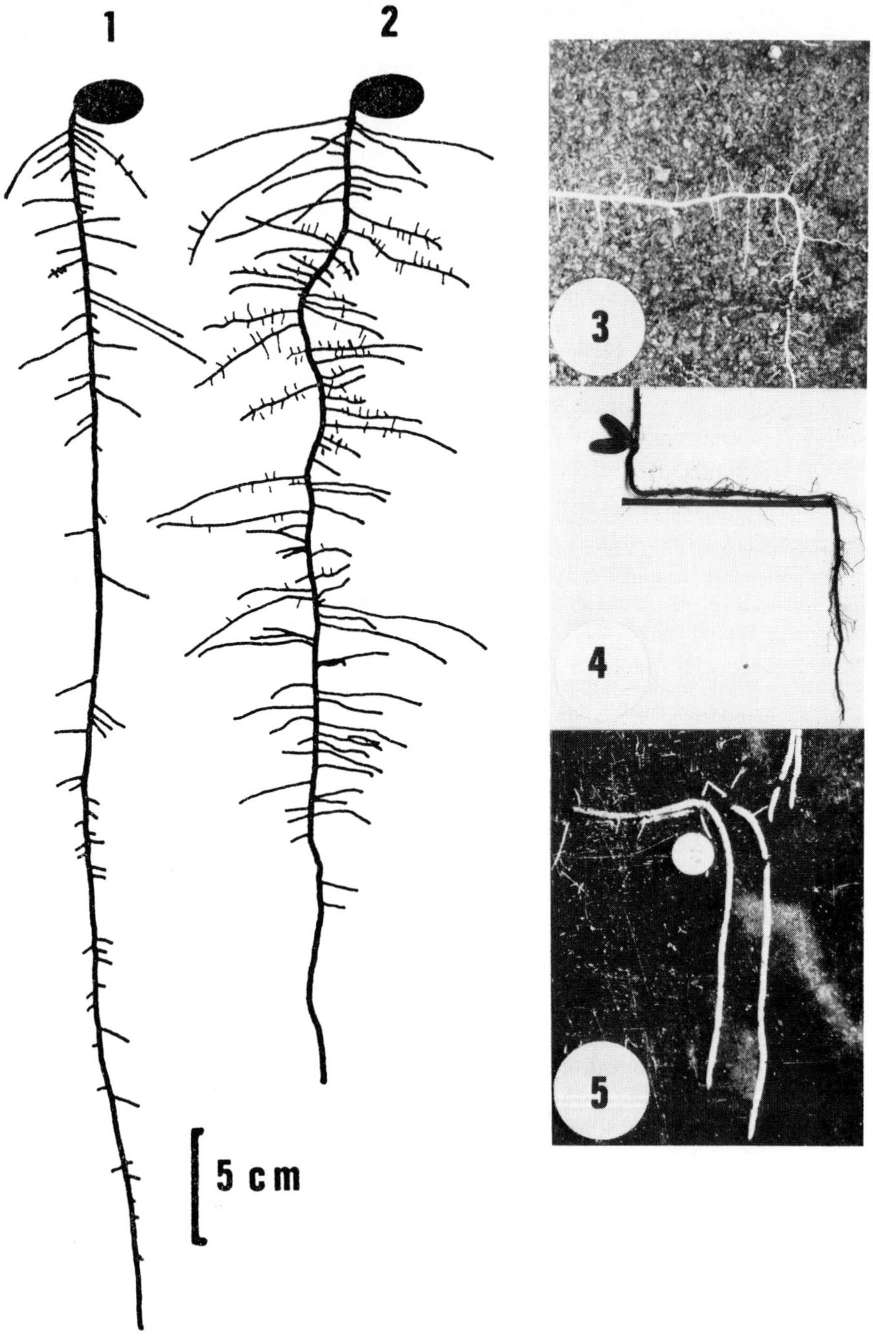
1
2
3
4
5
5 cm

apparaissent latéralement, d'autres sur le cal que forme parfois la bouture. Mais, aucune de ces racines, mêmes celles qui croissent verticalement dans la tourbe, n'est orthogéotrope. Disposées verticalement dans le sens contraire de la pesanteur, elles se réorientent généralement en deux étapes. Une première courbure à rayon relativement faible les amène à croître presque horizontalement. Une seconde courbure à très grand rayon les fait ensuite croître obliquement avec une pente variable, très souvent inférieure à 45° (Fig. 10). Lorsque'on les dispose verticalement dans le même sens que la pesanteur, elles ne se réorientent pas. Et, lorsqu'elles recontrent un obstacle plat, horizontal et long (4 à 5 cm), elles ne redeviennent pas verticales après l'obstacle. Elles adoptent une pente inférieure à 45° (Fig. 9). Quand l'obstacle rencontré est court (1 cm), elles peuvent redevenir verticales par suite d'une forte réaction exotropique.

Enfin, lorsqu'on fait subir une rotation de 90° à des boutures dont les racines croissaient verticalement, leurs pointes se réorientent et produisent une seule courbure, les faisant croître obliquement (Fig. 10), avec une pente inférieure ou égale à 45°. Et, lorsqu'on décapite ces racines, elles regénèrent 1 à 2 racines ayant la même pente (Fig. 11). Les racines principales des boutures issues de rejets de vieilles souches sont donc semi-plagiogéotropes et diffèrent en cela fortement des pivots des semis qui sont orthogéotropes. Cette différence de comportement se maintient pendant plusieurs années.

Aprés deux ans de pépinière, ou encore à un âge un peu plus avancé, après deux ans de pépinière et deux ans de croissance sur le terrain (Fig. 15), les racines principales sont encore généralement horizontales et régénèrent des racines semi-plagiogétropes. Elles diffèrent toujours des semis (Fig. 20).

Figures 1 à 5. Systèmes racinaires de semis de chênes.
1. Aspect d'un semis d'une trentaine de jours élevé en minirhizotron sur tourbe.
2. Aspect d'un semis d'une trentaine de jours élevé en minirhizotron sur un sol hydromorphe, mais sans nappe. Le pivot est moins long, et les racines secondaires et tertiaires sont plus développées.
3. Réaction géotropique d'un pivot de 30 cm de long après une rotation de 90° dans le plan vertical.
4. Pivot rencontrant un obstacle plat horizontal de 20 cm.
5. Réaction géotropique de pivots de remplacement après une rotation de 90° dans le plan vertical.

Figures 1 to 5. Root systems of oak seedlings.
1. Aspect of a 30 days old seedling raised on peat in a minirhizotron.
2. Aspect of a 30 day old seedling raised on a waterlogged soil in a minirhizotron but without a watertable. The taproot is shorter and lateral roots are longer.
3. Geotropic reaction of a 30 cm long taproot after a 90° rotation in a vertical plane.
4. Taproot encountering a flat horizontal, 20 cm long, obstacle.
5. Geotropic reaction of replacement taproots after a 90° rotation in a vertical plane.

6
7
8
9
10
11

Morphogénèse des systèmes racinaires de semis et de boutures lorsque la croissance des racines principales est freinée

Le freinage des racines principales conduit toujours à des modifications de comportement des racines secondaires. Sur tourbe, substrat facilement pénétrable par le pivot, les racines secondaires de semis ont une croissance faible (1,5 mm par jour) et peu durable. Elles restent fines (diamètre 0,1 à 0,2 mm), obliques ou subhorizontales et courtes. Quand le substrat se dessèche, la plupart d'entre elles, et tout particulièrement les plus fines, se dessèchent et disparaissent. Sur un sol, substrat moins facilement pénétrable que la tourbe, la croissance du pivot est plus lente (Fig. 1 et 2) et l'allongement des racines secondaires plus durable. Après blocage des pointes de pivot (Fig. 16), il apparaît des racines secondaires plus grosses, devenant orthogéotropes et à croissance indéfinie[12].

En revanche, quand le pivot se dessèche à l'air en se régénérant plusieurs fois, les racines secondaires s'allongent davantage, tout en restant fines et horizontales[20] (Fig. 17). Il apparaît également des racines secondaires tardives horizontales dont le diamètre de l'apex est dès la naissance environ le double de celui des racines secondaires apparues en séquence normale.

Figures 6 à 11. Systèmes racinaires de boutures de chênes issues de rejets de vieilles souches.
6. Racines obliques en minirhizotron.
7. Certaines boutures ont des racines verticales, d'autres des racines horizontales ou subhorizontales, d'autres encore ont les deux types de racines.
8. Courbures géotropiques d'une racine principale de bouture orientée en sens inverse de la pesanteur. La première courbure a un faible rayon. Elle fait pousser la racine horizontalement. La seconde courbure a un très grand rayon et entraîne une croissance oblique.
9. Une racine orientée verticalement (sens de la pesanteur) qui rencontre un obstacle plat et long la déviant horizontalement ne redevient pas verticale, mais reste oblique.
10. Racines orientées d'abord suivant la direction de la pesanteur; elles ne se réorientent pas. Dans un deuxième temps, la bouture a subi une rotation de 90° dans le plan vertical. On observe une seule courbure amenant les racines à croître obliquement.
11. Racines décapitées ayant régénéré deux racines symétriques par rapport à l'axe de la racine principale.

Figures 6 to 11. Root systems of oak cuttings taken from sprouts on old stumps.
6. Oblique roots in a minirhizotron.
7. Some cuttings produce vertical roots, some only horizontal roots, some produce both types.
8. Geotropic curvatures of a cutting main root oriented vertically in the opposite direction of the gravity. The first curvature has a small radius and orients the root horizontally. The second curvature has a great radius and induces an oblique growth.
9. A root oriented like gravity encountering a long horizontal and flat obstacle remains oblique after the obstacle.
10. Roots first oriented in the same direction as gravity. They produced no curvature. They have then undergone a 90° rotation in a vertical plane. A single curvature reorients the roots and results in an oblique growth.
11. Decapited roots having regenerated two roots symetrical to the axis of the decapited root.

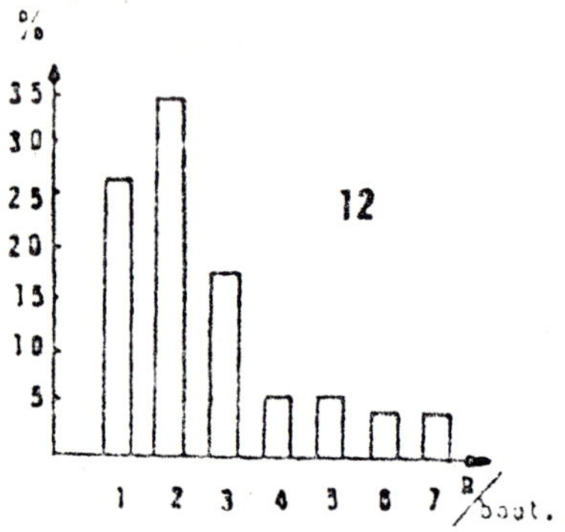

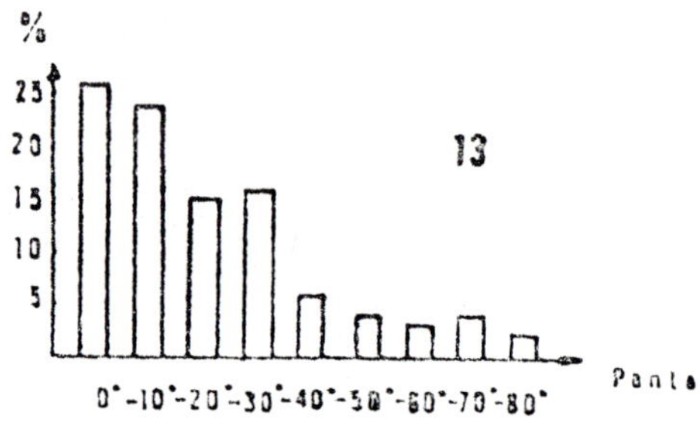

Figures 12 et 13. Systèmes racinaires de boutures de chênes issues de rejets de vieilles souches.
12. Pourcentage de boutures classées suivant le nombre de racines.
13. Pourcentage de racines classées suivant la pente en degrés.

Figures 12 and 13. Roots systems of cuttings taken from sprouts on old stumps.
12. Percentage of cuttings classified according to their number of roots.
13. Percentage of roots classified according to their slope in degrees.

Lorsqu'une racine principale de bouture est bloquée, les racines secondaires s'allongent également davantage, tout comme lorsqu'on bloque un pivot et qu'on le dispose horizontalement. Les racines secondaires supérieures s'allongent cependant davantage que chez les semis. Par ailleurs, une simple déviation d'une racine principale de bouture se traduit aussi par une croissance plus importante des racines secondaires que la déviation d'un pivot.

Ces racines secondaires ne deviennent toutefois pas orthogéotropes.

En présence d'une nappe à –8 cm de profondeur, la vitesse du pivot des semis est fortement ralentie. Sur un sol sans nappe, le pivot franchit les 8 premiers centimètres en 8 ± 1 jours, alors qu'en présence d'une nappe, il lui faut 19 ± 2 jours. Très souvent, il se courbe à environ 5 cm de la nappe dans une zone sans doute déjà fortement hypoxique. Une seconde courbure en sens inverse apparaît en général au moment où sa pointe pénètre dans la nappe (Fig. 19).

Sa vitesse se ralentit, puis devient nulle au bout d'un mois à 3 ou 4 cm en dessous du niveau de la nappe. Sa pointe noircit, mais ne régénère pas de racines. Les racines secondaires évoluent en revanche différemment des plants témoins. Celles situées près de la surface du sol s'allongent davantage, tout en restant fines et horizontales. Celles qui naissent dans la zone sans racines secondaires apparentes, lorsque la pointe du pivot pénètre dans la nappe, deviennent obliques (pente de 45°) et plus grosses (diamètre de 2 mm). Nous les appelons des *racines secondaires (ou latérales) transformées* (Fig. 19-22 j). Elles s'allongent plus vite (3,4 mm/jour) que les racines secondaires normales, mais sont à leur tour bloquées lorsqu'elles entrent dans la nappe (Fig. 19-60 j).

Après 50 jours d'ennoyage du pivot, il se forme au milieu des racines secondaires apparues normalement des *racines secondaires (ou latérales)*

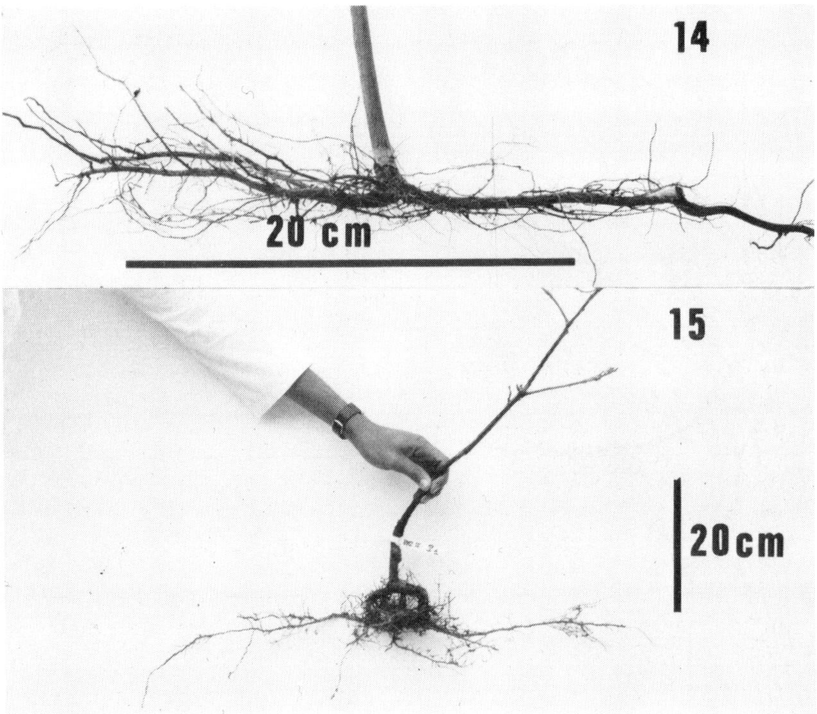

Figures 14 en 15. Systèmes racinaires de boutures de chênes issues de rejets de vieilles souches.
14. Bouture de 2 ans, avec trois racines horizontales.
15. Boutures de chêne de 6 ans. On notera que les racines sont toujours horizontales.

Figures 14 and 15. Root systems of cuttings taken from sprouts on old stumps.
14. 2 years-old cutting with 3 horizontal roots.
15. 6 year-old cutting. Lateral roots are still horizontal.

tardives horizontales, d'un diamètre encore plus grand que les racines latérales transformées (diamètre supérieur à 3 mm), mais dont la vitesse d'allongement est comparable à ces dernières. Elles restent blanches beaucoup plus longtemps que les racines secondaires normales. La porosité des racines secondaires transformées (14,3%) et tardives (14%), ainsi que du pivot (14 à 20%) est bien supérieure à celle des racines des plants témoins sans nappe (3%). La morphologie des systèmes racinaires de semis de chêne élevés en présence d'une nappe superficielle est en définitive bien différente de celle des semis témoins sans nappe (Fig. 2 et 19).

Lorsque la nappe d'eau est supprimée après 2 mois d'ennoyage de la pointe du pivot (Fig. 19-90 j.), cette dernière recommence à s'allonger. Sa vitesse d'allongement peut atteindre 16 mm/jour durant les quatre premiers jours suivant le drainage. Il n'apparaît pas plusieurs pivots de remplacement comme après une dessication de la pointe ou comme après une décapitation.

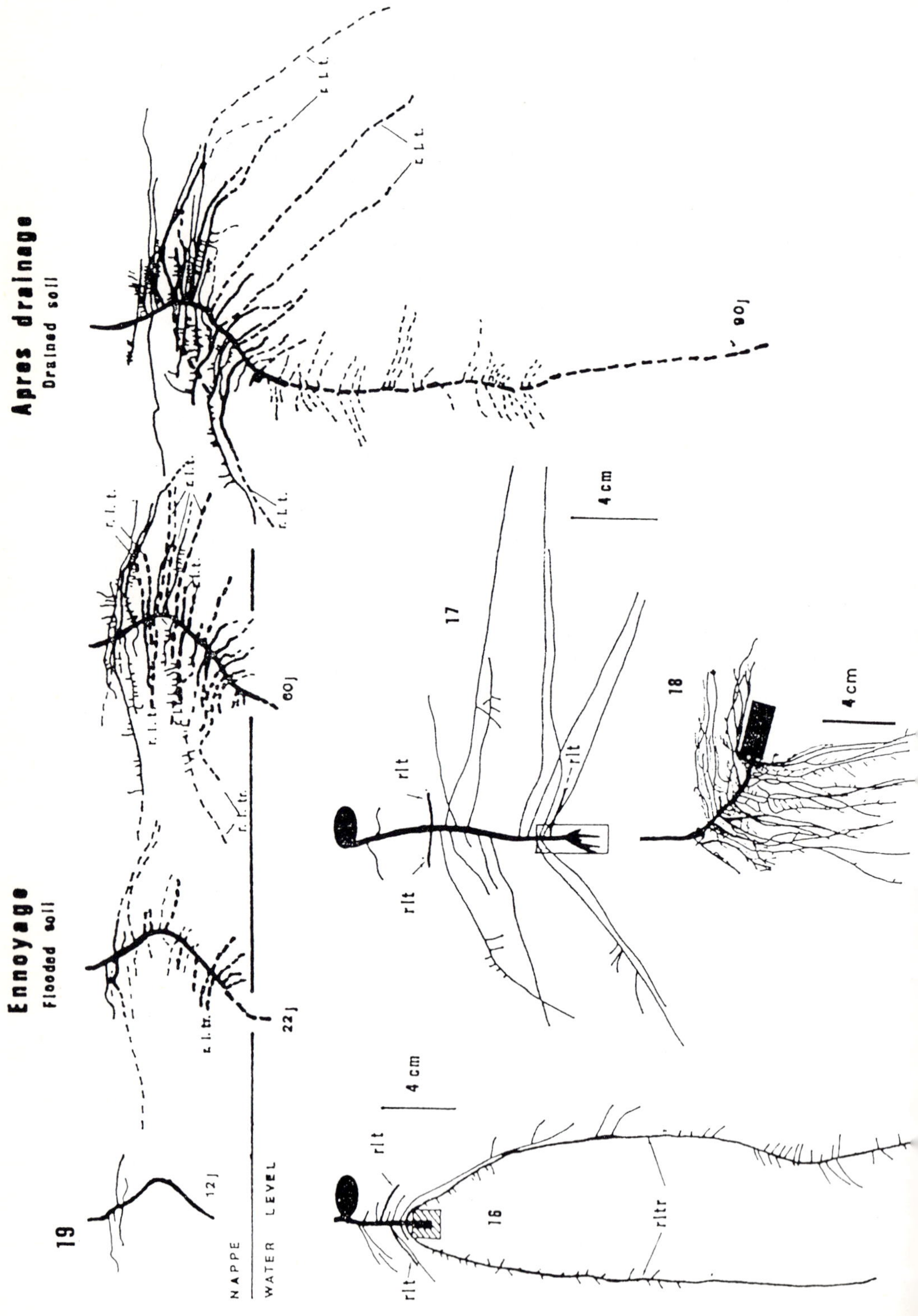
19
Ennoyage
Flooded soil
Apres drainage
Drained soil
NAPPE
WATER LEVEL
12 j
22 j
60 j
90 j
r.l.t.
r.l.tr.
16
17
18
rlt
rltr
4 cm

Pendant cette période, les racines latérales transformées s'allongent également à une vitesse élevée (9 mm/jour). Puis, la vitesse d'allongement du pivot redevient comparable à celle des témoins sur sol (10 mm/jour), l'allongement des racines secondaires transformées (rlt) s'annule, et des racines secondaires normales s'allongeant de 1,5 mm/jour naissent sur le pivot en croissance.

Lorsqu'on observe après 4 ans des systèmes racinaires de semis et de boutures implantés dans une station à nappe printanière superficielle, on s'aperçoit que les plants issus de semis ont des racines verticales et obliques qui cherchent à coloniser le sol en profondeur, comme l'ont révélé les observations décrites ci-dessus sur semis élevés en minirhizotron (Fig. 21). En revanche, les boutures issues de rejets de souche, quoique présentant une partie aérienne souvent orthotrope et bien développée, n'ont qu'un enracinement superficiel (Fig. 22).

Déterminisme du géotropisme des racines principales Nous avons vu que les racines principales des boutures issues de rejets de souche étaient toujours semi-plagiogéotropes, contrairement aux pivots des semis qui sont orthogéotropes. Deux hypothèses peuvent être faites:

1) cette propriété est due à l'âge du matériel végétal utilisé.

2) c'est la méthode de bouturage qui induit des racines semi-plagiogéotropes, tout comme la croissance du pivot induit ce caractère chez les racines secondaires.

Aussi, avons-nous bouturé simultanément dans les mêmes conditions des rameaux issus soit de rejets de vieilles souches, soit de semis. Dans ces conditions, on obtient un bien meilleur taux d'enracinement des boutures issues des semis:

Figures 16 à 19. Modification de la morphogénèse des systèmes racinaires en minirhizotrons.
16. Blocage du pivot par un bloc de plâtre. Les racines latérales les plus distales se transforment (r l tr) et deviennent orthogéotropes. Il apparaît aussi des racines latérales tardives (rlt).
17. Dessications et régénérations successives de la pointe d'un pivot arrivant à l'air (fenêtre rectangulaire). Les racines latérales restent fines et subhorizontales, mais s'allongent davantage. Il apparaît également des racines latérales tardives subhorizontales.
18. Blocage de la racine principale d'une bouture. Les racines latérales s'allongent davantage.
19. Freinage du pivot par une nappe d'eau à – 8 cm dans un sol hydromorphe. Des racines latérales se transforment (rltr) et des racines latérales tardives apparaissent (rlt). A droite morphogénèse du système racinaire lorsqu'on supprime la nappe après 2 mois d'ennoyage de la pointe du pivot.

Figures 16 to 19. Modification of root morphogenesis in minirhizotrons.
16. Tip of the taproot blocked with plaster. The most distal lateral roots are transformed (r l tr) and become orthogeotropic. Thick late lateral roots appear also (rlt).
17. Air pruning and regeneration of the tip of a taproot arriving in the air (rectangular window). Lateral roots remain thin, subhorizontal but become longer. Late lateral subhorizontal roots appear also.
18. Tip of a main root of a cutting blocked with plaster. Lateral roots become longer.
19. Taproot blocked by a watertable (– 8 cm) on a waterlogged soil. Lateral roots are transformed (rltr) and late lateral roots appear (rlt). On the right hand side morphogenesis of the root system when the watertable is removed, the tip of the taproot having been in the watertable for 2 months.

Fig. 20. Système racinaire d'un semis âgé de 8 ans sur sol drainé. Noter le pivot et le système racinaire superficiel.

Root system of an 8 year old seedling on a drained soil. Notice the taproot and the superficial root system.

Fig. 21. Système racinaire d'un plant âgé de 7 ans sur sol brun lessivé à pseudogley et à nappe hivernale et printanière superficielle (– 20 cm). Noter la présence de racines obliques à forte pente et les racines verticales.

Root system of a 7 year old seedling on a brown soil with pseudogley a winter and spring watertable at –20 cm. Notice the oblique and vertical roots.

Fig. 22. Système racinaire de bouture issues de rejet d'une vieille souche âgée de 7 ans sur le même sol que le plant 21. Toutes les racines sont horizontales.

Root system of a 7 year old cutting taken from a sprout on an old stump on the same site as for fig. 21. All the roots are horizontal.

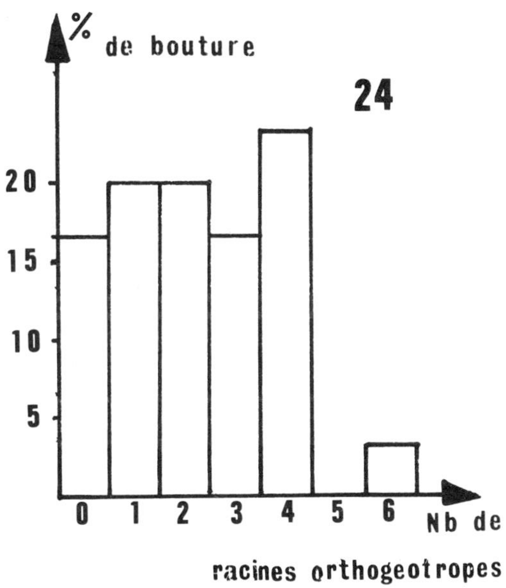

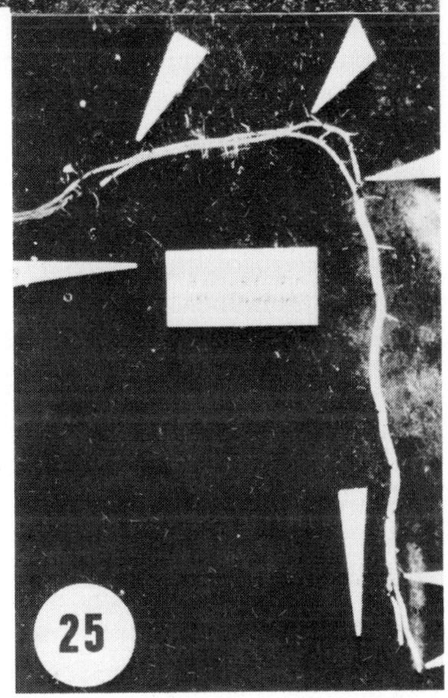

Figures 23 à 25. Systèmes racinaires de boutures issues de semis.
23. Système racinaire étalé sur la tourbe d'un minirhizotron, présentant 6 racines redevenant verticales, dont orthogéotropes.
24. Fréquence des boutures présentant un nombre donné (0 à 6) de racines orthogéotropes.
25. Réaction géotropique d'une racine verticale après une rotation du minirhizotron de 90° dans le plan vertical.

Figures 23 tot 25. Root systems of cutting taken from seedlings.
23. Root system with 6 orthogeotropic roots which became vertical after having been displayed on the peat of the minirhizotron.
24. Percentage of cuttings with a given number (0 to 6) of orthogeotropic roots.
25. Geotropic reaction of a vertical root after a 90° rotation in a vertical plane.

97% contre 50%, un nombre moyen de racines par bouture plus élevé: 3,3 contre 2. Enfin, 83% des boutures de semis présentent au moins une racine orthogéotrope, alors que les boutures issues de rejets n'en ont jamais. Certaines boutures présentent même jusqu'à 6 ou 7 racines principales orthogéotropes (Fig. 23 et 24). Celles-ci redeviennent verticales après rotation avec un rayon de courbure relativement faible (Fig. 25) et, lorqu'elles sont décapitées, elles régénèrent une nouvelle racine orthogéotrope. Notons aussi que les quelques boutures issues de semis ne présentant pas de racines orthogéotropes avaient en général une croissance très médiocre.

Le résultat de cette expérience est donc en faveur de la première hypothèse. Le plagiogéotropisme pourrait être dû à une épigénèse particulière. Dès lors, on peut se demander s'il n'est pas possible de revenir aux stades primaires et en particulier d'obtenir des racines orthogéotropes, en effectuant certaines manipulations. Des décapitations effectuées quelques semaines après l'apparition des extrémités des racines principales de boutures placées sur minirhizotron en position verticale ou oblique n'ont jusqu'ici permis d'obtenir que des régénérations de racines semi-plagiogéotropes. De même, l'élevage sur caisson à aérosol, qui permet d'obtenir des racines verticales parce qu'il n'y a pas de support, ne permet pas – tout au moins au cours des mois qui suivent le bouturage, et en l'absence d'autres manipulations – d'obtenir des racines orthogéotropes. Mais sur 4 des 12 boutures de 4 ans (lot 1) déterrées durant l'hiver 1981–82 nous avons pu observer des racines verticales. Une rotation de 90° dans le plan vertical a permis demontrer que les racines verticales de ces boutures placées en minirhizotron étaient aussi orthogéotropes. Notons que la partie aérienne des boutures possédant des racines orthogéotropes était également bien orthotrope.

Conclusions et discussions

Les pointes des racines primaires des semis et des boutures de semis de chêne sont orthogéotropes, tandis que celles des boutures issues de rejets de vieilles souches sont semi-plagiogéotropes. Il en est de même chez l'Hevéa[23]. Ces propriétés géotropiques paraissent difficilement modifiables: en particulier la décapitation des racines primaires entraîne généralement la régénération de racines ayant le même géotropisme. L'orthogéotropisme des racines principales pourrait donc figurer parmi les critères de juvénilité au même titre que l'aptitude à l'enracinement, l'inaptitude à la floraison et d'autres critères morphologiques (Schaffalitzky de Muckadell).

Le développement des racines secondaires est chez les boutures comme chez les semis sous le contrôle de la pointe de la racine principale. Lorsque la croissance axiale de cette dernière est annulée par un blocage quelconque (soit par un bloc de plâtre, soit par une nappe d'eau dans le sol, soit par décapitation répétée[20]) les racines secondaires acquièrent en général le géotropisme de la racine principale: les racines secondaires agéotropes des boutures issues de rejets de souches deviennent semi-plagiogéotropes, et les racines secondaires semi-plagiogéotropes des semis deviennent orthogéotropes.

Lorsque la pointe de la racine principale est seulement freinée (par exemple quand la pointe du pivot pousse dans des conditions hypoxiques[20]) ou lorsque la pointe du pivot se régénère régulièrement (par exemple quand la pointe du pivot arrive à l'air) les racines secondaires restent horizontales. Il est d'ailleurs possible qu'après un certain temps de fonctionnement semi-plagiogéotrope, ce fonctionnement s'automaintienne tout comme chez les racines secondaires prégerminatives de cacaoyer lorsque la racine principale est décapitée trop tardivement[5].

Il semble donc que le géotropisme d'une racine soit sous la dépendance à la fois:

– de la pointe de la racine d'ordre supérieur lorsqu'elle existe;

– des propriétés du tissu (bouture de semis, ou de rejet de souches, pivot ou racine primaire de bouture...) qui donne naissance à cette racine;

– et peut-être de la partie aérienne qui pourrait agir à la fois sur les corrélations entre racines primaires et secondaires et sur les propriétés du tissu qui donne naissance aux racines primaires.

Ce dernier point mérite toutefois encore d'être élucidé. Les observations réalisées suggèrent que des tailles sévères de racines de boutures de rejets de souche après la première ou la deuxième saison de végétation permettraient peut-être d'obtenir la régénération de racines orthogéotropes. Les essais réalisés jusqu'ici n'ont cependant pas encore permis d'obtenir des résultats positifs. Sur le plan pratique, il serait pourtant utile de pouvoir obtenir des racines orthogéotropes. L'enracinement des boutures issues de rejets de souches est en effet généralement très superficiel. La stabilité des boutures actuellement produites pourrait devenir insuffisante et leur sensibilité à la sécheresse trop grande. Les horizons profonds sont en effet pas ou mal colonisés. Or les sols à hydromorphie temporaire présentent souvent une sécheresse estivale. Même en sol hydromorphe, les semis de chêne, contrairement aux boutures, présentent en effet des racines obliques et verticales susceptibles de coloniser les horizons profonds après disparition de la nappe.

Remerciements Ce travail a été réalisé en collaboration avec Messieurs Muller, Gioria et Courrier, que nous tenons à remercier.

Bibliographie

1 Champagnat M, Baba J and Delaunay M 1974 Corrélations entre le pivot et ses ramifications dans le système racinaire de jeunes chênes cultivés sous brouillard nutritif. Rev. Cytol. Biol. Vég. 37, 407–418.

2 Cornu D, Garbaye J and Le Tacon F 1975 Résultats d'un essai préliminaire sur le bouturage du Chêne et du Hêtre. Rev. For. Fr. XXXIII, 2, 75 (139–140).

3 Cornu D, Delran S, Garbaye J and Le Tacon F 1977 Recherches des meilleures conditions d'enracinement des boutures herbacées de Chêne rouvre et de Hêtre. Ann. Sci. Forest., vol 34, n° 1.

4 Coutts M P and Philipson J J 1978 Tolerance of tree roots to water logging. II. Oxygen transport in Lodgepole pine and Sitka spruce roots of primary structure. New Phytol. 80, 341–349.

5 Dyanat-Najed 1970 Contrôle de la plagiotropie des racines latérales chez *Theobroma cacao* L. Bull. Soc. Bot. Fr. Mem. 117, 183–192.

6 Hook D D, Brown C L and Kormanik P P 1970 Lenticel and water root development of swamp tupelo under various flooding conditions. Bot. Gaz. 131, 217–224, 1970.

7 Hook D D and Scholtens J R 1978 Adaptations and flood tolerance of tree species. pp 299–331. In Plant Life in Anaerobic Environments. Eds. D D Hook and R M M Crawford. Ann. Arbor Sciences (Michigan).

8 Jenik Y 1957 Root system of *Quercus robur* L. and *Quercus petraea* Liebl. Rozprary čsl. Akad. Ved. 67, 1–85.

9 Jensen C R, Luxmoore R J, Gundy S D and Stolzy L H 1969 Root air space measurement by a pycnometer method. Agric. Jal. 61, 474–475.

10 Kawase M 1978 How plants adapt to water logging. Ohio report on research and development. Vol. 63, n° 1, 14–15.

11 Köstler J N, Bruckner E and Bibelriether H 1968 Die Wurzeln der Waldbäume. Verlag Paul Parey Hamburg-Berlin, 284 p.

12 Lamond M, Tavakol R and Riedacker A 1983 Influence d'un blocage de l'extrémité du pivot de semis de Chêne sur la morphogénèse de son système racinaire. Ann. Sci. For. *Sous presse.*

13 Levandowsky D W 1959 Propagation of clonal *Hevea brasiliensis* by cuttings. Trop. Agric Trinidad 36, 247–257.

14 Lyford W H L 1980 Development of the root system of northern red oak (*Quercus rubra* L.) Harvard forest paper n° 21, 31 p. Petersham Massachusetts.

15 Mingeau M 1977 Porosité racinaire et tolérance à l'ennoyage. Ann. Agron. 28, 117–184.

16 Pochet P D 1965 Le système radiculaire de boutures d'*Hevea brasiliens* Muell ARG. Dans le cadre des recherches sur l'amélioration des méthodes culturales en hévéa culture. Agricultura, Juin 1965, vol. XIII, 2ème série, n° 2. Laboratoire de Phytotechnie Tropicale de l'Université de Louvain.

17 Raymond Ph, Bruzeau F and Pradet A 1978 Etude du transport d'oxygène des parties aériennes aux racines à l'aide d'un paramètre du métabolisme: la charge énergétique, C.R. Acad. Sci. Paris, t. 286, série D- 1061–1063.

18 Riedacker A 1974 Un nouvel outil pour l'étude des racines et de la rhizosphère: le minirhizotron. Ann. Sci. Forest. 31, 129–134.

19 Riedacker A 1978 Etude de la déviation des racines horizontales ou obliques de boutures de peuplier qui rencontrent un obstacle; application à la conception de conteneurs. Ann. Sci. Forest 35, 1–18.

20 Riedacker A, Tavakol R, Dexheimer J and Alaoui H 1982 Modifications expérimentales de la morphogénèse et des géotropismes dans le système racinaire de jeunes Chênes. Can. J. Bot. *Sous presse.*

21 Schaffalitzky de Muckadell J 1954 Juvenile stages in woody plants. Physiol. Plant. 7, 782–796.

22 Tinley G H 1960 Vegetative propagation of clones of *Hevea brasiliensis* by cuttings. Proc. Nat. Rubb. Res. Conf. Kuala Lumpur, 409 p.

23 Yoon P K, and Leong S K 1975 Induction of pseudo taproots of cuttings and production of clonal root stocks in Hevea. Proceedings of the International Rubber Conference 1975, Kuala Lumpur, 85–108.

Plant and Soil 71, 147–153 (1983). A-14

Some aspects of the development of the endodermis and cortex of *Tilia cordata* and *Picea sitchensis*

K.A.D. MacKENZIE
East Malling Research Station, Maidstone, Kent, ME19 6BJ, UK

Key words Cortex Endodermis *Picea sitchensis* *Tilia cordata* Transfer cells Ultrastructure

Summary The endodermis of both *Tilia cordata* and *Picea citchensis* progressess through 3 characteristic phases of development. These developments are delayed somewhat in the xylem pole endodermis of *T. cordata*, while in *P. sitchensis* 3–5 passage cells are found. The cortex of *T. cordata* is characterised by very thick walls, while that of *P. sitchensis* is characterised by a thick walled layer just outside the endodermis and by 2–3 outer layers of transfer cells.

Introduction

The structure and development of roots, in particular the endodermis, has been studied in detail for many annual species[2], but that of tree roots has been relatively neglected[9]. Certainly, the roots of annual species are more readily cultured for experimental purposes, but since trees are among the most long lived and economically important crops, this neglect is unfortunate. Many trees, most notably the Gymnosperms, grow on nutrient deficient soils; it is important, therefore, to know something of their structure and function in order to understand how they cope with such adverse conditions. This short paper deals with the development of the endodermis in both *Tilia cordata* and *Picea sitchensis*, and some aspects of cortical anatomy which may be important in nutrient uptake.

Materials and methods

Roots from pot grown specimens of *Tilia cordata* and *Picea sitchensis* were used. Lengths of root up to 100 mm were cut into 2–4 mm pieces and fixed and embedded in Araldite[9]. Sections for light microscopy were cut at 0.25, 2.5, and 5.0 μm and stained with periodic acid-Schiff reagent and toluidine blue either separately or in combination. Ultrathin sections for electron microscopy were stained with uranyl acetate and lead citrate for 15 min each.

Results and discussion

Endodermis

Identification In *T. cordata*, as in apple[9], the presence of vacuolar phenolics (Fig. 1) is a useful feature by which the endodermal layer can be identified at its earliest stages of development. At later stages, histochemical tests for suberin are

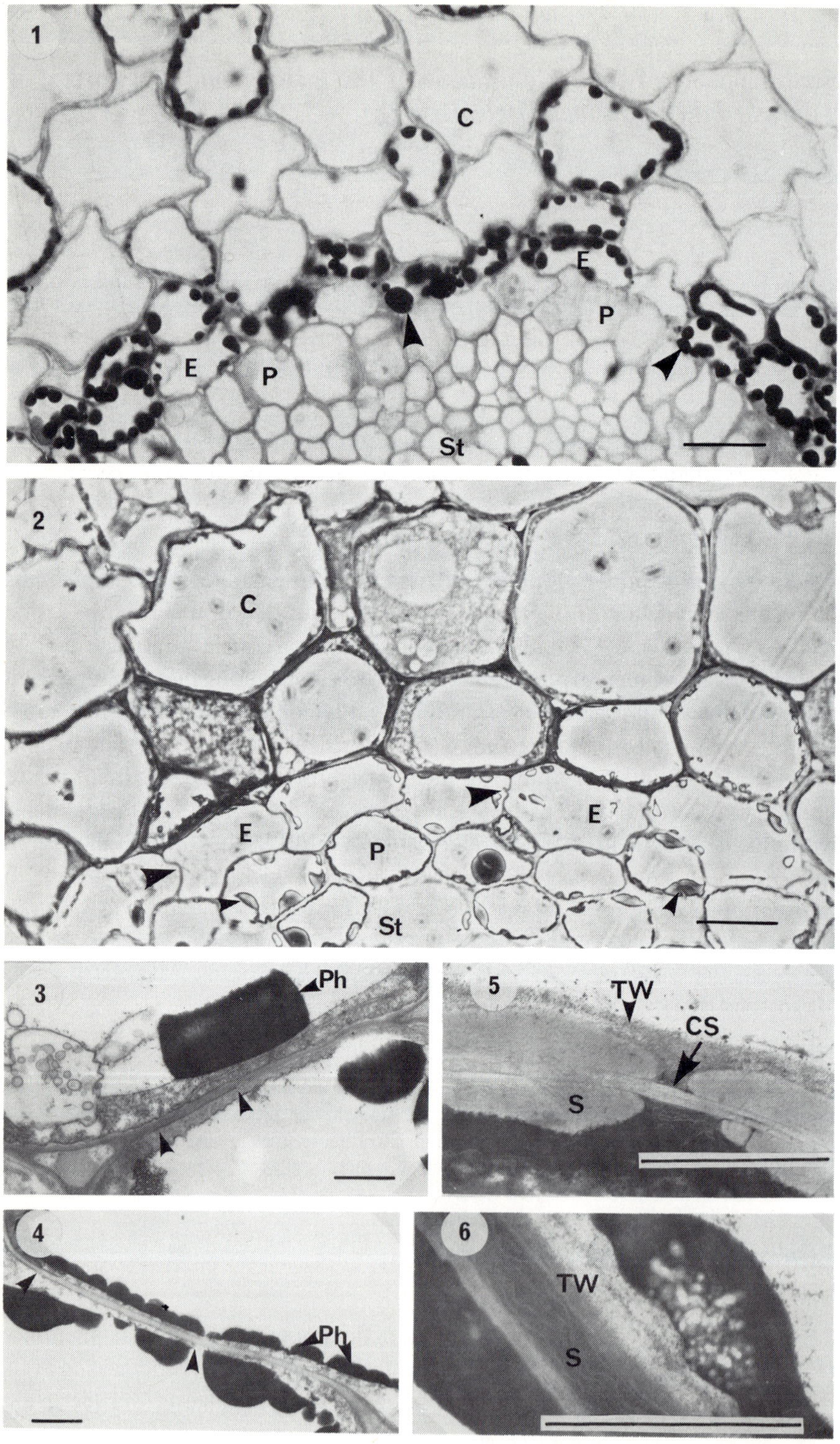
1
C
E
P
E
P
St
2
C
E
E
P
St
3
Ph
5
TW
CS
S
4
Ph
6
TW
S

a useful alternative. In *P. sitchensis*, however, both the endodermis and pericycle accumulate vacuolar phenolics, though to a lesser degree than *T. cordata*. In this case the endodermis is more accurately identified by histochemical means or with the electron microscope. The thin, often convoluted anticlinal walls of the endodermis can also be a useful distinguishing feature (Fig. 2).

Structure Both *T. cordata* and *P. sitchensis* progress through the 3 characteristic stages of endodermal development[2,9]. Much is made in the literature about distances from the root tip at which these stages occur, but these varying distances seem to depend as much on the growth rate of the root concerned as on the species. Little reference will, therefore, be made here to such distances.

Stage I. The Casparian strip in both species always develops less than 5 mm from the root tip and shows the characteristic fine granularity of the suberised anticlinal wall and the close apposition of the plasmalemma to that part of the wall (Fig. 3, 4). It occupies a rather longer portion of the wall in *P. sitchensis* than in *T. cordata*.

Stage II. Suberin lamellae are laid down inside the primary wall in both species, but in *T. cordata* the Casparian strip is most often not covered (Fig. 5).

Stage III. Still further from the root tip, a tertiary cellulosic layer is laid down inside the suberin lamellae in both species (Fig. 5, 6). Externally, the root still appears white at this stage. In *T. cordata* this stage frequently occurs before the completion of Stage II; as a result the cellulosic wall comes in contact with the Casparian strip itself.

In *T. cordata* Stages II and III are delayed somewhat outside the xylem poles; consequently, for a certain length of primary root, 50% of the cells have suberin lamellae while 50% do not. In this respect *T. cordata* is similar to apple[9]. In a sense the xylem pole endodermis may be seen as 'passage cells' in which the plasmalemma is not isolated from the primary cell wall and is consequently accessible for the transport of water and nutrients between the cortex and stele.

Figures 1 to 6.

1. and 2. are light micrographs, scale bar = 10 μm. All other figures are electron micrographs, scale bar = 1 μm.

1. *T. cordata*. Bright field. Endodermis (E) with vacuolar phenolics (arrows). P-pericycle, St-stele, C-cortex (note thick cell walls).

2. *P. sitchensis*. Phase contrast. Endodermis with vacuolar phenolics and thin convoluted anticlinal walls (large arrows). Note phenolics also in pericycle (P, small arrows), and thick cell walls in layer outside endodermis. C-cortex, St-stele.

3. and 4. Stage I endodermis in *T. cordata* and *P. sitchensis* respectively. Casparian strip is between arrows. Note vacuolar phenolics (Ph).

5. and 6. Stage III endodermis in *T. cordata* and *P. sitchensis* respectively. Note that in *T. cordata* the suberin (S) does not cover the Casparian strip (CS). TW-tertiary cellulose wall.

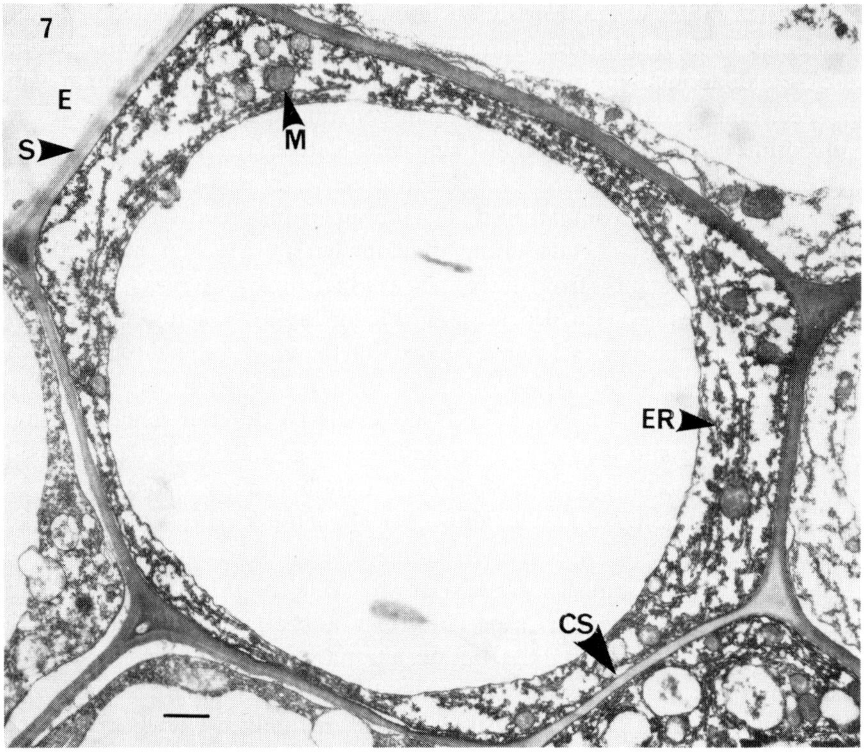

Fig. 7. *P. sitchensis*. Passage cell. Note abundance of endoplasmic reticulum (ER) and mitochondria (M) in the cytoplasm. E- normal endodermal cell with suberin (S). CS-Casparian strip (scale bar = 1 μm).

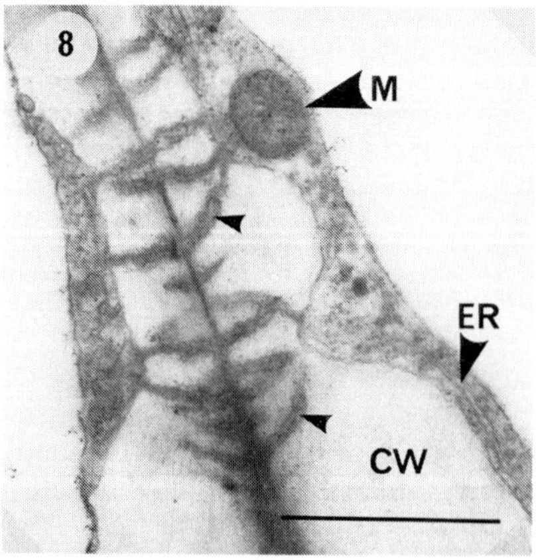

Fig. 8. *T. cordata* cortex. Note very thick cell wall (CW). Thinner pit areas are traversed by many plasmodesmata (arrows). Note endoplasmic reticulum (ER) and mitochondrion (M). (Scale bar = 1 μm).

In *P. sitchensis*, the situation is more complex. In early season (April) samples conventional passage cells are found. Three to 5 cells of the endodermis have no suberin lamellae. The positioning of these cells bears no relationship to the xylem or phloem poles. In samples taken from the same plants in September 2 unsuberised passage cells were found at each of the two xylem poles. No explanation can be offered here for this change in distribution. Ultrastructurally, all these cells are similar; the cytoplasm occupies much of the cell and contains many mitochondria and lengths of endoplasmic reticulum (Fig. 7), in contrast to the thin peripheral cytoplasm of the Stage II endodermis, suggesting a high degree of metabolic activity. This characteristic is not shared by the xylem pole endodermis of *T. cordata* or apple[9].

The cortex

The main feature of interest in *T. cordata* is the thick walls of the cortex (Fig. 1, 8). These appear to be cellulosic, showing a positive PAS reaction and a negative phloroglucinol-HCl reaction for lignin. Thinner pit areas are traversed by numerous plasmodesmata and the cytoplasm in these regions is particularly rich in mitochondria and endoplasmic reticulum (Fig. 8).

In *P. sitchensis*, the cortex is more complex. The layer of cells just outside the endodermis has very thick walls (Fig. 2). These walls appear to be cellulosic, giving an intense red colouration with the PAS reaction towards the middle lamella. This suggests a particularly dense packing of cellulose microfibrils in that area, an observation confirmed with the electron microscope. There is a negative phloroglucinol-HCl reaction indicating the absence of lignin.

The outermost layers are particularly interesting in *P. sitchensis*, though again there are inconsistencies with sampling time. In the April samples the outer tangential walls of the outermost 2–3 cells have the characteristic wall ingrowths of transfer cells (Fig. 9, 10) while the inner wall remains of normal thickness. There is a gradient of increasing degradation of both walls and cytoplasm towards the outer layer, apparently caused by the many micro-organisms impinging on the root surface (Fig. 9). The wall ingrowths are of a different texture from the main body of the wall, suggesting formation at a later stage of development. At the light microscope level, the wall ingrowths stain pale pink with the PAS reaction, suggesting a cellulose wall, in contrast to the darker pink of the rest of the wall. In this sense, the wall ingrowth are similar to, though not as well developed as, those recently reported to occur in the outer cell layer of Pisonia roots in contact with an ectomycorrhiza[1].

In later samples, however, no wall ingrowths are found in the outer cell layers. Instead, there is a layer inside the primary wall with similar textural characteristics (Fig. 11) and PAS reaction to the wall ingrowths. This layer is traversed by electron-dense strands which may be plasmodesmata, these being continuous with the plasmalemma and endoplasmic reticulum in places (Fig. 12). In Fig. 12 there is some evidence of localised wall-building activity which

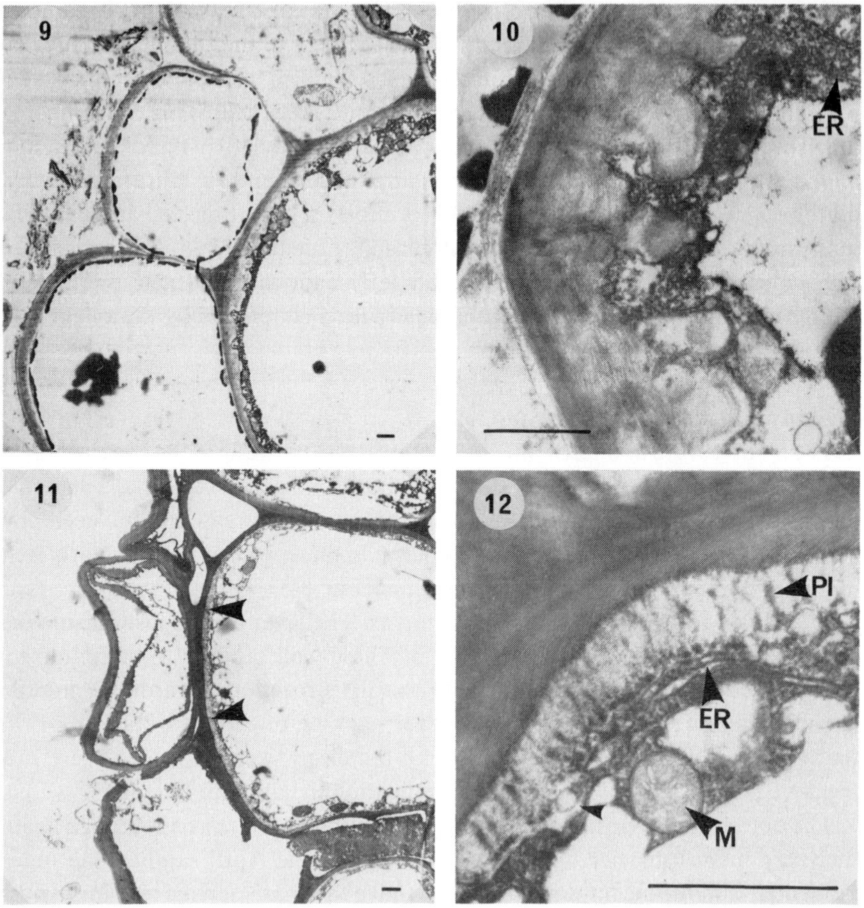

Figures 9–12.
9. *P. sitchensis* outer cortical layers. Note thickenings in the wall of the next to outer cell. The outer layer is partially degraded by soil micro-organisms.
10. Higher magnification of ingrowths in transfer cell. Note dense cytoplasm with lengths of endoplasmic reticulum (ER).
11. *P. sitchensis.* Thickened walls in next to outer layer of cortex. Note difference in density between primary wall and secondary wall layer (arrow).
12. Higher magnification of Fig. 11 showing what may be plasmodesmata (Pl). ER-endoplasmic reticulum, M-mitochondrion. There is a suggestion of further wall building activity at the point marked by the arrow. Scale bar = 1 μm.

suggests that these cells may be transfer cells at an early stage of development, or whose development has been altered due perhaps to changing seasonal conditions.

There does not appear to be any previous reference to transfer cells in the roots of *P. sitchensis*. Transfer cells have been found in other roots, often forming in response to some external stimulus, such as the presence of ectomycorrhizas[1], iron deficiency[8], high salt concentrations[5,6,7], or nematode activity[4]. The greater-than-normal area of plasmalemma in transfer cells is generally considered to render these cells highly efficient in membrane mediated transport. It is not possible to say from the available material whether or not they are a normal occurence in the roots of *P. sitchensis*, but it seems a strong possibility that they are formed in response to some form of nutrient deficiency, similar to the manner in which transfer cells are induced to form in the roots of *Helianthus annuus* grown in an iron deficient medium[8]. This would be an invaluable ecological adaptation for a plant which frequently grows on nutrient deficient soils, enabling it to become more efficient in the uptake of the nutrients in short supply. Further experimental work is needed to determine whether this is indeed a deficiency induced phenomenon, and if so, how widespread it is in a range of taxa.

Acknowledgements My thanks to Hugh Shepherd for supplying the plant material and to Marion Brookfield for photographic assistance.

References

1 Ashford A E, and Allaway W G 1982 A sheathing mycorrhiza on *Pisonia grandis* R.Br. (Nyctaginaceae) with development of transfer cells rather than a Hartig net. New Phytol. 90, 511–519.
2 Clarkson D T, Robards A W and Sanderson J 1971 The tertiary endodermis in barley roots: fine structure in relation to radial transport of ions and water. Planta Berlin. 96, 292–305.
3 Gunning B E S 1977 Transfer cells and their roles in transport of solutes in plants. Sci. Progr. Oxford 64, 539–568.
4 Jones M G K and Northcote D H 1972 Multinucleate transfer cells induced in Coleus roots by the root-knot nematode *Meloidogyne arenaria*. Protoplasma 75, 381–395.
5 Kramer D 1978 Transfer cells in the epidermis of roots of the halophyte *Atriplex hastata* L. Naturwissenschaften 65, 339–474.
6 Kramer D, Anderson W P and Preston J 1978 Transfer cells in the root epidermis of *Atriplex hastata* L. as a response to salinity: a comparative cytological and X-ray microprobe investigation. Aust. J. Plant Physiol. 5, 739–747.
7 Kramer D, Läuchli A, Yeo A R and Gullasch J 1977 Transfer cells in roots of *Phaseolus coccineus*: ultrastructure and possible function in exclusion of sodium from the shoot. Ann. Bot. 41, 1031–1041.
8 Kramer D, Römheld V, Landsberg E and Marschner H 1980 Induction of transfer-cell formation by iron deficiency in the root epidermis of *Helianthus annuus* L. Planta Berlin 147, 335–339.
9 MacKenzie K A D 1979 The development of the endodermis and phi layer of apple roots. Protoplasma 100, 21–32.

Plant and Soil 71, 155–160 (1983). A-15

When is the structural root system determined in Sitka spruce?

M. P. COUTTS and G. J. LEWIS
Forestry Commission, Northern Research Station, Roslin, Midlothian, UK

Key words *Picea sitchensis* Root morphogenesis Sitka spruce Woody root

Summary Growth ring analysis was carried out on root systems of Sitka spruce trees which had been planted for 8 and 34 years. Retrospective measurements were made on cross-sectional area increment near the root base. Differentiation of the main laterals into roots of widely different radial growth rates took place mainly during the first eight years, resulting in 3–11 'major' woody roots and a larger number of small 'minor' ones, with some of intermediate vigour. The major roots established during the first few years constituted the main structural root system at 34 years. Many of the minor roots stopped growing in diameter after a few years, but were still alive and extending at 34 years. The differentiation into major and minor roots is discussed with reference to their origins and the local environment.

Introduction

The number and spatial distribution of large woody roots is important for tree stability (see Refs. in this volume). Little is known about the origin and development of such structural roots although plant root systems exhibit 'compensatory growth' whereby those roots in a favourable environment grow more rapidly than, and even at the expense of less favoured roots[7].

In plantation-grown Sitka spruce (*Picea sitchensis* Bong. (Carr)) large numbers of small roots are present on the root system of the transplant one year after planting whereas the older tree usually has 3–11 large woody roots radiating from the stump, together with a number of smaller ones (Fig. 1A and B). This paper describes the uneven development which gives rise to roots of contrasting size on the root system.

Materials and methods

Root systems of Sitka spruce of Queen Charlotte Islands seed origin were excavated on a peaty gley site in Scotland. The trees had been established from 1 + 1 year transplants 34 years previously at 1.8 m spacing on turfs cut from shallow ploughing. The four root systems sampled were from subdominant to co-dominant trees with a mean height of 18.0 m. Additional trees on a deep peat site were sampled 8 years after planting but results for these are not presented in detail. All of the trees had shallow root systems varying in depth from 15–40 cm.

Secondary growth of the 16–23 largest roots on each root system was measured at a point 20 cm from the base of the original seedling. Photographs were made of each root in transverse section and photomicrographs were made of transverse sections cut from the small inner-most growth rings. Primary xylem diameter was measured and each years growth increment was cut from photographic enlargements and the cross-sectional area (CSA) measured on an optical planimeter. On roots of

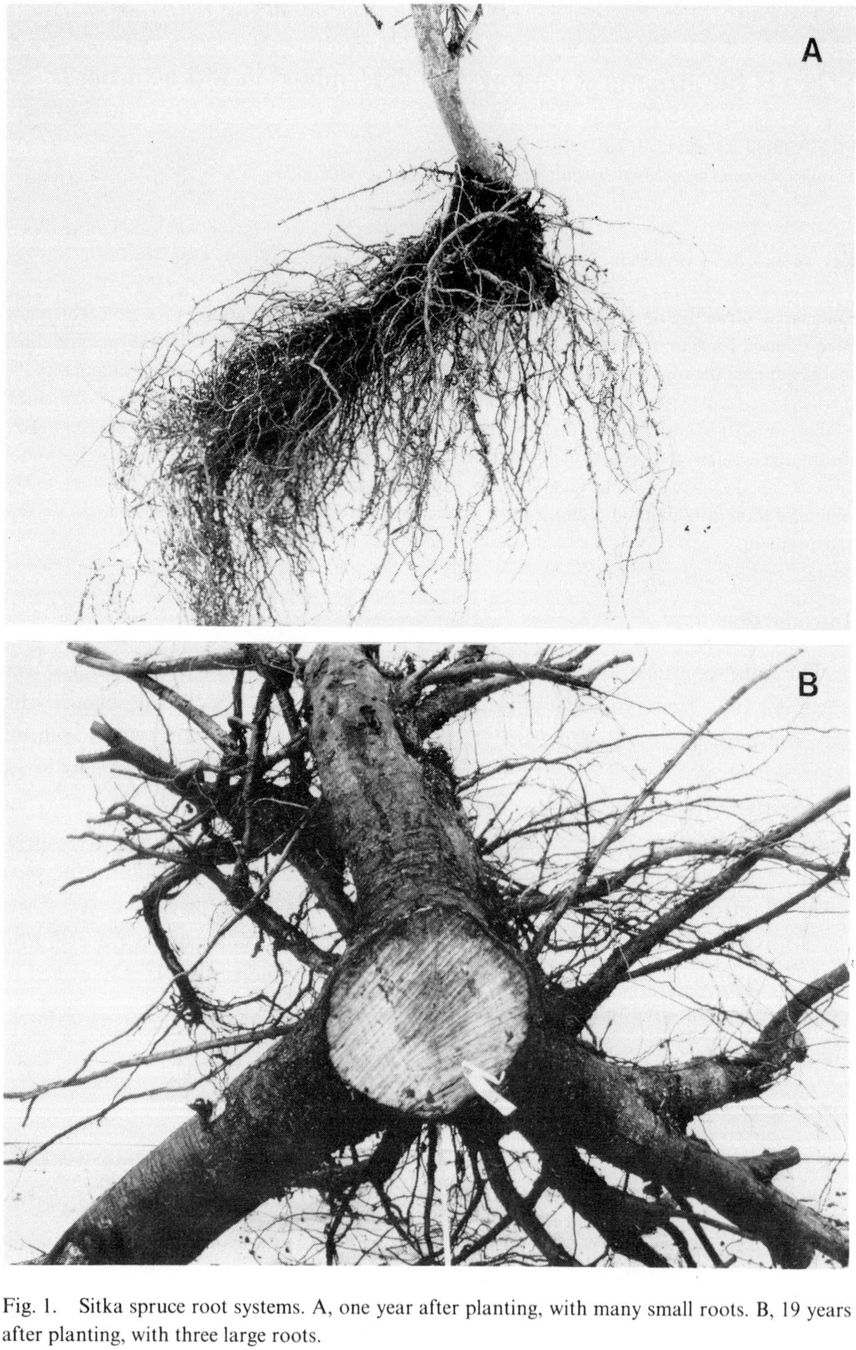

Fig. 1. Sitka spruce root systems. A, one year after planting, with many small roots. B, 19 years after planting, with three large roots.

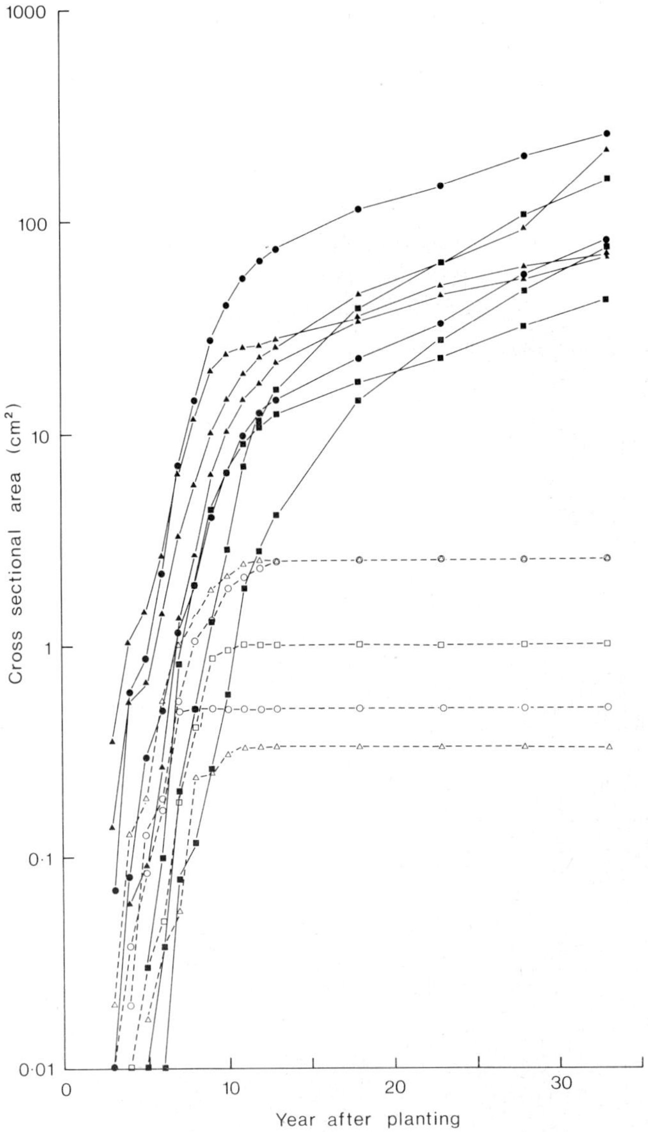

Fig. 2. Semi-log plot of cumulative cross sectional area of woody roots on a Sitka spruce tree 34 years after planting, measured at a point 20 cm from the centre of the stump. Each line represents a single root. Solid lines, the eight largest roots. Dashed lines, minor roots; the uppermost dashed lines represent the 9th and 10th largest roots on the tree.

small diameter some of the growth rings were missing and the age of the root was determined by tracing growth rings back to their connection with rings of known age in the stump.

Results and discussion

The results are illustrated by data from an individual tree which was typical in many respects. Thirty-four years after planting, the tree had eight large roots, which will be termed 'major' roots, ranging in CSA from 42–660 cm^2 at the 20 cm sampling point (Fig. 2). The remainder of the roots were much smaller (2.5 cm^2 or less in CSA) and are termed 'minor' roots. The differentiation into major and minor roots took place over a number of years. The roots took 3–6 years to reach the 20 cm sampling point, but on most of the 34 year and 8 year trees one or more roots had reached 20 cm by the first or second year after planting.

Three of the major roots were among the largest on the root system from years

Fig. 3. Transverse section of a 1 cm diameter Sitka spruce root cut 20 cm from the centre of the stump. The root is 30 years old and has seven growth rings.

3 to 34. The five other major roots grew at rates comparable to those for the minor roots during the early years. During this early phase of development both major and minor roots showed a logarithmic rate of increase in CSA (Fig. 2). However, 7–10 years after planting, the growth of the minor roots slowed down and it eventually stopped (Figs. 2 and 3). By contrast, the major roots continued to grow in CSA, so that by 8 years all but two of the roots classed as major roots at 34 years were already among the largest on the root system. The two remaining major roots became distinguishable from minor roots when they outgrew the latter 10 and 12 years after planting. At about that time canopy closure usually begins and the resulting competition from adjacent trees probably accounts for the change in the slope of the logarithmic growth curves for major roots in Fig. 2. The growth curves for major roots frequently cross one another, so that the ranking of the roots on a size basis changed many times during the 34 years. However, the curves for major roots remained separated from those for the minor roots after the initial sorting out period, so that the main structural root system at 34 years was largely differentiated between years 3 and 8. Other trees showed similar patterns of growth, but did not always show the clear distinction between major and minor roots shown in Fig. 2. and examples were found with a continuum of root sizes.

Missing growth rings in roots are well documented[4] but it is not generally recognised that the xylem can stop growing for many years near the root base while the root remains alive with actively extending tips, as found in minor roots in the present study. Growing mycorrhizas were found on distal parts of minor roots in which no secondary xylem had been produced for 20 years at the 20 cm sampling point. It would be of interest to know whether the cambium continues to produce new phloem in these roots because sieve tubes are believed to survive for little more than a year[11].

Our observations on Sitka spruce, where the dominance of a few major roots was sustained for a long period, differs from reports on other species. Successive changes over periods of many years brought about by a vigorous outgrowth of adventitious roots from the base of the tree have been reported in certain broad-leaved species[8] and in White spruce (*P. glauca* Moench (Voss))[10]. In the latter, the vigorous new roots emerged from the upper part of the root system or stem base in response to increased soil wetness and the accumulation of organic matter, and such large changes in site conditions did not occur in the Sitka spruce plantations in the present study.

The predominant factors which determine the vigour of an individual root in the field are not known, but the original nature of the root when it is in the primary stage, its position on the root system, and the local environment may all contribute. The mean primary xylem diameter (pxd) of the minor roots at the 20 cm sampling point was 0.45 ± 0.87 mm, and spanned the entire range encountered (0.22–1.16 mm). The major roots, many of which were adventitious in origin, had pxds in the upper part of the range with a mean of 0.65 ± 0.50 mm.

Measurement at a single point along a root does not take into account the changes in pxd which take place with age and season[3,9], but the data suggest that the possession of a large pxd is a prerequisite for development into a major root, although it does not guarantee such development and other factors are clearly required. In *Betula papyrifera* Marsh the possession of a relatively large pxd was found to be necessary if a branch on a root was itself to become a permanent long root[5].

The local soil environment can have a strong influence on root development. Experiments with Sitka spruce seedlings with roots divided between wet and dry soils have shown that radial growth of the woody root is related to the soil moisture environment of the primary roots[1]. Comparable experiments in which Sitka spruce root systems were split between different nutrient environments also gave rise to compensatory growth, with most growth occurring in the part of the root system well supplied with certain mineral nutrients[6]. After differential root growth had been induced by contrasting nutrient regimes, the growth re-adjusted when gradients in nutrient supply were removed[2]. Root systems evidently have the capacity to respond to the local environment and localized changes in soil conditions, and the exploration of new areas of soil by root tips may help to account for differentiation into roots of varying vigour, and for the changes in ranking of the major roots.

References

1 Coutts M P 1982 Growth of Sitka spruce seedlings with roots divided between soils of unequal matric potential. New Phytol. 92, 349–361.

2 Coutts M P and Philipson J J 1977 The influence of mineral nutrition on the root development of trees. III. Plasticity of root growth in response to changes in the nutrient environment. J. Exp. Bot. 28, 1071–1075.

3 Fayle D C F 1975 Archy and diameter of primary xylem in horizontal and vertical roots of Red pine. Can. J. For. Res. 5, 122–129.

4 Fayle D C F 1975 Distribution of radial growth during the development of Red pine root systems. Can. J. For. Res. 5, 608–625.

5 Horsley S B and Wilson B F 1971 Development of the woody portion of the root system of *Betula papyrifera*. Am. J. Bot. 58, 141–147.

6 Philipson J J and Coutts M P 1977 The influence of mineral nutrition on the root development of trees. II. The effect of specific nutrient elements on the growth of individual roots of Sitka spruce. J. Exp. Bot. 28, 864–871.

7 Scott Russell R 1977 Plant Root Systems: their Function and Interaction with the Soil. McGraw-Hill Book Company (UK) Limited, London. 298 p.

8 Stout B B 1956 Studies of the root systems of deciduous trees. Black Rock For. Bull. 15.

9 Wilson B F and Horsley S B 1970 Ontogenetic analysis of tree roots in *Acer rubrum* and *Betula papyrifera*. Am. J. Bot. 57, 161–164.

10 Wragg J W B 1967 Origin and development of White spruce root forms. Canada Department of Forestry and Rural Development. Depart. Publ. 1192.

11 Zimmermann M H and Brown C L 1971 Trees, Structure and Function. Springer-Verlag, Berlin.

Plant and Soil 71, 161–166 (1983). A-16

Differences between stem and root thickening at their junction in red pine

D. C. F. FAYLE
Ontario Tree Improvement and Forest Biomass Institute, Ministry of Natural Resources, Maple, Ont., Canada LOJ 1EO

Key words Annual ring area Crown size Red pine Root base Soil moisture Stem base

Summary The annual increment in cross-sectional area at the base of red pine roots increased and decreased periodically relative to that at the stem base during 4 decades analysed. This cyclical pattern was similar in both 1.8 by 1.8 m and 3.6 by 3.6 m spaced trees but was largely absent when crown size was substantially reduced by pruning. The increases and decreases in root thickening relative to the stem were associated with overall decreases and increases respectively in soil moisture deficits.

Introduction

At the stem-root junction, carbohydrates and growth regulators transported in the phloem pass from the single circumference of the stem into the separate circumferences of individual roots; conversely, water and dissolved substances travelling in the xylem pass from the separate cross sections of the roots into the single cross section of the stem. The number and distribution of roots that join the stem around and beneath its base affect the configuration of this junction. Physical stresses produced by the mass and sway of the aerial parts are concentrated here, and major environmental changes occur at the air-soil boundary.

Previously, I had suggested[2] that alternating periods of relative gain and loss in thickening at the root base compared to the stem base of red pine (*Pinus resinosa* Ait.) might be the result of changes with time in: the distance of the crown from the roots, the photosynthetic output of the crown, physical stresses and response to these stresses, and moisture availability. The latter was considered of secondary importance.

These points are examined further here.

Materials and methods

Four trees were selected from a spacing-pruning experiment planted in the fall of 1949 on a well-drained sandy loam in York Region, Ontario, to provide trees of the same age but with different crown development. Spacings were 1.8 by 1.8 m and 3.6 by 3.6 m, with pruned and unpruned plots. Pruning to produce a knot-free bole took place from year 5 to 27 (1954–76), during which the current, lowermost 3 live whorls were removed over a 2- to 4-year period. One tree per treatment was sampled; these are referred to as trees P1.8, P3.6, U1.8 and U3.6. The plantation design was such that P3.6 and U3.6 were adjacent, while P1.8 and U1.8 were within 30 m, all on similar soil conditions.

All branches in the original basal whorl of U1.6 were dead by age 13 (1962), after which 11 or 12 live whorls were retained for a decade, increasing to 14 by age 31 (1980); U3.6 retained 3–5 more live whorls than U1.8. Pruning substantially reduced the number of live whorls, but the number retained was similar in both pruned trees: 5–6 from age 5 to 12 (1954–61) increasing to 10 by age 27 (1976). Branch size (diameter, length) was greater in the 3.6-m than 1.8-m spaced trees.

Sections were cut from the middle of each stem internode, at 10 cm above the ground line, and at 30, 50 and 100 cm from the stem centre along roots that were at least 1.5 cm in diameter at 30 cm. These positions are referred to as S_{10} (stem base), R_{30} (root base), R_{50} and R_{100}. Ring widths were measured to the nearest 0.04 mm along 2 diameters at right angles to each other with a stereomicroscope fitted with a measuring eyepiece. Average ring width, circumference, and cumulative and annual cross sectional areas were calculated.

Additional information was obtained from similar measurements of other codominant red pines: 2 from a mixed (alternate row) plantation of red and white pine (*P. strobus* L.) established in 1956 (S_{10}, R_{30} and R_{50} positions only); 6 from a plantation established in 1937[1,2] (of which 2 trees were in a plot irrigated in 1964 and 1965); and 2 from a plantation established in 1939[1,2]. The spacing of the plantations was approximately 1.8 by 1.8 m and they were within 20–40 km of the spacing-pruning experiment. Soil moisture deficits for May-October were computed by Thornthwaite's method[7] using climatic data from stations within this area and a water holding capacity in the root zone of the soil of 100 mm. An additional 2 trees were sampled (S_{10}, R_{30} and R_{50}) in a plantation established in 1936 some 200 km north east of the other plantations, in Renfrew County, and thinned in 1960 and 1970.

Results

The total cross sectional area of the roots at 30 cm was similar to that of the stem at 10 cm by age 15 (1964) in the unpruned trees and age 30 (1979) in pruned trees. The circumference of the roots at 30 cm, however, was greater than that of the stem by the following factors at age 15 and 30 respectively: 2.3 and 3.5 for P1.8, 2.8 and 3.1 for U1.8, 2.7 and 3.8 for P3.6; and 4.8 and 6.1 for U3.6. The differences partly reflect the number of roots: 19 in P1.8, 12 in U1.8, 15 in P3.6 and 39 in U3.6.

The annual increments in cross sectional area at the stem and root bases differ with spacing and pruning in amount and in the patterns of growth, reflecting differences in crown development (Fig. 1, upper). When the ratios of the root to stem annual ring areas (R/S) are examined, however, differences exist with pruning, but are minimal with spacing (Fig. 1, lower). The overall increase with time in the ratios is partly due to the increasing diameter of the stem bringing the measured positions closer together.

Specifically, in the unpruned trees a cyclical pattern is evident in the R_{30}/S_{10} and R_{50}/S_{10} curves with highs about 1960, the late 1960s and late 1970s, and intervening lows in the early 1960s and 1970s. The highs represent periods of relative gain in increment at the root base compared to the stem base, and the lows a relative loss. The pattern is the result of the basal root increments fluctuating more than those at the stem base. Fluctuations in increment at 100 cm along the roots were less than at 30 and 50 cm, and the cyclical pattern is not noticeable in the R_{100}/S_{10} curves.

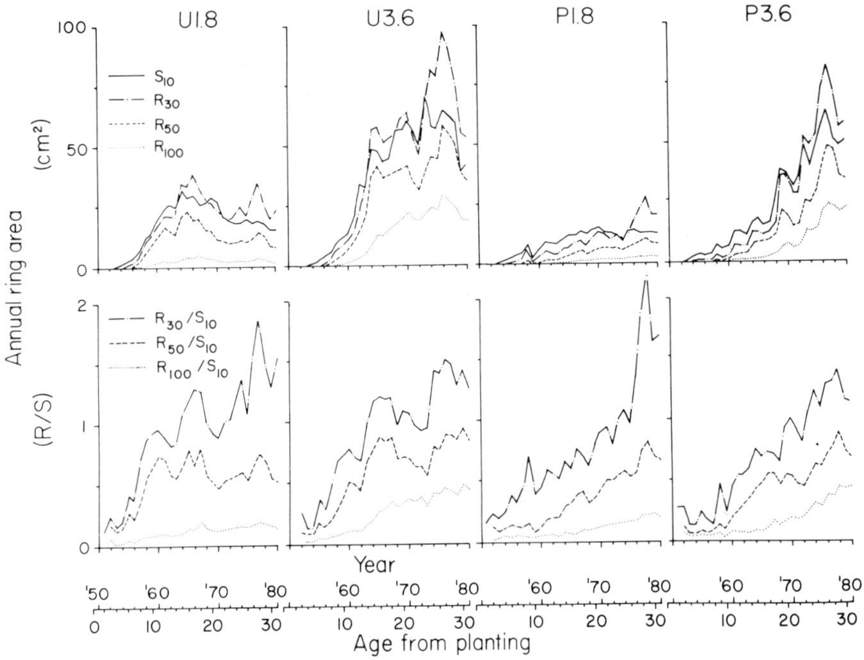

Fig. 1. (*Upper*) Annual ring areas at 10 cm height in the stem and at 30, 50 and 100 cm along the roots. (*Lower*) The ratio of annual ring areas at 30, 50 and 100 cm along the roots to the annual ring area at 10 cm height in the stem.

In the pruned trees the cyclical pattern is absent or weak in the R_{30}/S_{10} and R_{50}/S_{10} curves; a low does appear in the early 1970s in R_{50}/S_{10} for P3.6. No cyclical pattern is evident in the R_{100}/S_{10} curves.

No alternating highs and lows, similar to those in the R_{30}/S_{10} and R_{50}/S_{10} ratios, occurred in either unpruned or pruned trees in comparisons of annual ring areas between points within the basal 100 cm of the stem or within the basal 100 cm of the roots.

Exclusion of the vertical root component, *i.e.* those located under the stem, did not alter the R/S patterns, except for U3.6 where the first high was absent, being contributed primarily by the vertical roots. The annual ring area at the base of the vertical roots reached a maximum relative to the horizontal roots in the second decade, and occurred a few years earlier in the 3.6-m than 1.8-m spaced trees, and in U than P trees.

There appeared to be no effect of pruning or spacing in the time that eccentric growth started at 30 cm along the roots, i.e. when ring widths on the top plus bottom became greater than ring width at the sides.

The high and low periods in the R/S ratios of the unpruned trees match those found in the 1937, 1939 and 1956 plantations. Furthermore, the highs and lows correspond to periods of decreased and increased soil moisture deficits

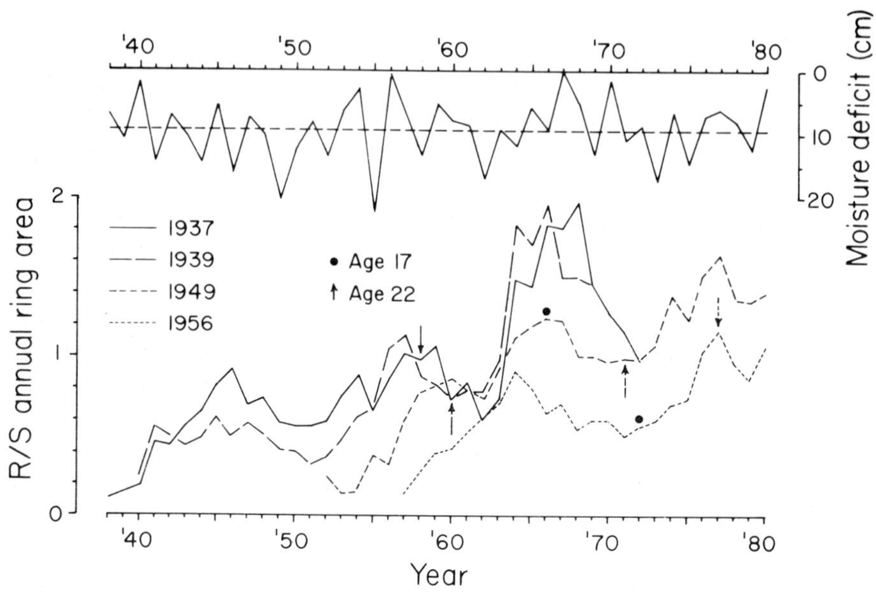

Fig. 2. The ratio of annual ring areas at 30 cm in the roots to annual ring areas at 10 cm height in the stem for trees planted in 1937, 1939, 1949 and 1956, compared to the annual May–October soil moisture deficit. Each R/S curve is the average of 2 trees, except the 1937 curve which is the average of 6 trees. The broken horizontal line with the soil moisture deficit curve represents the average deficit for the period.

respectively (Fig. 2). These changing deficits occurred at different tree ages and/or stages of crown development, whereas the highs and lows in the R/S curves relate more to changing deficits than to age.

The R_{30}/S_{10} ratios increased when soil moisture deficits decreased from 1962–63 to 1964–65 in both irrigated and unirrigated trees in the 1937 plantation, and there was a marginal shift in increment within the basal 100 cm of the roots towards their base. However, the R/S ratio of the irrigated trees increased less than that of the unirrigated trees as a result of a combined increase in stem thickening with a possible relative reduction in thickening at the root base.

Thinning in the Renfrew County plantation also appeared to alter thickening at the stem-root base. For several years prior to each thinning, annual ring areas of the stem and root bases followed the same trend in both trees. After thinning the annual ring area of the root bases increased immediately whereas that of the stem did not increase much until the 3rd year. Thus the R/S ratios increased after thinning, reaching a maximum within 4–6 years and then declining.

Discussion

Since shifts in thickening *within* the basal part of the stem or *within* the basal part of the roots were much less than those *between* the stem and root bases, some attribute of the stem-root junction appears to be significant. There was little change in total cross-sectional area at this junction, but the increase in circumference was substantial, particularly as the number of roots increased. The same has been reported for Picea [5,6], and between either side of a branching point in long roots of *Pinus strobus* [8]. In the latter case, there was little change in phloem thickness, thus the concentration of photosynthate per unit circumference would decrease beyond the junction [8]. Detailed measurements of phloem thickness in the red pine stems and roots were not made, but some incidental measurements indicated about a 50% reduction in phloem thickness from stem to root. Therefore, even when the circumference of the root bases was twice that of the stem, the concentration of photosynthate per unit circumference might be maintained. However, increases in circumference were more than double.

If the thicknesses of the phloem at the stem and root bases changed annually in relation to each other, there could be a change from stem to root in the concentration of photosynthate per unit circumference, affecting xylem production by the cambium. Differences in the rate and (or) duration of xylem production between the stem and root bases could produce a differential increment. Both rate and duration would be affected by internal moisture status, substrate availability and growth regulator balance. Annual cessation of cambial activity in the tree may occur last near the base of the roots, perhaps influenced by temperature.

The similarity in the R/S curves of the unpruned trees at the 2 spacings (Fig. 1) ınd the occurrence at the same age of both highs and lows in the R/S ratios (Fig. ?) suggest that the increasing distance of the crown from the roots and responses ɩo changing mechanical stresses at the stem-root base are not a major factor affecting thickening at the base of roots relative to that at the stem base. The correspondence of fluctuations in the R/S curves with those in the soil moisture deficit curves suggests that changes in moisture status have a greater influence. However, stem sway effects cannot be dismissed entirely. Restraining stems does reduce thickening at the stem-root base [3,8]; initially the reduction can be greater at the root base than in the stem [3]. The increase in the R/S ratios immediately after thinning may be a partial effect of increased sway. However, thinning would also improve the moisture supply by reducing root competition.

The relative change with time in moisture status appears to be more important than the absolute amount in affecting root-stem thickening, because the cyclical pattern of the R/S curves is equally well expressed on dry and very fresh sites [2]. Variation occurs between trees rather than sites, particularly in the first 1–2 decades, perhaps reflecting differences in individual crown-root relations.

If moisture availability is important, why was the cyclical pattern in the R/S curves absent or poorly expressed in the pruned trees? A possible explanation was the greatly reduced crown size, leading to a shortage of photosynthate for root growth. This is suggested by the much lower R/S ratios for 1959–1969 in the pruned trees: the maximum difference between pruned and unpruned trees in the number of live worls in the crown occurred within that period. At the same time internal moisture stress might have been reduced due to the smaller transpiring area of the crown. As a result, thickening would not be reduced in the roots relative to the stem when soil moisture deficits increased, but neither would there be sufficient photosynthate available for increased thickening when deficits decreased. Later, however, as the crown size increased, there is evidence of effects on thickening at the stem-root base of changes in the soil moisture deficits, *e.g.* the decline in the R_{50}/S_{10} ratios in P3.6 in the early 1970s (Fig. 1).

In conclusion, the cyclical changes in thickening between stem and root where they conjoin are brought about more changes in moisture availability than by other factors. Studies are needed to determine what attributes of the junction effect these growth shifts. Furthermore, these shifts should be related to those that can occur within the root system itself[3,5], to total annual wood production throughout the tree, and should be recognized, for example, when estimates are made of the annual biomass production or carbon balance by tree components if few locations only are sampled within the tree.

References

1 Fayle D C F and Pierpoint G 1975 Evolvement of root and shoot systems of red pine in soils of different moisture availability. *In* Forest Soils and Forest Land Management. Eds. B Bernier and C H Winget. Les Presses de l'Universite Laval, Que., pp 11–25.

2 Fayle D C F 1975 Distribution of radial growth during the development of red pine root systems. Can. J. For. Res. 5, 608–625.

3 Fayle D C F 1976 Stem sway affects ring width and compression wood formation in exposed root bases. For. Sci. 33, 193–194.

4 Fayle D C F 1978 The case of the thickening root. *In* IUFRO Sympos. Root Physiology and Symbiosis. Eds. A Riedacker and J Gagnaire-Michard. INRA-CNRF, Nancy, France, pp 216–227.

5 Riedl H 1937 Bau und Leistungen des Wurzelholzes. Jahrb. wiss. Bot. 85, 1–75. (Structure and function of root wood, English translation by M H Zimmermann, available from National Translation Center, Chicago, Ill.)

6 Rowe J S 1964 Studies in the rooting of white spruce. Can. Dep. For., For. Res. Br. Mimeo Rep. 64-H-13, 13 p.

7 Thornthwaite C W and Mather J R 1957 Instructions and tables for computing potential evapotranspiration and the water balance. Drexel Instit. Tech., Lab. Climat., Publ. Climat. 10, 198–311.

8 Wilson B F 1976 Distribution of secondary thickening in tree root systems. *In* The Development and Function of Roots. Eds. J G Torry and D T Clarkson. Academic Press, London, pp 197–219.

Plant and Soil 71, 167–170 (1983).

The development of root systems analysed by growth rings

EVAN R. C. REYNOLDS
Department of Forestry, University of Oxford, England

Key words Annual rings Coniferae Root system development

Summary Rings in root systems of 7 conifer species from 11–15yr plantations indicated primary roots were initiated 3–5yr after planting; horizontal roots extended at 20 cm yr^{-1} (*Tsuga heterophylla*) to 55 cm yr^{-1} (*Pseudotsuga menziesii*). Vertical roots grew slower. Systems varied from the flat root systems of *Picea sitchensis* or *P. omorika* to the heart root system of *Abies grandis*. *P. menziesii* favoured the growth of adjacent roots of some species. Planting produced asymmetric root systems.

Introduction

Retrospective analysis of the development of root systems by counting annual rings to assess radial or extension growth, is an attractive alternative[3] to *in vivo* observation, destructive sampling or experimental control of root morphogenesis. This 'ring analysis' is analogous to stem analysis. However, not all roots produce recognizable annual rings: each species requires critical investigation[11]. Sometimes ring analysis is only reliable in young roots[5]. It has provided root productivity estimates[3,6,11], dendrochronology[8], and, as herein, developmental information[5,9].

Information on the development of root systems is needed, firstly for water balance studies, where the change in rooted volume (particularly the depth in woodlands) and its available moisture is needed for the model. Secondly, urban arboriculturalists need to know if and when different tree species might be expected to interfere with neighbouring foundations or underground services. Lastly, further elucidation is desirable of the multifarious ways in which the physical and cultural environments fashion root systems.

Site and methods

Twelve conifer species were planted at 1.8 × 1.8 m in March 1967 on a site in the Forest of Dean (Gloucestershire, England) where the previous 45-yr crop of *Picea sitchensis* had been severely infected with butt-rotting *Polyporus schweinitzii*. Six species were planted in 40 × 40 m plots, the others (of these I investigated only *Picea omorika*) in irregularly shaped plots of c.405 m^2. Plots were triplicated. The loamy stony soil developed in sandstones and shales of the Upper Coal Series (Carboniferous) is somewhat compact between 30 and 60 cm, with pH 4.0 or less at the surface and up to pH 4.6 at depth. Internal soil drainage is impeded in much of that part of the experiment where I investigated the tree roots.

Pathologists pulled up edge trees from some plots, and occasionally the next row inwards, with a tractor-mounted winch. Extraction was of *P. sitchensis* (from all three plots) when plantations were 11- and 14.5-yr old and, from single plots at 15-yr of *P. sitchensis*, *Pseudotsuga menziesii*, *Pinus nigra maritima*, *Abies grandis*, *Tsuga heterophylla*, *Picea abies* and *P. omorika*.

Subjectively I assessed the development of six classes of root comprising the root systems[6,10], *viz*. (a) the tap root complex, (b) shallow (< 15 cm) horizontal primary roots, (c) deep horizontal primary roots, (d) deep horizontal secondary roots, (e) oblique (plagiogeotropic) roots (primary or secondary), and (f) sinker roots. The orientation of each root system was recorded relative to 8 compass points. The thickness of the extracted root plate was noted.

A shallow horizontal root was removed from each of two stumps of each species on the last sampling date. These roots were roughly radially directed and the straightest and longest pulled out. A few vertical roots were also collected. The roots were cut into 25 cm lengths from their bases. Two stumps of *P. menziesii* were collected and their roots removed near the origins. The cut surfaces of root segments and the stumps were sanded and the annual rings were counted under a lens or low power microscope.

Results

All seven species produced countable growth rings. Despite false rings[8], incomplete rings[8], small radial growth in early years[5], distal reduction in summer wood[8], and tapering radial growth, age was rarely as much as 2 yr in doubt, and probably more often underestimated than overestimated.

The long horizontal roots were almost all initiated in the 3rd to 5th growing seasons agreeing with others' findings[3,5]. *A. grandis* and *P. n. maritma* were, if anything, earlier than the other species (although beating up over the first 3 yr confuses the picture). Sinker roots, being branches from the horizontal roots, were naturally 3- or 4-yr younger, but roots of the tap root complex were as old as the horizontal primaries. Investigations of the root bases on the stumps of *P. menziesii* confirmed that major structural 'primary' roots (>2 cm 'horizontal' diameter under bark) were all at least 9-yr old (frequencies, sums of both stumps: 2(9 yr), 7(10 yr), 6(11 yr), 17(12 yr), 10(13 yr), 7(14 yr)). Only roots 11-yr or more achieved > 5 cm diameter, but not all roots of this age had become larger skeletal roots (26% achieved > 5 cm under bark diameter, 55% 2–5 cm, and 19% < 2 cm).

The mean extension rate for the first metre of horizontal roots was 30 cm yr^{-1}. Means (cm yr^{-1}) for each root (crudely in species order of increasing rapidity) were: *T. heterophylla* 19(0.75 m), 21(1.5 m); *P. omorika* 29 (2 m, age rather doubtful), 31(2 m); *A. grandis* 33(1 m), 33(1 m); *P. n. maritima* 25(1.25 m but a lateral substituted for the main axis at 0.75 m), 63(1.25 m); *P. sitchensis* 58(1.75 m), 30(1.5 m), crude preliminary field analyses 14.5-yr after planting suggested rates of 30 to 50; *P. menziesii* 63(1.25 m), 37(1.5 m); *P. abies* 50(1 m), 63(1.25 m). Figures in parentheses are the lengths analysed. Rates might have decreased as roots extended further[7]. Corresponding mean rates of extension of vertical roots were *P. menziesii* 10(0.5 m), 25(0.75 m); *P. sitchensis* 25(0.75 m), the crude field data indicated around 15 cm yr^{-1}. As in *Pinus resinosa*[7], the slower rates of vertical extension of the root system are depressed by frequent replacement of the growing point (presumably due to periodic waterlogging as in *P. resinosa*[7]).

The mean thicknesses (cm with SE) of the root plates were: *P. omorika** 29 ± 2.2; *P. menziesii** 36 ± 3.5; *T. heterophylla* 38 ± 2.6; *P. abies** 42 ± 6.2; *P. sitchensis* 51 ± 7.2; *P. n. maritima** 52 ± 3.4; *A. grandis* 59 ± 4.3. Species asterisked were growing on better drained soil; had the site been more uniform, differences might have been greater.

Symptoms of planting deficiencies were roots 'swept' in one direction, often tangential to the stem, and sometimes encircling or even girdling it (producing structural weakness). The base of the rootstock often had distorted grain[10] with inclusions of bark and soil. In the first sampling of *P. sitchensis* there was a significant preference to rooting eastwards and NE in the next sampling, presumably due to planting method.

Since the sampled trees were on plot edges or only one line inside, there could be an interaction with the neighbouring species[12]. Thus *P. sitchensis* had a significant preference for rooting into adjoining *P. menziesii* and slight indications of an avoidance of *T. heterophylla*. Effects of planting method and adjacent plantations were confounded in the sampling of other species, although *A. grandis* perhaps showed some tendency to root into *P. menziesii*.

Despite intraspecific differences, the experiment produced evidence of interspecific differences between root systems as shown by the development of each root class. The crude order of morphology from a heart root system[1] to a flat root system[1] is *A. grandis*, *P. n. maritima*, *P. abies*, *P. menziesii* (nearest the median type), *T. heterophylla*, *P. omorika*, *P. sitchensis*.

Conclusions

The species selected, the site conditions, and extractions by winching only edge trees from plots limit generalizations from my observations. Within these constraints, fir and pine had the largest root volumes and would be expected to develop the largest soil moisture deficits on this site, and *P. omorika* the least (but suffering more from water shortage). Lateral development of the root systems of *P. omorika*, *P. sitchensis*, and perhaps *T. heterophylla* could lead to greatest intraspecific competition[6].

Early initiation and rapid growth of long horizontal roots (as in *P. sitchensis*, *P. abies* and *P. menziesii*) would be expected to interfere with foundations and services, but, perhaps due to their shallowness on wetter sites, conifers are in general less damaging than broad-leaved species[2]. However, my results suggest that subterranean nuisance by urban trees probably relates closely to species.

Loss of central vertical roots at planting probably contributed to the limited rooting depths I observed. Asymmetry of the root systems and physical weaknesses at the root collar attributable to planting deficiencies have confirmed many previous observations[4]. It is probable that the development of roots is affected by which species are planted adjacently[12].

Acknowledgements The experiment was initiated by D K Barret of the Department of Forestry, University of Oxford, and B J W Grieg of the Forest Research Station, Alice Holt Lodge, Farnham, Surrey. The Forestry Commission also conducted the experiment and extracted the sample trees.

References

1 Büsgen M (3rd edition, translated) 1929 The Structure and Life of Forest Trees. Chapman and Hall Ltd, London, 436 p.

2 Cutler D F and Richardson I B K 1981 Tree Roots and Buidings. Construction Press, London, 94 p.

3 Deans J D 1981 Dynamics of coarse root production in a young plantation of *Picea sitchensis*. Forestry 54, 139–155.

4 Eerden E van, and Kinghorn J M (Ed.) 1978 Proceedings of the root form of planted trees symposium, held in Victoria, British Columbia, May 16–19, 1978. B.C. Ministry of Forests/Canadian Forestry Service. Joint Report No. 8, 357 p.

5 Fayle D C F 1975 Extension and longitudinal growth during the development of red pine root systems. Can. J. For. Res. 5, 109–121.

6 Fayle D C F 1975 Distribution of radial growth during the development of red pine root systems. Can. J. For. Res. 5, 608–625.

7 Fayle D C F and Pierpoint G 1975 Evolvement of root and shoot systems of red pine on soils of different moisture availability. *In* Forest Soils and Forest Land Management. Eds. B Bernier and C H Winget. Les Presses de l'Université Laval. pp 11–25.

8 Schulman E 1945 Root growth-rings and chronology. Tree Ring Bull. 12, 2–5.

9 Stout B B 1956 Studies of the root systems of deciduous trees. Bull. Black Rock Forest, Cambridge, Mass. 15, 45 p.

10 Somerville A 1979 Root anchorage and root morphology of *Pinus radiata* on a range of ripping treatments. N.Z.J. For. Sci. 9, 294–315.

11 Yamakura T, and Saito H 1972 Production and structure of under-ground part of Hinoki (*Chamaecyparis obtusa*) stand. I. Estimation of root production by means of root analysis. J. Jap. For. Soc. 54, 118–125.

12 Yeatman C W 1955 Tree root development on upland heaths. Bull. For. Comm. London 21, 49 p.

Plant and Soil 71, 171–188 (1983). A-18

Root architecture and tree stability

M. P. COUTTS
Forestry Commission, Northern Research Station, Roslin, Midlothian, UK

Key words Anchorage Root architecture Root strength Soil strength Stability Windthrow

Summary Root anchorage is discussed with a view to determining the optimum use of root material for enhanced stability. Field observations were made on Sitka spruce root systems while lateral forces were applied to the stem with a winch to pull the tree over. Measurements included the applied force, angles of inclination, soil and root movement, timing of the sound of root breakage using buried microphones, weight and shape of the root-soil plate and damage to the roots.

Components of anchorage include the dimensions and mass of the root-soil plate levered from the ground by the displaced stem, and tensile strength of roots and soil beneath the plate; root and soil tensile strength and root/soil resistance on the windward perimeter; and on the lee side the stiffness of the hinge at the fulcrum.

Strength properties of roots and soil are reviewed. Models devised for landslip are extended to consider behaviour under tension, of roots singly and in groups, and the concept is developed of a 'critical rooting density' at which root/soil resistance exceeds soil strength, giving rise to the characteric root-soil plate on uprooted trees. The lee side part of the root-soil plate acts as a cantilevered beam and determines the distance of the fulcrum from the tree. Physical laws defining the reduced stiffness of beams as a result of subdivision, indicate the importance of the number/size distribution of roots and weakening effects of branching.

On the windward side upward movement of the root-soil plate causes sequential breakage of soil and roots. Under an increasing applied load, failure occurs in parts of the soil-root system before the maximum force for uprooting is achieved. A preliminary approach is made to modelling where the changing contributions of the components of anchorage are allowed for throughout the uprooting process.

Introduction

Functions of tree roots which receive little attention from plant physiologists are those concerned with support and anchorage. At the juvenile stage failure due to bending at the root or stem base leads to 'toppling'. The tree continues to grow and the stem becomes permanently bowed reducing timber value[8]. At a later stage, when the stem base is more rigid, uprooting (windthrow) may occur. Windthrow is a serious problem in forest plantations[4,10,18,19,32,40,43] and orchards[30] and can cause extensive damage to amenity trees[7]. Uprooting is sometimes a sequel to the initial instability caused by toppling. Studies on windthrow in forest plantations include aerodynamic effects[5,6,13,21]; the movement and mechanical properties of stems[20,41]; effects of silvicultural treatments such as thinning[4], drainage and cultivation[29,31]; and measurements on root systems and the forces required to pull trees over by winching[12,14,15,31].

Substantial variation in stability has been recorded between different sites and species[15]. Associated measurements on root systems usually lack detail, and with one exception[31] have been confined to such features as the width and depth of the root system exposed by uprooting. The ideal root system for anchorage is not known for any site because root morphology tends to be dominated by site conditions, and the latter are often difficult to alter. For example, rooting depth is limited by seasonally high water tables[2] on soils which are notoriously difficult to drain[28,35]. However, if the components of anchorage can be elucidated a modelling approach can be adopted to study its sensitivity to root morphology and distribution, and to soil properties. This paper is addressed, not to the mathematical expression of such a model, but to the concepts and observations on which it can be based. These include the properties of the roots and soil, the mechanics and dimensions of the system under stress and the stages of structural failure. The field observations are confined to the uprooting of plate-like root systems found on exposed sites where Sitka spruce (*Picea sitchensis* Bong. Carr), the main commercial forest species in Great Britain, is widely grown.

Concepts

The root soil system

When a tree stem is moved laterally by the force of the wind the lee and windward sides behave differently (Fig. 1a). On the lee side, the root-soil system is subjected to bending and compressive forces against the bearing surface of the soil below, whereas on the windward side, where the root-soil plate is lifted, the system is subjected to tensile and possibly to shearing forces, and there is no bearing surface to counteract upward movement, so that bending takes place farther from the stem. Furthermore, the stiffening effect of the roots in the soil is like reinforcement in a beam. When a beam is forced to bend, tensile elements in it will stiffen it when they lie near the outer, convex edge. Main laterals lie near the soil surface in shallow root systems, and therefore stiffen the soil more efficiently on the windward than on the lee side. However, deeply located horizontal roots will have a reinforcing, stiffening effect on the lee side when they come under tension in the lower convex part of the bending root-soil plate. Thus when

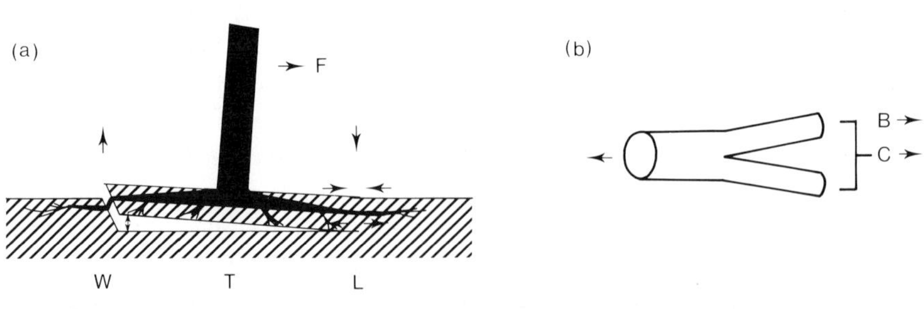

Fig. 1. **a**, Some of the forces acting on the root-soil system when the stem of a shallow rooted tree is moved horizontally. F, horizontal force of wind or winch; W, fracture of root-soil plate on windward edge; T, centre of tree stem; L, hinge on lee side. **b**, a root which has divided into two branches with total cross-sectional area equal to that proximal to the fork. See text.

modelling the lee side the contribution of the roots should be considered in the context of the root-soil interaction. Additional resistances to displacement arise from the mass of the root-soil plate, and from soil tensile strength round its edge and beneath it. The direction of movements in root forms other than plate-like systems are indicated by Hintikka[17].

Important properties of root architecture can be deduced from the physical properties of materials. The stiffness of a beam circular in cross-section is related to the fourth power of its diameter. Therefore if the beam forks into two of even size, and with a total cross-sectional area (CSA) equal to that proximal to the fork (Fig. 1b), then the *total* stiffness of the beam will be halved. Roots are often irregular in cross-section but can maintain an equal CSA either side of major root branches[42]. Branching on the lee side will thus cause a substantial reduction in stiffness, so that the fulcrum, on which the root-soil plate is levered up out of the soil, occurs closer to the stem and stability is reduced. Similarly, a root system containing only a few thick lateral roots at the stem base will be stiffer than one with many thin ones. However a wide angle between laterals shortens the lever arm, L, on the lee side ($L = R \cos (b/2)$, where R = root length proximal to the fulcrum where the roots bend and b = angle between roots). L becomes rather sensitive to angles $> 60°$, *e.g.* when fewer than six evenly spaced roots are present.

On the windward side, the strength of roots stressed under tension is not in theory subject to a reduction on subdivision, but in practice is reduced by unequal loading. If the anchorage of the forked root in Fig. 1b, is loaded under tension at B instead of at C, failure will occur at a much lower stress than if the load is applied at C, evenly to both parts of the fork.

Strength of the materials

The roots and soil differ enormously in their strength and physical properties (Table 1). Behaviour of a tree root under tension generally follows the curve in Fig. 2 with a linear relationship between stress and strain up to the elastic limit. The root then undergoes plastic deformation before failure. The elastic limit occurs at 2–4% strain under a load of 9 000–16 000 kPa for a number of broadleaved and coniferous species, with 11–20% strain under loads of 15 000–63 000 kPa at failure. Root diameter can influence tensile strength per unit of CSA[9,16,37,39]. Bark contributed little to strength[37]. Considerable variation was found in two types of Douglas fir growing in their native localities[9] and clonal variation in *Populus* spp.[16] was related to specific gravity and in some instances to cellulose content. Genetic differences are known to occur in stemwood strength[45] and variation in rootwood strength could have an important influence on anchorage.

The tensile strength of soil is rarely measured, but is 3–5 orders of magnitude weaker than that of roots under tension, and at failure strain is only about 1.8% (Table 1). Variation must be expected with wetness and organic matter content. The shear strength of soils decreases with increased moisture content and

Table 1 Tensile strength of roots and soils

Species	Proportional limit		Failure		Comments	Authority
	Stress (kPa)	Strain (%)	Stress (kPa)	Strain (%)		
					Roots*	
Picea abies	–	–	27800	–		Turmanina[37]
Picea sitchensis	15000	3	35000	13		Coutts
Pinus ponderosa	4300	4	9000	20	Seedlings, overbark	Waldron and Dakessian[39]
Pinus radiata	–	–	17600	11	Overbark	O'Loughlin and Watson[24]
Thuja plicata	–	–	57000	–		O'Loughlin[22]
Pseudotsugo taxifolia						
(Coastal)	–	–	63000	12		O'Loughlin
(Coastal)	–	–	45000			Burroughs and Thomas[9]
(Rocky Mountain)	–	–	15000			Burroughs and Thomas
Nothofagus sp.	–	–	33000			O'Loughlin and Watson[25]
Betula pendula	–	–	38200			Turmanina[37]
Tilia cordata	–	–	26500			Turmanina
Quercus rubra	–	–	32400			Turmanina
Populus deltoides	–	–	39100			Turmanina
Populus × eur-americana						
(clone 1–48)	8900	3	30900	16	Root air-dried, rewetted for measurement	Hathaway and Penny[16]
(clone 1–78)	24000	3	53900	16		Hathaway and Penny
Populus deltoides	8900	3	26600	12		Hathaway and Penny
Populus yunnanensis	12900	4	41000	18		Hathaway and Penny
Salix matsudana	12900	3	41000	18		Hathaway and Penny
Salix purpurea	15800	2	45300	18		Hathaway and Penny
					Soil	
Silt clay	–	–	7–19	1.8	Undisturbed	Thorne, Tuvey and Bryant[36]
Clay	–	–	4–7	0.1	Remoulded	Ajaz and Parry[1]

* Root strength is based on under-bark measurements unless otherwise stated.

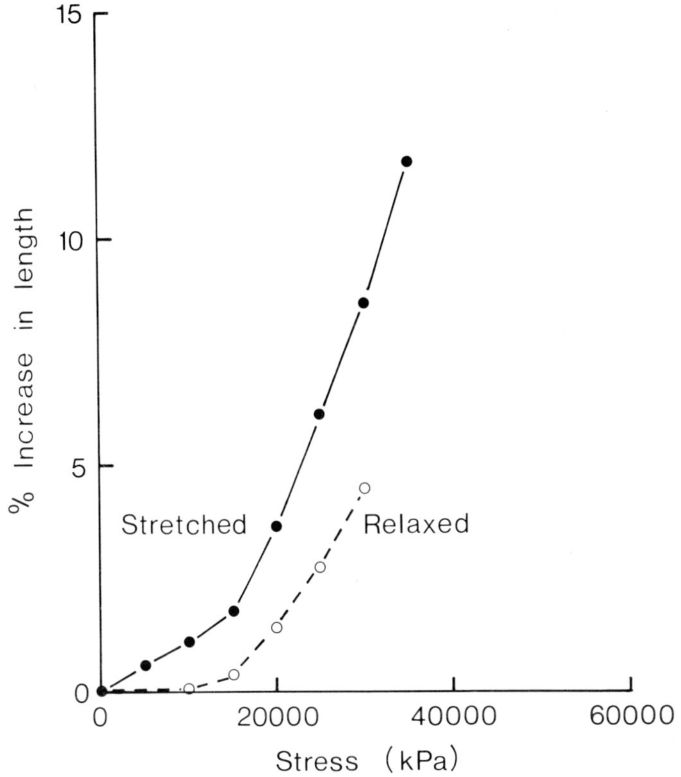

Fig. 2. Mean stress-strain curve for Sitka spruce rootwood stretched to failure under tension (——). The force was relaxed at intervals to determine the load causing permanent deflection (– – –). Roots were 4–10 mm diameter, and were taken from a position 80 cm from the stem base in a 34 year-old plantation.

increases with stress normal to the shear plane. Windthrow occurs during winter gales when the soil is wet so that shear values at or near saturation are of most interest. The shear resistance tends to be an order of magnitude larger than the tensile strength. At a depth of 30 cm, with a normal load (soil + crop[3]) of c 10 kPa, shear strength lies in the range 10–30 kPa[26,27]. An important aspect of shear strength is that after failure a substantial residual force is required to maintain movement in the shear plane.

Root/soil resistance

The behaviour of rooted soil under stress has been studied in relation to land slippage[9,11,23,33,34,38,39,44]. The increased soil shear strength due to tree roots was between 1 and 17 kPa. The most detailed analyses are those of Waldron[38] and Waldron and Dakessian[39] who used an experimental approach. Their model is relevant to aspects of root anchorage at the edge of the root-soil plate on the windward side. Plants were grown in columns of reconstructed soils which were then saturated with water and sheared at right angles to the main direction of the roots. Results were compared with those predicted from a mathematical model which included root tensile strength, soil shear strength and a calculated factor for root-soil resistance (R_{rs}). Largest effects of reinforcement were found in plants rooting through the (loam) soil into a compacted gravel[38], when shear, which occurred at a displacement of c 25 mm, was associated with a sound of roots breaking. A linear relationship was found between the total CSA of the roots prependicular to the shear plane and the shear strength of the column.

In clay soil the roots did not break but pulled out[38], and the equations indicate that in saturated fine textured soils slippage is the most important reinforcement factor provided by roots, and R_{rs} is determined by the soil rather than by root morphology or strength. However, the latter features may become important in soils with moisture content a little below saturation. The soil columns used[39] were 61cm long and were sheared at 30cm whereas roots of windthrown trees often extend several metres beyond the point at which the soil shears. Furthermore, roots are tapered and not evenly distributed. Modification to Waldron and Dakessian's conclusions are therefore required when considering anchorage.

Factors which determine the positions of root and soil breakage can be deduced from a simple theoretical model. Consider first an unbranched root of uniform diameter buried in the soil over an indefinite length under tensile load from one end A (Fig. 3a). The tensile force will stretch the root. R_{rs} will be proportional to the distance AB, and at some point along the root R_{rs} equals the applied load. No strain will occur beyond that point. Maximum strain, and therefore failure, will occur at A where there is no R_{rs}. If the root tapers from the point where the load is applied (Fig. 3b) the distribution of strain will be determined not only by R_{rs} but also by the root CSA. Under a given load, strain will increase with distance from A as the root decreases in diameter, but will then decrease again as R_{rs} increases with distance from A. There will be a point of maximum strain between A and B where the root will break. An increase in root thickness or a decrease in R_{rs} *(e.g.* wind loosening) will move the point of breakage outwards away from the tree, whereas factors which increase R_{rs}, such as drying of the soil, will cause the roots to break nearer to the base.

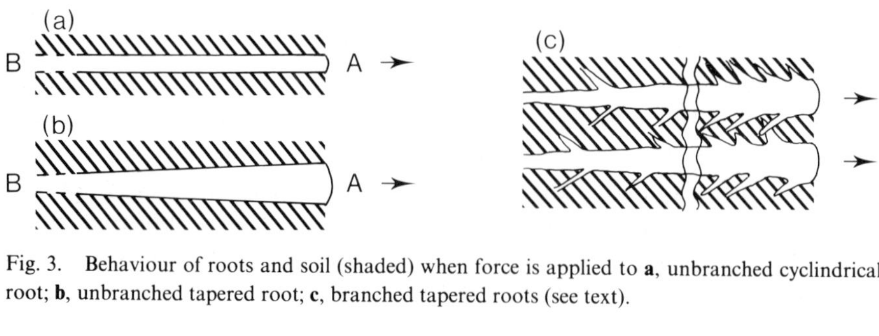

Fig. 3. Behaviour of roots and soil (shaded) when force is applied to **a**, unbranched cylindrical root; **b**, unbranched tapered root; **c**, branched tapered roots (see text).

Where a large amount of much-branched root material is present (Fig. 3c), R_{rs} will exceed the tensile strength of the soil. The applied load will cause the root soil mass in this reinforced region to behave as a unit. Where the amount of root material diminishes to give R_{rs} < soil strength, strain will first cause fracture of the soil, because of its low elasticity. After the soil has fractured the force may be considered to act on the roots as in Fig. 3b, so that roots will break distal to the soil fracture and project from the broken soil surface. The amount of root material in the soil required to increase R_{rs} to equal the soil strength will be termed the critical rooting density (CRD). The CRD will be modified by soil factors which influence the root-soil bond as described by Waldron and Dakessian. It will largerly determine the distance WT in Fig. 1a and the depth of the root-soil plate.

Observations

Root systems under stress

Measurements were made on trees subjected to mechanical stress by winching with a cable attached to the stem[31] at a height of 3.5 m. The trees were co-dominants in a 34-year-old Sitka spruce crop planted on spaced turves from shallow ploughing, on a peaty gley soil at Kershope Forest in Scotland, where sporadic windthrow had already occurred. Measured trees had a mean height of 20 m, mean diameter of 21 cm at 1.3 m. Movement of the soil surface was measured with dial gauges attached to rods hammered into the subsoil; the angle of the stem with a protractor and pendulum attached closed to the stem base[31]; and the applied force with a load cell at the point of attachment of the cable to the tree, and a chart recorder. The sound of roots breaking was recorded using a tape recorder, with microphones at a depth of 20 cm in 40 cm deep holes in the soil.

The first increment of force (Fig. 4b) caused upward movement of the root-soil plate on the windward side 25–75 cm from the stem centre, and a small depression on the lee side, as found in lateral roots by Hintikka[17]. Greater force moved soil which lay farther from the tree on the windward side, as surface roots extending horizontally beyond the edge of the root-soil plate were dragged upward. These are badly positioned for anchorage (Fig. 5), especially at small angles to the horizontal, *e.g.* in the tree illustrated in Fig. 4, only 14% of the stress on these shallow lateral roots was employed in anchorage. This weakening effect is enhanced by reducing R_{rs} before the root breaks, so that failure occurs further from the tree, where roots are thin. Soil at the base of the root-soil plate must have fractured during these early stages because of its low tolerance to tensile strain; the distribution of strain (Fig. 4b) suggests that the fracture occurred first under the stem and spread towards the edge of the root-soil plate. A series of clicking noises recorded on the tape recorder indicated that the first roots began to break when upward movement at the edge of the root-soil plate was 5 mm. The sounds increased in amplitude as the root-soil plate was lifted further, and

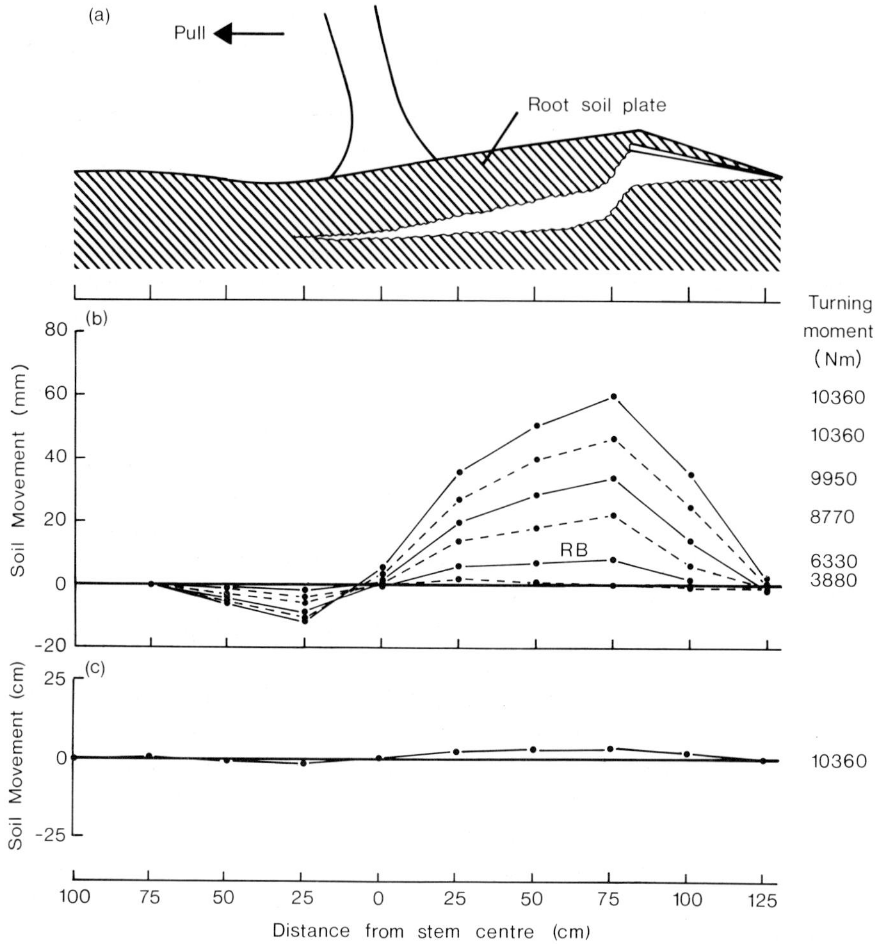

Fig. 4. Vertical movement of soil surface when horizontal force was applied to a 20 cm diameter Sitka spruce tree on a peaty gley soil. **a**, profile view, diagrammatic. **b**, movement on exaggerated vertical scale showing depression of soil surface on 'lee' side of tree and elevation on 'windward' side under increasing increments of force to a maximum at 10360 Nm. Upward movement on windward side spreads beyond edge of root-soil plate as long horizontal roots are dragged upward. **RB**, position of soil surface when roots are first heard breaking. **c**, soil surface at maximum applied force, drawn to true vertical scale.

continued with varying intensity throughout the uprooting process (Fig. 6). The maximum force (F max) was achieved when the root system had lifted 4.7 cm at 75 cm from the stem centre on the windward side and sunk 1.2 cm at a point 25 cm from the stem centre on the lee side. The root-soil plate curved very little even at F max (Fig. 4b), indicating the stiffness of the structure. The angle of the root-soil plate at F max was only 2°, about half that measured at the stem base; the discrepancy is accounted for by bending in the stem. Lateral movement was

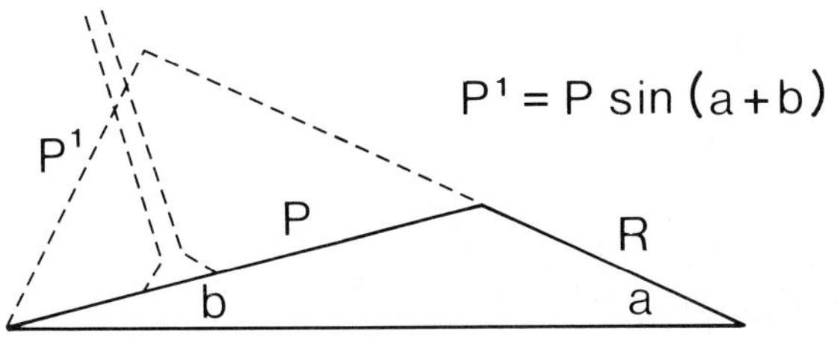

Fig. 5. When the tree is inclined to the left the root-soil plate (P) drags surface roots (R) which extend beyond the edge of the plate, upwards. The force applied to the stem then acts on the shorter lever arm P^1.

only 1–5 mm at F max, *i.e.* close to the value expected if the root-soil plate is considered to be the upwardly moving radius of a circle.

The amounts of movement associated with initial root breakage showed large between-tree variation. In some trees breakage occurred after an upward movement of 1–5 mm at 75 cm from the stem base on the windward side, associated with an angle of 0.5 to 1° at the stem base. In others a movement of up to 30 mm and an angle at the stem base of 1–2° was required; these had a smooth lower surface to the root-soil plate, indicating previous breakage by wind action and compression by repeated rocking. The relatively large amount of movement before roots began to break, when these rocking trees were pulled by the winch, indicates that many roots must have been broken previously by wind action.

Winching appeared to be a useful experimental technique provided that the most relevant measurements were made. The value of measuring the angle of the stem[15,17,31] is limited because the resistance produced by anchorage is confounded with stem stiffness, and even fastening the measuring device to the stem base did not entirely resolve this problem, whereas measurement of root-soil plate movement can be used in modelling.

The commonest objection to winching is that a static force is applied, whereas gusts of wind sway the tree and rock the root system. However, observation on damage to roots and soil did not distinguish between trees uprooted by wind or by winching, indicating that winching can give useful results. Wind action may reduce residual soil shear stress by abrasion of the root ball edge when the tree springs back between gusts, but preliminary experiments in which trees were uprooted by a series of pulls and relaxations on the winch, gave similar results to continuous pulling.

A problem in modelling is that roots of neighbouring trees are often interlocked. The effect of this can be large; one tree which was winched over uprooted four surrounding trees.

The acoustical method of recording root breakage is useful for detecting the onset of failure, and indicates when the elastic limit of the root-soil system has been exceeded. It also provides a vivid demonstration of the sequential nature of root breakage. As used in this study it is not a quantitative technique in that the

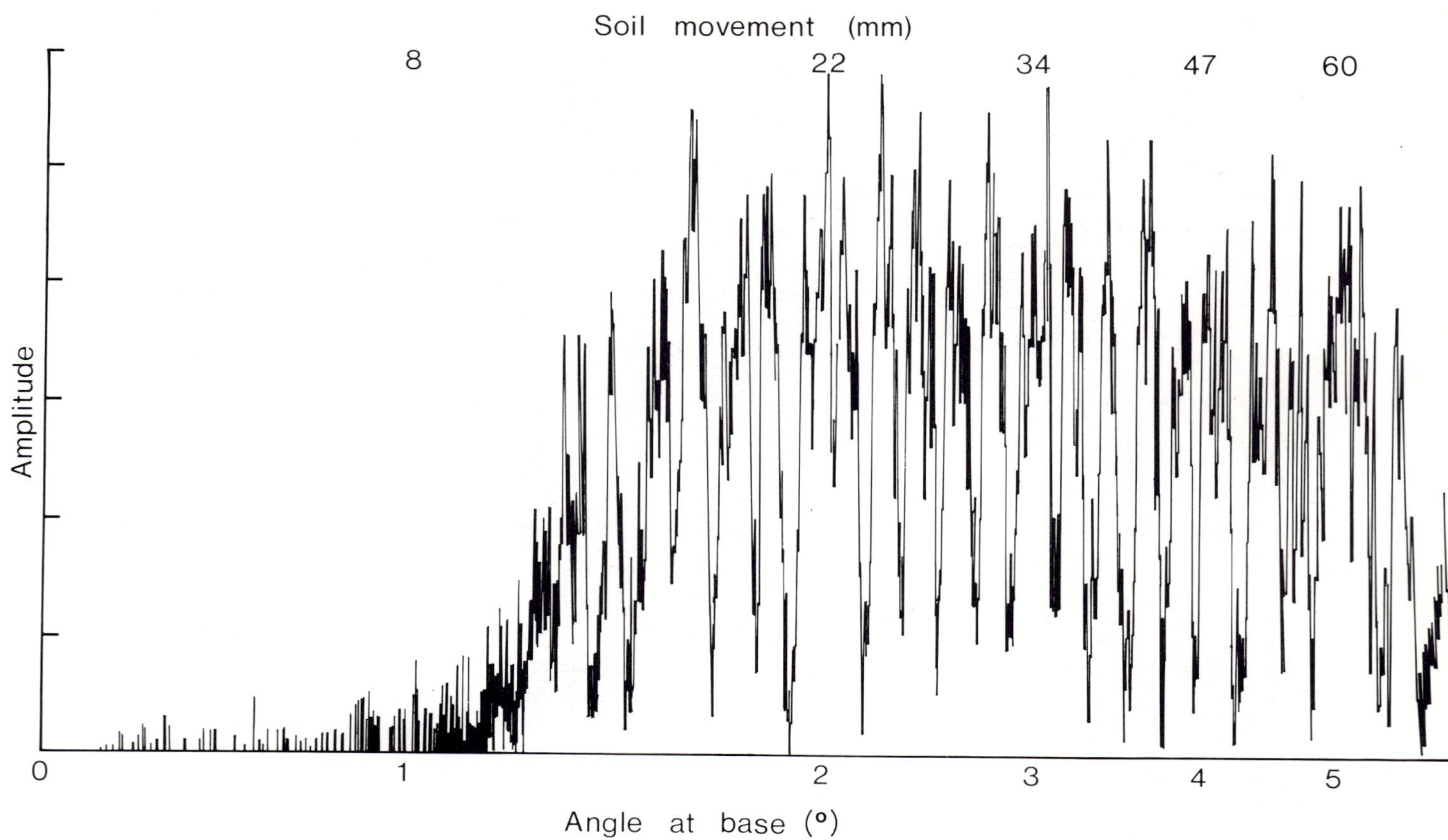

Fig. 6. Tracing from music analyser chart, on arbitrary scale of amplitude representing tape-recorded sounds registered by a microphone at a depth of 20 cm in the soil, 50 cm to the 'windward' side of a Sitka spruce tree being uprooted by winching.

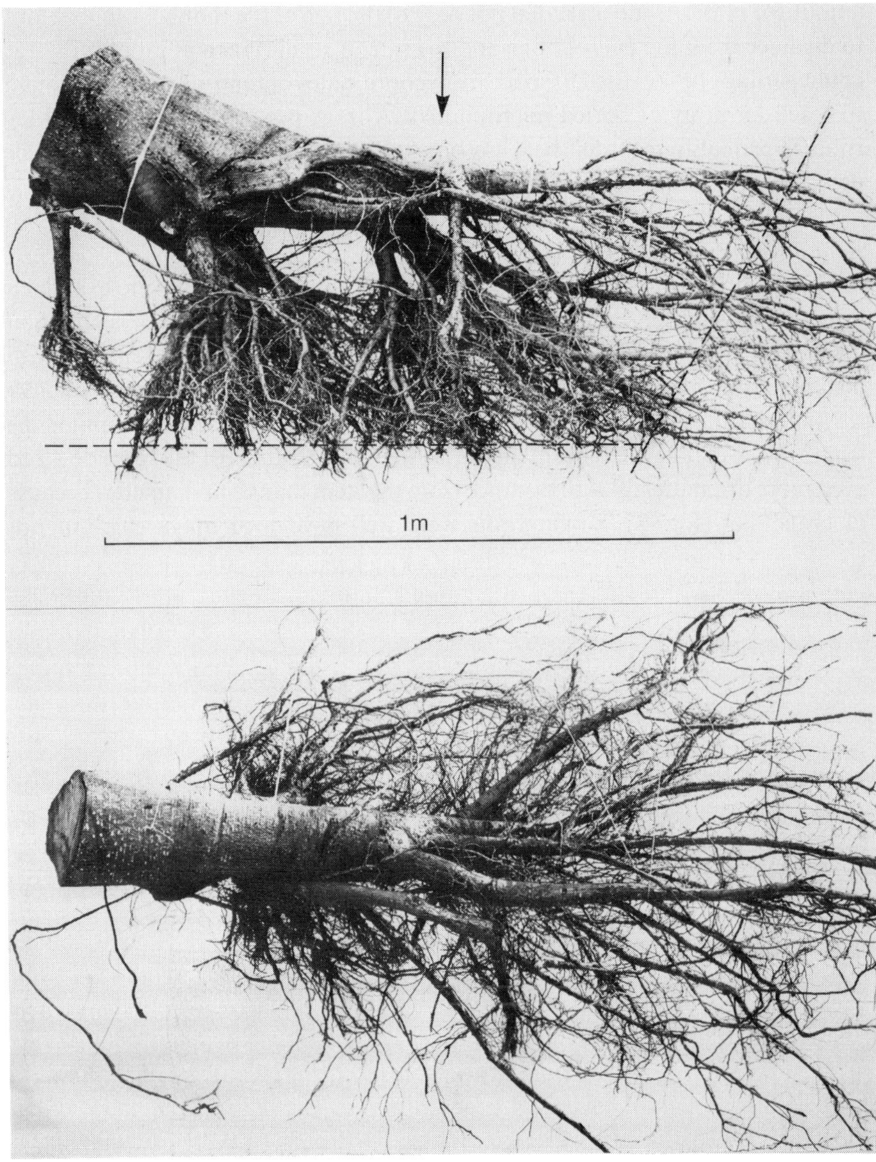

Fig. 7. A main lateral root from the windward side of an uprooted 34 year-old Sitka spruce tree on a peaty gley. *Above*, profile view. The dotted line marks the edge of the root-soil plate before soil was removed. The lateral roots mostly broke at a point just outside the right-hand edge of the photograph. The arrow indicates the characteristic point of bending for a root of this type when on the lee side. *Below*, plan view.

amplitude of the sound is related not only to the size of the root broken, but also to distance from the microphone and the rate of strain. Each point of breakage could perhaps be localised in space from coordinates of time taken for the sound to reach an array of buried microphones. A large proportion of the breaking roots is probably recorded because breakage of a 1 mm diameter root can be detected in soil at a distance of >1 m.

Root distribution and stiffness

The main framework of the Sitka spruce root systems which were examined typically consisted of large branched laterals bearing sinkers (Fig. 7). The total CSA of all the material in the lateral framework decreased exponentially with distance from the stem base (Fig. 8), as found for ring width in roots of *Acer rubrum* seedlings[42]. It was partly this steep exponential decline which limited the length of the cantilevered beam on the lee side. The stiffness of the roots declined even more dramatically with distance from the stem than their diameter, because of branching (Fig. 8). Sinker roots were well developed only near, but not

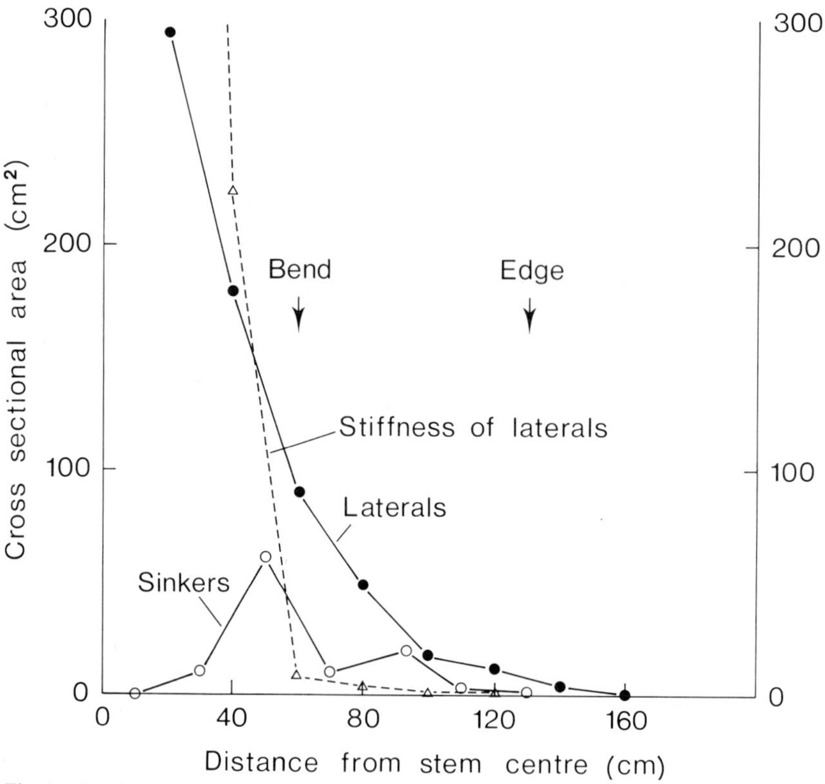

Fig. 8. Total cross-sectional area of the lateral (0–45°) and sinker (45–90°) roots on the portion of Sitka spruce root system shown in Fig. 10, and the calculated relative stiffness of the laterals. The arrows at 'edge' and 'bend' represent the edge of the root-soil plate on the windward side and the characteristic point of bending of such a root on the lee side, respectively.

immediately under, the stem base. They decreased in diameter beyond c 40 cm from the stem giving the erroneous visual impression that they were absent from the peripheral root system.

The root-soil plate

The shape of the root-soil plate was irregular in plan view but measurements of the 14 radii averaged for 32 root systems gave an eliptical shape (Fig. 9). The radius in line with the winch on the lee side of the tree is less than half as long as that to windward. Similar observations were made on windblown trees. The root-soil plate of an uprooted tree overlaps part of a similar area around adjacent trees, especially at the sides. Root-soil plates >40 cm deep of trees pulled away from the ploughed furrow were similar in shape to those pulled towards the furrow, but in shallower root systems the root-soil plate of trees pulled away from the furrow broke straight along the far edge of the furrow to give a more rectangular shape. In profile, root-soil plates usually taper toward the edges.

After uprooting, roots projected from the sheared edge of the root-soil plate and roots from adjacent trees projected from the soil in the crater, as predicted from theory. Most of the projecting main roots and laterals on them were broken but surface roots were sometimes pulled out with only minor breakage. Support on the windward side is supplied by a large number of roots of small diameter

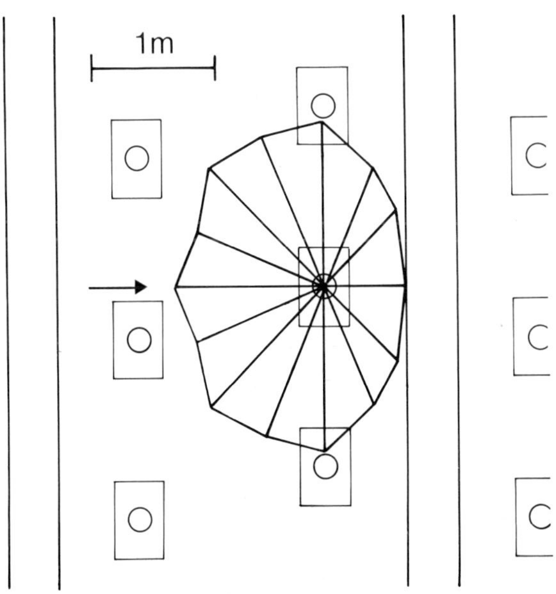

Fig. 9. Plan view of the Sitka spruce plantation showing mean shape of root-soil plates of trees pulled in the direction of the arrow. Note double rows of rectangular turves on which trees were planted between plough furrows.

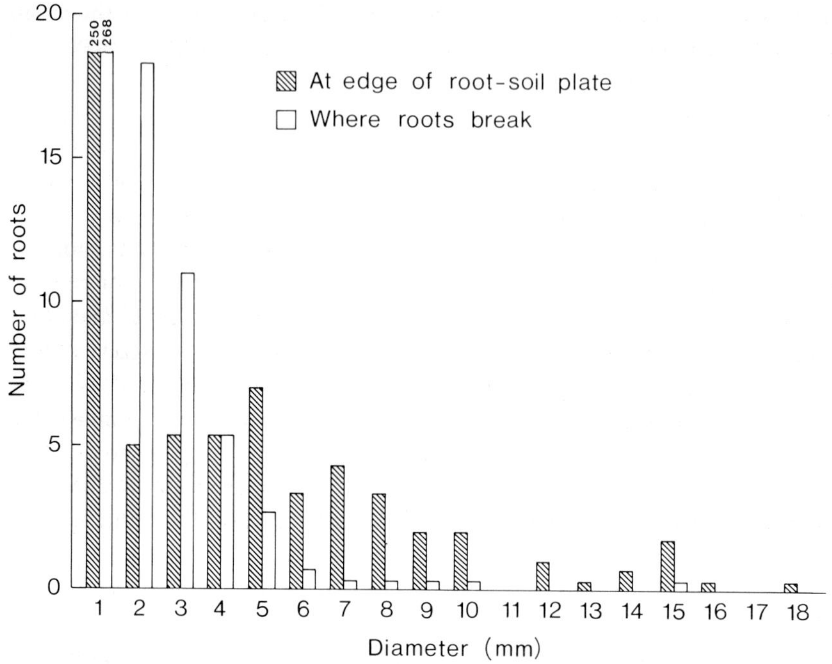

Fig. 10. Size distribution of roots projecting from the edge of the root-soil plate on the 'windward' side of a Sitka spruce tree uprooted by winching. Note the large number of small roots broken.

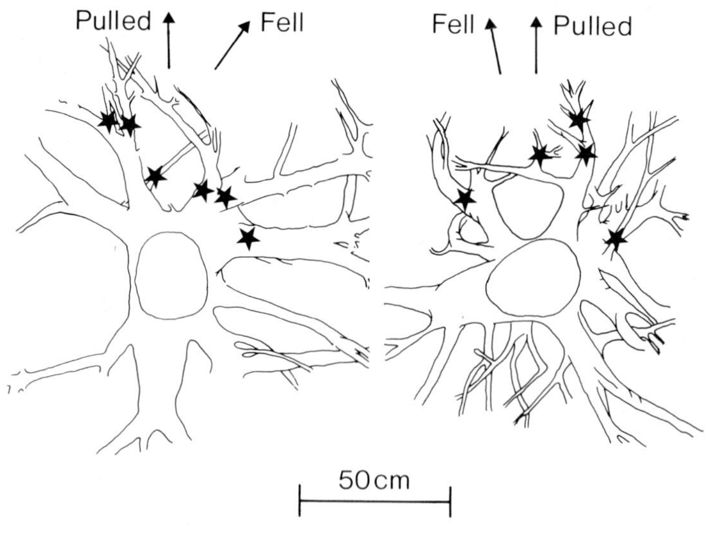

Fig. 11. Plan view of two Sitka spruce systems uprooted by winching, showing association between points of breakage (indicated by stars) and branching on 'lee' side. Note small number of relatively large roots broken.

(Fig. 10) and accords with the impression, gained from the tape recorder, of many small breakages. The number of points of breakage is larger than that shown in Fig. 10 because only data on main axes are presented.

In contrast to the windward side, damage on the lee side occurred to a few large roots near the stem base. Failure was largely at major points of branching (Fig. 11) as expected from theory. However, observations on damage were made only after trees had been uprooted and the soil removed, and Fig. 4c indicates that the angle of the root system at F max is too small to fracture roots under bending stress; indeed, loud sounds of breakage on the lee side, sometimes accompanied by suddenly decreased load on the winch, tended to occur after F max was exceeded. The contribution of roots on the lee side to stability concerns both their role in extending leverage to enhance the mechanical advantage of the root-soil system, and their resistance to bending at the hinge. Roots near the hinge and at right angles to the direction of windthrow are subjected to twisting forces, but torsion is probably unimportant under the small angular displacement at F max unless confined to the thick part of the root by peculiarities of morphology.

Components of anchorage

Contributions of roots and soil to stability in a Sitka spruce stand were estimated using values on soil strength from the literature and on root strength from measurement, and calculating the turning moment. The total force calculated to fracture the root-soil system was six times larger than the observed force for uprooting, although the contributions of torsion at the sides of the root-soil plate and bending at the hinge were ignored. The values for soil strength from the literature may have been too high, but even if these were reduced by an order of magnitude, the calculated force for uprooting still exceeded the observed.

The discrepancy between calculated and measured values reflects conclusions from modelling landslip[38], and highlights the importance of sequential breakage. The probable behaviour of the components of the root-soil system during displacement (Fig. 12) was deduced from mechanics and the physical properties of the materials. The turning moment of the root-soil plate mass follows a cosine curve and scarcely changes at the angles given. By contrast, the four other contributions to anchorage change markedly. Soils fails under tension soon after the system is displaced because of its low elasticity. If soil shear occurs at the edges of the root-soil plate, its residual component will assist anchorage throughout the uprooting process, but will decrease as the root-soil plate is levered clear of surrounding soil. Resistance of roots on the windward side takes longer to reach a maximum because greater movement is permitted by their higher elasticity and slippage. The lee side hinge has the rising curve characteristic of a cantilevered beam. The integrated curve of total resistance has the typical shape encountered in tree pulling studies[15,31]; but component curves

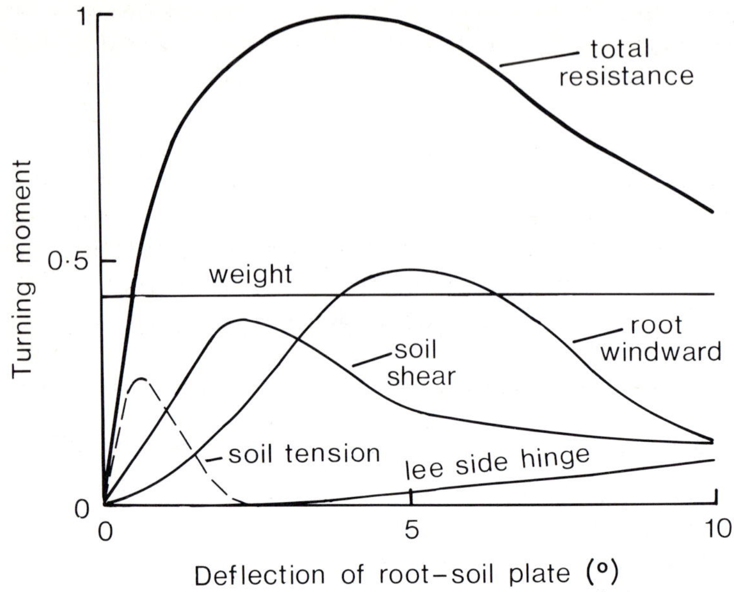

Fig. 12. Schematic representation of the contribution of components of anchorage to the total turning moment during uprooting.

will vary with the shape of the root-soil plate and the distribution, strength and elastic properties of the materials.

An important part of the integrated curve is where tensile failure occurs in the lower surface of the root-soil plate. This failure initiates rocking of the tree. Wind energy is then absorbed by damping in the root system rather than by bending of the stem[20]. F max occurs at a greater displacement of the root-soil plate and indicates the ultimate strength of the anchorage.

The observations reported indicate certain approaches to modelling root architecture by intimating the processes involved in anchorage. The limitations of a totally static model are evident, and attempts are being made to allow for the elastic and plastic resistances and sequential failure in different parts of the irregular root-soil system. The techniques for studying anchorage require further development and more data are needed on the strength and elastic-plastic properties of roots and soil, and on R_{rs}. It will be difficult to manipulate root growth to improve anchorage, and the first step is to determine the sensitivity of the system to any changes which might be attempted. The modelling approach outlined here may accomplish the latter, so long as the roots are considered in conjunction with soil properties.

Acknowledgements I thank Dr D. G. Pyatt and Dr D. C. F. Fayle for their useful discussion and help with the manuscript, Mr K. Clifford for help throughout the investigation and Margaret Brown for drawing the figures.

References

1 Ajaz A and Parry R H G 1974 An unconfined direct tension test for compacted clays. J. Test. Eval. 2, 163–172.

2 Armstrong W, Booth T C, Priestly P and Read D J 1976 The relationship between soil aeration, stability and growth of Sitka spruce (*Picea sitchensis* (Bong.) Carr) on upland peaty gleys. J. Appl. Ecol. 13, 585–591.

3 Bishop D M and Stevens M E 1964 Landslides on logged areas in southeast Alaska. U.S. Dept. Agr. For. Serv. Res. Pap. NOR-1.

4 Booth T C 1974 Silviculture and management of high-risk forests in Great Britain. Irish For. 31, 145–153.

5 Booth T C 1979 Wind flow over topography. For. Comm. Rept. For. Res. 1979, HMSO, London 19–20.

6 Booth T C and Mayhead G J 1972 Crop stability: tree pulling. For. Comm. Rept. For. Res. 1972, HMSO, London. 65–66.

7 Burdekin D A 1977 Gale damage to amenity trees. Arb. J. 3, pp 181–189.

8 Burdett A N 1979 Juvenile instability in planted pines. Irish For. 36, 36–47.

9 Burroughs E R and Thomas B R 1977 Declining root strength in Douglas fir after felling as a factor in slope stability. USDA For. Ser. Res. Pap. INT-190, 27p.

10 Cremer K W, Myers B J, van der Duys and Graig I E 1977 Silvicultural lessons from the 1974 windthrow in radiata pine plantations near Canberra. Aust. For. 40, 274–292.

11 Endo T and Tsuruta T 1969 On the effect of tree's roots upon the shearing strength of soil. Ann. Rep. For. Exp. Sta. Hokkaido 1968, 167–182.

12 Fraser A I 1962 The soil and roots as factors in tree stability. Forestry 35, 117–127.

13 Fraser A I 1963 Wind tunnel studies of the forces acting on the crowns of small trees. For. Comm. Rept. For. Res. 1962. HMSO, London. 178–183.

14 Fraser A I 1969 Tutkimuksia puiden latvuksen, rungon ja juuriston kehityksesta ohutturpeisilla soilla Britanniassa. Suo 20, 1–8.

15 Fraser A I and Gardiner J B H 1967 Rooting and stability in Sitka spruce. For. Comm. Bull. 40, HMSO, London. 28p.

16 Hathaway R L and Penny D 1975 Root strength in some Populus and Salix clones. N.Z. J. Bot. 13, 333–344.

17 Hintikka V 1972 Wind-induced root movements in forest trees. Comm. Inst. For. Fenniae 76, 56p.

18 Holtam B W 1971 Windblow of Scottish forests in January 1968. For. Comm. Bull. 45. HMSO Edinburgh, 53p.

19 MacKenzie R F 1976 Silviculture and management in relation to risk of windthow in Northern Ireland. Irish For. 33, 29–38.

20 Mayhead G J, Gardiner J B H and Durrant D W 1975 A report on the physical properties of conifers in relation to plantation stability. Int. Rept. U.K. For. Comm. Res. Dev. Div. 38p.

21 Miller K 1981 Topographic influences of wind structure. For. Comm. Rept. For. Res. 1981. HMSO, London. 23p.

22 O'Loughlin C L 1974 A study of tree root strength deterioration following clear felling. Can. J. For. Res. 4, 107–113.

23 O'Loughlin C L 1974 The effect of timber removal on the stability of forest soils. J. Hydrol. (N.Z.) 13, 121–134.

24 O'Loughlin C L and Watson A 1979 Root-wood strength deterioration in radiata pine after clear felling. N.Z. J. For. Sci. 9, 284–293.

25 O'Loughlin C L and Watson A 1981 Note on root-wood strength deterioration in *Nothofagus fusca* and *N. truncata* after clear felling. N.Z. J. For. Sci. 11, 183–185.

26 Parry R H G 1972 Stress-strain Behaviour of Soils. G T Foulis and Co Ltd. Henley-on-Thames. 752p.

27 Payne P C J and Fountaine E R 1952 A field method of measuring the shear strength of soils. J. Soil Sci. 3, 136–144.

28 Pyatt, D G 1973 Physical and mechanical properties of soil types. For. Comm. Rept. For. Res. 1973, HMSO, London. 62–66.

29 Pyatt D G and Booth T C 1973 Crop stability: tree pulling. For. Comm. Rept. For. Res. 1973, HMSO, London, 72–74.

30 Rogers W S and Parry M S 1968 Effects of deep planting on anchorage and performance of apple trees. J. Hortic. Sci. 43, 103–106.

31 Somerville A S 1979 Root anchorage and root morphology of *Pinus radiata* on a range of ripping treatments. N.Z. J. For. Sci. 9, 294–315.

32 Somerville A 1980 Wind stability: forest layout and silviculture. N.Z. J. For. Sci. 10, 476–501.

33 Swanston D N 1970 Mechanics of debris avalanching in shallow till soils in Southeast Alaska. USDA For. Ser. Res. Pap. PNW-103 17p.

34 Takahasi T 1968 Studies of the forest facilities to prevent landslides. Bull. Fac. of Agric., Shiznoka University, 18, 85–101.

35 Trafford B D 1974 A guide to drainage design technique based on scientific methods. Agr. Dev. Adv. Serv. Min. Agr. Fish Food, Tech. Bull. 74/7. 15p.

36 Thorne C R, Tovey N K and Bryant R 1980 Recording unconfined tension tester. Proc. Am. Soc. Civ. Eng. 106, 1269–1273.

37 Turmania V I 1965 On the strength of tree roots. Bjull. Mosk. Obsc. Ispty. Prir. (Otd. Biol.) 70, 36–45.

38 Waldron L J 1977 The shear resistance of root-permeated homogenous and stratified soil. Soil Sci. Soc. Am. J. 41, 843–849.

39 Waldron L J and Dakessian S 1981 Soil reinforcement by roots: calculations of increased soil shear resistance from root properties. Soil Sci. 132, 427–435.

40 Wrangler F 1976 The susceptibility of Norway spruce to windthrow as affected by site, stand type and stand heigth. Forst und Holz. 31, 220–222.

41 White R G, White M F and Mayhead G J 1976 Measurement of the motion of trees in two dimensions. Tech. Rept. Inst. Sound Vib. Res. Univ. Southampton 82, 22p.

42 Wilson B F 1975 Distribution of secondary thickening in tree root systems. *In* The Development and Function of Roots. Eds. J G Torrey and D T Clarkson. Academic Press,
43 London. 618p.
Wilson H H 1976 The effect of the gale of August 1975 on forests in Canterbury. N.Z. J. For. 21, 133–140.

44 Wu T H, McKinnell W P and Swanston D N 1979 Strength of tree roots and landslides on Prince of Wales Island, Alaska. Can. Geotech. J. 16, 19–33.

45 Zobel B 1961 Inheritance of wood properties in conifers. Silv. Genet. 10, 65–70.

Plant and Soil 71, 189–195 (1983). A-19

Modelling root structure and stability

J. D. DEANS and E. D. FORD
Institute of Terrestrial Ecology, Bush Estate, Penicuik, Midlothian, EH26 OQB, Scotland

Key words Roots Root system modelling Sitka spruce Windthrow

Summary Seven fully excavated 16 year-old root systems of Sitka spruce were analysed. All roots in excess of 1 cm diameter at their origin on the stumps were analysed, data being collected until root diameter declined to less than 0.5 cm.

Root morphology and distribution was identified as a balance between systematic biological mechanisms and their disruption by environmental factors, particularly changes of soil density and soil surface contours. The biological mechanisms have been modelled and the model is capable of simulating root systems in response to a few simple input variables *e.g.* the number of roots originating at stumps, stem ratius, total number of roots of all orders *etc.*

Additonally equations have been developed to estimate the distribution of root diameter, and root weight at all distances from tree stems and a similar equation permits the estimation of tree diameter at chosen heights. These latter estimates being utilized to calculate the turning moment of stems when blown by the wind.

The influence of the wind on turning moment is explored for simulated root systems of differing strength and gross morphology.

Introduction

Among the densely populated areas of the world the British Isles have one of the windiest climates[6]. Winter gales with mean hourly wind speeds of *c.* 20 m s^{-1} gusting up to *c.* 30 m s^{-1} cause damage almost every year in British forests[2]. Currently at least 50% of the forests around the Scottish/English border could be damaged by gales of this strenght. Several workers attempted to predict the tree height at which forests will be damaged by wind using static tree pulling techniques to estimate the turning moment[1], or by examining the topographic exposure of sites[10]. Although estimates made using these techniques have predicted wind damage reliably at some sites, they are insufficiently reliable for general application[10].

The principal limitation of the tree pulling technique is that it is applied statically, whereas in reality trees sway dynamically, furthermore in the interpretation of results scant attention has been given to the detailed architecture of the root systems. In this paper we illustrate the use of a path reconstruction model to produce simulated root systems and examine some aspects of the resistance such root systems offer to overturning when trees are swayed by the wind.

Methods

The root systems of 7 16 year-old Sitka spruce trees (*Picea sitchensis* (Bong.) Carr.) were excavated in a forest near Moffat in southern Scotland. While some root measurements were taken on site[3,4,7], the intact root systems were transported to the laboratory to facilitate more accurate and detailed analysis[7,8]. Three dimensional co-ordinates of root origins, branching points and bends were recorded on 4 root systems, while branching angles, angles of descent, azimuths of root origins, numbers of roots, branching points and bends and root lengths were recorded on all 7 root systems. Measurements were taken on all roots exceeding 5 mm in diameter; thinner roots were insufficiently rigid to maintain their *in vivo* positions after excavation.

Results of root system analysis

From detailed analyses of the root systems[7,8], it was clear that their geometry was more systematic than at first apparent. First order roots tended to be regularly spaced around stems, and all had small angles of descent into the soil. Also, lateral branches did not occur randomly; conditional upon the number of lateral roots initiated, their origins were regularly spaced along their parent roots. Fig. 1 illustrates the sites of lateral root origins for parent roots with 3 lateral branches, their origins are clustered and separated by bands lacking lateral root origins.

Lateral roots subtended large angles to parent roots thereby exploiting different soil regions, whereas forks which frequently result from root tip injury[11], subtended small angles to the truncated parent roots and maintained azimuths similar to that of the parent roots. Where roots changed direction at bends, they did so in a systematic manner such that they tended to keep to their original compass directions. That is, sequences of alternately clockwise and anticlockwise azimuth changes which maintained the overall azimuth of the roots similar to that at their stem origins occurred too frequently to be explained by chance.

However, these systematic processes were disrupted by changes in the soil environment. In particular, angles of descent became more horizontal where roots encountered denser substrates and frequently roots bifurcated into forks or changed azimuth abruptly where they encountered obstructions such as stones or impenetrable layers or where they protruded into ditches or above soil level on steep slopes. Consequently, forks and bends were distributed more randomly than lateral roots.

Root system modelling

From the 7 excavated root systems, Henderson[7,9] developed a computer simulation model utilising the method of path reconstruction *i.e.* having estimated the positions and azimuths of first order roots at the base of the stem, the model simulates entire root systems by estimating first the total length of each root, secondly the number of bends and branching points distributed along its

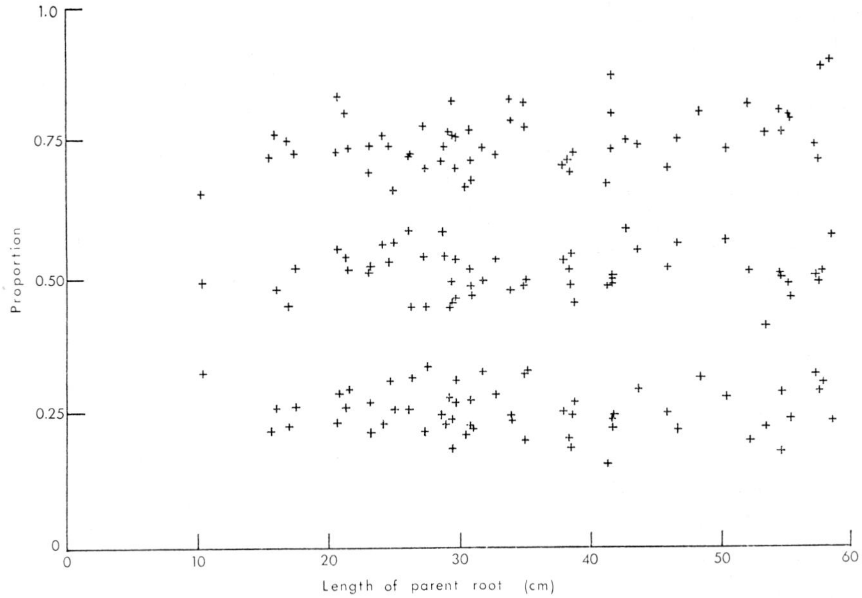

Fig. 1. Distribution of lateral root origins along roots of Sitka spruce trees for parent roots with three lateral roots only. Locations of origins are expressed as proportions of parent root length.

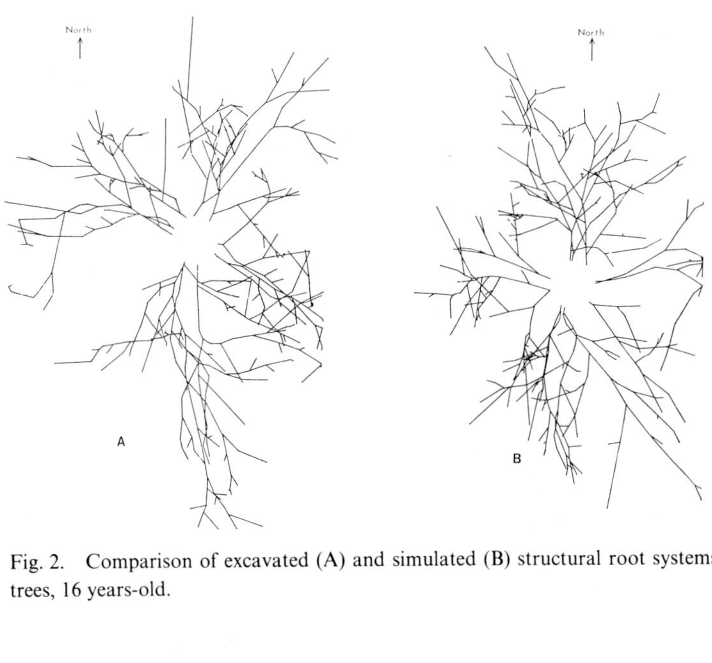

Fig. 2. Comparison of excavated (A) and simulated (B) structural root systems of Sitka spruce trees, 16 years-old.

length, and lastly the direction changes at bends. Figs. 2A and B present plan views of excavated and simulated root systems respectively. Clearly, the simulated root system is realistic in general appearance and bears many similarities with excavated roots. Various simulations were produced by altering the values of the input variables, *e.g.* tree diameter, number of roots originating on stems, total number of roots, overall dimensions of the root system *etc.* Sub routines were included to simulate root behaviour at the soil surface, ditches and indurated layers to ensure that root development proceeded within the soil. Output from the program included a list of 3 dimensional co-ordinates of root origins, branching points and bends, and the length of root between these points was calculated separately.

Mid-point diameters and weights of separate root lengths were then estimated. A root taper curve (polynomial of 4th degree) developed from the 7 excavated root systems provided accurate estimates of root diameters at all distances from stems while a 2nd degree polynomial provided good estimates of root weight cm^{-1} of root length for roots of different diameter. These latter estimates were necessary to estimate some of the resistances to uprooting when trees are swayed by the wind.

Modelling the response of crown and stem to wind

Wind forces acting on trees may be assumed to act at the centre of pressure of the stem/crown unit. The above ground height of the centre of pressure was calculated using published data from this site[5], while stem diameter at that height was estimated by a polynomial of 4th degree calculated from height/diameter measurements from 20 trees. A theoretical tree, typical of those being windthrown at this site was selected for study. It was 15 cm diameter at a height of 1.3 m with centre of pressure at 10 m. The effects of wind force on a tree of this size were then modelled for an unidirectional wind swaying the tree in the same direction. The model works by continuously redistributing energy between wind, momentum in the moving crown and spring energy stored in the stem at time intervals of 0.1s. Fig. 3 illustrates the calculated dynamic response of a stem, and drag on the tree crown for successive gusts of wind varying in speed and duration,the natural sway period of the tree being set at 2s. The centre of pressure of the tree is progressively displaced from vertical (0) in the direction of the wind (+ve values) as wind speed increases. As wind speed declines, the centre of pressure progressively returns towards 0 under the influence of spring energy stored in the stem. Thus the stem is deflected in the direction of the wind until stored spring energy exceeds applied wind energy, at this point the stem accelerates towards and then passes through 0 against the direction of the wind, the centre of pressure being displaced to the windward side of the tree (−ve values). The number and extent of stem sways of this type is dependent on the wind speed and duration of gusts causing the initial displacement of the stem, and current wind energy.

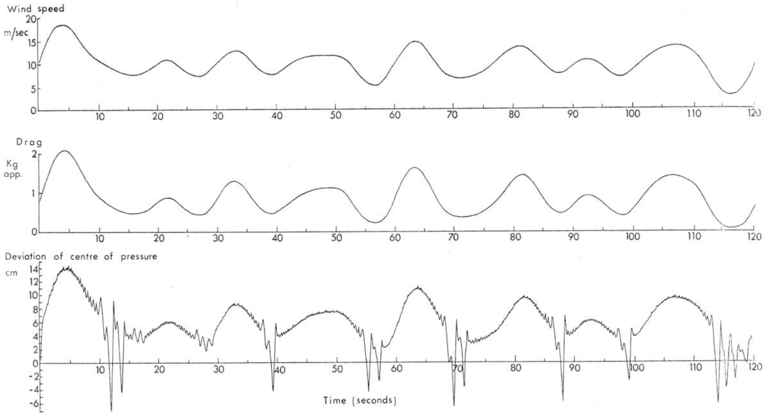

Fig. 3. Gusts of wind, the drag created by these gusts on a Sitka Spruce tree of 15 cm dbh and centre of pressure located 10 m above ground level, and deviations of the tree's centre of pressure from vertical (0) caused by drag. Positive deviations occur when the stem moves in the same direction as the wind, negative deviations occur when the stem swings back through 0 travelling in the opposite direction to the wind.

It is the capacity of the stem to store energy which buffers it's movement in relation to the 'movement' of wind and this is controlled by properties of the stem and wood, *e.g.* size of stem, taper and modulus of elasticity. *E.g.* if diameter at 1.3 m is increased from 15 cm to 18.1 cm in a tree of the same height, as could be expected for an open grown, rather than a stand grown tree, then deviations of the centre of pressure given in Fig. 3 are reduced by *c.*50%.

Interaction between stem sway and root resistance

Table 1 lists values of turning moment for a wind speed of 30 m s^{-1} acting on a stem of 15 cm dbh and centre of pressure 10 m above ground level and where the fulcrum point around which the tree moves is set at different distances from the stem centre. In these conditions the centre of pressure is displaced *c.* 30.5 cm leeward. Where the fulcrum is close to stem most of the crown and a substantial portion of the slender upper stem are displaced to leeward of the fulcrum and

Table 1. Turning moments resulting from a wind speed of 30 m s^{-1} acting on a tree of 15 cm dbh for fulcrums at different distances from the stem centre

	Distance from stem to fulcrum (cm)						
	10	15	20	25	30	35	40
Turning moments (kgf. m)							
Stem	−5.0	−9.3	−13.5	−17.7	−21.9	−26.1	−30.3
Crown and Stem	4.5	0.2	−4.0	−8.2	−12.4	−16.6	−20.8

Table 2. Root resistance to compression (kN) at different distances from the stem centre

	Distance from stem centre (cm)								
	10	15	20	25	30	35	40	45	50
Windward	1.67	223.4	140.6	0	46.8	0	0	0	0
Leeward	265.4	277.7	271.1	204.8	290.3	162.5	662.1	267.5	167.3

therefore apply +ve turning moments (Table 1) which exceed the −ve resistive moment applied by the remainder of the stem and crown. The net effect being a +ve turning moment acting to overturn the stem. Where fulcrums occur at distances >*c*. 20 cm from stems a larger proportion of the stem and crown is located windward of the fulcrum, the net effect being −ve turning moments which resist overturning. Clearly if fulcrums can be set at *c*. 20 cm from stem centres, stems are stable at this windspeed without additional support. We have examined the distribution of resistance to compression that root systems offer, an attribute of roots which could be important in forcing fulcrums out from stems. In this example, the simulated root system comprises 8 horizontal roots with root diameters at origin in the range 1.5–16 cm. Table 2 lists values of compressive resistance necessary to fracture these roots at different distances from stems. To leeward resistances are large although rather irregularly distributed with roots located 40 cm from tree centres combining to from the greatest total resistance. To windward in this example the distance over which roots are distributed is less, similarly the resistance to fracture offered by these roots is generally less with zero resistance occurring 25 cm from the centre of the stem and only 1.67 k Newtons of resistance being offered at 10 cm. At this level of resistance roots would probably fracture during normal winter gales. This result is typical where roots are *unevenly* distributed as frequently found and where thin roots arise directly from stems. In calculating these results roots descending into soil at angles >45° were assumed to have no resistance to compression.

Table 3 illustrates turning moments for the roots in this example assuming

Table 3. Turning moments (kgf. m) caused by the root system of a tree 15 cm dbh where the fulcrum occurs at different distances from the stem centre

	Distance from stem centre to fulcrum (cm)						
	10	15	20	25	30	40	50
Fulcrum to Leeward	6.2	7.9	9.7	11.8	13.8	18.5	23.6
Fulcrum to Windward	12.8	15.0	17.3	19.7	22.1	27.7	33.4

differing fulcrum points. Although these TMs increase with distance from the fulcrum, an additional advantage of moving fulcrums out from stems is that roots to leeward of the stem but windward of the fulcrum, provide an additional turning moment and since this would also apply to the soil mass, the effect of moving fulcrums out from stems is to considerably reduce the risk of overturning. Work is progressing towards characterising the gross morphology of root systems which offer optimum resistance to overturning. Consideration of the soil resistance to compression will be included.

References

1 Booth T C and Mayhead G J 1972 Rep. Forest Res. Lond. 66 p.

2 Booth T C 1974 Silviculture and management of high-risk forests in Great Britain. Ir. For. 31, 145–153.

3 Deans J D 1981 Dynamics of coarse root production in a young plantation of *Picea sitchensis*. Forestry 54, 139–155.

4 Deans J D 1983 Distribution of thick roots in a Sitka spruce plantation, 16 years after planting. Scott. For. (*In press*).

5 Ford E D 1982 High productivity in a polestage Sitka spruce stand and it's relation to canopy structure. Forestry 55, 1–17.

6 Gloyne R W 1968 Supplement to Forestry 'Wind effects on the Forest' 57 p.

7 Henderson R 1981 The structural root systems of Sitka spruce and related stochastic processes. Unpublished Ph. D. Thesis, Dept. of Statistics, University of Edinburgh.

8 Henderson R, Ford E D, Renshaw E and Deans J D 1983 Morphology of the structural root system of Sitka spruce. I. Analysis and quantitative description. *In preparation.*

9 Henderson R, Ford E D and Renshaw E 1983 Morphology of the structural root system of Sitka spruce. II. Computer simulation of rooting patterns.

10 Miller K F 1980 Rep. Forest Res. Edin. 21 p.

11 Wilson B F 1970 Evidence for injury as a cause of tree root branching. Can. J. Bot. 48, 1497–1498.

Plant and Soil 71, 197–209 (1983). A-20

The physiology of vesicular-arbuscular mycorrhizal roots

V. GIANINAZZI-PEARSON and S. GIANINAZZI
Station d'Amélioration des Plantes, INRA, BV 1540, F-21034 Dijon-Cedex, France

Key words Carbohydrate Disease resistance Hormones Micronutrients Phosphate Physiology VA mycorrhizae Water uptake

Summary The fungi of vesicular-arbuscular mycorrhizae colonize considerable portions of the root system and in spite of the carbon drain they impose on the host plant, their presence within the root tissues can positively influence several aspects of the host plant's physiology. In the majority of cases, improved phosphate uptake is the primary cause of growth and yield enhancements in the mycorrhizal plants. Mycorrhizal roots have different phosphate absorption kinetics and lower threshold values than nonmycorrhizal roots. The external hyphae developing around mycorrhizae explore a large volume of soil and absorb available phosphate beyond the depletion zone at the root surface. Phosphate accumulating in the external fungal hyphae is translocated to the internal mycelium by a well-developed transport system and transferred to the host tissues mainly across the intracellular arbuscules. Certain specialized enzyme activities are specifically associated with this alternative pathway of phosphate nutrition in mycorrhizal plants. Improved phosphate nutrition is not always sufficient to explain the observed effects of vesicular-arbuscular mycorrhizae on the host plant's physiology.

Introduction

The underground organs of a very large number of plants growing under natural conditions do not exist simply as roots but as complex vesicular-arbuscular (VA) endomycorrhizal associations. In these the mycorrhizal fungi invade the host cells and the host root tolerates and maintains the presence of what can be a large volume of fungal tissue. In spite of the repercussions that this must have on root metabolism together with the carbohydrate drain on the plant, the formation of VA mycorrhizae can positively influence several aspects of plant physiology: phosphate nutrition, trace element and water uptake, hormone production, nitrogen fixation and resistance to root disease.

VA endomycorrhizae form an integral part of many agricultural, horticultural and tree crops and certain plants, notably some tree species, appear to have an obligate physiological requirement for these associations[35,54]. However, considerable gaps still exist in our knowledge concerning the biology and the physiology of VA mycorrhizal systems. The present article summarizes the progress that has been made in recent years towards understanding the physiology of VA mycorrhizae and underlines the necessity for further research if benefits are to be obtained from the management of these symbiotic associations.

Carbohydrate physiology

The carbohydrate physiology of VA mycorrhizae has not been studied in great detail in the past but reports of reduced growth and yield in mycorrhizal plants[11,20,46,65,95] are now stimulating interest in this field. There is clear evidence from experiments in which mycorrhizal plants are allowed to photosynthesize in presence of $^{14}CO_2$ that the carbon requirements of VA mycorrhizal fungi are supplied by the host plant[12,25,51]. The form of carbon and its mode of transfer from plant to fungus are as yet unknown but since the host cells always outlive the fungus in VA mycorrhizas, transfer must be biotrophic and take place across the living interface between the two organisms. Low irradiation, short day lengths and defoliation, which all decrease photosynthesis, reduce the spread of VA mycorrhizal fungi in roots[29,50] and the development of mycorrhizal infection has been correlated with the exudation of both sugars and organic acids[44,89]. It is possible therefore that not only sugar levels but also other major components, such as organic acids, may be important in controlling the development of VA mycorrhizal infections in roots.

VA mycorrhizal fungi rapidly convert the transferred host photosynthates into specific fungal carbon compounds which cannot be readily utilized by the host plant and of which the major part appear to be as lipids or glycogen[12,21,25,41,59]. This rapid conversion of carbon compounds will have the effect of maintaining a concentration gradient of photosynthate from host to fungus so that net movement continues in the direction of the heterotroph. The way in which carbon compounds are transported to other parts of the fungal mycelia developing in soil around mycorrhizal roots has not been studied but one possibility is that protoplasmic streaming (which can be bidirectional in VA fungal hyphae[72,79]) could carry glycogen particles or lipid globules down a carbohydrate concentration gradient towards the external hyphae.

The magnitude of the drain on photosynthates arising from the development of a VA fungal mycelium within roots and soil is difficult to estimate. Host-fungus competition for carbon has been suggested as a cause of growth depressions in VA mycorrhizal plants[16,96]. Mycorrhizae incorporate greater amounts of carbon derived from photosynthate than nonmycorrhizal roots [59], starch grains disappear from cells during active fungal development[41] and carbon loss through respiration can be considerably higher[68]. This may result partly from the carbon demand of the fungus itself (about 10% of the mycorrhizal biomass[104]) and partly from the increased metabolic activity of root cells following infection[24,41,60]. However, the mycorrhizal plant can compensate for this enhanced carbon demand of the infected roots by increased photosynthesis[3,4] and, for example, in legumes plant growth is greatly improved when both *Rhizobia* and VA mycorrhizal fungi invade root tissues in spite of the high carbon demand of both microbial partners[5,28,91,93]. Whilst the use of photosynthate by actively developing mycorrhizal fungi may adequately explain temporary growth reductions in young plants[16,20,91], this alone does not seem sufficient in the case of ageing infections and the physiological basis of prolonged negative mycorrhizal effects[11,46,95] need to be examined further.

Phosphate physiology

In the majority of cases, improved phosphate uptake is the primary cause of growth and yield enhancement in VA mycorrhizal plants and most physiological studies have therefore been focussed on this aspect. Recent research has begun to provide information concerning the absorption efficiency, partitioning and biochemical fate of phosphate in VA mycorrhizal systems.

Mycorrhizae can absorb several times more phosphate than uninfected roots from soils[40,66,83] and from solutions[14,45,64]. Differences in phosphate absorption are most striking when soluble phosphate in the rooting medium becomes limiting for growth of the plant species involved, and Mosse[65] suggested that mycorrhizal roots can even absorb phosphate from soils that are so phosphate deficient that none is taken up by nonmycorrhizal roots. This greater phosphate absorption by VA mycorrhizae arises because of a superior efficiency of uptake from labile forms of soil phosphate and is not attributable to a capacity to mobilize phosphate sources unavailable to nonmycorrhizal roots[40,66,71,74]. Under certain conditions the mycorrhizae may, however, be able to absorb from fixed phosphate sources[99] or stimulate root phytase activities[3,40]. Under normal conditions where phosphate concentrations in the soil solution are low ($10^{-5}M$ to $10^{-6}M$) mycorrhizal roots have considerably greater phosphate inflow rates (moles P absorbed per unit root length per unit time)[84,85,92,94] and apparently possess a pathway of phosphate uptake with a much higher affinity for phosphate[27] than nomycorrhizal roots.

The most plausible explanation for this greater uptake efficiency and that for which experimental evidence is most convincing is that the fungal hyphae extending out from VA mycorrhizae are able to absorb and transport phosphate from the soil to host root. This hyphal pathway of phosphate supply has been repeatedly confirmed[22,49,70] and phosphate translocation by VA mycorrhizal fungi has been measured over distances up to 8 cm[76]. The absorbing surface offered by the ramifying external mycelia around mycorrhizal roots undoubtedly aids the plant to exploit phosphate in the soil beyond the depletion zone at the root surface[67]. The increase in surface area external hyphae represent for a mycorrhizal root is extremely difficult to measure for obvious reasons of fragility and inacessibility of soil mycelia but the values obtained of 0.8 to 1.34 m of hyphae per cm of infected root[85,100], and which must be considered as underestimations, would virtually double the potential absorbing surface of roots having a diameter of 0.5 to 1 mm. The role of the external mycelium depends, however, not only on its size and rate of spread but also on its uptake and translocation capacities.

Unfortunately very little is known about phosphate absorption by VA mycorrhizal fungi apart from the fact that it is temperature sensitive[14] and that it must occur against a concentration gradient in soil. There is no information concerning, for example, the uptake efficiency of different fungi, or the distribution and affinity of phosphate absorption sites along hyphae.

More attention has been given to the processes involved in the subsequent transport of phosphate to the host root by the fungi. Studies of P fluxes in hyphae of *Glomus mosseae* have shown that translocation of phosphate, measured either directly in sterile dual cultures[23,70] or derived theoretically in pot experiments[84], occurs at high rates ($0.1–3.8 \times 10^{-9}$ mol cm^{-2} sec^{-1}) and that it is driven by metabolically-dependent active processes[23] which probably involve protoplasmic streaming and to a lesser extent bulk flow of hyphal contents. However, translocation studies have up to now been limited to *G. mosseae* and they need to be extended to other VA mycorrhizal fungi in order to assess the relationship between translocation capacities and observed variations in VA mycorrhizal efficiency[19]. Accumulation and translocation of important amounts of phosphate by hyphae would be expected to cause serious disturbances in normal cell processes and there is increasing evidence that VA mycorrhizal fungi overcome this problem by sequestering phosphate in their vacuoles in the form of polyphosphate, seen as dense granules in the electron microscope[25,58,97,98,106]. When phosphate is supplied to phosphate-starved mycorrhizal plants, polyphosphate rapidly accumulates in the external hyphae then in the internal mycelium of infected roots[69]. As much as 40% of fungal P may be in the form of polyphosphate within hyphae[17] and ionic stabilization of the polyphosphate-rich fungal vacuoles appears to be maintained by the presence of calcium and other cations[98,106]. Flux values for P translocation calculated from the amount of P in the form of vacuolar polyphosphate and observed rates of protoplasmic streaming are in close agreement with those measured experimentally[26]. In the light of all these observations the most acceptable hypothesis to explain rapid phosphate translocation in VA mycorrhizal fungi is still that of Tinker[103] who proposed that the transport mechanism is cyclosis, plus bulk flow, with loading und unloading of polyphosphate into vacuoles as the method of varying the P_i concentration of the streaming protoplasm. Biochemical studies on VA mycorrhizae and their fungi have shown that they are equipped with enzymes necessary for polyphosphate synthesis and degradation[18] and that alkaline phosphatase is, as in other fungi storing polyphosphate[31,61], characteristically associated with the fungal vacuoles[32,34,42]. Although the rôle of these enzyme activities in the biochemical control of phosphate metabolism in VA mycorrhizae is still a matter of speculation[36], some interesting observations can be made on their activities in relation to the phosphate nutrition of mycorrhizal plants.

The accumulation of polyphosphate in VA mycorrhizal roots transferred from phosphate deficient conditions to a sufficient phosphate supply is accompanied by large increases in polyphosphate kinase activity[18] suggesting that this enzyme, contrary to that in yeast cells[48], is inducible in VA mycorrhizal systems. The necessity for such enzyme induction to prime the phosphate transport system in the fungal hyphae could explain the initial lag period in phosphate transport observed in translocation experiments[22,23]. The alkaline phosphatase activity of VA mycorrhizae, which can be easily detected in root extracts[38,39,42] or in fungal extracts of mycelium released from roots by enzymic digestion (Gianinazzi-Pearson, Gianinazzi and Callow, unpublished results), is closely linked to early infection development and plant growth stimulation. Moreover, in plants growing at levels of phosphate fertilization which considerably diminish the mycorrhizal growth effect without significantly influencing infection, there is an important decrease in the alkaline phosphatase activities associated with the VA mycorrhizal fungi (Table 1).

The phosphate absorbed and accumulated by the external hyphae of VA mycorrhizal fungi is translocated to the internal mycelium and transferred to the root tissues, but the cause and the mechanisms controlling the release of phosphate by the fungi still have to be determined. Polyphosphatase activity is only associated with the internal mycelium[18] and polyphosphate granules disappear from vacuoles in the fine arbuscule branches[25,98] which suggests that unloading of polyphosphate from VA mycorrhizal fungi occurs within host cells;

Table 1. Alkaline phosphatase activities, mycorrhizal infection and growth effects of VA mycorrhizal fungi in roots of 50-day old soybeans growing in soil with and without phosphate fertilization

VA mycorrhizal fungus	E_3 (*Glomus* sp.)		*Glomus mosseae*	
KH_2PO_4 added to soil* (g.kg^{-1})	0	0.25	0	0.25
Relative activity (%) of fungal alkaline phosphatase detected in 0.2 g of root[39]	100_a**	38.5_b	100_a	36.5_b
Mycorrhizal infection:				
%root length	92.1_a	82.8_a	68.8_b	63.8_b
% arbuscule frequency	45.1_a	36.4_a	38.9_a	31.1_a
Shoot f. wt. (g. plant^{-1}):				
Mycorrhizal plant	13.6_a	18.3_b	13.9_a	18.1_b
Nonmycorrhizal plant	6.0_c	16.0_{ab}	5.8_c	17.0_b
Root f. wt. (g. plant^{-1}):				
Mycorrhizal plant	10.8_a	13.5_a	11.7_a	12.9_a
Nonmycorrhizal plant	7.0_b	12.4_a	6.9_b	12.4_a

* Clay loam, pH 7.2 (H_2O); 56 ppm available P (Olsen)

** Values with different subscripts are statistically different ($P = 0.05$)

what becomes of the associated cations in not known. Unloading of polyphosphate could be a rate-limiting step in the phosphate nutrition of mycorrhizal plants and its retention in the fungal tissue (storage rôle?) may contribute to the greater P concentrations sometimes found in roots of mycorrhizal plants[22,40,92,94]. It is highly probable that the preferential site for phosphate transfer from fungus to plant is across the living interface in the intracellular arbuscules[13,79,103]. There is evidence that phosphate released by the fungus into the interfacial matrix is actively transported across the host plasmalemma[24] and ATPase activities bound to this membrane are specifically concentrated around the fine, living arbuscule hyphae[60].

It is certain that high levels of soluble phosphate in soil decrease mycorrhizal infection but apparently it is the resulting high P concentrations in the root tissues that regulates infection development[8,52,63,82]. The physiological basis of this effect is not known although changes in plant membrane integrity have been proposed[44,75]. The viability of VA mycorrhizal fungi is little affected by high soil P[52,65] but effects of high soil and plant P on the processes of phosphate absorption, translocation and transfer to the host plant by the fungi have not been investigated. The observations discussed previously concerning fungal alkaline phosphatase activities (Table 1) suggest that increasing phosphate fertilisation may well act on mycorrhizal function before reaching sufficient levels to impede fungal development within roots. Furthermore, P tolerant mycorrhizal fungi exist which both infect and enhance plant growth a high levels of phosphate fertilization[36,73]. Further work is clearly needed in this field since the magnitude of the VA mycorrhizal effect in a given host-fungus combination will depend on the efficiency of each of the processes involved in the fungal pathway of phosphate nutrition.

Other physiological aspects

VA mycorrhizal infections can also increase the uptake by plants of soil nutrients other than phosphate. Mycorrhizal inoculation frequently increases foliar concentrations of zinc and copper[33,53,56] and deficiency of these micronutrients in fruit trees growing in nursery beds has been related to the absence of mycorrhizae following fumigation or to reduced infection due to high phosphate fertilization[43,57,101,102]. It is not known how VA mycorrhizae improve copper nutrition but studies on zinc uptake have shown that absorption rates of mycorrhizae are greater than those of nonmycorrhizal roots and that VA fungal hyphae can translocate small quantities of zinc to host roots[15,22]. Direct mycorrhizal effects on mineral nutrition may well be limited to those nutrient ions which have a poor mobility and are present in low concentrations in soil solutions (*e.g.* P, Zn, Cu). The higher mobility of ions such as potassium, nitrate and sulphate in soil solutions[67] makes it unlikely that depletion zones would form around plant roots and that these nutrients would move more rapidly through hyphae than through soil. For example, VA mycorrhizal infections can increase the sulphate absorbing power of roots but this appears to be a secondary effect brought about by improved P nutrition[78], and although VA mycorrhizal fungi can translocate sulphate[22,77], fluxes are small suggesting that hyphal translocation does not contribute greatly to sulphur nutrition in VA mycorrhizal plants.

There are several reports of VA mycorrhizal plants being less sensitive to temporary periods of water stress than nonmycorrhizal plants and that the mycorrhizae may play a rôle in the water economy of plants[37,47,62,80,81,90]. Higher hydraulic conductivities (or lower resistance) per unit length of mycorrhizal root may be the major contributing factor to this better use of soil water[47,81]. It seems unlikely that hyphal transport of water could sufficiently account for this[104] and the improved P nutrition of the mycorrhizal plants appears a more plausible explanation. Some evidence exists to support this hypothesis: firstly, drought tolerance in plants is related to their P status[6] and, secondly, important additions of nutrients to nonmycorrhizal plants can essentially eliminate the differences in resistance to water transport whilst the application of fungitoxicant 48 hours before measuring resistances has no effect on mycorrhizal plants[81]. Some authors have also proposed a mycorrhizal effect on root hydraulic conductivity and on stomatal regulation resulting from changes in phytohormone levels[4,47].

Other effects of VA mycorrhizae on photosynthesis and host morphology[3,4,30,47,55,105], for example, could be hormonal. VA mycorrhiza formation can increase the level of cytokinin in the host plant and change the level of abscissic acid and giberellin-like substances[1,2]. To what extent altered hormone production in mycorrhizal plants is due to an improved nutrient status still remains unclear. One VA fungus has been shown to synthesize phytohormones[9] but whether these can pass into the plant and affect its growth and physiology, or whether they are important in the infection process has yet to be determined.

Increased nodulation and nitrogen fixation in VA mycorrhizal legumes essentially results from the plant's improved P nutrition[93,94]. However, some special relationship may exist between the mycorrhizae and nodules since excessively high levels of phosphate fertilization are necessary to eliminate the mycorrhizal effect on nitrogen fixation[5]. Again it has been suggested that hormones may contribute to the interactions within dual symbiosis in legumes[7,9].

Whilst VA mycorrhizal infections can favour the colonization of roots by other symbiotic microorganisms, they often reduce the susceptibility, or increase the tolerance, of roots to soil-borne pathogenes like nematodes[87], *Phytophthora* sp., *Thielaviopsis basicola* and *Fusarium oxysporium*[10,86]. Increased resistance to fungal pathogen attack is often due to reduced host damage and less frequently to decreased pathogen infection; in the case of nematodes, although root penetration does not seem to be inhibited, subsequent development within mycorrhizal roots is restricted[87]. The reasons for increased resistance or tolerance to pathogens in VA mycorrhizal roots are not understood. A well-established mycorrhizal infection is a prerequisite for protection[10] and the protective effects are localized in the roots, the shoots of mycorrhizal plants being no more resistant to aerial pathogens than those of nonmycorrhizal plants[88]. Modifications in root physiology following mycorrhizal infection may explain the decrease in susceptibility to pathogens. These effects may arise through changes in lignin formation, sugar levels, phenol synthesis, amino acid concentrations, oxidative enzyme activities or ethylene production following VA infection[87]. This possibility of biological disease control by VA mycorrhizae underlines once more the need to more closely investigate the physiology of these associations if their potential in crop production is to be exploited.

References

1 Allen M F, Moore T S and Christensen M 1980 Phytohormone changes in *Bouteloua gracilis* infected by vesicular-arbuscular mycorrhizae: I. Cytokinin increases in the host plant. Can. J. Bot. 58, 371–374.

2 Allen M F, Moore T S and Christensen M 1982 Phytohormone changes in *Bouteloua gracilis* infected by vesicular-arbuscular mycorrhizae: II. Altered levels of giberellin-like substances and absissic acid in the host plant. Can. J. Bot. 60, 468–471.

3 Allen M J, Sexton J C, Moore T S and Christensen M 1981 The influence of phosphate source on vesicular-arbuscular mycorrhizae of *Bouteloua gracilis*. New Phytol. 87, 687–694.

4 Allen M J, Smith W K, Moore T S and Christensen M 1981 Comparative water relations and photosynthesis of mycorrhizal and nonmycorrhizal *Bouteloua gracilis* H.B.K. Lag ex Steud. New Phytol. 88, 683–693.

5. Asimi S, Gianinazzi-Pearson V and Gianinazzi S 1980 Influence of increasing soil phosphorus levels on interactions between vesicular-arbuscular mycorrhizae and *Rhizobium* in soybeans. Can. J. Bot. 58, 2200–2205.

6 Atkinson D and Davidson A W 1973 The effects of phosphorus deficiency on water content and response to drought. New Phytol. 72, 307–313.

7 Azcon R, Marin A D and Barea J M 1978 Comparative role of phosphate in soil or inside the host on the formation and effects of endomycorrhiza. Plant and Soil 49, 561–567.

8 Azcon G de Aguilar C and Barea J M 1978 Effects of interactions between different culture fractions of 'phosphobacteria' and *Rhizobium* on mycorrhizal infection, growth and nodulation of *Medicago sativa*. Can. J. Microbiol. 24, 520–524.

9 Barea J M and Azcon-Aguilar C 1982 Production of plant growth-regulating substances by the vesicular-arbuscular mycorrhizal fungus *Glomus mosseae*. Appl. Environ. Microbiol. 43, 810–813.

10 Bärtschi H, Gianinazzi-Pearson and Vegh I 1981 Vesicular-arbuscular mycorrhiza formation and root rot disease (*Phytophthora cinnamomi*) development in *Chamaecyparis lawsoniana*. Phytopathol. Z. 102, 213–218.

11 Bertheau Y, Gianinazzi-Pearson V and Gianinazzi S. 1980 Développement et expression de l'association endomycorhizienne chez le blé I. Mise en évidence d'un effet variétal. Ann. Amélior. Plantes 30, 67–78.

12 Bevege D I, Bowen G D and Skinner M F 1975 Comparative carbohydrate physiology of ecto- and endomycorrhizas *In* Endomycorrhizas. Eds. F E Sanders, B Mosse and P B Tinker. Academic Press, London and New York, pp 149–174.

13 Bonfante-Fasolo P, Dexheimer J, Gianinazzi S, Gianinazzi-Pearson V and Scannerini S 1981 Cytochemical modifications in the host-fungus interface during intracellular interactions in vesicular-arbuscular mycorrhizae. Plant Sci. Lett. 22, 13–21.

14 Bowen G D, Bevege D I and Mosse B 1975 Phosphate physiology of vesicular-arbuscular mycorrhizas. *In* Endomycorrhizas. Eds. F E Sanders, B Mosse and P B Tinker. Academic Press. London and New York, pp 241–260.

15 Bowen G D, Skinner M P and Bevege D I 1974 Zinc uptake by mycorrhizal and uninfected roots of *Pinus radiata* and *Araucaria cunninghammii*. Soil Biol. Biochem. 6, 141–144.

16 Buwalda J G and Goh K M 1982 Host-fungus competition for carbon as a cause of growth depressions in vesicular-arbuscular ryegrass. Soil Biol. Biochem. 14, 103–106.

17 Callow J A, Capaccio L C M, Parish G and Tinker P B 1978 Detection and estimation of polyphosphate in vesicular-arbuscular mycorrhizas. New Phytol. 80, 125–134.

18 Capaccio L C M and Callow J A 1982 The enzymes of polyphosphate metabolism in vesicular-arbuscular mycorrhizas. New Phytol. 91, 81–91.

19 Carling D E and Brown M F 1980 Relative effect of vesicular-arbuscular mycorrhizal fungi on the growth and yield of greenhouse soybeans. Soil Sci. Am. J. 44, 528–532.

20 Cooper K M 1975 Growth responses to the formation of endotrophic mycorrhizas in *Solanum*, *Leptospernum*, and New Zealand ferns. *In* Endomycorrhizas, Eds. F Sanders, B Mosse and P B Tinker, Academic Press, London and New York, pp 391–407.

21 Cooper K M and Losel D M 1978 Lipid physiology of vesicular-arbuscular mycorrhiza I. Composition of lipids in roots of onion, clover and ryegrass infected with *Glomus mosseae*. New Phytol. 80, 143–151.

22 Cooper K M and Tinker P B 1978 Translocation and transfer of nutrients in vesicular-arbuscular mycorrhizas II. Uptake and translocation of phosphorus, zinc and sulphur. New Phytol. 81, 43–52.

23 Cooper K M and Tinker P B 1981 Translocation and transfer of nutrients in vesicular-arbuscular mycorrhizas IV. Effect of environmental variables on movement of phosphorus. New Phytol. 88, 327–339.

24 Cox G and Tinker P B 1976 Translocation and transfer of nutrients in vesicular-arbuscular mycorrhizas I. The arbuscule and phosphorus transfer: a quantitative ultrastructural study. New Phytol. 77, 371–378.

25 Cox G, Sanders F E, Tinker P B and Wild J A 1975 Ultra-structural evidence relating to host-endophyte transfer in a vesicular-arbuscular mycorrhiza. *In* Endomycorrhizas. Eds. F E Sanders, B Mosse and P B Tinker. Academic Press, London and New York, pp 297–312.

26 Cox G, Moran K J, Sanders F, Nockolds C and Tinker P B 1980 Translocation and transfer of nutrients in vesicular-arbuscular mycorrhizas III. Polyphosphate granules and phosphorus translocation. New Phytol. 84, 649–659.

27 Cress W A, Throneberry G O and Lindsey D L 1979 Kinetics of phosphorus absorption by mycorrhizal and nonmycorrhizal tomato roots. Plant Physiol. 64, 484–487.

28 Crush J R 1974 Plant growth responses to vesicular-arbuscular mycorrhiza. VII. Growth and nodulation of some herbage legumes. New Phytol. 73, 743–749.

29 Daft M J and El-Giahmi A A 1978 Effects of *Endogone* mycorrhiza on plant growth VIII. Effects of defoliation and light on selected hosts. New Phytol. 80, 365–372.

30 Daft M J and Okusanya B O 1973 Effect of *Endogone* mycorrhiza on plant growth VI. Influence of infection on the anatomy and reproductive development of four hosts. New Phytol. 72, 1333–1339.

31 DeNollins S, Thone F and Borgers M 1975 Enzyme cytochemistry of *Candida albicans*. J. Histochem. Cytochem. 23, 758–765.

32 Dexheimer J, Gianinazzi-Pearson V, Gianinazzi S and Marx C 1982 Rôle possible de la vacuole du champignon symbiotique des mycorhizes VA dans la nutrition des plantes. Bull. Soc. Bot. France: Actual. Bot. (*In press*).

33 Gerdemann J W 1964 The effect of mycorrhizas on the growth of maize. Mycologia 56, 342–349.

34 Gianinazzi S, Gianinazzi-Pearson V and Dexheimer J 1979 Enzymatic studies on the metabolism of vesicular-arbuscular mycorrhiza III. Ultrastructural localization of acid and alkaline phosphatase in onion roots infected by *Glomus mosseae* (Nicol & Gerd.). New Phytol. 82, 127–132.

35 Gianinazzi S, Trouvelot A and Gianinazzi-Pearson V 1982 Les endomycorhizes: importance dans la croissance et le développement des arbres fruitiers. 2ème Colloque sur les Recherches Fruitières, Bordeaux 73–81.

36 Gianinazzi-Pearson V 1982 Physiologie des endomycorhizes et perspectives offertes par leur utilisation. C.R. Acad. Agric. France 68, 380–389.

37 Gianinazzi-Pearson V and Diem H G 1982 Endomycorrhizae in the Tropics. *In* Microbiology of Tropical Soils. Implications in Soil Management, Eds. Y R Dommergues and H G Diem. Martinus Nijhoff, The Hague, pp 209–251.

38 Gianinazzi-Pearson V and Gianinazzi S 1976 Enzymatic studies on the metabolism of vesicular-arbuscular mycorrhiza I. Effect of mycorrhiza formation and phosphorus nutrition on soluble phosphatase activities in onion roots. Physiol. Vég. 14, 833–841.

39 Gianinazzi-Pearson V and Gianinazzi S 1978 Enzymatic studies on the metabolism of vesicular-arbuscular mycorrhiza II. Soluble alkaline phosphatase specific to mycorrhizal infection in onion roots. Physiological Plant Pathology 12, 45–53.

40 Gianinazzi-Pearson V, Fardeau J C, Asimi S and Gianinazzi S 1981 Source of additional phosphorus absorbed from soil by vesicular-arbuscular mycorrhizal soybeans. Physiol. Vég. 19, 33–43.

41 Gianinazzi-Pearson V, Morandi D, Dexheimer J and Gianinazzi S 1981 Ultrastructural and ultracytochemical features of a *Glomus tenuis* mycorrhiza. New Phytol. 88, 633–639.

42 Gianinazzi-Pearson V, Gianinazzi S, Dexheimer J, Bertheau Y and Asimi S 1978 Les phosphatases alcalines solubles dans l'association endomycorrhizienne à vésicules et arbuscules. Physiol. Vég. 16, 671–678.

43 Gilmore A E 1971 The influence of endotrophic mycorrhizae on the growth of peach seedlings. J. Am. Soc. Hortic. Sci. 96, 35–37.

44 Graham J H, Leonard R T and Menge J A 1981 Membrane-mediated decrease in root exudation responsible for phosphorus inhibition of vesicular-arbuscular mycorrhiza formation. Plant Physiol. 68, 548–552.

45 Gray L E and Gerdemann J W 1969 Uptake of phosphorus-32 by vesicular-arbuscular mycorrhizae. Plant and Soil 30, 415–422.

46 Hall I R, Scott R S and Johnstone P D 1977 Effect of vesicular-arbuscular mycorrhizae on response of 'Grasslands Huia' and 'Tamar' white clovers to phosphorus. N.Z. J. Agric. Res. 20, 349–355.

47 Hardie K and Leyton L 1981 The influence of vesicular-arbuscular mycorrhiza on growth and water relations of red clover. I. In phosphate deficient soil. New Phytol. 89, 599–608.

48 Harold F 1966 Inorganic polyphosphates in biology: structure, metabolism, and function. Bacteriol. Rev. 30, 722–794.

49 Hattingh M J, Gray L E and Gerdemann J W 1973 Uptake and translocation of ^{32}P-labelled phosphate to onion roots by endomycorrhizal fungi. Soil Sci. 116, 383–387.

50 Hayman D S 1974 Plant growth responses to vesicular-arbuscular mycorrhiza VI. Effect of light and temperature. New Phytol. 73, 71–80.

51 Ho I and Trappe J M 1973 Translocation of ^{14}C from *Festuca* plants to their endomycorrhizal fungi. Nature London 244, 30–31.

52 Jasper D A, Robson A D and Abbot L K 1979 Phosphorus and the formation of vesicular-arbuscular mycorrhizas. Soil Biol. Biochem. 11, 501–505.

53 Kleinschmidt G D and Gerdemann J W 1972 Stunting of citrus seedlings in fumigated soils related to the absence of endomycorrhizae. Phytopathology 62, 1447–1453.

54 Kormanik P P, Bryan W C and Schultz R C 1977 Quality hardwood seedlings require early mycorrhizal development in nursery beds. Proc. 14th South. For. Tree Improv. Conf., 289–293.

55 Krishna K R, Suresh H M, Syamsunder J and Bagyaraj D J 1981 Changes in the leaves of finger millet due to VA mycorrhizal infection. New Phytol. 87, 717–722.

56 Lambert D H, Baker D E and Cole H 1979 The role of mycorrhizae in the interactions of phosphorus with zinc, copper and other elements. Soil Sci. Soc. Amer. J. 43, 976–980.

57 LaRue J H, McClellan W D and Peacock W L 1975 Mycorrhizal fungi and peach nursery nutrition. Calif. Agric. 29, 6–7.

58 Ling-Lee M, Chilvers G A and Ashford A E 1975 Polyphosphate granules in three different kinds of tree mycorrhiza. New Phytol. 75, 551-555.

59 Lösel D M and Cooper K M 1979 Incorporation of ^{14}C-labelled substrates by uninfected roots of onion. New Phytol. 83, 415–426.

60 Marx C, Dexheimer J, Gianinazzi-Pearson V and Gianinazzi S 1982 Enzymatic studies on the metabolism of vesicular-arbuscular mycorrhizas IV. Ultrastructural evidence (ATPase) for active transfer processes in the host-arbuscule interface. New Phytol. 90, 37–43.

61 Matile Ph and Wiemken A 1976 Interactions between cytoplasm and vacuole. *In* Transport in Plants III. Encyclop. Plant Physiol. (NS) 3, pp 255–287.

62 Menge J A, Davis R M, Johnson E L and Zentmyer G A 1978 Mycorrhizal fungi increase growth and reduce transplant injury in avocado. Calif. Agric. April, 6–7.

63 Menge J A, Steirle O, Bayaraj D J, Johnson E L V and Leonard R T 1978 Phosphorus concentrations in plants responsible for inhibition of mycorrhizal infection. New Phytol. 80, 575–578.

64 Morrison T M and English D A 1967 The significance of mycorrhizal nodules of *Agathis australis*. New Phytol. 66, 245–250.

65 Mosse B 1973 Advances in the study of vesicular-arbuscular mycorrhiza. Annu. Rev. Phytopathol. 11, 171–196.

66 Mosse B, Hayman D and Arnold D J 1973 Plant growth responses to vesicular-arbuscular mycorrhiza V. Phosphate uptake by three plant species from P-deficient soils labelled with ^{32}P. New Phytol. 72, 809–815.

67 Nye P H and Tinker P B 1977 Solute movement in the soil-root system. Studies in Ecology, 4, Blackwell Scientific Publications, Oxford and London.

68 Pang P C and Paul E A 1980 Effects of vesicular-arbuscular mycorrhiza on ^{14}C and ^{15}N distribution in nodulated fababeans. Can. J. Soil. Sci. 60, 241–250.

69 Parish G R, Sanders F E and Tinker P B 1977 Polyphosphate translocation in VA mycorrhizas. 5th North Amer. Conf. on Mycorrhizae, p. 84 (Abstracts).

70 Pearson V and Tinker P B 1975 Measurement of phosphorus fluxes in the external hyphae of endomycorrhizas. *In* Endomycorrhizas. Eds. F E Sanders, B. Mosse and P B Tinker. Academic. Press, London and New York, pp 277–287.

71 Pichot J and Binh T 1976 Action des endomycorhizes sur la croissance et la nutrition phosphate de l'agrostis en vases de végétation et sur le phosphate isotopiquement diluable du sol. Agron. Trop. 31, 375–378.

72 Pons F and Gianinazzi-Pearson V 1982 Observations on vesicles produced *in vitro* by the vesicular-arbuscular mycorrhizal fungus *Gigaspora margarita*. Trans. Brit. mycol. Soc. (*Submitted*).

73 Porter W M, Abbot L K and Robson A D 1978 Effect of rate of application of superphosphate on populations of vesicular-arbuscular endophytes. Aust. J. Exp. Agric. Anim. Husb. 18, 573–578.

74 Powell C Ll 1975 Plant growth responses to vesicular-arbuscular mycorrhiza. VII. Uptake of P by onion and clover infected with different Endogone spores types in ^{32}P labelled soils. New Phytol. 75, 563–566.

75 Ratnayake M, Leonard R T and Menge J A 1978 Root exudation in relation to supply of phosphorus and its possible relevance to mycorrhizal formation. New Phytol. 81, 543–552.

76 Rhodes L H and Gerdemann J W 1975 Phosphate uptake zones of mycorrhizal and nonmycorrhizal onions. New Phytol. 75, 555–561.

77 Rhodes L H and Gerdemann J W 1978 Hyphal translocation and uptake of sulfur by vesicular-arbuscular mycorrhizae of onion. Soil Biol. Biochem. 10, 355–360.

78 Rhodes L H and Gerdeman J W 1978 Influence of phosphorus nutrition on sulfur uptake by vesicular-arbuscular mycorrhizae of onion. Soil Biol. Biochem. 10, 361–364.

79 Rhodes L H and Gerdemann J W 1980 Nutrient translocation in vesicular-arbuscular mycorrhizae. *In* Cellular Interactions in Symbiosis and Parasitism. Eds. C B Cooke, P W Pappas and E D Rudolph. Ohio State University Press, pp 173–195.

80 Safir G R, Boyer J S and Gerdeman J W 1971 Mycorrhizal enhancement of water transport in soybean. Science 172, 581–583.

81 Safir G R, Boyer J S and Gerdemann J W 1972 Nutrient status and mycorrhizal enhancement of water transport in soybean. Plant Physiol. 49, 700–703.

82 Sanders F E 1975 The effect of foliar-applied phosphate on the mycorrhizal infections of onion roots. *In* Endomycorrhizas. Eds. F E Sanders, B Mosse and P B Tinker, Academic Press, London and New York, pp 261–276.

83 Sanders F E and Tinker P B 1971 Mechanism of absorption of phosphate from soil by *Endogone* mycorrhizas. Nature London 233, 278–279.

84 Sanders F E and Tinker P B 1973 Phosphate flow into mycorrhizal roots. Pest. Sci. 4, 385–395.

85 Sanders F E, Tinker P B, Black R L B and Palmerley S M 1977 The development of endomycorrhizal root systems I. Spread of infection and growth-promoting effects with four species of vesicular-arbuscular endophyte. New Phytol. 78, 257–268.

86 Schenk N C and Kellam M K 1978 The influence of vesicular arbuscular mycorrhizae on disease development. Bull. 799 Florida Agric. Exp. Station No 798, 1–16.

87 Schönbeck F 1979 Endomycorrhiza in relation to plant disease. *In* Soil borne Plant Pathogens. Eds. B Schippers and W Gams. Academic Press, London, pp 271–280.

88 Schönbeck F and Dehne H W 1979 The influence of endotrophic mycorrhiza on plant diseases 4. Fungal parasites on aerial plants parts, *Olpidium brassicae*, TMV. J. Plant Dis. Prot. 86, 103–112.

89 Schwab S M, Johnson E L V and Menge J A 1982 Influence of simazine on formation of vesicular-arbuscular mycorrhizae in *Chenopodium quinona* Willd. Plant and Soil 64, 283–287.

90 Sieverding E 1981 Influence of soil water regimes on VA mycorrhiza I. Effect on plant growth, water utilization and development of mycorrhiza. J. Agron. Crop. Sci. 150, 400–411.

91 Smith F A and Smith S E 1981 Mycorrhizal infection and growth of *Trifolium subterraneum:* use of sterilized soil as a control treatment. New Phytol. 88, 299–309.

92 Smith S E 1982 Inflow of phosphate into mycorrhizal and nonmycorrhizal plants of *Trifolium subterraneum* at different levels of soil phosphate. New Phytol. 90, 293–303.

93 Smith S E and Daft M J 1977 Interactions between growth, phosphate content and N_2 fixation in mycorrhizal and nonmycorrhizal *Medicago sativa*. Aust. J. Plant Physiol. 4, 403–413.

94 Smith S E, Nicholas D J D and Smith F A 1979 The effect of early mycorrhizal infection on nodulation and nitrogen fixation in *Trifolium subterraneum*. Aust. J. Plant Physiol. 6, 305–311.

95 Sparling G P and Tinker P B 1978 Mycorrhizal infection in Pennine grassland II. Effects of mycorrhizal infection on the growth of some upland grasses on γ-irradiated soils. J. Appl. Ecol. 15, 951–958.

96 Stribley D P, Tinker P B and Rayner J H 1980 Relation of internal phosphorus concentration and plant weight in plants infected by vesicular-arbuscular mycorrhizas. New Phytol. 86, 261–266.

97 Strullu D G, Gourret J P, Garrec J P and Fourcy A 1981 Ultrastructure and electron-probe microanalysis of the metachromic vacuolar granules occurring in *Taxus* mycorrhizas. New Phytol. 87, 537–545.

98 Strullu D G, Gourret J P and Garrec J P 1981 Microanalyse des granules vacuolaires des ectomycorhizes, endomycorhizes et endomycothalles. Physiol. Vég. 19, 367–378.

99 Swaminathan V 1979 Nature of the inorganic fraction of soil phosphate fed on by vesicular-arbuscular mycorrhizal potatoes. Proc. Indian Acad. Sci. 88B, 423–433.

100 Tisdall J M and Oades J M 1979 Stabilization of soil aggregates by the root system of ryegrass. Aust. J. Soil Res. 17, 429–441.

101 Timmer L W and Leyden R F 1978 Stunting of citrus seedlings in fumigated soils in Texas and its correction by phosphorus fertilization and inoculation with mycorrhizal fungi. J. Am. Soc. Hortic. Sci. 103, 533–537.

102 Timmer L W and Leyden R F 1980 The relationship of mycorrhizal infection to phosphorus-induced copper deficiency in sour orange seedlings. New Phytol. 85, 15–23.

103 Tinker P B 1975 Effects of vesicular-arbuscular mycorrhizas on higher plants. Symp. Soc. Exp. Biol. 29, 325–349.

104 Tinker P B 1978 Effects of vesicular-arbuscular mycorrhizas on plant nutrition and plant growth. Physiol. Vég. 16, 743–751.

105 Wallace L L 1981 Growth, morphology and gas exhance of mycorrhizal and nonmycorrhizal *Panicum coloratum* L., a C_4 grass species, under different clipping and fertilization regimes. Oecologia 4, 272–278.

106 White J A and Brown M F 1979 Ultrastructural and X-ray analysis of phosphorus granules in a vesicular-arbuscular mycorrhizal fungus. Can. J. Bot. 57, 2812–2818.

Plant and Soil 71, 211–215 (1983). A-21

Role of the host-arbuscule interface in the VA mycorrhizal symbiosis: ultracytological studies of processes involved in phosphate and carbohydrate exchange

S. GIANINAZZI, J. DEXHEIMER*, V. GIANINAZZI-PEARSON and C. MARX*
Station d'amélioration des Plantes, INRA, BV 1540, F-21034 Dijon-Cedex, France

Key words ATPase Interfacial matrix Neutral phosphatase Onion Polysaccharides Sycamore VA mycorrhizas

Summary The host-arbuscule interface does not vary greatly between different VAM associations of herbaceous and woody plants. The presence in the interfacial matrix of living fine arbuscule branches, where the host wall material is reduced or nonexistent and the endophyte wall is thin, of host membranous vesicles, polysaccharides, ATPase and neutral phosphatase activities, clearly indicates that this structure is not an inert space, but a physiologically complex zone. Possible roles for these interfacial matrix components in host-fungus nutrient exchange is discussed.

Introduction

The interface formed by living host and fungus in the fine arbuscule branches of vesicular-arbuscular mycorrhizae (VAM) is considered to be the site of nutrient exchange between the two symbionts[1,3]. In the electron microscope the interfacial matrix appears as a space between the two organisms, but recent detailed cytochemical and cytoenzymological studies, mainly in herbaceous plants, have shown that it is not in fact empty but contains membranous vesicles, polysaccharides, proteins and ATPase activities[1,5,9].

In the present paper similar observations are made on the VA mycorrhizal association of a woody plant species, sycamore, and the presence of another enzyme activity (neutral phosphatase) in the interfacial matrix is described. The possible role of these components of the interfacial matrix in the nutrient exchanges between the plant and the VA fungus is discussed.

Materials and methods

Onion (*Allium cepa*, F_1 hybrid Hyper) and sycamore (*Acer pseudoplatanus* L.) plants were infected with *Glomus mosseae* as previously described[7,9]. Samples of infected roots were taken from 6-week old onions or 13-week old sycamore seedlings and prepared for electron microscope observations following methods described by Gianinazzi *et al.*[6] (acid or alkaline phosphatases), Dexheimer[4] (PATAg reaction for polysaccharides) and Marx *et al.*[9] (ATPase activity). The experimental procedure for neutral phosphatase localization was similar to that used for acid phosphatase except that specimens were incubated in Novikoff and Goldfischer's[10] medium containing either inosine

* *Laboratoire de Botanique II, Université de Nancy I, F-54037 Nancy-Cedex, France*

diphosphate (IDP), thiamine pyrophosphate (TPP) or β-glycerophosphate. Controls consisted of sections incubated in the medium without substrate or in the complete reaction medium with either 20 m*M* NaF (acid phosphatase inhibitor) or 4 m*M* KCN (alkaline phosphatase inhibitor) for β-glycerophosphatase activity. All enzyme activities were revealed by the deposition of a black precipitate formed by the product of the reaction (lead phosphate).

Results

As in herbaceous plants[1,5], the presence of membranous vesicles of host origin in the interfacial matrix of sycamore mycorrhiza appeared to be a general characteristic (Plate 1a). These structures, which reacted to the PATAg test in the same way as the host plasmalemma, were particularly abundant around the living fine arbuscule branches where the endophyte wall was thin and host wall material extremely reduced (scattered fibrils only) or absent. The PATAg reaction also revealed the presence of scattered polysaccharide fibrils in the interfacial matrix which were attached to the surface of either the host plasmalemma or the membranous vesicles. As previously reported[9], the latter possessed an intense ATPase activity similar to that observed along the invaginated host plasmalemma extending around the fine arbuscule branches whilst the non-invaginated membrane showed only a weak activity (Plate 1b).

In sycamore, as in onion, acid or alkaline phosphatase activity could not be detected in the interfacial matrix around the living arbuscule branches when β-glycerophosphate ar α-naphthyl phosphate were used as substrates (Plate 1c). However, at neutral pH, intense TPPase, IDPase and β-glycerophosphatase activities could be detected all along the host plasmalemma in both onion and sycamore mycorrhizae, although they appeared to be particularly intense along the invaginated membrane around the finer arbuscule branches (Plate 1d, e and f). Contrary to ATPase[9], these enzyme activities persisted around the collapsed hyphal branches. In the control preparations no precipitate was observed in the absence of the substrates and β-glycerophosphatase activity was not inhibited by KCN or NaF. Furthermore, preliminary results from more detailed inhibitor studies indicate that the IDPase activity is not sensitive to diethylstilbestrol (0.3 m*M*), a specific inhibitor of plasmalemma ATPase activity. These observations indicate that another group of enzymes, neutral phosphatases, are present in the VAM interfacial matrix. Whether these consist of one enzyme with a large substrate spectrum or of several enzymes has still to be determined.

Discussion

During intracellular development of the fungal symbiont in VA mycorrhizas the plasmalemma of the host cell proliferates and extends to surround the entire arbuscule. As the present observations indicate the structure of the host-fungus interface does not vary greatly between different VAM associations of

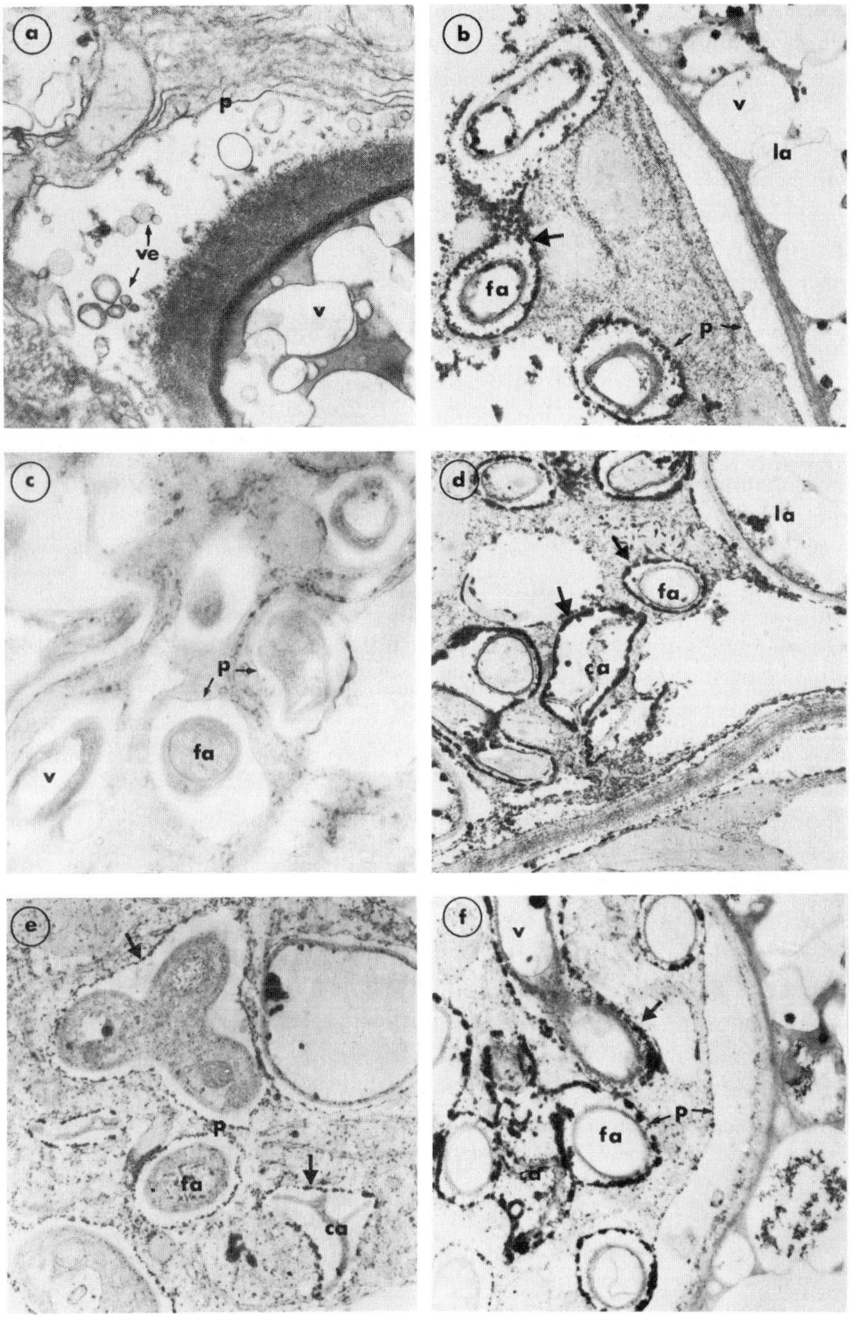

Plate 1. Electron micrographs of host-arbuscule interfaces in sycamore plants infected by *G. mosseae:* a) PATAg reaction for polysaccharide (× 19000), b) ATPase activity (× 17000), c) acid (pH 5.0) β-glycerophosphatase activity (× 20500) d) TPPase activity (× 13700) e) IDPase activity (× 14200) and f) neutral (pH 7.0) β-glycerophosphatase activity (× 17000).

Arrows indicate enzyme activity; 1a, large arbuscule branch; fa, fine arbuscule branch; p, host plasmalemma; v, fungal vacuole; ca, collapsed arbuscule branch and ve, membranous vesicles.

herbaceous and woody plants. The interfacial matrix is not an inert space or 'espace membranaire'[3], but rather a physiologically complex zone which shows certain fundamental differences from that occurring in pathogenic associations. In contrast to plant-fungal pathogen systems[2,8], the invaginated host plasmalemma in VA mycorrhizas does not differ morphologically or cytochemically from the uninvaginated membrane[1]. Furthermore, host plasmalemma-bound ATPase activity is specifically localised along the invaginated membrane around the fine arbuscule branches[9], whereas this enzyme activity is inhibited in fungal pathogen infections[12]. In contrast, the fungi causing both symbiotic and pathogenic infections show marked ATPase activity along their plasmalemma. Plasmalemma-bound ATPase activity is considered to play a role in active transport processes at this membrane[11]. These observations underline the fundamental difference that exists between symbiotic VA mycorrhiza and pathogenic associations. In the former the host plasmalemma conserves its normal function of bidirectional active nutrient transfer, phosphate passing from fungus to plant and carbohydrates in the reverse direction, whilst in pathogenic relationships this transfer is unidirectional the fungus being the only partner to benefit from the association.

It appears from the present study that another group of enzymes, neutral phosphatases, are also produced by the host plasmalemma in the arbuscule interfacial matrix. The exact function of neutral phosphatases in plants remains obscure, but it has been proposed that they may participate in the processes of cell wall formation in plants[13]. However, in VAM despite the presence of intense neutral phosphatase activity very little or no polysaccharide material is deposited in the interfacial matrix around the fine arbuscule branches. Sparse polysaccharide fibrils can be seen attached to the host plasmalemma[5] and membranous vesicles. These observations suggest that the invaginated host plasmalemma retains its normal capacity for cell wall synthesis but that subsequent polymerisation of the host-produced polysaccharides is somehow inhibited. A possible explanation for this is that VA fungi are able to lyse the cell wall precursors produced by the host cell, perhaps using them as a carbon source. The persistence of host neutral phosphatase activities and the accumulation of disorganised wall material in the interfacial matrix around dead collapsed hyphae[1,5] lends supports to this hypothesis.

References

1 Bonfante-Fasolo P, Dexheimer J, Gianinazzi-Pearson V and Scannerini S 1981 Cytochemical modifications in the host-fungus interface during intracellular interactions in vesicular-arbuscular mycorrhizae. Plant Science Letters 22, 13–21.

2 Bracker C E and Littlefield L J 1973 Fungal Pathogenicity and the Plant's Response. Academic Press. pp 159–313.

3 Cox G and Sanders F 1974 Ultrastructure of the host-fungus interface in a vesicular-arbuscular mycorrhiza. New Phytol. 73, 901–912.

4 Dexheimer J 1976 Etude de la sécrétion de mucilage par les cellules des glandes digestives de

Drosera (*D. rotundifolia* L. et *D. capensis* L.). Application de quelques techniques cytochimiques. Cytobiologie 13, 307–321.

5 Dexheimer J, Gianinazzi S and Gianinazzi-Pearson V 1979 Ultrastructural cytochemistry of the host-fungus interfaces in the endomycorrhizal association *Glomus mosseae*/*Allium cepa*. Z. Pflanzenphysiol. 92, 191–206.

6 Gianinazzi S, Gianinazzi-Pearson V and Dexheimer J 1979 Enzymatic studies on the metabolism of vesicular-arbuscular mycorrhiza. III. Ultrastructural localization of acid and alkaline phosphatase in onion roots infected by *Glomus mosseae* (Nicol. & Gerd.). New Phytol. 82, 127–132.

7 Gianinazzi-Pearson V and Gianinazzi S 1976 Enzymatic studies on the metabolism of vesicular-arbuscular mycorrhiza. I. Effect of mycorrhiza formation and phosphorus nutrition on soluble phosphatase activities in onion roots. Physiol. Vég. 14, 833–841.

8 Gil F and Gay J L 1977 Ultrastructural and physiological properties of the host interfacial components of *Erysiphe pisi in vitro* and *in vivo*. Physiol. Plant Pathol. 10, 1–12.

9 Marx C, Dexheimer J, Gianinazzi-Pearson V and Gianinazzi S 1982 Enzymatic studies on the metabolism of vesicular-arbuscular mycorrhiza. IV. Ultracytoenzymological evidence (ATPase) for active transfer processes in the host-arbuscule interface. New Phytol. 90, 37–43.

10 Novikoff A B and Goldfischer S 1961 Nucleoside diphosphatase activity in the Golgi apparatus and its usefulness for cytological studies. Proc. Nat. Acad. Sci. 47, 802–810.

11 Raven J A 1976 Transport in algal cells *In* Transport in Plants II, Encyclopedia of Plant Physiology. New series 2, 129–188.

12 Spencer-Phillips P T N and Gay J L 1981 Domains of ATPase in plasma membranes and transport through infected cells. New Phytol. 89, 393–400.

13 Zerban H and Werz G 1975 Localization of nucleoside diphosphatases and thiamine pyrophosphatase in various stages of the life cycle of the green algae *Acetabularia cliftonii* and *Acetabularia mediterranea*. Cytobiologie 12, 13–27.

Plant and Soil 71, 217–221 (1983). A-22

Studies of VA mycorrhizae *in vitro*: mycorrhizal synthesis of axenically propagated wild cherry *(Prunus avium* L.) plants

F. PONS, V. GIANINAZZI-PEARSON, S. GIANINAZZI and J. C. NAVATEL*
Station d'Amélioration des Plantes, I.N.R.A., B.V. 1540, F-21034 Dijon-Cedex, France

Key words Cherry rootstocks *Gigaspora margarita* *In vitro* mycorrhizae Micropropagation *Prunus avium* L.

Summary An increasing number of woody plant species can be axenically propagated using plant tissue culture methods. These techniques produce high quality healthy plants but transplant problems often arise due to their being susceptible to different environmental stresses. Mycorrhizae can play a vital role in plant nutrition and survival and their introduction during the production of axenically propagated plants could contribute to the success of this technique. Using a simple method for axenic establishment and direct observation of VA mycorrhizal cultures, *in vitro* mycorrhizal synthesis of a woody plant species has been obtained for the first time using axenically propagated plants of *Prunus avium* L. (wild cherry) and the VA fungus *Gigaspora margarita*.

Introduction

Vesicular-arbuscular (VA) mycorrhizae are widely represented in numerous types of ecological systems and concern many woody species[3,7]. The development of certain plants, especially tree species like Liquidambar[10], Citrus[12] or Malus[4,21], depends directly on the presence of VA fungi to such an extent that they exhibit an important physiological requirement for VA mycorrhizae.

Nowadays, *in vitro* micropropagation techniques are being increasingly applied to the production of fruit trees[16,17]. This procedure results in the disappearance of the natural microflora of the micropropagated plants. Thus during relatively long periods of their development the plants are without VA mycorrhizal fungi, a condition which is prolonged with the subsequent transfer of such plants into inert potting media or fumigated nursery beds. Several studies have demonstrated the importance of infecting plants as early as possible if adverse effects such as stunting of seedlings in fumigated soils or poor growth and development after transplanting[9,11,13,17] are to be avoided.

Wild cherry plants serving as cherry rootstocks are currently propagated *in vitro*[16] and in preliminary experiments these plants have shown a marked improvement in growth when inoculated with a VA mycorrhizal fungus at the time of transplanting into a peat : gravel : soil potting mixture (Plate 1a). The aim of the present work was to see whether a VA mycorrhizal infection could be

* *C.T.I.F.L., Centre de Balandran, F-30127 Bellegarde, France*

established in micropropagated cherry rootstocks *in vitro* in order to anticipate the necessity for inoculation during transplanting. *In vitro* VA mycorrhizae have up to now been obtained using herbaceous species[1,8,22] and this is the first report of their culture in both a micropropagated and a woody species.

Material and methods

VA mycorrhizal inoculum

Gigaspora margarita[2] was chosen for these experiments because of its large and easily manipulated spores, but since the optimum pH for spore germination and mycorrhiza development of this fungus were unknown, and this factor is often important in determining patterns of VA infections, some preliminary germination and growth experiments were carried out. Spores of *G. margarita* were surface sterilized using Mosse's method[14] with 400 ppm streptomycin and germinated at 24°C in the dark either on 0.5% water agar (pH 6.8) or on media of pH 5.8, 6.3, 6.8, 7.5 and 7.8 obtained by mixing solutions of 1.6×10^{-3} *M* KH_2PO_4 and $K_2HPO_43H_2O$.

Micropropagation of cherry rootstocks

Cherry rootstocks (wild cherry, clone F 12.1) were multiplied and elongated on Lepoivre's agar medium[23] containing 2.2×10^{-6} *M* BA and 5.8×10^{-7} *M* IBA during two to three weeks. Rooting was then induced on Németh's medium[18] using 5.8×10^{-7} *M* IBA. Both media contained Walkey's vitamins[24].

In vitro mycorrhiza inoculation

Young wild cherry plants 3 cm high with roots 1–2 cm long were transplanted into the experimental system shown in Plate 1b. Crush and Hay's medium[5] modified by the addition of agar (1.5%) and either K_2HPO_4 (184 mg/l) or Ca phytate (150 mg/l) was used with or without sterilized soil (clay loam pH 7.2, 20 ppm available P (Olsen)). The wild cherry plants were inoculated with pregerminated surface sterilized spores of *G. margarita* placed close to the roots. Petri dishes were covered in the lower half with black strips to avoid excess illumination and incubated vertically in a growth cabinet (16–19°C, 16 h day, 12000 lux).

Assessment of infection

Nine weeks after inoculation, the presence of a mycorrhizal infection was determined by clearing the roots in 10% KOH and staining them in Trypan blue[20].

Table 1. *In vitro* germination and hyphal development of *Gigaspora margarita* at different pHs

	pH of medium ($KH_2PO_4 + K_2HPO_43H_2O$)					
	5.8	6.3	6.8	6.8 (water agar)	7.5	7.8
Average % spores germinating	91.7	92.7	85.0	90.8	82.9	91.7
Average number of germ tubes/spore	4.6	3.8	5.9	3.7	7.1	4.9
% of viable germ tubes	13.9	26.4	25.6	31.4	9.4	15.2

Results and discussion

Spores of *G. margarita* germinated after 3 days on all media and germ tubes continued to grow for 3 to 4 weeks (Table 1). After 12 days the percentage spore germination was comparable on all media which suggested that the fungus was not sensitive to a variation in pH between 5.8 and 7.8. However, the number and

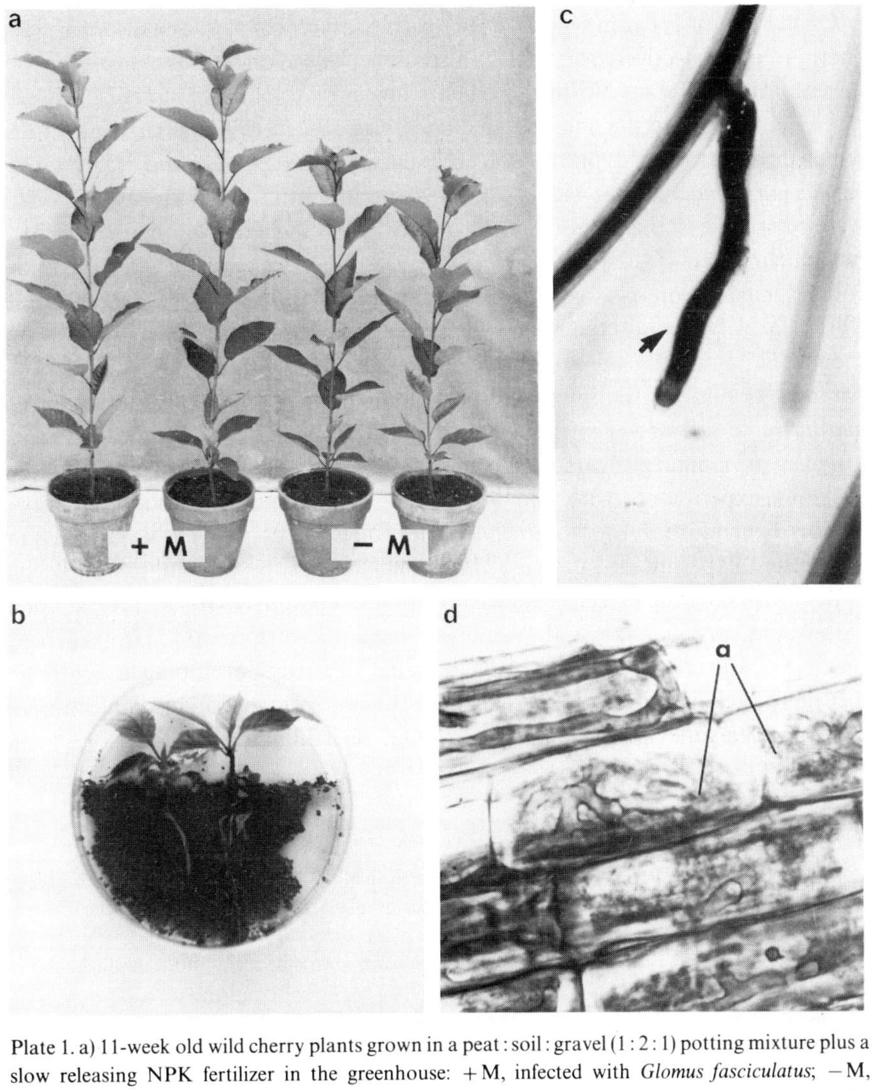

Plate 1. a) 11-week old wild cherry plants grown in a peat : soil : gravel (1 : 2 : 1) potting mixture plus a slow releasing NPK fertilizer in the greenhouse: +M, infected with *Glomus fasciculatus*; −M, uninoculated. b) Experimental system for *in vitro* mycorrhizal inoculation of micropropagated wild cherry plants. c) Lateral root of a wild cherry plant infected *in vitro* by *Gigaspora margarita* (arrow) compared with uninfected roots (× 20). d) Portion of a wild cherry root infected *in vitro* with *G. margarita* and showing typical arbuscules (a) (× 1200).

the appearance of the germ tubes produced by each spore were modified by the composition of the media. There was an increase in the number of germ tubes produced by each spore as pH increased up to 7.5 and there was an inverse relationship between number and viability of germ tubes; in general the greater the number of germ tubes, the smaller was the proportion of germ tubes which remained viable. Since germ tube viability was greatest on water agar and on media with pH in the range 6.3 to 6.8, inoculum was produced on the former and Crush and Hay's medium was adjusted to pH 6.8 with KOH when necessary.

Of the six different media used for *in vitro* mycorrhizal inoculation of wild cherry plants, infections occurred in the presence of soil on the basic medium with or without Ca phytate. No enhancement of the growth of the mycorrhizal cherry plants, compared to the uninoculated ones, was observed. The *in vitro* infections which developed were typical of VA endomycorrhizae (Plate 1c and d) formed by *G. margarita*, arbuscules but no vesicles within roots. Mycorrhizal infections were restricted to the young lateral roots (Plate 1c) where no suberization had begun; this type of distribution has also been observed in field infected apple roots (Dr Atkinson, personal communication). Because of this, infection levels were relatively low and the good root development of the wild cherry plants on the media containing soil must have contributed to the success of the infection. Another reason for the low infection levels could be related to the use of *G. margarita*; in soil slower mycorrhizal establishment has been observed with this fungus[19] (Gianinazzi-Pearson, Pons and Gianinazzi unpublished results). In previous experiments with the herbaceous plants red clover and ryegrass, mycorrhizal infection was obtained without soil using the same basic media as in the present experiments but using *Glomus mosseae* as the VA fungus[22].

In conclusion, the present results show that *in vitro* mycorrhizal infection of a woody species is possible and that mycorrhizal micropropagated plants can be produced. Future research is being directed towards developing a synthetic medium which favours both lateral root production and mycorrhizal infection of *in vitro* propagated woody plants inoculated with different VA fungi.

References

1 Allen M F, Moore Jr T S, Christensen M and Stanton N 1979 Growth of vesicular-arbuscular mycorrhizal and nonmycorrhizal *Bouteloua gracilis* in a defined medium. Mycologia 71, 666–669.

2 Becker W N and Hall I R 1976 *Gigaspora margarita*, a new species in the Endogonaceae. Mycotaxon 4, 155–160.

3 Boullard B 1968 Les Mycorhizes. Masson and Co. Paris, 135 p.

4 Covey R P, Koch B L and Larsen H J 1981 Influence of vesicular-arbuscular mycorrhizae on the growth of apple and corn in low-phosphorus soil. Phytopathology 71, 712–715.

5 Crush J R and Hay M J M 1981 A technique for growing mycorrhizal clover in solution culture. N. Z. J. Agric. Res. 24, 371–372.

6 Green N E, Graham S O and Schenck N C 1976 The influence of pH on the germination of vesicular-arbuscular mycorrhizal spores. Mycologia 68, 929–934.

7 Harley J L 1969 The Biology of Mycorrhiza. 2nd edition Leonard Hill, London, 334 p.

8 Hepper C M 1981 Techniques for studying the infection of plants by vesicular-arbuscular mycorrhizal fungi under axenic conditions. New Phytol. 88, 641–647.

9 Kleinschmidt G D and Gerdemann J W 1972 Stunting of citrus seedlings in fumigated soils in relation to the absence of endomycorrhizas. Phytopathology 62, 1447–1453.

10 Kormanik P P, Bryan W C and Schultz R 1976 Endomycorrhizae: their importance in nursery production of hardwood seedlings. Proc. SE Area Nurserymen's Conf. Eastern Session, Charleston, USA pp 18–21.

11 Lambert D H, Stouffer R F and Cole Jr H 1979 Stunting of peach seedlings following soil fumigation. J. An. Soc. Hortic. Sci. 104, 433–435.

12 Menge J A, Johnson E L V and Platt R G 1978 Mycorrhizal dependency of several citrus cultivars under three nutrient regimes New Phytol. 81, 553–559.

13 Morandi D, Gianinazzi S and Gianinazzi-Pearson V 1979 Intérêt de l'endomycorrhization dans la reprise et la croissance du Framboisier issu de multiplication végétative *in vitro*. Ann. Amélior. Plantes 29, 623–630.

14 Mosse B 1959 The regular germination of resting spores and some observations on the growth requirements of an Endogone sp. Trans. Brit. Mycol. Soc. 46, 273–286.

15 Mosse B, Stribtley D P and Le Tacon F 1981 Ecology of mycorrhizae and mycorrhizal fungi. *In* Advances in Microbiol. Ecology. Ed. Alexander, Plenum Press, N. Y. and London pp 137–210.

16 Navatel J C 1980 L'utilisation des cultures *in vitro* pour la multiplication de quelques espèces légumières et fruitières. C. R. Acad. Agric. France 66, 681–696.

17 Navatel J C 1982 Multiplication végétative *in vitro*: intérêt et problèmes liés peut-être au défaut de mycorhization du matériel obtenu. Les Colloques de l'I.N.R.A. 13, 339–343.

18 Németh G 1981 Milieux de multiplication végétative *in vitro* Table Ronde sur la Multiplication *in vitro* d'Espèces Ligneuses 6–7 June 1978. C.R.A.-Gembloux, B. 39–42.

19 Ollivier B, Diem H G, Pinta M and Dommergues Y R 1982 Effet de l'endomycorhization sur la concentration en phosphore et en zinc des parties aeriennes de *Vigna unguiculata* cultivé dans un sol Dek. Les Colloques de l'I.N.R.A. 13, 155–161.

20 Phillips J M and Hayman D S 1970 Improved procedures for clearing and staining parasitic and vesicular-arbuscular mycorrhizal fungi for rapid assessment of infection. Trans. Brit. Mycol. Soc. 55, 158–161.

21 Plenchette C, Furlan V and Fortin J A 1981 Growth stimulation of apple trees in unsterilized soil under field conditions with VA mycorrhiza inoculation. Can. J. Bot. 59, 2003–2008.

22 Pons F, Gianinazzi-Pearson V and Gianinazzi S 1982 Synthèse *in vitro* des endomycorhizes ericoïdes et VA: complément à la micropropagation. Les Colloques de l'I.N.R.A. 13, 345–349.

23 Quoirin M, Lepoivre Ph and Boxus Ph 1977 Un premier bilan de 10 années de recherches sur les cultures de méristèmes et la multiplication *in vitro* de fruitiers ligneux. C. R. Rech. 1976–1977. Rapp. Synth. Stn. Cult. fruitières maraîchères Gembloux, B. 93–117.

24 Walkey D G 1972 Production of apple plantlets from axillary-bud meristems. Can. J. Plant Sci. 52, 1085–1087.

Plant and Soil 71, 223–246 (1983). A-23

The development of vesicular-arbuscular mycorrhizal infection in plant root systems

F. E. SANDERS and N. A. SHEIKH
Department of Plant Sciences, University of Leeds, Leeds LS2 9JT, UK

Key words Growth response Infection spread Models Mycorrhizae VA-mycorrhizae

Summary Many crop plants carry VA mycorrhizal infection. The fungi benefit their host principally by increasing the rate of phosphorus uptake from soils low in available phosphorus.

Infection of seedling root systems occurs from soil-borne propagules and/or from neighbouring mycorrhizal roots. When, under the right conditions, the propagules germinate, a mycelium is produced which cannot grow extensively until mycorrhizal infection is established.

The level of initial (primary) infection is related to propagule density. The relation can be described by a mathematical model which takes account of propagule germination rates, rates of hyphal growth through the soil and rates of root growth.

Subsequent spread of secondary infection is by growth, along and between roots, of hyphae originating from established infection units and by the formation of new infection units. This spread can be modelled.

If certain simplifying assumptions are made, the effects on the host of developing mycorrhizal infection can also be modelled and satisfactory predictions of infection spread and the mycorrhizal growth response obtained.

In conclusion, some implications for agriculture and forestry are briefly discussed.

Introduction

The principal way in which vesicular-arbuscular (VA) mycorrhizal fungi benefit plant growth is now clearly understood. Where rates of phosphorus uptake by non-mycorrhizal roots from soil are limited by slow rates of diffusion, rates of uptake by mycorrhizal roots may be substantially greater, leading to higher rates of growth in mycorrhizal than in non-mycorrhizal plants. Although both mycorrhizal and non-mycorrhizal roots take up phosphate ions from the same soil fraction, mycorrhizal roots possess, by virtue of their external mycelium, a larger and better distributed absorbing surface. Transport of phosphorus along the external hyphae into the root then occurs in parallel with transport to the root surface by diffusion through the soil[39,42].

Rates of uptake of phosphorus can be measured by nutrient flow analysis[7] and expressed per unit of root weight (specific absorption rate – SAR) or per unit of root length (inflow – I). Experiments indicate that inflows of phosphorus to mycorrhizal roots can be greater than inflows to comparable non-mycorrhizal roots by up to one order of magnitude. Under conditions where growth responses to inoculation with VA mycorrhizal fungi are obtained, average inflows to mycorrhizal root systems are typically 2–5 times greater than inflows to the root systems of non-mycorrhizal controls. The rate of uptake of phosphorus by a whole root system may thus be greatly influenced by the proportion of roots that are mycorrhizal[36].

In annual crops in field soils mycorrhizal infection originates from propagules present in the soil at the time of sowing. Development of mycorrhizal infection can be followed by frequent sampling of roots and assessment[17] of the proportion of mycorrhizal root in the samples after clearing and staining[17,30]. Typically, an initial lag phase, where the fraction (%) of root mycorrhizal remains low, is followed by a rapid rise towards a relatively constant level or plateau (Fig. 1). The per cent mycorrhizal infection reached at the plateau seems to depend on the nutrient status of the soil, particularly the phosphorus status. Thus increasing rates of phosphorus fertiliser addition may give rise to decreasing plateau levels of mycorrhizal infection (Fig. 1).

Development of mycorrhizas and/or changes in phosphorus nutrition may be associated with changes in root:shoot ratio[39]. In addition, the growth and maintenance of the obligately-biotrophic mycorrhizal fungus may have an energy cost which is not negligible. Recent work at Rothamsted and elsewhere[28,29,33,41] indicates that as much as 10% of that host photosynthate translocated below ground may be expended in fungal growth and respiration.

In summary, the growth of a VA mycorrhizal plant in phosphorus-deficient soil is the result of complex dynamic interactions between shoot, root, mycorrhizal fungus and soil. These interactions are not, at present, clearly understood.

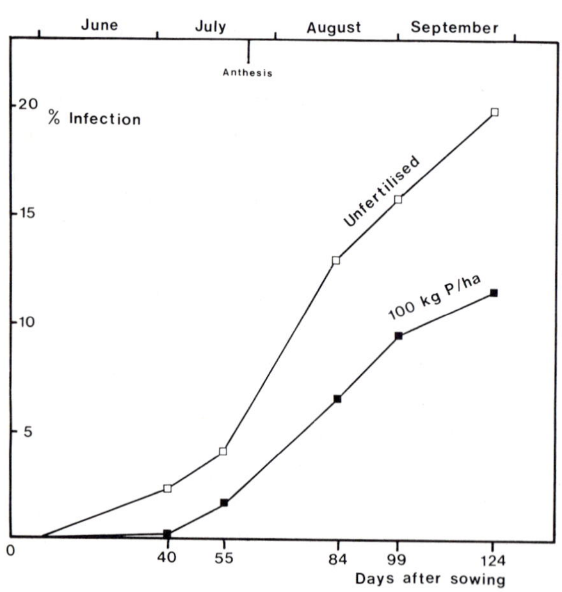

Fig. 1. The progress of mycorrhizal infection in field-grown barley.

In this paper we discuss some of these interactions and propose models for a) the spread of mycorrhizal infection in root systems and b) consequent effects on plant growth.

The infection process

Forms and numbers of propagules in field soils and their germination

The soil inoculum of VA mycorrhizal fungi consists of spores and sporocarps, mycelium, or intact or fragmented mycorrhizal roots. Where propagules can be identified visually, *e.g.* spores wet-sieved from field soils, they can be counted directly. Alternatively, densities of viable propagules can be estimated by procedures based on calculations of most probable number (MPN) from the results of experiments where host plants are grown in field soil, serially diluted with sterile soil[1,31]. Although MPN estimates of propagule density are subject to wide error, they usually exceed estimates made from direct counts of spores. MPN estimates of propagule densities and direct counts of spores in a number of arable topsoils near Leeds have been made on several occasions since 1977. Most MPN estimates lay in the range 0.1–10 viable propagules g^{-1} soil while direct spore counts were up to ten times smaller[33].

There is some information available on the factors influencing germination of spores of VAM fungi in the laboratory but little on germination behaviour in field soils. Laboratory experiments indicate the importance of temperature, pH and ionic environment[13,15,18,37]. Spores can be stored in a viable condition for long periods under dry conditions[43]. Germination appears not to be influenced by the presence of host roots[26,32]. Spores are frequently found to contain hyperparasites which may prevent germination[14,35]. In general, after the onset of favorable conditions of temperature and moisture, germination starts after a few days, germination rate reaches a maximum a few days later and then declines as nearly all the viable spores have completed germination. Spores of some species may exhibit dormancy.

Data from an experiment[33] in which the germination of spores in soil was followed are shown in Fig. 2. Known numbers of spores of *Glomus clarus* Nicholson & Schenck were mixed into sterile soil in small pots which were watered, planted with pregerminated maize seedlings and placed in a controlled environment chamber at 25°C. Pots were harvested at intervals and numbers of spores that had *not* germinated were counted after quantitative wet-sieving. Germination was essentially complete after 6–8 days.

To our knowledge, there is no experimental work on spore germination under field conditions, but there is no reason to suppose behaviour greatly different from that in the laboratory. In the UK spore germination probably occurs in response to increasing soil temperature in spring. Our observations in Nigeria indicated that spore germination in the field occurred rapidly in response to increased soil moisture at the onset of the rainy season.

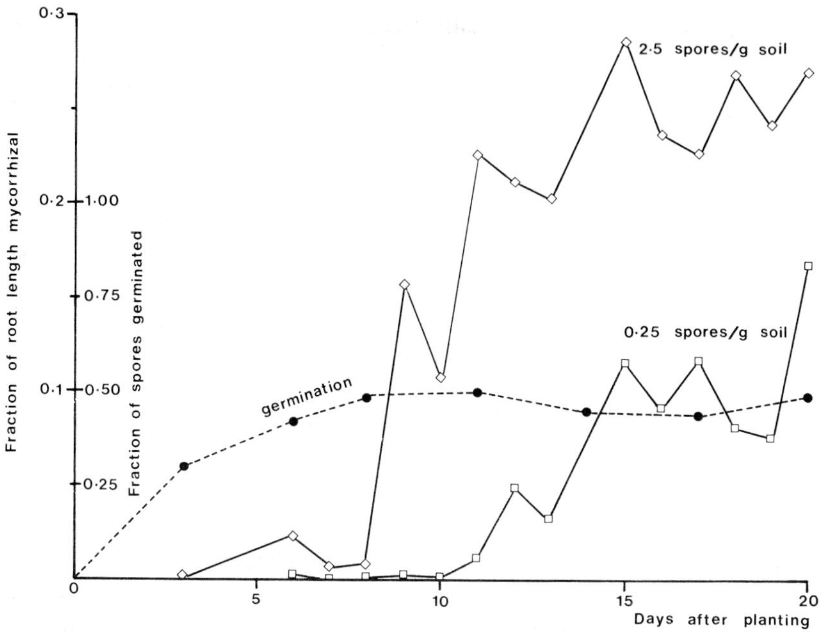

Fig. 2. Spore germination in the soil and progress of mycorrhizal infection in maize seedlings at two inoculum densities.

Germination of most spores of VA mycorrhizal fungi is marked by the appearance of a simple germ tube, although some may have the capacity for multiple germination. The germ tube subsequently branches in a rather characteristic way[26,32]. There are no published observations of hyphal growth following germination in undisturbed soil but the hyphae probably spread out radially to occupy a spherical region of wide extent compared to spore diameter but containing a low hyphal density. Hyphae do not appear preferentially to grow towards roots[27,32]. Nevertheless, the form of hyphal growth following germination must influence, in part, the probability that an individual propagule causes a primary infection point on a host root system.

The relation between propagule density and early levels of mycorrhizal infection

Fig. 2 shows also the effect of a ten-fold decrease in spore density on the development of mycorrhizal infection in the roots of maize seedlings. In the pots containing spores at a density of $0.25\,g^{-1}$ infection was apparently delayed. Later, infection remained lower than in the pots containing 2.5 spores g^{-1}.

Another series of plants was raised under similar conditions in soils with a range of spore densities. The soils were prepared by serially diluting soil containing 25 spores g^{-1} with sterile soil. The dilution factor was 10. Pots (five replicates at each dilution) were harvested after 14 days. Fig. 3 shows $\log_{10}$ inoculum dilution plotted against $\log_{10}$ infected root length. With a spore density

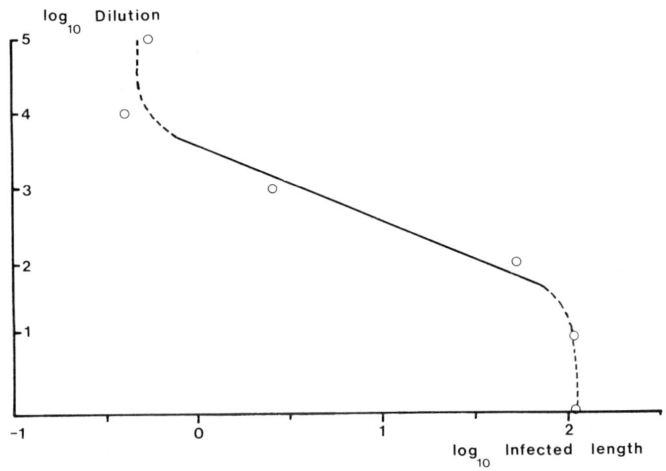

Fig. 3. The relation between inoculum dilution and infected root length in mycorrhizal maize seedlings.

of 25 spores g^{-1} the roots were clearly 'saturated'. At dilutions of 10^4 and 10^5, not all of the replicates became mycorrhizal as the spores were 'diluted out'. At intermediate dilutions, the plot is linear with a slope close to unity, indicating that infected root length was proportional to the propagule number in the pot.

Quantitative relations between propagule density and the establishment and development of VA mycorrhizal infection are clearly of considerable interest. Recently, there have been several attempts to derive theoretical relationships for soil-borne plant pathogens[3,16,19,23]. These attempts have engendered some controversy[2]. The ideas of Gilligan[16] seem most applicable to VA mycorrhizal fungi. Smith and Walker[40] have developed a simple mathematical model for infection from soil-borne propagules of VA mycorrhizal fungi but this model only covers in detail events within the root.

In approaching this problem, we have considered the following events: germination, hyphal growth, the encounter of host roots by hyphae, formation of appressoria and penetration of the root cortex. The symbols used hereafter are listed in Table 1.

Consider a unit volume of soil, containing N propagules, into which grow L cm of host roots. The latter are assumed to consist of a randomly-distributed array of straight cylindrical segments with radius r_o, see[4,24]. The probability that a single propagule lies closer to a root than a distance r is then given approximately by

$$1 - \exp[\frac{\pi}{\lambda} L(r + r_o)^2]$$

For N propagules, the number that lie within a distance r from the root surface will then be

$$N\{1 - \exp[\frac{\pi}{\lambda} L(r + r_o)^2]\}$$

Table 1. List of symbols

Symbol	Description
t	time, days
W	plant dry weight, g
ΔW	change in W in one time increment
P	plant phosphorus content, mg
ΔP	change in P in one time increment
S	root length per g root dry weight, cm g^{-1}
R	root : total plant dry weight ratio
r_0	root radius, cm
L, L_0, L_t	root length (cm) at time = any, 0, t
F, F_0, F_t	fraction of total root length that is mycorrhizal
L_i, L_{i0}, L_{it}	infected (mycorrhizal) root length, cm
n, n_0, n_t	number of entry points or infection units
Δn	change in n in one time increment
b	relative rate of increase of root length, day^{-1}
k	relative rate of increase of entry points, day^{-1}
RMR	relative multiplication rate, day^{-1}
RGR	relative growth rate, g dry matter g^{-1} day^{-1}
U	specific utilisation rate, g dry matter day^{-1} mg^{-1} phosphorus
I_m, I_c	inflow of phosphorus to mycorrhizal and non-mycorrhizal root segments respectively, mg cm^{-1} day^{-1}
N	propagule number per unit volume
r	radius of competence, cm
hgr	rate of hyphal growth in soil, cm day^{-1}
a	rate of growth of infection units, cm day^{-1}
$l, \bar{l}$	length of an infection unit, mean length of units, cm
n	number of size class
f_n	relative frequency in class n
f_i	fraction of infectable root length infected
f_s	fraction of total root length susceptible to infection

If it is assumed that a propagule has to be closer to the root than the distance r to cause an infection (r is now the 'radius of competence' of the propagule), it is possible to estimate the number of primary infection points that could potentially arise from a given propagule density at a given root density. Suppose, for example, that there is one germinated spore per g of soil and that the root density is 1 cm^{-2}. If each propagule that is 'in range' may give rise to only one primary infection point, for $r_o = 0.025$ cm and values of r of 0.1, 0.2 and 0.3 cm, potentially there will be in one litre of soil 48, 147 and 282 primary infection points.

We have attempted to measure r in an experiment carried out in tubes where inocula containing many spores of *Glomus caledonius* were separated from host roots by layers of soil of different thicknesses. Roots were prevented from growing into the separating layer of soil by barriers of finely woven terylene cloth (Fig. 4). Tubes were sampled at frequent intervals. Infection was first observed at

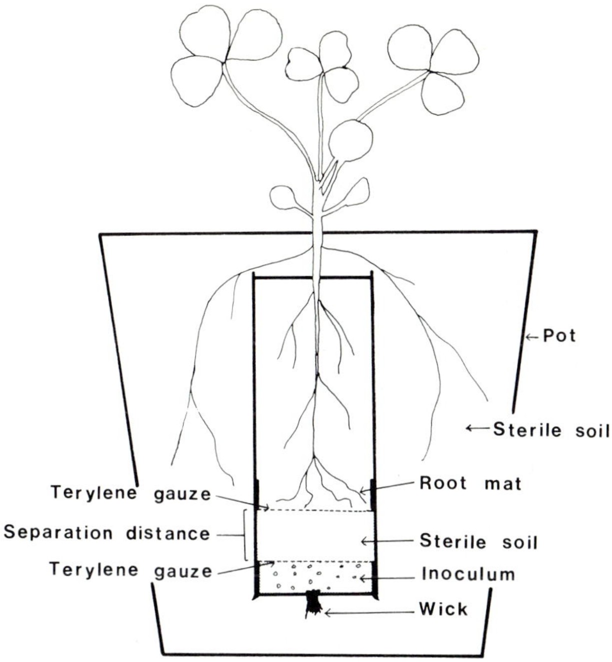

Fig. 4. Schematic diagram of the tubes used to measure the radius of competence of spores. For explanation – see text.

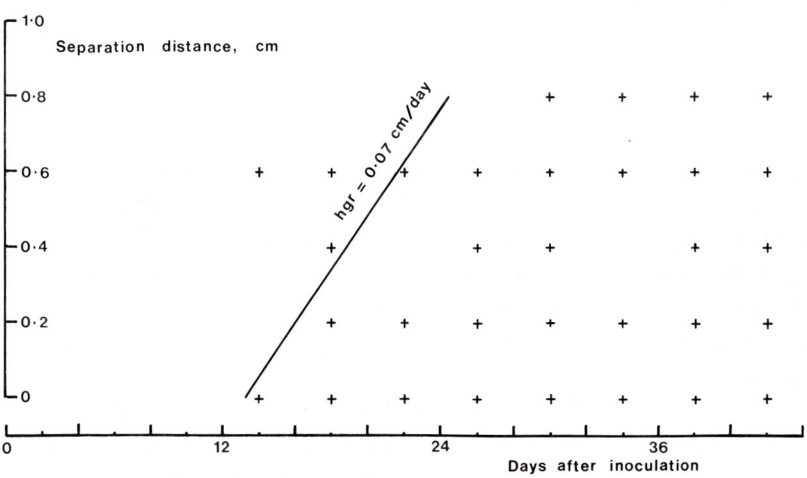

Fig. 5. The occurrence of infection with time at different separation distances. + = mycorrhizal infection recorded in the root mat. For explanation – see text. The solid line, the slope of which represents hyphal growth rate (hgr) was obtained by fitting a regression of separation distance on time at which infection was first observed at each separation distance.

a separation distance of 0 cm after 14 days. Thereafter it was observed successively at greater and greater separation distances, to the maximum of 0.8 cm at 30 days (Fig. 5). The rate of hyphal growth away from the spores was estimated from the data to be approximately 0.07 cm day^{-1}. Similar results were obtained with *Glomus clarus*.

This hyphal growth rate can be included in the theoretical example above. If all the spores in the litre of soil germinate together at time zero, there will be 15, 60 and 129 potential infection points after 1, 2 and 3 days respectively. 95% of the spores will be 'in range' after 15 days.

The infection data shown in Fig. 2 can be analysed in the same manner. When spore density was 2.5 spores g^{-1}, it can be calculated that approximately 60 spores per pot would have germinated and been 'in range' after 8 days when root density was 4 cm g^{-1}. At 0.25 spores g^{-1}, only 6 spores per pot would have been 'in range' after 8 days. This may be the explanation of the apparent lag in infection at the higher dilution.

Thus, in a system of roots extending in unit volume of soil, the observed rate of appearance of primary infection points will depend on the density of viable propagules and their rate of germination, on the rate of root growth, on the rate of radial growth of hyphae through the soil and on the probability of infection after a hypha has encountered a root surface. The latter probability may decline as the distance from propagule to root increases. The pursuit of such questions is, however, academic until more information from experiments has accumulated.

Growth of infection units

VA mycorrhizal infections observed in roots seem to be built of discrete infection units[9] which may or may not overlap longitudinally. Each infection unit consists of an entry point or group of entry points, deriving from an external hypha and associated internal mycelium (see Fig. 6). The following general picture of development is supported by observations on mycorrhizal roots of onion and red clover[9,10,36].

A hypha arriving at the epidermal surface of the root forms an appressorium from which hyphae penetrate the root cortex, usually between the epidermal cells and then between or through the subepidermal cells where hyphal coils may be formed. On reaching the inner layers of the cortex, the penetrating hyphae branch and grow out longitudinally in both directions usually along the intercellular spaces, but sometimes passing from cell to cell. As the hyphae grow out along the cortex, arbuscules are formed on side branches. These side branches penetrate the walls of the cortical cells and then branch repeatedly within the lumen. Arbuscule formation first occurs in cortical cells located near the entry point but later at progressively greater distances away along the cortex. In older infection units it is possible to observe a sequence of arbuscule formation and senescence with the youngest stages behind the growing tips of the longitudinal hyphae and the oldest near the entry point[9]. Arbuscules are

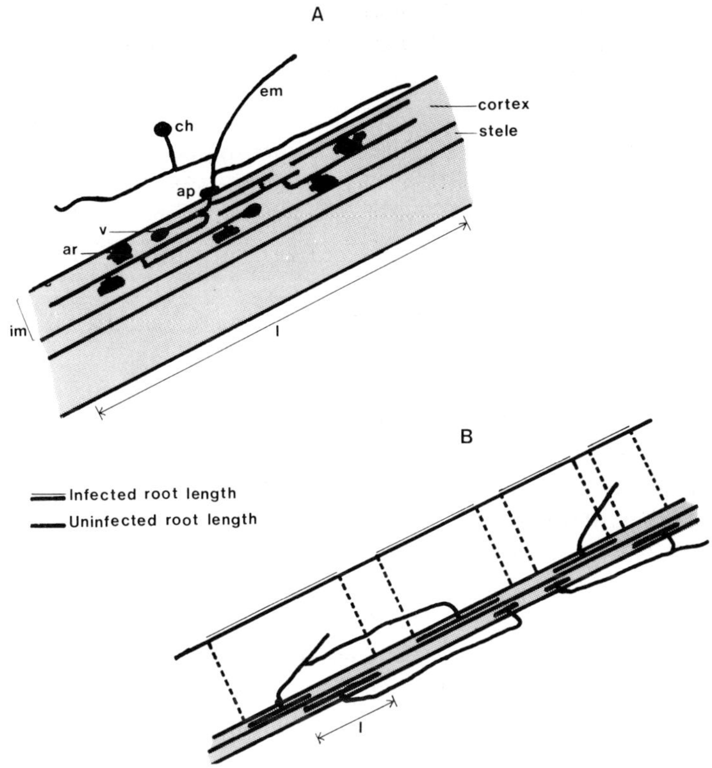

Fig. 6. Schematic diagram of mycorrhizal roots. A. Single infection unit: em, im – external and internal mycelium; ar – arbuscule; v – vesicle; ap – appressorium; ch – chlamydospore; *l* – length of infection unit. B. Multiple overlapping infection units.. ═══, ——— – portions recorded by the random intercept method as mycorrhizal or non-mycorrhizal respectively.

considered to be the main site of transfer of metabolites between host and fungus and it is likely that the infection unit becomes self-supporting through carbon compounds derived from the host only after the formation of arbuscules[10]. Until this time the growth of the infection unit is probably 'fuelled' *via* its connection to the external mycelium. In the case of a primary infection point initial growth is presumably at the expense of reserves within the spores or other propagules. Secondary infection units must depend for their initial growth mostly on carbon compounds translocated from those formed previously.

Information on rates of growth of individual infection units is clearly desirable but not readily obtainable because infection units cannot be observed *in vivo*. Experimental approaches to this problem involve frequent sequential harvests of replicate plants, observation of developmental stages and the fitting of mathematical models to infection data.

Efforts have been made[6,11] to determine the duration of arbuscules but neither paper considered the rate of longitudinal growth of infection units. Cox and Sanders[9] found that the length of infection units of *Glomus mosseae* in the roots of onion did not appear to exceed 1.0 cm but again did not consider rates of longitudinal growth.

Following the methodology of Tinker[42], Smith and Walker[40] considered the principles likely to govern the process of infection from soil-borne propagules and formulated two differential equations. They were then able, by fitting the equations to their data, to calculate values of two parameters. These were the rate of longitudinal extension of infection units and the rate of formation of infection units per unit length of root.

We have recently attempted to measure rates of longitudinal growth of infection units of *Glomus caledonius* in the roots of red clover. Seedlings were grown in tubes without a separating soil layer (Fig. 4). They were harvested as before at intervals of two days, starting at 10 days. Root systems from the tubes were cleared, stained and mounted on microscope slides. Entry points were counted and the lengths of every discrete infection unit measured.

When the logarithms of the numbers of entry points were plotted against time in days from the first observation of infection, a straight line fit was obtained with a slope of 0.22. Thus the increase in entry point numbers with time could be described by the equation $n_t = n_0 \exp(0.22\,t)$. The frequencies of infection units in different length classes for a root system harvested 16 days after the first observation of infection are shown in Table 2.

We propose the following simple model to explain these results. Let the population of infection units increase with time exponentially, such that $n_t = n_0 \exp(kt)$. Let the rate of growth in length of the infection units be a cm day^{-1}.

Table 2. Frequencies of infection units in different length classes

Class* n	Observed frequency	Observed rel. frequency	Calculated rel. frequency, f_n
1	158	0.612	0.612
2	46	0.178	0.237
3	20	0.078	0.092
4	13	0.050	0.036
5	6	0.023	0.014
6	5	0.019	0.005
7	5	0.019	0.002
8	2	0.008	0.001
9	2	0.008	<0.001
10	1	0.004	<0.001
Total	258	0.999	0.999

* The class interval was 1.75 mm.

If, after a period of time, the population is classified according to the length of the infection units, it is easy to show that the relative frequency in class n will be given by

$$f_n = [\exp(-(n-1)kc/a)][1 - \exp(-kc/a)]$$

where c is the class interval and $n = 1$ corresponds to the length class whose lower boundary is zero.

In the data given in Table 2, $f_1 = 0.612$ and the class interval is 1.75 mm. For $k = 0.22$, a may be estimated from the equation to be 0.4 mm day^{-1}. Also shown in Table 2 are predicted values of f_n which are in reasonable agreement with the observed values up to 8.75 mm (class 5). For greater lengths the predicted relative frequencies are lower than the observed. There are several possible explanations for this discrepancy of which the simplest is that we have occasionally measured as single units, what were, in fact, groups of overlapping infection units. This mistake can easily be made if appressoria become detached during preparation of the samples.

Our estimate of the longitudinal growth rate of infection units, 0.4 mm day^{-1}, is close to the figure of 1.00 mm day^{-1} for infection units in subterranean clover calculated by Smith and Walker[40].

Spread of secondary infection in root systems

After the phase of primary infection, secondary infection units are produced at an ever-increasing rate. These arise from the spread, along or between roots, of the abundant external hyphae which grow out from previously formed infection units. The ratio of external to internal fungal biomass may be as great as 1 : 1[20]. External mycelium appears to be produced in proportion to the length of root infected[36] and may grow away from the root for considerable distances. Spread of infection is therefore unlikely to be limited by the growth of external mycelium unless root densities are very low.

A theoretical relationship between numbers and lengths of infection units and the fraction of total root length that is mycorrhizal may be derived as follows. Let it be assumed that infection units develop at random along susceptible root length and that they have a length frequency distribution given by $dF(l) = f(l)\,dl$, $0 < l < \infty$, where l is the length of an infection unit. If there are n infection units per unit length of root, the number of length $l' = nf(l')\,dl'$ and the probability, P′, that a point on the root is not overlapped by an infection unit of length l' is $\exp[-nl'f(l')\,dl']$ (see[34]). The probability, P, that a point is not overlapped at all is then given by

$$P = \prod_0^\infty \exp[-nl'f(l')\,dl'] = exp\left[-\int_0^\infty nl'f(l')\,dl'\right]$$

$$= \exp[-n\bar{l}]$$

where $\bar{l}$ is the mean length of the infection units. The proportion of root length covered by infection units is therefore $1 - \exp[-n\bar{l}]$. It is this quantity, the fraction of root length mycorrhizal, that is easily, and hence usually, measured (see[17]).

Assuming that infection is equally probable along the whole of the root, the length of infected root $L_i = L(1 - \exp[-n\bar{l}])$. In fact, suberised root and meristematic and elongating regions may not be susceptible.

If infection units are propagated exponentially according to the equation $n_t = n_0 \exp(kt)$, then the fraction of root length infected at time t, F_t, will be given by

$$F_t = 1 - \exp[-n_0 \bar{l} \exp(kt)]$$

In an exponentially increasing population, the mean age approaches a constant value. If a, the rate of longitudinal growth of infection units in the cortex does not vary, then $\bar{l}$ will also approach a constant value.

If root length per unit volume is also increasing exponentially (which may often occur) according to the equation $L_t = L_0 \exp(bt)$, then F_t is given by

$$F_t = 1 - \exp[-n'_0 \bar{l} \exp((k - b)t)/L_0]$$

where n'_0 is the starting number of infection units per unit volume of soil. This Gompertz function produces a S-shaped curve when F_t is plotted against time. It is interesting to note that curves of the progress of plant disease are often better fitted by Gompertz than logistic functions[5].

The dynamics of infection spread in root systems

Computer models

When applied to whole plants growing in soil, the simple model of infection spread outlined above gives rise to very complex calculations, but these are readily handled on the computer, using the technique of reiterative calculation of changes taking place in the system during successive time steps.

We have adopted the following general procedure. Firstly, we have collected sets of data from experiments on the spread of mycorrhizal infection in red clover and other plants grown in controlled environments. Measurements made at successive times included root and shoot dry weight, root length, numbers of entry points and fraction of root length mycorrhizal, lengths of isolated infection units, phosphorus contents *etc.* Secondly, we have smoothed the data by fitting polynomial functions. Thirdly, we have constructed computer models using the principles described above, which simulate the behaviour of the real system when supplied with appropriate initial values and values of parameters.

A model of infection spread

Experimental procedure Red clover plants were grown from seed in a growth chamber in small pots of soil inoculated with spores of *Glomus caledonius*. Each pot received exactly the same dose of inoculum and there was no check to growth other than shortage of phosphorus. There was a similar series of uninoculated pots. Harvests of five replicate pots were made at intervals of 4 days, starting on day 12 and finishing on day 56 after emergence of the seedlings. A substantial positive growth response to inoculation was obtained.

The polynomial coefficients and r^2 for the equations fitted to the log-transformed data are given in Table 3.

Construction of the model Root length – this was deemed to increase at a rate given by the polynomial function derived from the measurements.

Increase in numbers of entry points – differentiation of the polynomial equation describing the relation between log n_t and time gave an equation for relative multiplication rate (RMR $= n^{-1}\, dn/dt$). RMR declined from 0.39 day^{-1} at 12 days to 0.05 at 56 days (Fig. 7). RMR is equivalent to *k*, but is clearly not constant as assumed previously. The decline in RMR accompanied an increase in the fraction of the root length mycorrhizal and may have been due to physiological factors determining susceptibility to infection and/or to an inability of new infection units to develop on regions of the root already mycorrhizal. In developing the model, these possibilities were taken into account by assuming that 1) a fixed fraction of the root was not susceptible to infection by the fungus and 2) new infection units could only develop in previously uninfected root.

Programming – a short programme was written in Basic incorporating the following initial values, parameters and calculation steps for time intervals of one day.

Table 3. Polynomial coefficients of the equations used in connection with the simulation of infection spread*

Variable $\log_e$ of	Coefficient of				r^2
	t^0	t^1	t^2	t^3	
Total root length	+2.0923	+0.24097	−0.00327	+0.0002197	0.993
Total entry points	−4.2449	+0.60307	−0.009784	+0.0000588	0.997
Infected root length	−3.5452	+0.40364	−0.003858	+0.00000749	0.993
% infection	−1.0324	+0.16267	−0.000588	−0.00001430	0.985
RMR**	+0.60307	−0.01957	+0.000176	–	–

* Polynomial equations were fitted to the log-transformed data using the MINITAB package implemented on the Amdahl computer installation at the University of Leeds.

** RMR, relative multiplication rate, new entry points per existing entry point per day.

Infection spread model:

Alterable parameters – a, rate of longitudinal growth of infection units, cm day^{-1}
k, RMR, day^{-1}
f_s, fraction of root length susceptible to mycorrhizal infection

Initial values – L_{i0}, infected root length, cm
n_0, number of infection units

Function determining total root length – $L_t = f(t)$, a polynomial function derived from the data

Calculation steps –

1) time t
2) $L_t = f(t)$
3) infectable root length $= L_t f_s$
4) fraction of infectable root length infected, $f_i = 1 - \exp[-n_t \bar{l}/L_t f_s]$
5) fraction of total root length infected $F_t = f_i f_s$
6) potential number of new entry points $= kn_t$
7) actual number of new entry points Δn = potential number $\times (1 - f_i)$
8) new total number of entry points $n_{t+1} = n_t + \Delta n$
9) total length of all infection units $\sum l = \sum n_t a(\mathrm{t} - t)$ where $\mathrm{t} - t$ is the age of units formed at time t
10) $\bar{l} = \sum l / n_{t+1}$
11) mean length of infected root per entry point $= F_t L_t / n_{t+1}$
12) return to 1) with $t = t + 1$

The programme was run on a CBM 8032 microcomputer with the initial values from the data set. When the parameters a and f_s were set at 0.1 cm day^{-1} and 0.5 respectively, and k was set at the starting value of 0.39 day^{-1} derived from the data, agreement between simulation and experiment was excellent (Figs. 7–9).

The value of a used was close to that estimated above and was therefore not unrealistic. In the real system, the maximum value of F, the fraction of root length mycorrhizal, is probably determined by a feedback loop. Here it was necessary arbitrarily to set f_s to 0.5 in order to obtain a good fit. This will be discussed further below.

The simulation also predicted $\bar{l}$, the mean length of infection units. In principle, $\bar{l}$ could have been measured but this proved impossible in practice because of overlapping. There were too few isolated infection units for measurements to be made on these except at very early times.

The close agreement possible between the simulation and the real data has increased our confidence that this line of attack may be a useful aid to understanding the biological mechanism of infection spread. For example, any

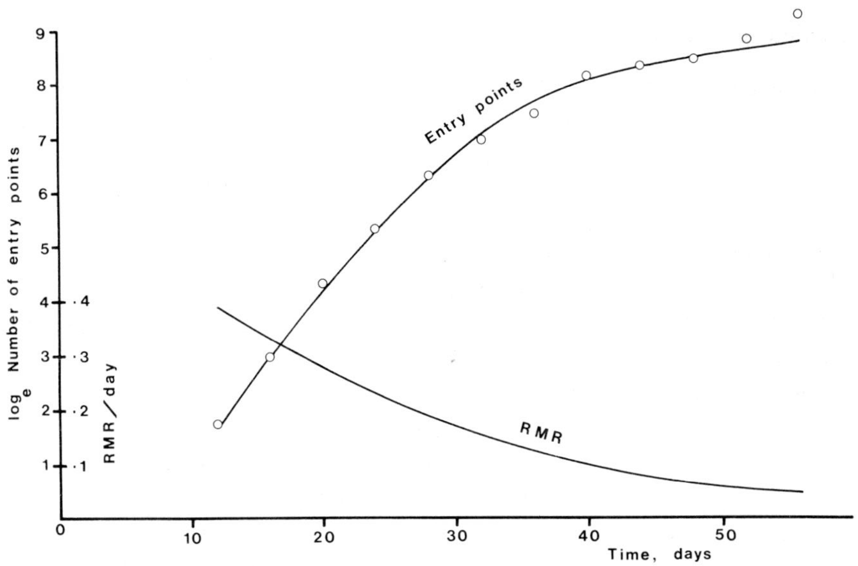

Fig. 7. Infection spread model. The increase in entry point numbers (n) and the change in relative multiplication rate (RMR) with time. 0 = data points; —— = simulation.

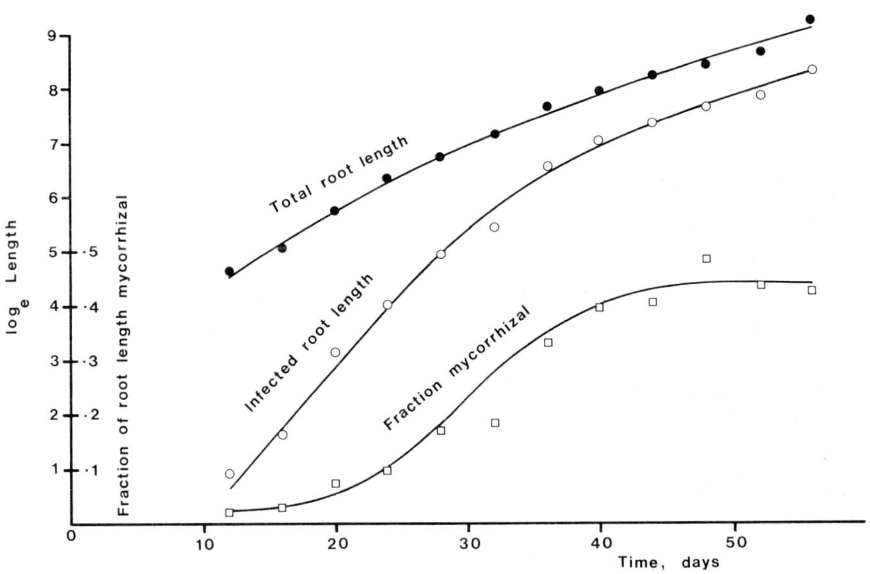

Fig. 8. Infection spread model. The increase of total root length (L cm), infected root length (L_i cm) and the fraction of root length mycorrhizal (F) with time. ●, ○, □ – data points; solid lines – simulation.

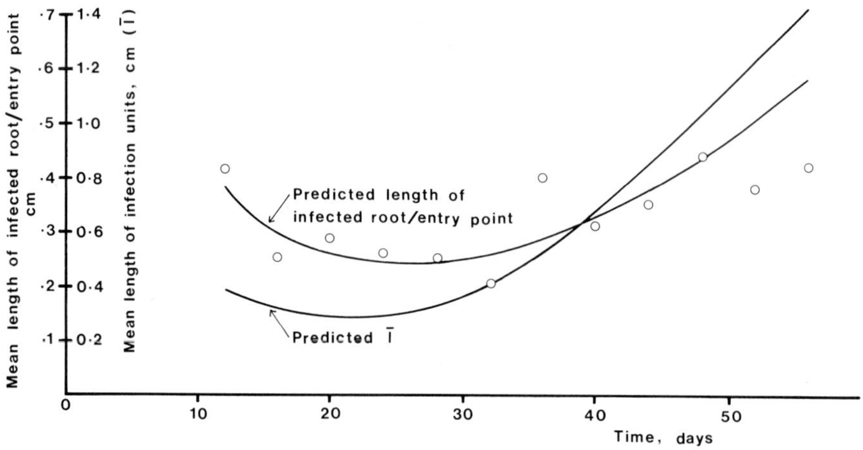

Fig. 9. Infection spread model. The changes in mean length of infected root per entry point (L_i), and mean length of infection units ($\bar{l}$) with time. ○ – data for mean length of infected root per entry point; solid lines – simulation.

factor which may interfere with the growth of external mycelium such as low soil moisture, grazing by soil fungivores[44], or ionic environment[8] should lead to a decrease in RMR and hence to an increase in $\bar{l}$ and the mean length of infected root per entry point.

Consequences of infection spread for the growth of the host

A whole-plant model

Our current version of this model contains the following two main assumptions: 1) that inflows of phosphorus to mycorrhizal root systems can be considered to have two distinct components, inflow to the root tissue itself and inflow via the fungus (hyphal inflow[36]); 2) that host growth is linked to the uptake of phosphorus such that a parameter determining the rate of dry matter increase can be calculated if we know how it is related to a parameter dependent on the uptake of phosphorus. Assumption 1 seems generally satisfactory although it can be criticised in detail. For example, it ignores possible competition for phosphorus between root and fungal hyphae. It may also be difficult to specify appropriate values for the two components which tend to decrease with time as the soil becomes depleted of phosphorus. Assumption 2 is more complex and difficult to justify. Utilisation of phosphorus within the plant may be subject to 'ontogenetic drift', while mycorrhizal and non-mycorrhizal plants may exhibit very different behaviour in terms of the efficiency with which they utilise phosphorus in dry matter production[41].

Experimental data We have taken for analysis original data[36] relating to the effects of the LAM endophyte (*Glomus caledonius*) on the growth of seedling

Table 4. Data on the growth of onions ± inoculation with *Glomus caledonius**

Harvest time days	Total dry weight** g		Calculated RGR*** day^{-1}	
	Mycorrhizal	Control	Mycorrhizal	Control
17	.014	.014	.105	.110
20	.027	.020	.104	.105
23	.031	.021	.103	.101
26	.030	.037	.101	.096
29	.053	.047	.100	.092
33	.081	.067	.098	.086
37	.130	.092	.096	.080
41	.170	.124	.094	.074
45	.257	.169	.092	.068
49	.348	.211	.090	.062

* Data additional to those provided by Sanders *et al.*[36]
** Dry weights are means for three replicate plants
*** RGR was calculated after fitting polynomials to log-transformed dry weight data

onions. Details of RGR were not included and are therefore given in Table 4.

RGR of the controls declined more rapidly than that of the mycorrhizal plants due probably to lower phosphorus inflows and an associated decline in internal phosphorus concentration (see Figs. 4 and 7 in ref. 36). When RGR was plotted against internal phosphorus concentration, separate curvilinear relations were obtained for the two types of plant. This was not convenient for the purpose of the model. There appeared to be only one parameter of growth which had a unique relation to plant phosphorus content irrespective of whether or not plants were mycorrhizal. This was specific utilisation rate, given by the equation $U = P^{-1}\,dw/dt$, where P is total phosphorus content and dw/dt is absolute growth rate[21,22]. Fig. 10 shows the relation between dw/dt and P for the onion data. The equation of the fitted curve was $P^{-1}\,dw/dt = 0.0483 - 0.0331\,P + 0.0138\,P^2$, $r^2 = 0.995$.

Values of inflow used in the model were close to the time average of mean values measured[36], namely, 0.5×10^{-4} mg phosphorus cm^{-1} day^{-1} for non-mycorrhizal root segments and 1.9×10^{-4} mg cm^{-1} day^{-1} for mycorrhizal segments. For simplicity, variation of inflow with time was ignored although this introduced inaccuracy.

The model takes no account of changes in root : shoot ratio and specific root length with time and phosphorus status.

Details of the model The whole-plant model again employs reiterative calculation of changes during time steps of one day. The calculations, programmed in Basic, are carried out as follows.

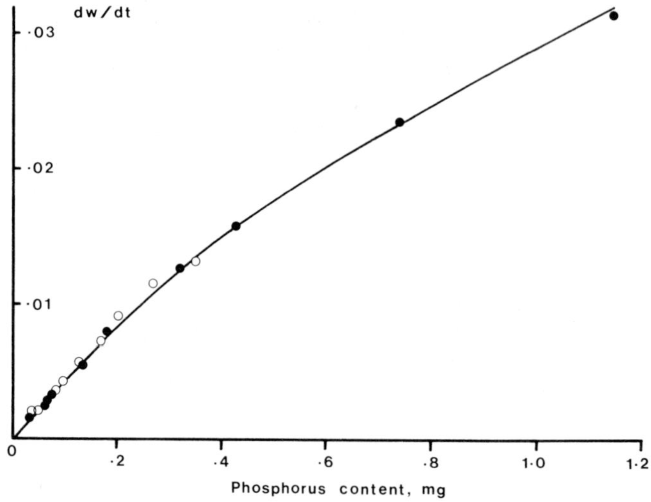

Fig. 10. The relation between absolute growth rate (dw/dt, g d^{-1}) and phosphorus content (P) used in the whole plant model. ●, ○ – data for mycorrhizal and non-mycorrhizal plants respectively; solid line – fitted polynomial function.

Whole-plant model:

Alterable parameters –	a, k, f_s – as before
	I_m, I_c inflows to mycorrhizal and non-mycorrhizal segments, mg cm^{-1} day^{-1}
	S specific root length cm g^{-1} dry weight
	R root : total dry weight ratio
Initial values –	L_{i0}, n_0 – as before
Function –	P^{-1} dw/dt = f(P)
Calculation steps –	1) time t
	2) dry weight increase $\Delta W = Pf(P)$
	3) RGR = $\Delta W/W$
	4) $W = W + \Delta W$
	5) $L_t = WRS$
	6) steps as in infection spread model, leading to L_{it}, F_t and n_{t+1}
	7) P uptake, $\Delta P = I_m \cdot L_{it} + I_c(1 - L_{it})$
	8) $P = P + \Delta P$
	9) P concentration = P/W
	10) calculation of $\bar{l}$, as before
	11) return to 1) with t = t + 1

Results of simulations Reasonable fits to the data were obtained with the following initial and parameter values: $L_{i0} = 1$ cm, $n_0 = 4$, $k = 0.27$ day^{-1}, $a = 0.1$ cm day^{-1}, I_m, I_c = 1.9, 0.5 × 10^{-4} mg cm^{-1} day^{-1} respectively, $f_s = 0.8$, R = 0.354, S = 8 × 10^3 cm g^{-1}.

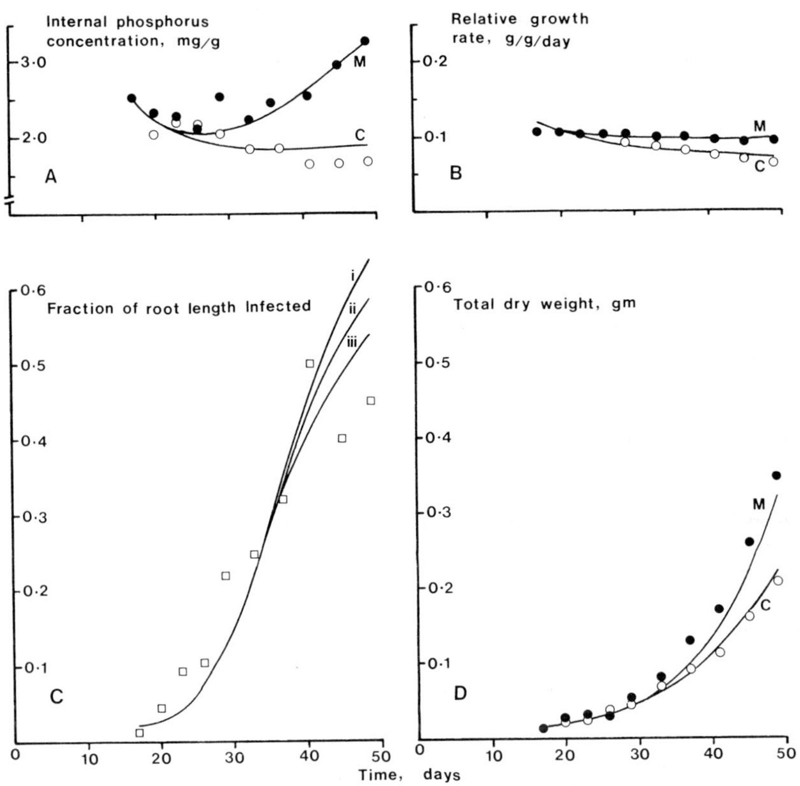

Fig. 11. Results of the simulation using the whole-plant model. Changes with time in: A – Phosphorus concentration in the dry matter; B – relative growth rate (RGR); C – fraction of root length mycorrhizal (F) [i, ii, iii – infection curves corresponding to values of f_s of 1.0, 0.9 and 0.8 respectively.] D – plant dry weight (W). ●, ○ – data from mycorrhizal and control plants; solid lines – simulation.

Since numbers of entry points were not recorded in the experiment, values of parameters and initial values relating to infection spread were arbitrarily adjusted to give satisfactory agreement between the predicted and measured values of F_t (Fig. 11c). Values of R and S were similar for mycorrhizal and control plants and did not vary greatly during the experiment.

f_s was set at 0.8 to improve the fit of the infection curve. When f_s was set at 1.0, the infection curve reached a plateau at 0.67 (Fig. 11c), but this change alone had little effect on predicted dry matter increase and phosphorus uptake.

The comparison between the predicted and observed dry weights (Fig. 11d) shows reasonable agreement although growth was underestimated in mycorrhizal plants. RGR, phosphorus concentration in the dry matter and phosphorus content were satisfactorily predicted (Figs. 11a, 11b, 12b). Root growth was overestimated (Fig. 12a).

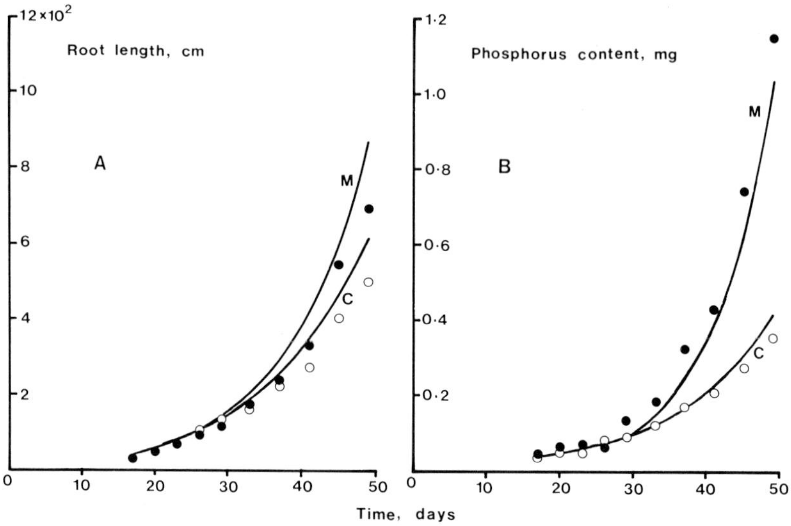

Fig. 12. Results of the simulation using the whole plant model. Changes with time in: A – root length (L); B – phosphorus content (P). Symbols as in Fig. 11.

Discussion

The whole-plant model is a crude over-simplification of the real system and it is surprising that it predicts the growth response arising from infection spread so well. The approach may therefore be useful in indicating future directions of experimentation.

An interesting feature of the model is its prediction that infection curves will reach a plateau at below $F = 1$ even if the entire root length is susceptible. This behaviour seems to result from feedback. At the plateau an equilibrium is established between root and fungal growth when the RGRs of total and infected root length become equal. Any factor which tends to change the RGR of the root will change the position of the equilibrium and hence the level of the plateau on the infection curve. The effect of phosphorus fertiliser addition to the soil on infection levels may be interpreted in this way. Increasing soil phosphorus should increase the value of I_c so that the relative importance of hyphal inflow will decrease[38]. An increase in root RGR will then reduce the level of the plateau on the infection curve (Fig. 13).

Extension of the model will depend on collection of data over a wider range of soil and environmental conditions.

Conclusion

Model building has greatly aided our understanding of the biology of the VA mycorrhizal symbiosis and is justifiable for this reason alone. There are however wider implications for the practical manipulation of VA mycorrhizal symbioses in the field.

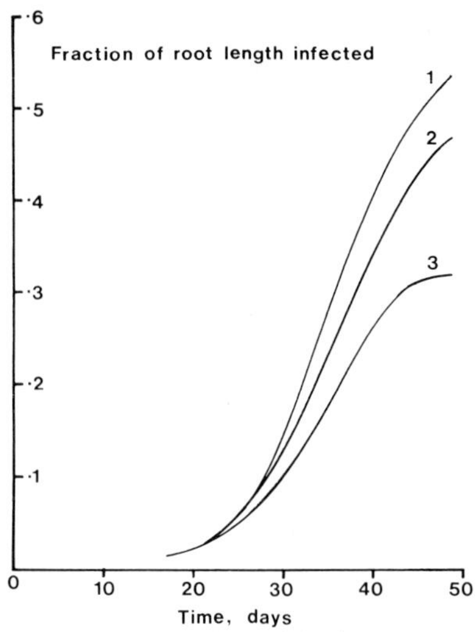

Fig. 13. Changes in the infection curve caused by alteration in the parameters I_m and I_c (mg cm^{-1} day^{-1}) to simulate the effect of added phosphorus fertiliser. Other parameters as for Fig. 12. 1. $I_m = 1.9 \times 10^{-4}$, $I_i = 0.5 \times 10^{-4}$ 2. $I_m = 2.3 \times 10^{-4}$, $I_c = 0.9 \times 10^{-4}$ 3. $I_m = 2.9 \times 10^{-4}$, $I_c = 1.5 \times 10^{-4}$.

On a world scale there is a need to develop sustainable agricultural systems with low fertiliser inputs. This can be achieved by breeding crop cultivars which produce more dry matter per unit of nutrient taken up, by technological improvements leading to greater recoveries of applied fertilisers and by the use of VA mycorrhizal associations, again to increase the recovery of current fertiliser applications and residues[1,12].

Practical problems in developing the use of VA mycorrhizas in agriculture and forestry have been recently reviewed[1,25]. Effective species of VA mycorrhizal fungi, suitable for use as inoculants on annual crops, must infect rapidly and extensively. Observed differences in effectiveness between species may be due largely to differences in infection rates. Our discussion has concluded that the latter will depend heavily on such factors as inoculum density and rates of propagule germination and hyphal growth. Perennial crops have received less attention than annual crops but ability of inoculated fungi to infect new roots rapidly must also be important if the fungi are to be effective.

Marx and Schenck[25] discuss the interesting example of a procedure used commercially in Mississippi for the inoculation of sweetgum. Cultures of selected VA mycorrhizal fungi are raised in pots in a glasshouse on the roots of a herbaceous host. The contents of the pots are then spread on to a recently fumigated nursery soil and incorporated. An annual agronomic crop is then grown to multiply the fungal inoculum in the soil before the tree seed is planted. Seedlings grown by this procedure are much superior to those seeded directly into the fumigated nursery. Procedures of this kind may well become more common in the future, particularly where crops are to be transplanted to infertile sites. A clear understanding of the factors affecting establishment and rates of spread of VA mycorrhizal infection will greatly facilitate management of inoculation processes as well as understanding of differences in rates of infection by indigenous fungi in field soils.

Acknowledgements The authors wish to thank the Agricultural Research Council and the Government of Pakistan for financial support, D J Pilbeam and L E Reed for valuable discussion and criticism of the manuscript and A. Khaliq for providing Fig. 1.

References

1 Abbott L K and Robson A D 1982 The role of vesicular-arbuscular mycorrhizal fungi in agriculture and selection of fungi for inoculation. Aust. J. Agric. Res. 33, 389–408.

2 Baker R and Drury R 1981 Inoculum potential and soilborne pathogens: the essence of every model is within the frame. Phytopathology 71, 363–372.

3 Baker R, Maurer C L and Maurer R A 1967 Ecology of plant pathogens in soil. VII. Mathematical models and inoculum density. Phytopathology 57, 662–666.

4 Barley K P 1970 The configuration of the root system in relation to nutrient uptake. Adv. Agron. 22, 159–201.

5 Berger R D 1981 Comparison of the gompertz and logistic equations to describe plant disease progress. Phytopathology 71, 716–719.

6 Bevege D I and Bowen G D 1975 Endogone strain and host plant differences in development of vesicular-arbuscular mycorrhizas. *In* Endomycorrhizas. Eds. F E Sanders, B Mosse and P B Tinker, Academic Press, London, pp 77–86.

7 Brewster J L and Tinker P B H 1972 Nutrient flow rates into roots. Soils Ferts. 35, 355–359.

8 Chambers C A, Smith S E and Smith F A 1980 Effects of ammonium and nitrate ions on the mycorrhizal infection, nodulation and growth of *Trifolium subterraneum*. New Phytol. 90, 293–303.

9 Cox G C and Sanders F E 1974 Ultrastructure of the host-fungus interface in a vesicular-arbuscular mycorrhiza. New Phytol. 73, 901–912.

10 Cox G C, Sanders F E, Tinker P B and Wild J 1975 Ultrastructural evidence relating to host-endophyte transfer in a vesicular-arbuscular mycorrhiza. *In* Endomycorrhizas. Eds. F E Sanders, B Mosse and P B Tinker. Academic Press, London, pp 297–312.

11 Cox G C and Tinker P B 1976 Translocation and transfer of nutrients in VA mycorrhizas. I. The arbuscule and phosphorus transfer: a quantitative ultrastructural study. New Phytol. 77, 371–378.

12 Cox W J and Robson A D 1980 The optimization of plant nutrition – improving the efficiency of fertiliser utilisation. Proc. Austr. Agron. Conf., 157–176.

13 Daniels B A and Graham S O 1976 Effects of nutrition and soil extracts on germination of *Glomus mosseae* spores. Mycologia 68, 108–116.

14 Daniels B A and Menge J A 1980 Hyperparasitization of vesicular-arbuscular mycorrhizal fungi. Phytopathology 70, 584–588.

15 Daniels B A and Trappe J M 1980 Factors affecting spore germination of the vesicular-arbuscular mycorrhizal fungus *Glomus epigaeus*. Mycologia 72, 457–471.

16 Gilligan C A 1979 Modelling rhizosphere infection. Phytopathology 69, 782–784.

17 Giovanetti M and Mosse B 1980 An evaluation of techniques for measuring vesicular-arbuscular mycorrhizal infection in roots. New Phytol. 84, 489–500.

18 Green W E, Graham S O and Schenck N C 1976 The influence of pH on the germination of vesicular-arbuscular mycorrhizal spores. Mycologia 68, 929–934.

19 Grogan R G, Sall M A and Punja Z K 1980 Concepts for modelling root infection by soilborne fungi. Phytopathology 70, 361–363.

20 Hepper C M 1981 Techniques for studying the infection of plants by vesicular-arbuscular mycorrhizal fungi under axenic conditions. New Phytol. 88, 641–647.

21 Hunt R 1978 Plant growth analysis. Studies in biology no. 58. Edward Arnold, London, 67 p.

22 Keay J, Biddiscombe E F and Ozanne P G 1970 The comparative rates of phosphorus absorption by eight annual pasture species. Aust J. Agric. Res. 21, 33–44.

23 Leonard K J 1980 A reinterpretation of the mathematical analysis of rhizoplane and rhizosphere effects. Phytopathology 70, 695–696.

24 Marriott F H C 1972 The distance distributions of a random point from randomly spaced lines. Biometrics, 874–875.

25 Marx D H and Schenck N C 1982 Potential of mycorrhizal symbiosis in agriculture and forest productivity. In Challenging problems in plant health. American Phytopathological Society. *In press.*

26 Mosse B 1959 The regular germination of resting spores and some observations on the growth requirements of an Endogone sp. causing vesicular-arbuscular mycorrhizae. Trans. Br. Mycol. Soc. 42, 272–286.

27 Mosse B and Hepper C M 1975 Vesicular-arbuscular mycorrhizal infections in root organ cultures. Physiol. Plant Pathol. 5, 215–223.

28 Pang P C and Paul E A 1980 Effects of vesicular-arbuscular mycorrhiza on ^{14}C and ^{15}N distribution in nodulated faba beans. Can. J. Soil Sci. 60, 241–250.

29 Paul E A and Kucey R M N 1981 Carbon flow in plant microbial associations. Science 213, 473–474.

30 Phillips J M and Hayman D S 1970 Improved procedures for clearing roots and staining parasitic and vesicular-arbuscular mycorrhizal fungi. Trans. Br. Mycol. Soc. 55, 158–160.

31 Porter W M 1979 The 'Most Probable Number' method for enumerating infective propagules of vesicular-arbuscular mycorrhizal fungi in soil. Austr. J. Soil Res. 17, 515–519.

32 Powell C Ll 1976 Development of mycorrhizal infections from Endogone spores and infected root segments. Trans. Br. Mycol. Soc. 60, 439–445.

33 Reed L E and Sanders F E Unpublished data.

34 Roach S A 1968 The Theory of Random Clumping. Methuen, London, 94 p.

35 Ross J P and Ruttencutter R 1977 Population dynamics of two vesicular-arbuscular endomycorrhizal fungi and the role of hyperparasitic fungi. Phytopathology 67, 490–496.

36 Sanders F E, Tinker P B, Black R L B and Palmerley S M 1977 The development of endomycorrhizal root systems. I. Spread of infection and growth-promoting effects with four species of vesicular-arbuscular endophyte. New Phytol. 78, 257–268.

37 Schenck N C, Graham S O and Green N E 1975 Temperature and light effects on contamination and spore germination of vesicular-arbuscular mycorrhizal fungi. Mycologia 67, 1189–1192.

38 Smith S E 1982 Inflow of phosphate into mycorrhizal and non-mycorrhizal plants of *Trifolium subterraneum* at different levels of soil phosphate. New Phytol. 90, 293–303.

39 Smith S E 1980 Mycorrhizas of autotrophic higher plants. Biol. Rev. 55, 475–510.

40 Smith S E and Walker N A 1981 A quantitative study of mycorrhizal infection in Trifolium: separate determination of the rates of infection and mycelial growth. New Phytol. 89, 225–240.

41 Stribley D P, Tinker P B and Rayner J H 1980 Relation of internal phosphorus concentration and plant weight in plants infected by vesicular-arbuscular mycorrhizas. New Phytol. 86, 261–266.

42 Tinker P B H 1975 Effects of vesicular-arbuscular mycorrhizas on higher plants. Symp. Soc. Exp. Biol. 29, 325–349.

43 Tommerup I C and Kidby D K 1979 Preservation of spores of vesicular-arbuscular endophytes by L-drying. Appl. Environ. Microbiol. 37, 831–835.

44 Warnock A J, Fitter A H and Usher M B 1982 The influence of a springtail *Folsomia candida* on the mycorrhizal association of leek *Allium porrum* and the vesicular-arbuscular mycorrhizal endophyte *Glomus fasciculatus*. New Phytol. 90, 285–292.

Plant and Soil 71, 247–256 (1983). A-24

The concept of succession in relation to the spread of sheathing mycorrhizal fungi on inoculated tree seedlings growing in unsterile soils

P. A. MASON, J. WILSON and F. T. LAST
Institute of Terrestrial Ecology, Bush Estate, Penicuik, Midlothian, EH26 OQB, Scotland

and C. WALKER
Forestry Commission, Northern Research Station, Roslin, Midlothian, EH25 9SY, Scotland

Key words Birch Mycorrhizal fruitbodies Mycorrhizal succession Sheathing mycorrhizas Sitka spruce

Summary Repeated annual assessments of the toadstools (fruitbodies) of mycorrhizal fungi associated with a mixed stand of *Betula* spp. indicated that they were produced in a pattern ordered in time and space, suggesting a succession with identifiable early- and late-stage fungi. This concept is supported by below-ground observations of mycorrhizas which, however, need to be augmented.

Both early- and late-stage mycorrhizal fungi form mycorrhizas on seedlings growing in axenic ('aseptic') conditions. In contrast, only early-stage fungi seem able to trigger mycorrhizal formation on seedlings growing in unsterile soils.

During axenic propagation, the early-stage *Hebeloma sacchariolens* and the late-stage *Amanita muscaria* formed similar numbers of mycorrhizas per root system. After being transplanted to a range of unsterile field soils, *A. muscaria* failed to keep pace with the spread of the developing root system: no more *A. muscaria* mycorrhizas were formed. On the other hand the continued development of *H. sacchariolens* mycorrhizas precluded, during the first season after transplanting, the development of mycorrhizas by fungi naturally occurring in field soils. In the second season, however, the development of *H. sacchariolens* mycorrhizas was restricted in acid peat but not in three other types of soil.

The development of *Laccaria* mycorrhizas after inoculating Sitka spruce with this fungus was associated, irrespective of soil type, with accelerated tree growth; with heights at the end of the first season being doubled.

Introduction

The benefits conferred on trees by sheathing mycorrhizas are not in dispute although the substantive field evidence leaves much to be desired. To ensure the early development of effective mycorrhizas, and hence successful plant establishment, forest nursery seedbeds and/or seedlings have been inoculated with naturally infested (i) soils, (ii) raw humus or (iii) leaf litter[17]. But the use of these usually bulky and unwieldy inocula has often incurred risks of introducing root pathogens and weeds. A possibly less hazardous approach, which is attracting increasing interest is the use of pure cultures. When testing pure cultures of mycorrhizal fungi, it was seen that different fungi possessed different *in vitro* cultural characteristics, growth rates and temperature optima as well as different host preferences[22]. This, and further evidence, now suggests that the (i) development and (ii) effectiveness of tree mycorrhizas depend, as in the legume root nodule complex, upon a range of genetically controlled host and fungal factors. But which fungus, or isolate of a fungus, should be used when attempting to forge beneficial associations which will persist after planting out?

Microbiologists are very much aware that the assemblages of bacteria and fungi colonising the surfaces of leaves and roots change in a systematic manner as their substrates get older. This being so, it should not be surprising if the fungi forming mycorrhizas with a tree, 40 years-old, differ from those associated with saplings – an age related succession. Although the concept of 'mycorrhizal succession' is only now being enunciated there are already several pieces of circumstantial evidence. How significant are they and how will the impact of mycorrhizal succession affect the choice of fungi for the controlled inoculation of tree seedlings?

Concept of succession

With some notable exceptions most of the lists associating fruitbodies of sheathing mycorrhizal fungi with trees have paid little or no attention to the age of the host[21]. But when the occurrence of fruitbodies was recorded during the first ten years after planting a mixed stand of birches (*Betula pendula* Roth and *B. pubescens* Ehrh.), clear evidence was obtained of a sequence in both time and space[15]. Fruitbodies of *Hebeloma crustuliniforme* (Bull. ex St. Amans) Quélet and a *Laccaria* sp. were observed within two years after planting. These fungi were joined by *Thelephora terrestris* Ehrenb. ex Fr. (in year 3), *Inocybe lanuginella* (Schroet.) Konrad and Maublanc and *Lactarius pubescens* (Fr. ex Krombh.) Fr. (year 4), *Hebeloma* spp. (year 5), species of *Cortinarius* and *Leccinum* in year 6 and of *Russula* in year 10.

In year 2, fruitbodies of *H. crustuliniforme* were concentrated in a 'ring', encircling stem bases, with a mean radius of 22 cm which by year 6 had increased to 73 cm. The fruitbodies of *Lactarius*, like those of *H. crustuliniforme*, tended to occur in zones of maximal density which first showed in year 4 with a radius of 40 cm and then progressively increased to 69 cm in year 6. In the same year, fruitbodies of the newly appearing *Leccinum* spp. were nearest the trees (27 cm from stem bases). Thus it was possible to discern the elements of a number of concentric rings with each new species initially appearing close to the bases of trees and then moving progressively outwards in successive years[6]. Interestingly, the fruitbodies of *Laccaria* sp., instead of being arranged in rings, seem to follow the lines of secondarily thickened roots.

In addition to being associated with birch, fruitbodies of *Hebeloma crustuliniforme*, *Laccaria* species, particularly *Laccaria laccata* (Scop. ex Fr.) Cooke, *Thelephora terrestris* and species of *Inocybe* have also been found in association with young conifers. *Hebeloma crustuliniforme* has been recorded in abundance with seedlings of *Pseudotsuga menziesii* (Mirb.) Franco[23] and *Pinus radiata* D. Don before, but not after, transplanting[2]. In contrast, fruitbodies of species of *Suillus* and *Inocybe* were associated with stands of *P. radiata* 5 or more years old but seldom with younger trees. Thus the sequences of mycorrhizal fruitbodies associated with birches and pines have many features in common. Furthermore, toadstools of *Amanita muscaria* (L. ex Fr.) Pers. ex Hooker were found in plantations of *P. radiata*[2] and *Pseudotsuga menziesii*[3] when more than 10 and 12 years old respectively. As yet they have not appeared in our experimental birch plot which is now 11 years old. This clear age dependence is supported by observations made in southern India where the annual production of *Amanita muscaria* toadstools increased from 32 to 3,720 per thousand trees in plantations of *Pinus patula* Schl. and Cham. 5 and 16 years-old respectively[10].

Together these fruitbody observations suggest that some mycorrhizal fungi (*e.g. Hebeloma, Laccaria* and *Inocybe*) are characteristic of young stands of trees whereas others *e.g.* species of *Leccinum, Russula* and particularly *Amanita* are characteristically associated with older trees. Although the occurrence of fruitbodies does not give a comprehensive insight of events below ground-level, the idea of mycorrhizal succession based on fruitbody data is given credence by the apparently close relation between the presence and position of (i) mycorrhizas and (ii) the fruitbodies of their associated fungi[8,20]. Furthermore, Warcup (pers. comm.) when working in the stand of *Betula* spp., already mentioned, was able to relate the fruitbodies to their own distinctive types of mycorrhizas (Table 1).

Root isolations by Chu-Chou and Grace[3] confirmed that *Hebeloma crustuliniforme* was only associated with young seedlings of Douglas fir and that *Amanita muscaria* could only be isolated from mycorrhizas when trees were 13 or more years old. Mycorrhizal root observations[9,13,19] on nursery and outplanted saplings have also shown that the range of mycorrhizas associated with forest trees can alter considerably as trees get older. Thus the evidence derived from root isolates and the appearance of mycorrhizas supports the concept of mycorrhizal succession.

We know that fungi found both early and late in the mycorrhizal sequence readily form mycorrhizas in axenic culture[14]. However, the functional importance of succession was recognised when birch seedlings were planted into soil cores sampled immediately beneath fruitbodies of species of *Hebeloma, Inocybe, Laccaria, Lactarius* or *Leccinum*; the seedlings developed numerous mycorrhizas of *Laccaria, Inocybe* and *Hebeloma*, a few with *Lactarius* and none with *Leccinum*[4]. Fox[7], similarly, found that birch seedlings planted into basidiospore amended soils readily developed mycorrhizas with *Hebeloma* and *Inocybe* species but not with *Lactarius* or *Leccinum*. Thus, it seems that early-stage fungi, *e.g. Hebeloma* and *Laccaria*, are able to form mycorrhizas on young seedlings growing in unsterile as well as sterile soils, whereas late-stage fungi will only form mycorrhizas on seedlings growing in axenic conditions. But is the

Table 1. The relation between (i) fungi isolated from mycorrhizas and (ii) the occurrence of fruitbodies when observations were made on 5 birch saplings of *Betula pendula* during their fourth season after being planted into a brown earth (Warcup, pers. comm., 1975)

Tree	Fungi isolated from different mycorrhizas	Fruitbodies recorded in association in the autumn
1.	*Thelephora terrestris*	None
2.	*Hebeloma* sp.	*Hebeloma* sp.
	An unidentified basidiomycete	
	An unidentified ascomycete	
3.	*Laccaria 'laccata'*	*Laccaria 'laccata'*
	An unidentified basidiomycete	
	A white ascomycete differing from the unidentified ascomycete ex tree 2.	
4.	*Hebeloma* sp.	*Hebeloma* sp.
	Lactarius pubescens	*Lactarius pubescens*
	Thelephora terrestris	
	An unidentified ascomycete	
5.	*Hebeloma* sp.	*Hebeloma* sp.
	Lactarius pubescens	*Lactarius pubescens*
	Laccaria 'laccata'	*Laccaria 'laccata'*
	An unidentified ascomycete	

development of mycorrhizas solely host/fungus dependent? Can the association be altered by other soil microbes or by different soil types?

The colonisation of birch seedlings by early-and late-stage fungi

To examine rates of spread and abilities of different mycorrhizal fungi to compete with naturally occurring microbes, birch seedlings were planted into a range of unsterile soils after being inoculated, during axenic propagation[16], with *Hebeloma sacchariolens* Quélet (an early-stage fungus), *Amanita muscaria* (a late-stage fungus) or *Paxillus involutus* (Batsch) Fr. (a fungus often associated with young birch trees growing on slag heaps and other sites of dereliction.)

At the end of the 8 week period of axenic propagation at least 65% of the root tips of inoculated seedlings were mycorrhizal. When a second sample was inspected 121 days after being transplanted into one of 4 different soils (2 peaty and 2 mineral) the distribution of mycorrhizas differed greatly. At this time, the end of the first growing season, mycorrhizas, derived from naturally occurring soil inocula and attributed to *Laccaria proxima* (Boud.) Pat., *Inocybe lanuginella* and *Thelephora terrestris*, had developed on the uninoculated controls. Mycorrhizas formed by a similar group of mycorrhizal fungi had developed on the seedlings inoculated with *Amanita muscaria*, this fungus appearing to be totally incapable of forming mycorrhizas on seedlings in unsterile conditions;

Table 2. Percentages of sheathing mycorrhizas, after growing seedlings of *Betula pendula* in a range of 4 unsterile soils, attributable to the fungi introduced two years earlier while propagating in axenic conditions

		Fungi introduced during propagation		
		Hebeloma saccharíolens	*Paxillus involutus*	*Amanita muscaria*
Peaty soils	A	0	44	0
	B	100	29	0
Mineral soils	C	93	14	0
	D	99	43	0

even the mycorrhizas formed during axenic propagation could not be traced. In contrast, *H. sacchariolens* and *P. involutus* both spread to form numerous mycorrhizas to the total exclusion of mycorrhizas formed by naturally occurring microbes. The latter were absent even though many roots were non-mycorrhizal.

Although the different soil types didn't significantly affect the development of *H. sacchariolens* and *P. involutus* during the first year, there were appreciable interactions in the second year (Table 2).

H. sacchariolens continued to spread and colonise new roots on seedlings in 3 of the 4 soils, with at least 90% of root fragments becoming mycorrhizal but in the fourth soil (A), an acid peat, it failed totally. Instead the mycorrhizas formed in the second year in this combination of treatments, were attributable to naturally occurring soil-borne inocula. Mycorrhizas of *Paxillus involutus* developed less prolifically than those of *H. sacchariolens*. They accounted for 40–50% of the mycorrhizas on seedlings in two of the four soils and less than 30% on seedlings in the other soils. At the end of the second year a detailed

Table 3. Percentages of mycorrhizas, two years after planting inoculated birch seedlings into a range of four unsterile soils, attributable to naturally occurring mycorrhizal fungi. (See Table 2)

		Naturally occurring mycorrhizal fungi*				
		*Hebeloma*** sp.	*Inocybe* sp.	*Thelephora terrestris*	*Cenococcum geophilum*	*Leccinum* sp.
Peaty soils	A	0	0	17	0	51
	B	0	5	66	<1	<1
Mineral soils	C	51	0	9	8	0
	D	49	0	31	0	0

* These figures show the percentages of mycorrhizas of naturally occurring mycorrhizal fungi on seedlings irrespective of their original treatment.

** These figures do not include mycorrhizas of *Hebeloma sacchariolens* (see Table 2).

assessment was made of the mycorrhizas formed by naturally occurring fungi irrespective of the inoculation treatment (Table 3). Interestingly, mycorrhizas were only formed by naturally occurring *Hebeloma* spp. when seedlings were growing in mineral soils. This suggests that the occurrence of this genus may be strongly controlled by soil type.

The inability of naturally occurring fungi, during the first year, to colonise seedlings inoculated with *H. sacchariolens* and *P. involutus* suggests that they can be kept in check by the use of judiciously selected fungi. At the end of the second year, it appears that the early advantage afforded by inoculation during axenic propagation may eventually be overtaken, albeit by isolates of other early-stage fungi.

Effects of *Laccaria* sp. and *Paxillus involutus* on the growth of Sitka spruce seedlings

The results of experiments done with birch encouraged the inoculation of Sitka spruce (*Picea sitchensis* (Bong.) Carr.) with *Laccaria* sp. or one of two isolates, 0086 and 16, of *Paxillus involutus*. As before seedlings were inoculated during axenic propagation and then transferred eight weeks later to pots containing two peat and two mineral soils.

Unlike the experiment done with birch there were significant differences in mycorrhiza formation during propagation (Table 4). These differences were maintained so that by the end of the first growing season, mycorrhizas attributable to *Laccaria* sp. were most numerous (67%) and those attributable to *P. involutus* 0086 the least numerous (5%) (Table 4). Moreover, growth appeared to be directly related to the abundance of mycorrhizas. Isolate 0086 of *P.*

Table 4. Percentages of Sitka spruce root fragments with mycorrhizas at the end of the first growing season after inoculating with three different fungi

Samples taken at	Inoculated during propagation with: Uninoculated controls	*Paxillus involutus* isolate 0086	*Paxillus involutus* isolate 16	*Laccaria* sp.	L.S.D. ($p = 0.05$)
End of propagation in axenic conditions (days 0–63)	0* (*0.0*)	2 (*7.0*)	12 (*20.0*)	76 (*60.7*)	(*27.72*)
End of first growing season in unsterile soils (days 63–221)	5 (*12.3*)	5 (*12.9*)	22 (*27.9*)	67 (*55.2*)	(*2.35*)

* For statistical analysis, percentages were transformed to angles which are italicised.

Table 5. Effects of inoculating with three mycorrhizal fungi on the heights of Sitka spruce seedlings.

Samples taken at	Inoculated during propagation with				LSD ($p = 0.05$)
	Uninoculated controls	*Paxillus involutus* Isolate 0086	*Paxillus involutus* Isolate 16	*Laccaria* sp.	
End of propagetion in axenic conditions (days 0–63)	2.8 cm	2.9 cm	2.9 cm	3.0 cm	1.02
End of first growing season in unsterile soils (days 63–221)	3.1 cm	3.6 cm	4.8 cm	6.5 cm	0.17

involutus was associated with a 58% height increase whereas *Laccaria* sp. stimulated a greater than 100% increase in shoot height (Table 5). Similarly inoculation with *Laccaria* trebled numbers of root fragments whereas *P. involutus* isolate 0086 increased them by merely 71%.

These results emphasise the importance of inoculating tree seedlings with 'effective' isolates of selected mycorrhizal fungi. Nevertheless, although not too

Table 6. Effects of inoculating Sitka spruce seedlings with three mycorrhizal fungi on subsequent root growth expressed as numbers (n) of root fragments*

Samples taken at	Inoculated during propagation with				LSD ($p = 0.05$)
	Uninoculated controls	*Paxillus involutus* Isolate 0086	*Paxillus involutus* Isolate 16	*Laccaria* sp.	
End of propagation in axenic conditions (days 0–63)	7** (*0.91*)	14 (*1.17*)	16 (*1.23*)	13 (*1.15*)	(*0.43*)
End of first growing season in unsterile soils (days 63–221)	52 (*1.72*)	89 (*1.96*)	130 (*2.12*)	150 (*2.18*)	(*0.04*)

* For statistical analysis numbers (n) of root fragments per plant were transformed to log (n + 1). Transformed data are in italics.

** For the sample taken at the end of propagation, (n) represents the total number of roots on each seedling, whereas for the end of season sample, (n) represents the total number of roots counted in zones 1 cm wide and spaced 2 cm apart.

Table 7. The characteristics of early- and late-stage fungi associated with birch saplings planted in a brown earth at Bush Estate, Edinburgh in 1971

	Laccaria spp. (early -stage)	*Lactarius pubescens*	*Amanita muscaria* (late -stage)
Years after planting in which fruitbodies of mycorrhizal fungi were first observed	2	4	>10
Years after planting in which fungi were first isolated from mycorrhizas	2	4	>10
Ability (+ or −) to form mycorrhizas when			
a) basidiospores were added to unsterile soil[7]	+	−	N.t.
b) mycelial cultures were added to unsterile soil[4]	+	+/−*	−
Ability (+ or −) of mycorrhizal fungus to spread after transferring inoculated seedlings from axenic propagation to unsterile soil	+	+	−

N.t. = Not tested.
* +/− = Mycorrhizas very seldom formed.

much should be drawn from the use of one isolate, it seems that the use of *Laccaria* species might prove to be rewarding[18] – this fungus seems easy to establish and appears to be a strong competitior, conferring, at the same time measurable benefits on its hosts.

Critique

The results in this other papers[4,5,7,11,12] suggest that the concept of 'succession' has an important bearing on the selection of fungi for inoculating young tree seedlings. Although fruitbody observations by themselves do not tell the whole story, they have highlighted the occurrence of a sequence of early-, followed by late-stage fungi (Table 7).

Early-stage fungi, such as *Laccaria* spp., are associated with young tree seedlings; they have the ability to colonize roots and develop mycorrhizas in unsterile soil, irrespective of naturally occurring inocula. In contrast, late-stage fungi, such as *Amanita muscaria* are unable to spread and form mycorrhizas on young seedlings growing in unsterile soils. No doubt, a range of intermediates exist and will be identified in due course. *Lactarius pubescens*, for example, possesses little or no ability to infect seedlings from inoculum added to unsterile soils yet can spread from mycorrhizas, pre-formed in axenic conditions (Table 7).

The concept of mycorrhizal succession seems to provide a rational basis for the selection of fungi for inoculating tree seedlings. However much more needs to be known about the sequence of fungi associated with tree saplings growing in a range of soils, *e.g.* the peats which occupy a central position in British forestry. Knowledge of this sort may help explain the recently reported failure of *Laccaria laccata* and *Hebeloma crustuliniforme* to infect new roots of Douglas fir[1]. It may also enable us to understand the role of *Paxillus involutus* which appears to dominate from the outset on sites of dereliction yet can be replaced by a range of fungi in mineral and peaty soils within 1–2 years.

On balance, the evidence to date suggests, when considering controlled inoculations, that the choice of fungus should, in the first instance, be controlled by its position within the mycorrhizal succession; secondly by its ability to stimulate the growth of its host, some isolates being more beneficial than others; and thirdly it is clear that the persistence of the association will depend not only on the host × mycorrhizal fungus interplay but on a three membered system host × mycorrhizal fungus × other soil microbes, with each component being subject to environmental influences.

References

1 Bledsoe C S, Tennyson K and Lopushinsky W 1982 Survival and growth of outplanted Douglas-fir seedlings inoculated with mycorrhizal fungi. Can. J. For. Res. 12, 720–723.

2 Chu-Chou M 1979 Mycorrhizal fungi of *Pinus radiata* in New Zealand. Soil Biol. Biochem. 11, 557–562.

3 Chu-Chou M and Grace L J 1981 Mycorrhizal fungi of *Pseudotsuga menziesii* in the North Island of New Zealand. Soil Biol. Biochem. 13, 247–249.

4 Deacon J W, Donaldson S J and Last F T 1983 Sequences and interactions of mycorrhizal fungi on birch. Plant and Soil 71, 257–262.

5 Fleming L V 1983 Succession of mycorrhizal fungi on birch: infection of seedlings planted around mature trees. Plant and Soil 71, 263–267.

6 Ford E D, Mason P A and Pelham J 1980 Spatial patterns of sporophore distribution around a young birch tree in three successive years. Trans. Br. mycol. Soc. 75, 287–296.

7 Fox F M 1983 Role of basidiospores as inocula of mycorrhizal fungi of birch. Plant and Soil 71, 269–273.

8 Laiho O 1970 *Paxillus involutus* as a mycorrhizal symbiont of forest trees. Acta For. Fenn. 106, 1–72.

9 Lamb R J 1979 Factors responsible for the distribution of mycorrhizal fungi of *Pinus* in Eastern Australia. Aust. For. Res. 9, 25–34.

10 Last F T, Mason P A, Smith R, Pelham J, Bhoja Shetty K A and Mahmoud Hussain A M 1981 Factors affecting the production of fruitbodies of *Amanita muscaria* in plantations of *Pinus patula*. Proceedings of the Indian Academy of Science (Plant Science) 90, 91–98.

11 Last F T, Mason P A, Wilson J, Deacon J W and Walker C 1983 Succession: its relevance when selecting sheathing mycorrhizal fungi for inoculating young trees. Nature London *in litt.*

12 Last F T, Mason P A, Wilson J and Deacon J W 1983 Fine roots and sheathing mycorrhizas: their formation, function and dynamics. Plant and Soil 71, 9–21.

13 Marks G C and Foster R C 1967 Succession of mycorrhizal associations on individual roots of Radiata pine. Aust. For. 31, 193–201.

14 Mason P A 1980 Aseptic synthesis of sheathing (ecto-) mycorrhizas. *In* Tissue Culture Methods for Plant Pathologists. Eds. D S Ingram and J P Helgeson. pp 173–178 Blackwell Scientific, Oxford.

15 Mason P A, Last F T, Pelham J and Ingleby K 1982 Ecology of some fungi associated with an ageing stand of birches (*Betula pendula* and *B. pubescens*). For. Ecol. Manage. 4, 19–39.

16 Mason P A, Dighton J, Last F T and Wilson J 1983 Procedure for establishing sheathing mycorrhizas on tree seedlings. Forest Ecol. Manage. 5, 47–53.

17 Mikola P 1980 Mycorrhizae across the frontiers. *In* Tropical Mycorrhiza Research. Ed. P Mikola. pp 3–10. Clarendon Press, Oxford.

18 Molina R 1982 Use of the ectomycorrhizal fungus *Laccaria laccata* in forestry. 1. Consistency between isolates in effective colonisation of containerized conifer seedlings. Can. J. For. Res. 12, 469–473.

19 Riffle J W and Tinus R W 1982 Ectomycorrhizal characteristics, growth and survival on artificially inoculated ponderosa and Scots pine in a greenhouse and plantation. For. Sci. 28, 646–660.

20 Tominaga Y 1975 Studies on the mycorrhiza of 'fairy ring' of *Tricholoma matsutake* (S. Ito et Imai) Sing. VI. On the mycorrhiza of *Pinus densiflora*. Bull. Hiroshima Agric. Coll. 5, 159–163.

21 Trappe J M 1962 Fungus associates of ectotrophic mycorrhizae. Bot. Rev. 28, 538–606.

22 Trappe J M 1977 Selection of fungi for ectomycorrhizal inoculation in nurseries. Annu. Rev. Phytopathol. 15, 203–222.

23 Trappe J M and Strand R F 1969 Mycorrhizal deficiency in a Douglas-fir region nursery. For. Sci. 15, 381–389.

Plant and Soil 71, 257–262 (1983). A-25

Sequences and interactions of mycorrhizal fungi on birch

J. W. DEACON, S. J. DONALDSON
Microbiology Department, School of Agriculture, West Mains Road, Edinburgh, EH9 3JG, Scotland

and F. T. LAST
Institute of Terrestrial Ecology, Bush Estate, Penicuik, Midlothian, EH26 0QB, Scotland

Key words *Betula pubescens* Birch Fungi Mycorrhizae Mycorrhizal interactions Mycorrhizal sequences Sheathing mycorrhiza

Summary Soil cores collected under a birch tree (*Betula pubescens*) on an experimental plot showed a progressive change in types of sheathing mycorrhiza with distance from the tree base. Seedlings grown in cores in a glasshouse also developed different mycorrhizal types depending on distance from the tree at which the cores were taken, but the types on seedlings were often different from those in the parent cores. When cores were taken directly beneath fruitbodies and sown to birch in a glasshouse, seedlings developed mycorrhizas of Laccaria, Inocybe and Hebeloma in cores from beneath these fruitbodies, but they seldom developed Lactarius mycorrhizas and never developed Leccinum mycorrhizas in cores taken beneath these fruitbodies. Similarly, when seedlings were grown in soils supplemented with vermiculite-peat inocula in a glasshouse, Laccaria and Hebeloma readily formed mycorrhizas, but *Lactarius pubescens* seldom did so and Leccinum and *Amanita muscaria* never dit so. Yet all these fungi form mycorrhizas on birch seedlings in aseptic conditions.

The results suggest a distinction between 'early stage' and 'late stage' mycorrhizal fungi of birch. Early stage fungi readily infect seedlings from resident or introduced inoculum in normal, unsterile soil, whereas late stage fungi do not readily form mycorrhizas in these conditions.

Introduction

A succession of mycorrhizal fungi has been demonstrated by mapping fungal fruitbodies around birch trees (*Betula pendula* Roth. and *B. pubescens* Ehrh.) in an experimental plot at Bush Estate near Edinburgh[2,7]. The succession is seen both with respect to time, because fruitbodies of different species first appear after different numbers of years, and with respect to space, because the mean positions of fruitbodies of different species from the tree bases change in successive years[2,7]. A preliminary survey of soil cores beneath fruitbodies showed the presence of mycorrhizas belonging to the fruiting species[2].

Since 1977, collaborative work between the Institute of Terrestrial Ecology at Bush Estate and the Microbiology Department of Edinburgh University has focused on the reasons for mycorrhizal successions and the consequences for tree growth. This paper outlines experimental evidence for a distinction between 'early stage' and 'late stage' mycorrhizal fungi of birch, based on behaviour in unsterile soil. Papers by Fleming[1] and Fox[3] also relate to this subject.

Is the succession real?

Harper and Webster[4] showed the danger of interpreting fungal successions solely on the basis of fruiting, because different fungi fruit after different periods of vegetative growth. For mycorrhizal fungi it can be expected that species with small fruitbodies will fruit sooner than species with large fruitbodies, because small fruitbodies require correspondingly lower mycelial nutrient reserves or access to less host-derived carbohydrate. The observed succession of fruitbodies on the experimental plot at Bush Estate (reference 7 and P. A. Mason, personal communication) has included the following species, with representative air-dry weights of fruitbodies collected from the plot in parentheses: *Laccaria* sp. (0.03 g) and *Hebeloma* spp. (*e.g. H. crustuliniforme*, 0.35 g) in the early years after saplings were planted; then *Inocybe lanuginella* (0.16 g) and *Lactarius pubescens* (2.23 g) in the fourth and subsequent years, and *Leccinum* spp. (5.03–10.59 g), *Russula* spp. and *Cortinarius* sp. in the later years. These weights, which followed a systematic trend, suggest that mycorrhizas *per se* need to be analysed so as to confirm succession and at the same time to counter reservations expressed by Harper and Webster[4].

The distributions of mycorrhizas around an 8-year-old tree of *B. pubescens* chosen at random in the experimental plot at Bush Estate are shown in Figure 1. Three radii were marked out from the base of the tree, and batches of ten cores (5 × 2.5 cm diam) were taken at 25 cm intervals along them. Half of the cores from each position were examined for the mycorrhizas they contained – *dissection cores* – and half were sown to birch in a glasshouse at 18°C, the seedling roots being examined for mycorrhizas 8 to 10 weeks later – *seedling cores*. Results for the separate radii were very similar so they are combined in Fig. 1.

Four main mycorrhizal types were seen in dissection cores: Hebeloma-type occurred mainly in the outer sampling positions, Leccinum- and an unknown mycorrhizal type occurred mainly in the inner positions and Lactarius-type was found at all positions though it tended to 'peak' in the middle. These patterns closely reflect the distributions of fruitbodies around trees in the experimental plot and they strongly support the concept of succession[2,5,7]. The seedling assay gave different results. Neither Lactarius nor Leccinum infected seedling roots; instead, unknown type '2' occurred on seedlings in cores from near the tree, unknown type '3' at 50 cm, unknown type '4' mainly at 75 cm and Hebeloma-type mainly at 100 cm (Fig. 1). The similar results for all radii showed the validity of this pattern . So in addition to a 'true' succession, there are other changes that occur with distance from the tree base and that are relevant to the establishment of different mycorrhizal types on seedlings.

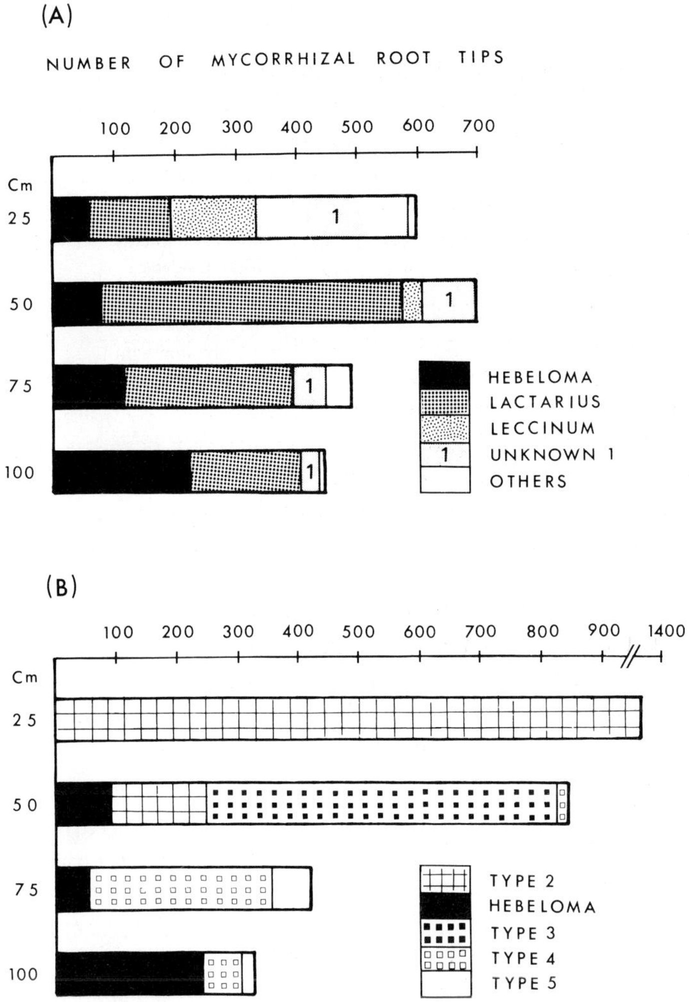

Fig. 1. Numbers of mycorrhizas of different types (A) in 15 soil cores or (B) on 15 birch seedlings grown in soil cores taken at distances of 25, 50, 75 and 100 cm from the base of a tree of *Betula pubescens*.

Assay of soil cores beneath fruitbodies

Mycorrhizas of particular fungi are likely to be found near the fruitbodies of those species, but it is interesting to know if they occur to the exclusion of other mycorrhizal types and if they can provide an inoculum source for infection of seedlings. To investigate these points, soil cores were taken directly beneath fruitbodies on the experimental plot at Bush Estate and used for dissection or seedling assay, as before (Table 1).

Table 1. Summary of occurrence of mycorrhizal types in soil cores beneath fruitbodies, assessed by microscopical examination of cores (dissection cores) or by means of seedling assay (seedling cores)

Fruitbody	Mycorrhizal types	
	Dissection cores**	Seedling cores***
Laccaria 'laccata'	6/10 Laccaria; never dominant	10/10 Laccaria; always dominant
Inocybe lanuginella	8/10 Inocybe; never dominant	10/10 Inocybe; dominant but few
Hebeloma sacchariolens	10/10 Hebeloma; co-dominant	8/10 Hebeloma; usually dominant
Hebeloma populinum	5/5 Hebeloma; dominant	5/5 Hebeloma; dominant
Lactarius pubescens	10/10 Lactarius; dominant, abundant	3/10 Lactarius*; very few, never dominant
Leccinum spp. (*subleucophaeum* + *rigidipes*)	10/10 Leccinum; dominant, abundant	0/10 Leccinum

* The 3 seedlings bearing Lactarius-type mycorrhizas were incubated for 26 weeks instead of the usual 10–12 weeks.

** Numbers of replicate cores, of 10, in which mycorrhizas attributable to the different fruitbodies were found

*** Numbers of replicate seedlings, of 10, on which mycorrhizas, attributable to the different fruitbodies, developed.

A mixture of mycorrhizal types was found beneath fruitbodies of Laccaria and Inocybe, and the fruitbody types were never dominant. Yet seedlings grown in these soil cores always developed mycorrhizas typical of the fruiting species. Beneath Hebeloma fruitbodies there was usually a predominance of Hebeloma-type mycorrhizas, and seedlings in the soil cores usually developed Hebeloma-type mycorrhizas (exclusively so in the case of *H. populinum*). Of interest, the mycorrhizas of *H. sacchariolens* were consistently associated with sclerotium-like bodies. Beneath Lactarius and Leccinum fruitbodies there were very large numbers of mycorrhizas of these fungi, to the exclusion of all other types. Yet seedlings seldom (Lactarius) or never (Leccinum) developed mycorrhizas of the fruitbody type when grown in the cores. Instead, Hebeloma-type or unknown types of mycorrhiza developed on the seedling roots.

These results suggest that fungi that occur early in the fruitbody sequence can readily initiate infection of seedling roots – even in soil cores containing few mycorrhizas of the appropriate type – whereas fungi that occur later in the fruitbody sequence cannot readily initiate infection of seedling roots.

Establishment of mycorrhizas from inoculum added to soil

Inocybe spp. have not been obtained in pure culture, but all other mycorrhizal fungi discussed here can form mycorrhizas on birch seedlings in aseptic conditions (*e.g.* by the method in reference 6). So they are compatible with *seedling* root systems even though they may not occur on seedlings in nature. An attempt was made to establish them from artificial inoculum (colonized vermiculite-peat) added to soil (1:3, w/w) in glasshouse conditions, and for this purpose two soils were used: a 'mycorrhizal' soil taken from the root zone of trees within the experimental plot at Bush Estate, and a 'non-mycorrhizal' soil taken from outside the boundary of the plot. Seedlings were sampled from 10 to 26 weeks after the seed was sown.

Seedlings in control pots of non-mycorrhizal soil bore a mixture of Hebeloma- and unknown mycorrhizal types. Supplementation of this soil with inoculum of Laccaria gave rise mainly to Laccaria-type mycorrhizas, and supplementation with *Hebeloma sacchariolens* or *H. crustuliniforme* gave rise only to Hebeloma-type mycorrhizas (those of *H. sacchariolens* being associated with sclerotium-like bodies). These inoculant fungi could thus establish on seedling roots and at least partly exclude the normal range of colonists in glasshouse conditions. In contrast, *Lactarius pubescens* colonized the seedling roots very poorly from added inoculum (and only on seedlings incubated for 20 to 26 weeks) and *Leccinum roseofracta* and *Amanita muscaria* did not colonize the seedling root systems; instead, either Hebeloma-type or a mixture of Hebeloma- and unknown types occurred on the seedling roots in these treatments.

The pattern of results for 'mycorrhizal' soil was similar to that for 'non-mycorrhizal' soil in all important respects, though a distinctive mycorrhizal type know to occur in this soil (termed 'brown with cream tips') was never completely excluded by the added inocula.

'Early-stage' and 'late-stage' mycorrhizal fungi

The work outlined here supports the concept of a mycorrhizal succession on birch, first recognized on the basis of fruitbody occurrence[2,5,7]. The work suggests further that a broad distinction can be made between 'early-stage' and 'late-stage' mycorrhizal fungi, based on differences in behaviour in unsterile soil. Early-stage fungi can readily infect seedling roots from resident sources of inoculum in soil or from added vermiculite-peat inoculum, whereas late-stage fungi cannot do so. Yet both types can infect seedling roots in aseptic conditions. Based on evidence presented by Fox[3] and Fleming[1] it can be suggested that early-stage and late-stage fungi differ in the relative inoculum potentials needed to initiate mycorrhizal infection.

References

1 Fleming L V 1983 Succession of mycorrhizal fungi on birch: infection of seedlings planted around mature trees. Plant and Soil 71, 263–267.

2 Ford E D, Mason P A and Pelham J 1980 Spatial patterns of sporophore distribution around a young birch tree in three successive years. Trans. Br. mycol. Soc. 75, 287–296.

3 Fox F M 1983 Role of basidiospores as inocula of mycorrhizal fungi of birch. Plant and Soil 71, 269–273.

4 Harper J E and Webster J 1964 An experimental analysis of the coprophilous fungus succession. Trans. Br. Mycol. Soc. 47, 511–530.

5 Last F T, Mason P A, Wilson J, Deacon J W and Walker C 1983 Succession: Its relevance when selecting sheathing mycorrhizal fungi for inoculating young trees. Nature London (*in press*).

6 Mason P A, Dighton J, Last F T and Wilson J 1983 Procedure for establishing sheathing mycorrhizas in tree seedlings. For. Ecol. Manage. 5, 47–53.

7 Mason P A, Last F T, Pelham J and Ingleby K 1982 Ecology of some fungi associated with an ageing stand of birch (*Betula pendula* and *B. pubescens*). For. Ecol. Manage. 4, 19–39.

Plant and Soil 71, 263–267 (1983). A-26

Succession of mycorrhizal fungi on birch: infection of seedlings planted around mature trees

L. V. FLEMING
Microbiology Department, School of Agriculture, West Mains Road, Edinburgh, EH9 3JG, Scotland and Institute of Terrestrial Ecology, Bush Estate, Penicuik, Midlothian, EH26 0QB, Scotland

Key words Betula Ectomycorrhiza Mycelial strands Mycorrhiza Succession

Summary Non-mycorrhizal seedlings of *Betula pendula* were planted around an 11-year old tree of *B. pubescens* in an experimental plot at Bush Estate, south of Edinburgh. Half (23) of the seedlings were in untreated planting positions and half (24) in positions that had been cored to sever connections of the roots with the parent tree. After 17 weeks, seedlings in the non-cored positions bore mycorrhizas mainly of *Lactarius pubescens* whereas mycorrhizas of this species were uncommon on seedlings in cored positions. Much smaller and usually non-significant differences were found with respect to other mycorrhizal types in cored and non-cored positions. It is concluded that *L. pubescens* infected seedling roots by means of mycelial strands which must remain attached to the parent tree (food base) in order to infect. The significance of the results for the concept of early-stage and late-stage mycorrhizal fungi is discussed.

Introduction

A succession of mycorrhizal fungi has been reported on roots of birch (*Betula pendula* Roth. and *B. pubescens* Ehrl.)[2,6,7]. In this respect a distinction has been made between early-stage and late-stage fungi, because the former can infect seedlings grown in cores of soil taken beneath their fruitbodies whereas the latter cannot do so[2,6]. A common feature of the reported late-stage fungi, like *Lactarius pubescens* Fr. and *Leccinum* spp., is that they form conspicuous mycelial strands. Most work on such mycelial strands has concentrated on their abilities to absorb and translocate materials from the external environment to the plant root (*e.g.* reference 3). But Garrett[4] considered that the major role of mycelial strands and rhizomorphs in root-infecting fungi as a whole was as agents of infection. The possibility that strands serve this role in mycorrhizal fungi has been recognized by several workers (*e.g.* reference 1) but never conclusively demonstrated.

Because the infectivity of mycelial strands and rhizomorphs depends on nutrients supplied by a food base[5] it was thought possible that the failure of late-stage mycorrhizal fungi to infect seedlings in soil cores was a result of the severence of inoculum in these cores from the food base provided by the parent tree. An experiment was done to investigate this.

Experimental

An 11-year old tree of *B. pubescens* was selected in an experimental plot at Bush Estate, south of Edinburgh, where the distributions of mycorrhizas and fruitbodies had been recorded in previous years[7]. The tree had mycorrhizas of *L. pubescens* more or less evenly distributed around it to a distance of at least 1 metre from the trunk. The ground beneath the tree was covered with black polyethylene sheeting from December 1980 until May 1981, to suppress growth by herbaceous plants. Meanwhile, non-mycorrhizal seedlings of *B. pendula* were grown in vermiculite-peat in laboratory conditions[8], hardened-off in a cold-frame and prepared for transplanting into the field when they were 55 days old. They were then 20.1 ± 1.75 (S.E.) mm tall, weighed 3.6 ± 0.51 mg (oven-dry) and bore 49.3 ± 5.26 uninfected root tips.

Forty-eight planting positions were chosen, in four annuli around the tree, at 20, 35, 50 and 75 cm radii from the trunk. The planting positions were evenly spaced within the annuli, so there were progressively more seedlings further out from the tree. Alternate positions were untreated and received one seedling each – 'non-isolated'. At other positions a core of soil 8 cm deep and of 10 cm diameter was taken with a corer and then replaced with minimal disturbance, knowing that roots within the core had been severed from the parent tree. Seedlings were then planted – 'isolated'. All plantings were done on 3 June 1981 by removing a narrow core of soil with a cork-borer (1 cm diameter) and inserting into the hole a root system of a seedling grown in a 1 cm diameter plastic sleeve (the sleeve being removed at planting)[8].

The seedlings were watered as required in the summer of 1981, and a 2 cm deep mulch of sphagnum peat was applied to the site to help conserve moisture. At intervals the cores containing the 'isolated' seedlings were carefully lifted and replaced immediately, to prevent re-invasion by roots or mycelial strands. Light intensities at seedling level, and temperatures and rainfall were recorded, and the positions of fungal fruitbodies were recorded as they appeared.

Seedlings were sampled 17 weeks after they had been planted into the field, when the isolated seedlings were both taller (39.1 ± 1.42 mm) and heavier (37.6 ± 3.93 mg dry wt.) than the non-isolated ones (34.1 ± 1.43 mm and 27.5 ± 3.53 mg). The root systems were washed and assessed for mycorrhizal types, distinguished on the basis of colour, gross morphology and microscopical features, using as standards mycorrhizas formed in bixenic culture or consistently associated with fruitbodies in the field (Table 1).

Results

There were no significant differences in (a) total numbers of root tips and (b) numbers of uninfected root tips between isolated and non isolated seedlings (Table 1). Four predominant mycorrhizal types were seen on the seedling root systems. Unidentified types '4' and '5' were found on similar numbers of isolated and non-isolated seedlings, and the numbers on these seedlings did not differ significantly (Table 1). By contrast, Lactarius-type mycorrhizas were present on significantly more non-isolated than isolated seedlings (Table 1) and they were frequently the dominant type on non-isolated seedlings (in 11 out of 23 instances) but were never dominant on isolated seedlings. The difference in the case of Lactarius-type mycorrhizas can be illustrated in another way: a total of 73 mycorrhizas of this type was recorded on the 24 isolated seedlings, compared with 1469 on the 23 non-isolated seedlings (one of the original 24 seedlings in this category died).

Hebeloma-type mycorrhizas were present on many non-isolated seedlings but

Table 1. Occurrence of mycorrhizal types on birch seedlings grown for 17 weeks in isolated (cored) * and non-isolated positions around a mature birch tree

Mycorrhizal type	No. seedlings bearing each mycorrhizal type (no. with > 20% root tips of that type in parentheses)		Mean no. root tips of each mycorrhizal type (means of log x + 1 in parentheses)		
	Isolated (Max. 24)	Non-isolated (Max. 23)	Isolated	Non-isolated	LSD (P = 0.05)
Hebeloma	6 (3) **	19 (4)	16 (0.35)	18 (0.89)	(0.39)
Lactarius	3 (2) ***	22 (17)	3 (0.15)	64 (1.66)	(0.32)
Type 4	17 (16)	11 (11)	70 (1.33)	34 (0.87)	(0.50)
Type 5	19 (13)	17 (9)	54 (1.28)	47 (1.12)	(0.49)
Total no. mycorrhizal root tips/seedling			143	166	
Total no. uninfected root tips/seedling			11	7	
Total no. root tips/seedling			154	173	

* For explanation see text
** Difference significant at $P = 0.01$
*** Difference significant at $P = 0.001$

not on the isolated seedlings. In most cases, however, they were present in very low numbers and if the value of 20% of total root tips represented by Hebeloma-type mycorrhizas is taken as an indication of good establishment on a seedling (even this is much less than would be true of a dominant mycorrhizal type) then Table 1 shows that few seedlings in either category were heavily colonized by it. In fact, there was no difference between treatments in the number of Hebeloma-type mycorrhizas: 389 were recorded on the 24 isolated seedlings, compared with 408 on the 23 non-isolated seedlings.

The distribution of mycorrhizal types on the parent tree root system was examined early in November 1981 by taking small cores of soil (8 × 2 cm diameter) at 25 cm distance intervals on ten radiating transects from the tree trunk. Lactarius-type mycorrhizas represented only 12% of the total number of mycorrhizal root tips, the highest recorded percentage being 31.0 for the ten cores at 100 cm from the trunk. Type '5' was the dominant type, representing 48.8% of root tips overall. Hebeloma-type represented 26.1% type '4' 9.9% and other types 3.2%. Thus, comparing the results for seedlings and the parent tree root system, Lactarius was markedly over-represented on the non-isolated seedlings and under-represented on the isolated seedlings, Hebeloma and type '5' were under-represented on the seedling roots in general, and type '4' was over-represented on the seedling roots in general.

Most fruitbodies that appeared around the parent tree in autumn 1981 were of *L. pubescens*, *Hebeloma velutipes* Bruchet and *Cortinarius* sp. Three species of *Russula* were also seen (*R. versicolor* J. Sch., *R. betularum* Hora and *R. grisea* (Secr.) Gill) but neither they nor the *Cortinarius* sp. seemed to represent the unidentified mycorrhizal types on the roots.

Discussion

L. pubescens was the only one of the mycorrhizal types observed that formed mycelial strands; indeed, it was the only late-stage type (in the sense defined in reference 2) that occurred on the seedlings, *Russula* and *Cortinarius* spp. being excluded from consideration because they apparently did not infect seedling roots.

The results show clearly that Lactarius can establish mycorrhizas on seedlings in soil, provided that the fungus is not separated from the parent tree; furthermore, it can colonize the seedling roots to a greater degree than its level of establishment on the parent tree root system would suggest. This is consistent with the view that Lactarius colonized the seedling roots by means of mycelial strands or, at least, by means of hyphae that need to remain attached to the food base provided by the parent tree. In this respect the results parallel those previously obtained for root-infecting pathogens, like *Armillaria mellea* (Vahl ex Fr.) Kummer[5].

Supporting evidence for the view that mycelial strands are important in colonization of seedling roots by late-stage mycorrhizal fungi has recently been obtained. Naturally occuring birch seedlings were collected in a chestnut (*Castanea sativa* Mill.) coppice with oak standards near Rochester, Kent, late in July 1981. The seedlings were less than one year old but 73% of them bore a late-stage mycorrhizal type associated with thick yellow mycelial strands and resembling some mycorrhizas formed by the Boletaceae. Soil in which these seedlings were growing was taken to a glasshouse and sown to birch, but only the usual range of early-stage mycorrhizal types occurred on the seedling roots. This could be interpreted as evidence that seedlings developing around parent trees or their living coppice stools in the field are infected by late-stage mycorrhizal fungi but that isolation of these fungi from the parent tree leads to failure to infect seedlings.

The results of this study raise interesting questions about the respective roles of early- and late-stage mycorrhizal fungi. It is unclear at present if early-stage mycorrhizal fungi have a significant role in the establishment of naturally regenerating tree seedlings in the woodland environment.

Acknowledgements I am very grateful to Dr J W Deacon and Professor F T Last for supervising this work, and to the Natural Environment Research Council of Great Britain for the award of a Research Studentship. I thank also Dr P A Mason for help in various ways.

References

1 Chilvers G A and Gust L W 1982 The development of mycorrhizal populations on pot-grown seedlings of *Eucalyptus st-johnii* R T Bak. New Phytol. 90, 677–699.

2 Deacon J W, Donaldson S J and Last F T 1983 Sequences and interactions of mycorrhizal fungi on birch. Plant and Soil 71, 257–262.

3 Duddridge J A, Malibari A and Read D J 1980 Structure and function of mycorrhizal rhizomorphs with special reference to their role in water transport. Nature London 287, 834–836.

4 Garrett, S D 1951 Ecological groups of soil fungi: a survey of substrate relationships. New Phytol. 50, 149–166.

5 Garrett S D 1956 Rhizomorph behaviour in *Armillaria mellea* (Vahl) Quel II. Logistics of infection. Ann. Bot. 20, 193–209.

6 Last F T, Mason P A, Wilson J, Deacan J W and Walker C 1983 Succession: its relevance when selecting sheathing mycorrhizal fungi for inoculating young trees. Nature London *in litt.*

7 Mason P A, Last F T, Pelham J and Ingleby K 1982 Ecology of some fungi associated with an ageing stand of birch (*Betula pendula* and *B. pubescens*). For. Ecol. Manage. 4, 19–39.

8 Mason P A, Dighton J, Last F T and Wilson J 1983 Procedure for establishing sheathing mycorrhizas on tree seedlings. For. Ecol. Manage. 5, 47–53.

Plant and Soil 71, 269–273 (1983). A-27

Role of basidiospores as inocula of mycorrhizal fungi of birch

FRANCES M. FOX
Microbiology Department, School of Agriculture, West Mains Road, Edinburgh, EH9 3JG, Scotland and Institute of Terrestrial Ecology, Bush Estate, Penicuik, Midlothian, EH26 0QB, Scotland

Key words Basidiospore Betula Ectomycorrhiza Mycorrhiza Succession

Summary Fruitbodies of sheathing mycorrhizal fungi collected under birch (*Betula pendula* and *B. pubescens*) were suspended over pots of soil and the resulting spore-supplemented soils were planted with seedlings of *B. pendula*. *Inocybe lacera*, *I. lanuginella*, *Hebeloma sacchariolens* and *H. leucosarx* formed mycorrhizas readily. *Lactarius pubescens* and *Leccinum roseofracta* did not form mycorrhizas from basidiospore inocula, even after prolonged periods of seedling growth. *Paxillus involutus* gave equivocal results, perhaps because the soil was unsuitable for this species. Storage of the basidiospore-supplemented soils for 6 months in outdoor conditions or in a growth room at 18°C did not materially alter the results.

The results are discussed in relation to the concept of mycorrhizal succession.

Introduction

Basidiospores are generally assumed to be the main means by which ectomycorrhizal fungi are dispersed and initiate infections, but there is only limited experimental evidence to support this view[9,11]. A general difficulty in this respect concerns the low degree of germinability of the spores, though this can be overcome to some extent by adding activated charcoal or the yeast *Rhodotorula glutinis*[2] or germinating in the presence of living roots[3] or mycelium of the appropriate type[7].

A succession of mycorrhizal types has been demonstrated on roots of birch (*Betula pendula* Roth. and *B. pubescens* Ehrh.) in collaborative studies at Edinburgh, and a general distinction has been made in this respect between 'early stage' and 'late stage' mycorrhizal fungi[1,5,8]. The late-stage types do not readily infect seedling roots in unsterile soil, though they can do so in aseptic conditions[1]. As part of this programme the infectivity of basidiospores of early- and late-stage mycorrhizal fungi was investigated.

Materials and methods

Fruitbodies of seven sheathing mycorrhizal basidiomycetes were collected beneath birch trees at three sites in autumn 1981. The stipes were partly removed and the caps were then suspended individually over 50 cm^3 plant pots containing brown earth soil, by pressing the remaining part of the stipe into the soil. After 24 h in dark humid conditions the fruitbodies were removed and discarded, and the spores that had fallen onto the soil were buried by lightly mixing the top 1 cm of soil in each pot. No attempt was made to count the number of spores released from each fruitbody, but in each case the corner of a glass coverslip was placed beneath part of the cap to ensure that spores had been released in large numbers. The soil had been taken from a tree-less site south of Edinburgh and had been stored air-dry for 21 days. It was sieved <2 cm, mixed with autoclaved grit (3 : 1, v/v), watered to 40% saturation and left in pots for 4 days before use. The soil was unsterile.

Variable numbers of replicates were prepared, depending on availability of fruitbodies. In all treatments at least twelve replicate pots were prepared and these were sown to birch immediately as described below. In some instances further replicates were available; these pots were placed in plastic plant propagators which were sealed in polyethylene bags and then wrapped in aluminium foil. The propagators were stored for 6 months either in a growth room at constant 18°C or in outdoor conditions where they were exposed to winter temperatures (which dropped as low as −20°C).

To test the infectivity of spores in soil, seeds of *B. pendula* were surface-sterilized in H_2O_2 and placed on tap-water agar, on which the resulting seedlings were allowed to grow for 20 days at room temperature. Then two seedlings were transferred to each pot and the pots were incubated in a growth room at 18°C, with a 16 h light/8 h dark cycle. The pots were arranged in pairs with the different pairs being spaced in a randomized layout. They were sprayed daily with water. The seedlings were thinned to one per pot after 6 weeks and harvested at 14 to 16 weeks, though in many cases a few pots were incubated for longer to see if the pattern of results changed on prolonged incubation of seedlings. For assessment, the root systems were washed free from soil, placed on a grid and cut into 1 cm sections. Alternate sections were scored for numbers of root tips and numbers of tips with different mycorrhizal types, these being compared with reference material. Throughout the experiment, pots of unsupplemented soil were used for comparison with the treatment series.

Results

Results of the main part of the experiment, involving the immediate planting of seedlings, are shown in Table 1. Of the 6 control seedlings growing in unsterile soil and assessed after 16 weeks, two had Hebeloma-type mycorrhizas, three had Laccaria-type mycorrhizas and two had Inocybe-type; it is notable that each mycorrhizal type occurred almost to the exclusion of other types on

Table 1. Establishment of mycorrhizas on seedlings of *Betula pendula* planted in soil recently supplemented with basidiospore inocula of different fungi

	Basidiospore inoculum*							
	None	*I. lac*	*I. lan*	*H. sac*	*H. leu*	*P. inv*	*L. pub*	*L. ros*
No. seedlings examined	6	6	6	6	6	6	6	6
No. with inoculant-type mycorrhizas	—	6	6	5	6	4	0	0
Mean no. root tips	379	264	410	526	622	346	684	635
Mean % root tips:								
(a) uninfected	63	46	70	48	34	37	44	34
(b) with inoculant type mycorrhizas	—	52	30	49	59	20	0	0

* I. lac, *Inocybe lacera*; I. lan, *Inocybe lanuginella*; *H. sac*, *Hebeloma sacchariolens*; H. leu, *Hebeloma leucosarx*; P. inv, *Paxillus involutus*; L. pub, *Lactarius pubescens*; L. ros, *Leccinum roseofracta*.

Table 2. Establishment of mycorrhizas on seedlings of *Betula pendula* planted in soil supplemented with basidiospore inocula of different fungi and stored for 6 months

	Basidiospore inoculum* and conditions of storage							
	Growth room (18°C)					Outdoor (winter)		
	None	*I. lac*	*H. leu*	*P. inv*	*L. pub*	None	*P. inv*	*L. pub*
No. seedlings examined	6	5	5	6	8	6	8	6
No. with inoculant-type mycorrhizas	–	4	5	2	0	–	3	0
Mean no. root tips	182	65	148	168	169	193	209	192
Mean % root tips:								
(a) uninfected	87	70	54	31	43	61	92	74
(b) with inoculant-type mycorrhizas	–	30	45	3	0	–	1	0

* See Table 1.

individual seedlings. Treatments involving spores of *Inocybe lacera* (Fr.) Quel., *I. lanuginella* (Schroet.) Konrad & Maublanc, *Hebeloma sacchariolens* Quelet and *H. leucosarx* Orton gave successful mycorrhizal establishment: the inoculant types developed on most or all seedlings and in relatively large numbers on those seedlings. It is notable that these fungi are in the early-stage category of Deacon *et al.*[1]. In contrast, *Lactarius pubescens* Fr. and *Leccinum roseofracta* Watling did not establish on seedling roots from spore inocula; instead, the seedlings bore Laccaria-type or Thelephora-type or, less commonly, Paxillus-type mycorrhizas. The inoculant fungi in these cases are late-stage[1]. The results for *Paxillus involutus* (Fr.) Karsten spore inocula were equivocal: some seedlings were infected but few mycorrhizas were attributable to *P. involutus*, the predominant type on most replicate seedlings being Laccaria (Table 1). Indeed the overall mean of 20% root tips attributable to Paxillus in this treatment was similar to that in the treatment involving *Lactarius pubescens* (results not presented in detail in Table 1).

Some of the remaining seedlings in replicate pots from this part of the experiment were incubated for 26 weeks because it was thought that mycorrhizas of Lactarius or Leccinum might develop after longer incubation periods. But mycorrhizas of these fungi were still not present after 26 weeks.

When soils that had been stored for 6 months were planted, the development of mycorrhizas was generally similar to that described above (Table 2). Mycorrhizas of Inocybe and Hebeloma developed in pots supplemented with spores of these fungi; Paxillus gave few mycorrhizas and Lactarius gave none.

Discussion

Results in this paper suggest that early-stage mycorrhizal fungi of birch can establish mycorrhizas on seedlings from spores in soil but that late-stage mycorrhizal fungi cannot do so. The failure of some spores to initiate infections does not seem to alter during 6 months' storage in soil – even over winter – but the effects of prolonged storage are being studied further. It is notable that the spores of early-stage mycorrhizal fungi remained infective for at least 6 months at 18°C. This seems to be ecologically relevant because spores released from fruitbodies in the autumn in Britain would presumably need to persist in soil until the following growing season.

The results for Paxillus are difficult to interpret. This fungus is commonly found as a mycorrhizal symbiont on very poor soils, including coal-spoil heaps, but it has not been found in the experimental plot of birches at Bush Estate, where the soil is richer[8]. Possibly the spores of *P. involutus* would initiate infections in coal spoil, and this is currently under study. The fungus has been found to produce mycorrhizas when vermiculite-peat inocula are added to coal spoil in glasshouse conditions (P A Mason, personal communication).

My work has helped to reinforce the view that a broad distinction is possible between early- and late-stage mycorrhizal fungi of birch, and it has identified one of the important differences in this respect, *i.e.* that of infectivity of spores in soil. The failure of some previous attempts to use spore inocula might now be explained by the choice of fungi used. But it is possible also that the success of this work was due to the fact that spores were not allowed to dry: they were released into a humid environment and immediately mixed into moist soil. Simon[10] has reviewed the effects of drying and re-wetting fungal spores and other plant propagules, a consistent finding being that substantial loss of nutrients occurs if rehydration takes place too rapidly. Perhaps any such loss of nutrients significantly reduces the germinability or infectivity of spores[4] or predisposes them to microbial antagonism[6].

Acknowledgements I am very grateful to Dr J W Deacon and Professor F T Last for suggesting and supervising this work, and to the Science Research Council of Great Britain for the award of a Research Studentship. I also thank Dr P A Mason for valuable help and discussion.

References

1 Deacon J W, Donaldson S J and Last F T 1983 Sequences and interactions of mycorrhizal fungi on birch. Plant and Soil 71, 257–262.

2 Fries N 1978 Basidiospore germination in some mycorrhiza-forming Hymenomycetes. Trans.Br. mycol. Soc. 70, 319–324.

3 Fries N and Birraux D 1979 Spore germination in *Hebeloma* stimulated by living plant roots. Experimentia 36, 1056–1057.

4 Last F T 1960 Longevity of conidia of *Botrytis fabae* Sardiña. Trans. Br. Mycol. Soc. 43, 673–680.

5 Last F T, Mason P A, Wilson J, Deacon J W and Walker C 1983 Succession: its relevance when selecting sheathing mycorrhizal fungi for inoculating young trees. Nature London (*in litt.*).

6 Lockwood J L 1977 Fungistasis in soils. Biol.Rev. 52, 1–43.

7 Losel D M 1964 The stimulation of spore germination in *Agaricus bisporus* by living mycelium. Ann Bot. 28, 541–554.

8 Mason P A, Last F T, Pelham J and Ingleby K 1982 Ecology of some fungi associated with an ageing stand of birches (*Betula pendula* and *B. pubescens*). For. Ecol. Manage. 4, 19–39.

9 Robertson N F 1954 Studies on the mycorrhiza of *Pinus sylvestris*. I. The pattern of development of mycorrhizal roots and its significance for experimental studies. New Phytol. 53, 253–283.

10 Simon E W 1974 Phospholipids and plant membrane permeability. New Phytol. 73, 377–420.

11 Theodorou C and Bowen G D 1973 Inoculation of seeds and soil with basidiospores of mycorrhizal fungi. Soil Biol. Biochem. 5, 765–771.

Plant and Soil 71, 275–284 (1983). A-28

Methods for synthesizing ectomycorrhizas and their effect on mycorrhizal development

J. ANDRÉ FORTIN, YVES PICHÉ and CHRISTIAN GODBOUT
Département d'Ecologie et Pédologie, Faculté de Foresterie et Géodésie, Université Laval, Ste-Foy, Québec, Canada G1K 7P4

Key words Alnus Ectomycorrhizas Mycelial strands Mycorrhizas Pinus Populus Sclerotia Synthesis

Summary The different methods used for synthesizing ectomycorrhizae since Melin's first experiments were reviewed and evaluated. A non-aseptic growth pouch method was proposed that allowed for the formation of ectomycorrhizae on several species of Pinaceae and of actinorhizal and non-actinorhizal angiosperms including woody cuttings. More than 30 fungal species thus tested formed ectomycorrhizae. In most cases full ectomycorrhiza formation required 3–10 days after inoculation. The method allowed for chronosequential destructive (fixation) and non-destructive (photography) observations. It was particularly useful for the description of extramatrical phase development including hyphal strands, sclerotia and fruit body primordia formation.

Introduction

Melin[1] devised the first equipment to apply Koch's postulate and to define the nature of the causal organisms of ectomycorrhizae. Sand was used as a substrate and cotton stoppers maintained the aseptic condition within the flasks. It took several months for the associations to take place and bulky apparatus was needed to deal with watering problems during the experimental period. This equipment limited the number of simultaneous tests that could be carried out.

Hacskaylo[2] introduced the use of expanded mica as a substrate instead of sand; this substitution increased the water holding capacity of the substrate and improved aeration of the root. Seedlings were maintained in aseptic conditions in one liter flasks for up to three months during ectomycorrhizal formation, and supplementary watering was unnecessary. Over the following 20 years, this method and various modifications were widely used and applied to a large number of host tree species and several dozen species of ectomycorrhizal fungi[3]. Some physiological studies were facilitated such as the translocation of radioactive tracers through the fungal mycelia[4]. Marx and Zak[5] improved the usefulness of Hacskaylo's method by introducing the use of peat moss to counteract the high buffering capacity of expanded mica (pH – 6.5). This made it possible to investigate acidophilic species such as *Picea mariana* (Mill.) BSP that can hardly grow in presence of pure expanded mica[6].

The flask method has some major disadvantages. Under greenhouse conditions high temperatures sometimes occur within the flasks and cooling baths are necessary[5]. The growth of control plants within a cotton stoppered flasks can be limited by the availability of CO_2 after a few hours of active photosynthesis[7]. Moreover the addition of glucose is necessary, supposedly to help the growing hyphae to reach the roots, thus creating even more artificial conditions. The assimilation of glucose by the fungus increases the level of CO_2 in the flasks. Therefore the observed growth stimulation of inoculated hosts compared with controls is probably more the result of increased CO_2 concentration than improved mineral nutrition. In our experiences the growth of contaminated controls was often stimulated compared with uncontaminated ones.

To overcome some of these difficulties and especially the access to CO_2, Trappe[8] used food jars with the aerial part of the plants in normal air conditions, (*i.e.* outside of the jar) and the root in aseptic conditions within the bottle in presence of a potential ectomycorrhizal fungus. Expanded mica, peat moss and sucrose were still used. However, all the methods discussed up to now still have many disadvantages. The equipment used in these methods remain bulky and do not permit investigation with a large number of experimental units at the same time. They do not readily allow for the observation of ectomycorrhiza genesis and the development of the extramatrical phase. Also the maintenance of aseptic conditions for long periods of time is not always easy.

Pachlewska[9] used a test tube technique to assess the compatibility of ectomycorrhizal hosts and fungi. Seedlings were grown aseptically on water agar containing thiamine but no minerals. Ectomycorrhiza formation was obtained with a large number of fungi[10]. The lack of nutrients raises doubts about the negative results and the method is not suitable for developmental or physiological studies.

Pachlewska's method was improved by Mason[11,12] by adding minerals, glucose and thiamine. With these modifications ectomycorrhiza formation took place 4–6 weeks after inoculation. It was successfully used by Giltrap[13] on *Betula* spp.

Several other devices were developed for special physiological and structural experiments on ectomycorrhizae under aseptic conditions. Sohn[14] introduced the use of ion exchange resins. Biggs and Alexander[15] used a sophisticated apparatus for the study of nutrient uptake by ectomycorrhizal plants and Nylund[16] worked with special glass chambers to investigate structural modifications of roots during ectomycorrhizal formation. Moreover, Skinner and Bowen[17,18] observed the development of ectomycorrhizae and hyphal strand formation by using glass sided boxes containing sterilized soil.

In an attempt to design equipment that would require limited space and yet permit the observation of ectomycorrhiza formation, we have developed a root-hypocotyl method[19,20]. This method enables the pine roots to grow well in absence of their photosynthetic apparatus which is replaced by agar-filled glass shell vials containing sucrose, thiamine and auxin. The root part extends within a

Petri dish under conditions similar to those used by Marx and Zak [5] for whole seedlings in flasks. Typical ectomycorrhizae were obtained.

Axenic conditions are usually considered essential for synthesis of ectomycorrhizae under laboratory conditions. However Mullette [22] using Leonard jars observed no differences in the formation of ectomycorrhizae under axenic and non-axenic conditions.

For several years we have searched for a system that would not be bulky, would place the seedling under optimum air conditions, would function without addition of sugars or vitamin and would allow for rapid formation and direct observation of ectomycorrhiza. It became obvious that the growth pouch technique used in studies of nitrogen-fixing root symbiosis [23] could meet all these objectives.

General description of the growth pouch method

Synthesis of ectomycorrhizae using growth pouches was first mentioned by Fortin *et al.* [24], and fully described by Piché and Fortin [25]. Essentially it uses flat plastic polyester pouches containing a thin paper pad and nutrient solution; the root system grows in one plane on the surface of the pad. With this equipment ectomycorrhiza formation took place quickly, 3–10 days from inoculation. The method allows all phases of development to be observed and makes it possible to conduct structural and physiological experimentations simultaneously on a large number of seedlings using limited growing spaces.

With this method ectomycorrhizae were readily obtained on all species of tree tested up to now. They are *Pinus strobus* L., *Pinus sylvestris* L., *Pinus banksiana* Lamb., *Abies balsamea* (L.) Mill., *Picea glauca*, (Moench) Voss, *Picea mariana* (Mill.) B.S.P., *Betula glandulosa* Michx., *Alnus crispa* (Ait.) Pursh, *Alnus rugosa* (Du Roi) Spreng., *Alnus glutinosa* (L.) Gaertn., *Populus tremuloides* Michx., all of which were grown from seeds. Woody cuttings of *Salix planifolia* Pursh. and *Populus* x *interamericana* hybrids also formed ectomycorrhiza under such conditions. Up to now, more than 30 species of fungi have been induced to form ectomycorrhizae on different species of trees by this method.

Some researchers might be sceptical about the value of this method but it must be said that in all the unsuccessful cases we have seen, proper precautions had not been taken. Although non-aseptic, the growth pouch method requires a maximum of cleanliness. The pouches come already sterilized and do not require autoclaving; indeed autoclaving promotes hemicellulose hydrolysis and encourages the development of contaminants. Excess of nitrogen is also likely to promote the development of cellulolytic organisms. Mites should also be strictly controlled to prevent spreading contaminants that always develop to some extent. Up to now, in our experiences ectomycorrhiza formation was never observed among several hundreds of control seedlings used for different experiments.

Since successful ectomycorrhiza formation have been readily obtained without preventing light reaching the root system, no effort is being made to cover the pouches. However placing them close one to each other reduces the amount of light reaching the roots and the development of algae.

Synthesis of ectomycorrhizae in Pinaceae using the growth pouches

Most ectomycorrhizal Pinaceae enter the autotrophic phase of growth by their 4–6th week after onset of germination depending on the species. At this stage the primary needles start to form and the first lateral roots emerge from the tap root. A few more days are required for these lateral roots to develop receptive short roots. When using the pouch method for the first time with a species, it is recommended that seedlings be pregerminated aseptically on agar and grown under sterile conditions in test tubes[20] until the 5th or 6th week before their transfer into the pouches. This will reduce substantially the number of undesirable propagules. After introducing the seedlings into the pouches partial wilting may occur. Reducing light intensity and increasing air moisture for 24–48 h promotes rapid recovery. Inoculation must be postponed for a few more days until the short receptive rootlets begin to form.

Inoculation should be made with actively growing hyphae. Plugs of inoculum cut with a cork borer must be placed for a few days on a fresh medium to encourage the regeneration of hyphal tips. The introduction of pieces of styrofoam into the pouches maintains an air space by preventing the front wall of the pouches from contacting the upper surface of the root. These conditions favor root development, mycelium growth and mycorrhiza formation more than maintaining a film of nutrient solution between the pad and the pouche's front wall.

After the formation of some ectomycorrhizae the addition of fresh nitrogen-containing nutrient solution encourages further formation of ectomycorrhizae and promotes the development of the extramatrical phase including hyphal strands and sclerotia formation with *P. tinctorius* on *P. strobus*[25].

Synthesis of ectomycorrhizae on angiosperms

The method was successfully used, with adaptations, for the small-seeded *P. tremuloides*, for actinorhizal *Alnus* spp. and for woody cuttings of willow and poplar.

Handling the tiny seeds of *P. tremuloides* required special treatments. The seeds were first germinated on 70 mesh silica sand in Petri dishes and transferred on the same substrate into open containers under non-aseptic conditions. Complete Long Ashton solution[26] was used. At the four-leaf stage the sand was flooded and the seedlings could be lifted without harming the root system, and placed into the pouch. Inoculation was made a few days later and successful ectomycorrhiza formation was observed with several fungal species.

Surface sterilized seeds of alder were germinated for 5 days in Petri dishes and transferred directly into the pouches. Inoculation with pure culture of Frankia [27] was applied 7 days later in presence of Crone's mineral solution. The solution was supplemented with 5 mg/l of nitrate to sustain the growth of the seedlings until nodules became effective. The seedlings were succesfully colonized with several ectomycorrhizal fungal species applied 5–8 weeks later.

Cuttings of willows and of hybrid poplars were collected at the end of the winter, surface sterilized with 70% alcohol and placed directly into the pouches in presence of water. Inoculation with ectomycorrhizal fungi was applied 2 weeks later and ectomycorrhizae formed readily with several species of fungi without any addition of nutrients.

Potential and limitations of the growth pouch method for the study of ectomycorrhizae

The outstanding advantage of the growth pouch method is that it allows direct visual observations (Fig. 1, 5) and examination under stereoscopic lenses (Fig. 2–4 and 7–10) of the ectomycorrhiza development at all stages including the extramatrical phase. This advantage is in common with the method of Pachlewska [9] and of Mason [12]. The shape (Fig. 3), the colour, and the texture (Fig. 7, 5) [28] can be observed at any stage of the ectomycorrhiza genesis with destructive (fixation) as well as non destructive (photography) techniques. Moreover the observation of the extramatrical phase development including mycelium patterns (Fig. 6), hyphal strands (Fig. 3), sclerotia (Fig. 4, 9) and sporophore primordium (Fig. 10) becomes possible.

The method will facilitate description of the characteristics of host fungus pairs as suggested by Dominik [29], Zak [30] and Giltrap [13].

Another advantage of the growth pouch technique is that it allows for the inoculum to be placed precisely in the vicinity of selected root parts and permits time lapse observation to be made at the level of individual ectomycorrhiza [24, 30]. The minimum time required for ectomycorrhiza formation was reduced from 4–6 weeks [12] to 3–5 days after inoculation. This was achieved without addition of glucose or vitamins. Moreover the minimum space required makes it possible to conduct exhaustive studies on the compatibility between all the members of genera or families of potential host plants and fungi in order to detect whether specificity or phylogenetical relationships exist or not as attempted earlier by Molina [32]; Gaie and Heinemann [33] and Giltrap [13].

When the purpose of a synthesis experiment is merely to demonstrate the compatibility between a given fungal isolate and a potential ectomycorrhizal plant, the method could probably be much simplified; surface sterilized seeds of the Pinaceae could possibly be transferred to the pouch immediately after germination.

It should be possible to inoculate the seedlings with suspension of spores as

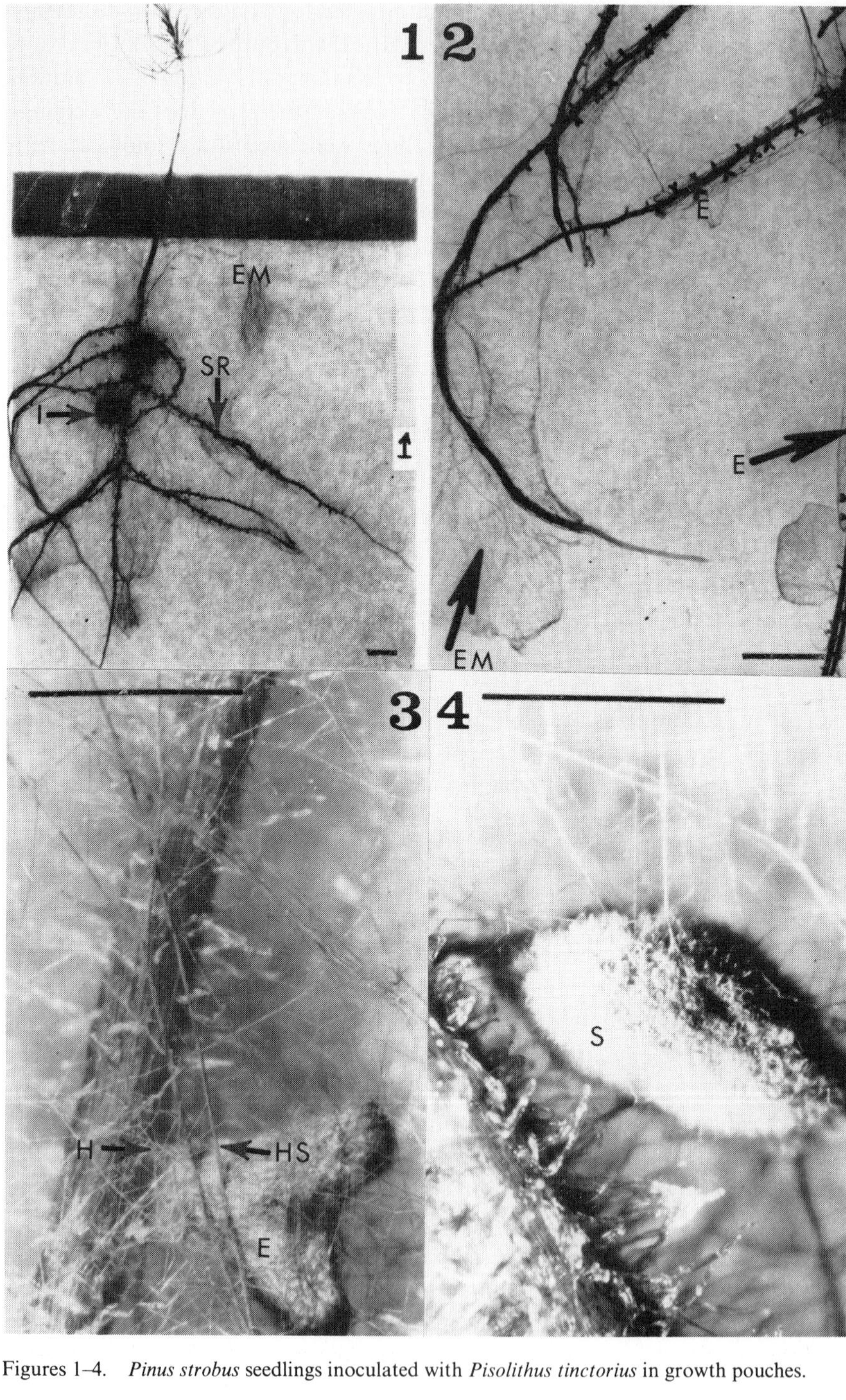

Figures 1–4. *Pinus strobus* seedlings inoculated with *Pisolithus tinctorius* in growth pouches.

Fig. 1. General view of a seedling. Note the ectomycorrhizal short roots (S.R.), the inoculum plug (I), and the extramatrical mycelium (E.M.). Bar = 1 cm.

Fig. 2. Region of root system showing dichotomous ecto mycorrhizae (E) and mycelium growth along the long lateral root forming extramatrical mycelium (E.M.). Bar = 1 cm.

Fig. 3. Nine day old ectomycorrhiza (E) on a long lateral root surronded by hyphae (H) and hyphal strands (H.S.) undistrubed as they formed. Bar = 1 mm.

Fig. 4. Sclerotium (S) formed around 25 days after inoculation. The rind at that stage retined a fuzzy surface. Bar = 1 mm.

successfully obtained by Blakeslee[21] and Marx and Ross[34], pieces of carpophores or sclerotia and even surface sterilized ectomycorrhizae from the field. If possible this would be helpful in studying suspected ectomycorrhizal species that can be hardly cultivated in the absence of host plants such as *Gomphidius* and *Russula* spp.

The fact that this method allows for the first time the observation of sclerotia formation by *P. tinctorius* suggests that these structures might exist on several other species of ectomycorrhizal fungi. If such is the case the pouch method is more likely to permit their observation than any of the flask techniques.

Competition between species of ectomycorrhizal fungi on the roots of a given seedling is another subject that could be studied with the growth pouch.

The method proposed here has some limitations. This is not an aseptic method and the specific composition and the abundance of the microorganisms is not controlled and may vary from one experiment to another. The method also requires daily attendance, high moisture level in the growth chambers and precludes the use of optimum nitrogen concentrations with non-nitrogen-fixing host plants.

References

1 Melin E 1922 Untersuchungen über die Larix-Mykorrhiza. I. Synthese der Mykorrhiza in Reinkultur. Svensk Bot. Tidskr. 16, 161–196.

2 Hacskaylo E 1953 Pure culture synthesis of pine mycorrhizae in terralite. Mycologia 45, 971–975.

3 Trappe J M 1962. Fungus associates of ectotrophic mycorrhizae. Bot. Rev. 28, 538–606.

4 Melin E and Nilsson H 1950 Transfer of radioactive phosphorus to pine seedlings by means of mycorrhizal hyphae. Physiol. Plant. 3, 88–92

5 Marx D H and Zak B 1965 Effect of pH on mycorrhizal formation of slash pine in asepticulture. For. Sci. 11, 66–75.

6 Richard C and Fortin J A 1970 Les mycorhizes du *Picea mariana* (Mill.) BSP.: Aspects morphologiques, anatomiques et systématiques. Nat. Can. 97, 163–173.

7 Richard C and Fortin J A 1975 Rôle protecteur du *Suillus granulatus* contre le *Mycelium radicis atrovirens* sur des semis du *Pinus resinosa*. Can. J. For. Res. 5. 452–456.

8 Trappe J M 1962 *Cenococcum graniforme*, its distribution, ecology, mycorrhiza formation, and inherent variation. Ph.D. Diss. Univ. Washington, Seattle. USA 148 p.

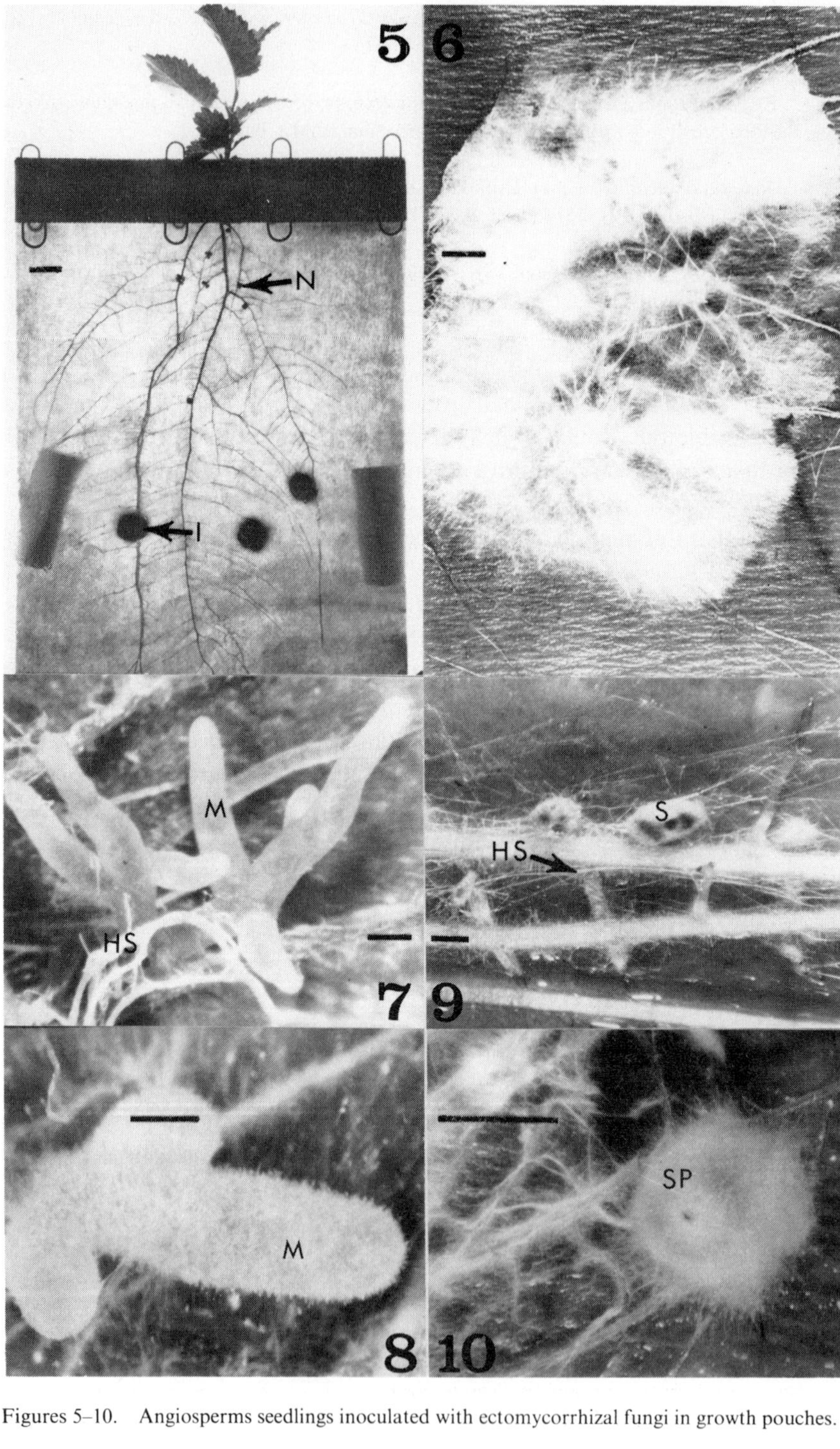

Figures 5–10. Angiosperms seedlings inoculated with ectomycorrhizal fungi in growth pouches.

Fig. 5. *Alnus crispa* seedling showing roots nodulated (N) with Frankia and inoculum plug (I). The small but abundant ectomycorrhizae are barely visible. Bar = 1 cm.

Fig. 6. Mycelial growth pattern of *Hebeloma crustuliniforme* inoculated on *Populus tremuloides* root system. Bar = 1 cm.

Fig. 7. Ectomycorrhizae formed on *Populus tremuloides* by *Lactarius pubescens var. pubescens* showing smooth surface of the mantle (M) hyphal strands (H.S.). Bar = 1 mm.

Fig. 8. Ectomycorrhiza formed by *Amanita muscaria* on *Populus tremuloides* showing asperulated surface of the mantle (M). Bar = 1 mm.

Figl. 9. Ectomycorrhizae (E) formed by *Paxillus involutus* on *Alnus rugosa* showing hyphal strands (H.S.) and sclerotia (s). Bar = 1 mm.

Fig. 10. *Cortinarius* cf. *subporphyrus* sporophore (SP) primordium formed on the extramatrical phase of *Populus tremuloides* ectomycorrhizae. Bar = 1 cm

9 Pachlewska J 1968 Investigations on mycorrhizal symbiosis of *Pinus sylvestris* in pure culture on agar. Prace Inst. Badaw. Lesn. 345, 3-76.

10 Pachlewski R and Pachlewska J 1974 Studies on symbiotic properties of mycorrhizal fungi of pine (*Pinus sylvestris* L.) with the aid of the method of mycorrhizal synthesis in pure cultures on agar. Forest Research Institute, Warsaw, 228 p.

11 Mason P A 1975 The genetics of mycorrhizal associations between *Amanita muscaria* and *Betula verrucosa*. The Development and Function of Roots. Eds. J G Torrey and D T Clarkson. Academic Press, New-York and London, pp 567–574.

12 Mason P A 1980 Aseptic synthesis of sheathing ectomycorrhizas. *In* Tissue Culture Methods for Plant Pathologists. Eds. D S Ingram and J P Helgeson. Blackwell Scientific publications, Oxford, London, Edinburgh Boston Melbourne, pp 173–178.

13 Giltrap N J 1979 Experimental studies on the establishment and stability of ectomycorrhizas. Ph.D. Thesis. Univ. of Sheffield, UK 196 p.

14 Sohn R F 1981 *Pisolithus tinctorius* forms long ectomy corrhiza and alters root development in seedlings of *Pinus resinosa*. Can. J. Bot. 59, 2129–2134.

15 Bigg W L and Alexander I J 1981 A culture unit for the study of nutrient uptake by intact mycorrhizal plants under aseptic conditions. Soil. Biol. Biochem. 13, 77–78.

16 Nylund J E 1981 The formation of ectomycorrhiza in conifers: Structural and physiological studies with special reference to the mycobiont, *Piloderme bicolor*. Ph.D. Thesis, Univ. Uppsala, Sweden, 34 p.

17 Skinner M F and Bowen G D 1974 The uptake and translocation of phosphate by mycelial strands on pine mycorrhizas. Soil Biol. Biochem. 6, 53–56.

18 Skinner M F and Bowen G D 1974 The penetration of soil by mycelial strands of ectomycorrhizal fungi. Soil Biol. Biochem. 6, 56–61.

19 Fortin J A 1965 Synthesis of mycorrhizae on explants of the root-hypocotyl of *Pinus sylvestris* L. Can. J. Bot. 44, 1087–1092.

20 Fortin J A and Piché Y 1979 Cultivation of *Pinus strobus* root-hypocotyl explants for synthesis of ectomycorrhizae. New Phytol. 83, 109–119.

21 Blakeslee G M 1977 Communication presented during a special session. Third North American Conference on Mycorrhizae.

22 Mullette K J 1976 Studies of Eucalypt Mycorrhizas. I. A method of mycorrhiza induction in *Eucalyptus gummifera* (Gaertn. and Hochr.) by *Pisolithus tinctorius* (Pers.)Coker and Couch. Aust. J. Bot. 24, 193–200.

23 Lalonde M 1979 Techniques and observations of the nitrogen-fixing Alnus root nodule symbiosis. *In* Recent Advances in Biological Nitrogen Fixation. Ed. N. S. Subba Rao. Oxford and I. B. H. Publication, New-Delhi, pp 421–434.

24 Fortin J A, Piché Y and Lalonde M 1980 Technique for the observation of early morphological changes during ectomycorrhiza formation. Can. J. Bot. 58, 361–365.

25 Piché Y and Fortin J A 1982 Development of mycorrhizae, extramatrical mycelium and sclerotia on *Pinus strobus* seedlings. New Phytol. 91, 211–220.

26 Hewitt E J 1966 Sand and Water Culture Methods used in the Study of Plant Nutrition. Tech. Comm. 22 (2nd edition revised). Commonwealth Agricultural Bureaux, London, 431 p.

27 Normand P and Lalonde M 1982 Evaluation of Frankia strains isolated from provenances of two *Alnus* species. Can. J. Microbiol. (*In press*).

28 Warrington S J, Black H D and Coons L B 1981 Entry of *Pisolithus tinctorius* hyphae into *Pinus taeda* roots. Can. J. Bot, 59, 2135–2139.

29 Dominik T 1969 Key to ectotrophic mycorrhizae. Folia For. Pol. Ser. A. 15, 309–328.

30 Zak B 1973. Classification of ectomycorrhizae. *In* ectomycorrhizae. Eds. G C Marks and T T Kozlowski. Academic Press, New-York, pp 43–78.

31 Piché Y, Fortin J A and Lafontaine J G 1981 Cytoplasmic phenols and polysaccharides in ectomycorrhizal and non-mycorrhizal short roots of pine. New Phytol. 88, 695–703.

32 Molina R 1981 Ectomycorrhizal specificity in the genus Alnus. Can. J. Bot. 59, 325–334.

33 Gaie W and Heinemann P 1980 Mycorhization du genre *Betula* par *Pisolithus arhizus*, en conditions axéniques. Bull. Jard. Bot. Nat. Belg. 50, 375–383.

34 Marx D H and Ross E W 1970 Aseptic synthesis of ectomycorrhizae on *Pinus Taeda* by basisdiopores of *Thelephora terrestris*. Can J. Bot. 48, 197-198.

Plant and Soil 71, 285–291 (1983). A-29

Apical meristems in mycorrhizal and uninfected roots of *Calluna vulgaris* (L.) Hull

G. BERTA
Istituto Botanico dell' Università di Torino

and P. BONFANTE-FASOLO
Centro di Studio sulla Micologia del Terreno del CNR, I-10125, Turin, Italy

Key words Apical meristem Ericales Mycorrhizas Root

Summary *Calluna vulgaris* root meristems, either uninfected or infected by *Pezizella ericae*, were compared using light and transmission electron microscopy. In lightly infected 'hair' roots, active tips contained initial cells and derivates similar to those of the uninfected roots. The meristem region of most of the heavily infected 'hair' roots was strongly reduced. The relationship between the fungus and the root is complex, and the presence of the fungus may influence the organization of the apical meristem.

Introduction

Ericoid mycorrhizal fungi can improve the survival and growth of many Ericaceae, especially those living in extremely nutrient-poor habitats[11]. Recently the ultrastructural organization of naturally and axenically infected mycorrhizal 'hair' roots has been described in different ericaceus genera[1, 2, 3, 10]. However, the effects of endomycorrhizal fungi on root tip organization have not been reported. Such studies have so far been limited to ectomycorrhizas where the apical root meristems, covered by a fungal sheath, are restricted and show low rates of mitosis except in the cap initials[5].

Moreover, results on the uninfected 'hair' root meristem organization are scanty and limited to light microscopical features[4].

The aim of the present paper is to compare, using light and electron microscopy, the organization of *Calluna vulgaris* (L.) Hull apical root meristems (the initials and immediately derived cells), either uninfected or infected by *Pezizella ericae* Read.

Materials and methods

Calluna vulgaris seeds were germinated aseptically on water-agar. Seedlings were transferred to a sterile soil-water agar mixture in tubes and some were inoculated with *Pezizella ericae* by Dr V. Gianinazzi-Pearson, who kindly provided materials. After three months, 'hair' roots of infected and uninfected seedlings were excised and prepared for light and transmission electron microscopy[2]. Flat-embedded samples were observed, counted and measured at low magnification with an ocular micrometer. Serial sections (1 μm thick) of 30 apices were prepared and stained with toluidine blue.

Only median longitudinal sections were used and were examined with a Zeiss Universal microscope. Thin sections, obtained from the same areas, and stained with uranyl acetate and lead citrate, were examined in a Philips EM 300 electron microscope.

Results

Longitudinal median sections of 'hair' roots (more than 1000 μm long) revealed a typical dermatocalyptrogen in their active tips. Initial cells dividing transversely and longitudinally, gave rise to two types of cells (Fig. 1). The first type led, by anticlinal divisions, to the epidermis consisting of cells rich in dense deposits. The second type produced, by anticlinal and periclinal divisions, a well differentiated cap, with large and sloughing cells. Epidermal and cap cells were surrounded by an extraradical slime layer.

The dermatocalyptrogen was clearly separated from the underlying meristematic area, where different initial cells gave rise to the cortex and central cylinder.

After several anticlinal divisions, the cortical cells divided periclinally, giving rise to two layers. The outer layer soon vacuolized and sometimes collapsed, whilst the inner one filled up with dense material.

In the inner layer of the meristem, central cylinder initial cells divided anticlinally and periclinally.

With the rare exception of some long 'hair' roots (about 1000 μm long), most of the infected 'hair' roots were short, with a mean length about 200 μm (range 100–400 μm). Two types of histological organization were observed. The first type, characteristic of the shortest 'hair' roots, occurred in a few lightly ones. Initial cells were observed originating epidermal and cap cells, whilst other initials originated central cylinder cells (Fig. 2). The apical organization was similar to that of the uninfected hair roots. Fungal mycelium generally surrounded these 'hair' roots, but penetrated only basal areas.

The second type of 'hair' root tip organization occurred in the heavily infected 'hair' roots (200-400 μm long). Initial and derivates cells with meristematic

---

Fig. 1. Median longitudinal section of a root tip from an uninfected *Calluna vulgaris* hair root. Initials (i) of epidermal (e) and cap (c) cells are evident, as well as the cortical (ct) cells and other initials (ii) giving rise to vascular cells (cc).

Fig. 2. Median longitudinal section of a root tip from a hair root, lightly infected by *Pezizella ericae*. External hyphae (ef) surround the hair root, without penetrating it.

Fig. 3. Median longitudinal section of a root tip from an infected hair root. Meristematic region is reduced and vascular cells are filled up with dense materials (d).

Fig. 4. Median longitudinal section of a root tip from a strongly infected hair root. Fungal coils (f) are present inside epidermal and cap cells.

Bar represents 10 μm.

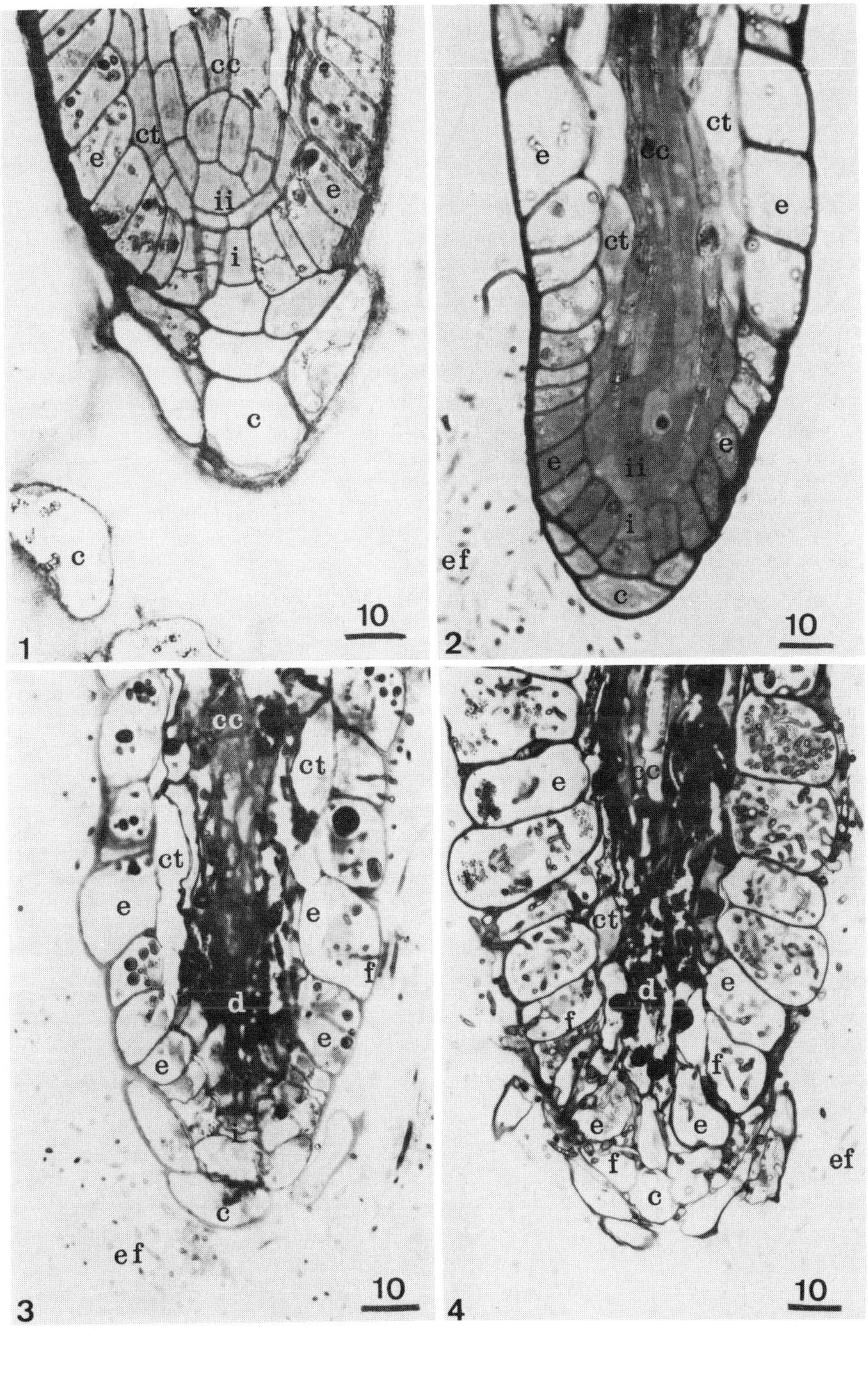
cc
ct
e
ii
e
i
c
c
10
1
e
ct
ec
e
ct
e
e
ii
i
ef
c
10
2
cc
ct
ct
e
e
f
d
e
e
i
c
ef
10
3
e
cc
ct
d
e
f
f
e
e
f
c
ef
10
4

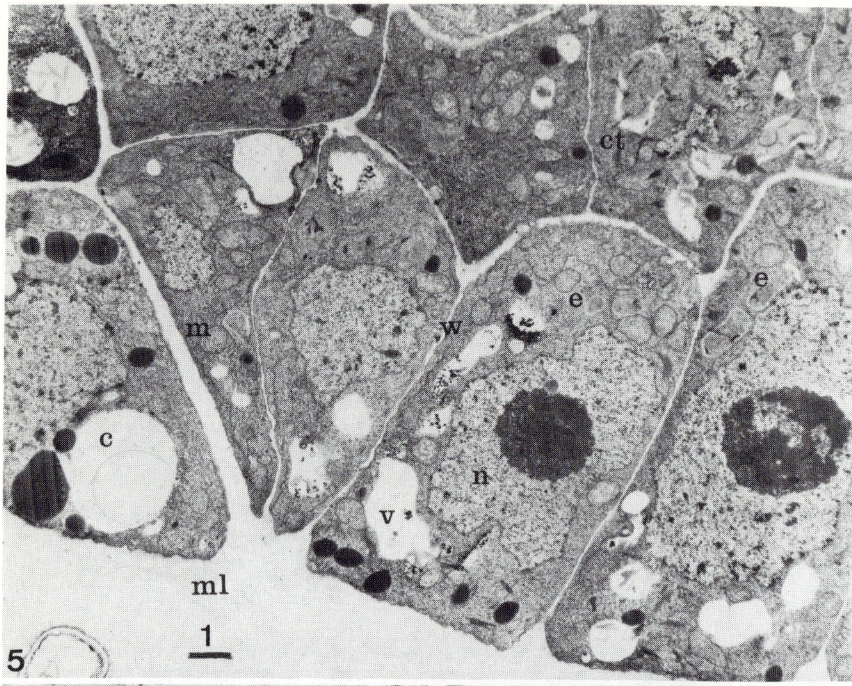

ct
e
e
m
w
c
n
v
ml
1
5

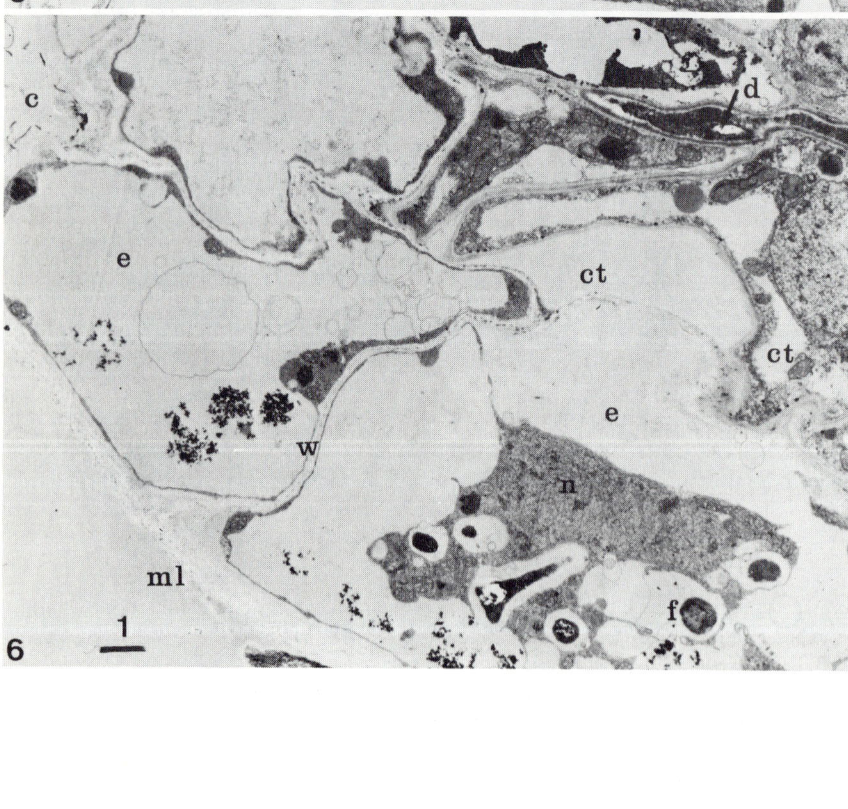

c
d
e
ct
ct
e
w
n
ml
f
1
6

characteristics of dermatocalyptrogen were no longer observed: there were a few highly vacuolated cells, often heavily infected by the fungus. In the area corresponding to the cortical and stelar initials and derivates, elongated cells were present rich in dense material (Fig. 4).

During the mycorrhizal tip analysis, other patterns of organization were observed, characterized by intermediate features. In some cases, meristematic regions were reduced but still present. A few uninfected derivate cells were found giving rise to epidermal and cortical cells (Fig. 3).

Uninfected and lightly-infected apical meristems displayed similar ultrastructural features: irregularly shaped nuclei, containing prominent nucleoli; small proplastids; abundant perinuclear mitochondria; peripheral rough endoplasmic reticulum; infrequent starch grains; minute vacuoles; thin electron-transparent cell walls containing plasmodesmata; periplasmic areas characterized by electron transparent vesicles (Fig. 5). The derivates quickly differentiated into mature cells.

All cells in the heavily infected 'hair' root tips were strongly modified. The cells corresponding to the dermatocalyptrogen region were highly vacuolated with a thin peripheral layer of cytoplasm, where unusual plastids and mitochondria were observed. Fine electron-dense deposits occurred inside the large vacuoles. Isolated hyphae were sometimes observed in cap cells, while well organized coils occurred in epidermal cells, close to the vacuolated region (Fig. 6). In the region of the cortex and stelar initials, irregularly shaped cells were observed. They were filled with electron-dense materials and were collapsed towards the tip.

Discussion

Several differences in the organization of the meristematic region were noted, between infected and uninfected roots. Uninfected 'hair' roots had active tips with an apical meristem pattern that is very common among Dicotyledons: *i.e.* a central cylinder and cortex deriving from specific initials, and epidermal and cap cells arising from common initials [7]. Moreover the presence of a clear boundary between cap and cortex suggests that the apical meristem of *C. vulgaris* can be considered as a closed meristem according to Clowes [6].

By contrast, the more vigorous mycorrhizal plants of the same age have a more branched and abundant root system, but the latter was formed of shorter hair

---

Fig. 5. Electron micrograph of the meristematic region of an uninfected hair root of *C. vulgaris*, showing cap (c), epidermal (e) and cortical (ct) cells. (n, nucleus; m: mitochondrion; v: vacuole; w: wall; ml: mucigel layer).

Fig. 6. Electron micrograph of the tip region of an infected hair root. The fungus (f) colonizes the apical epidermal cells. Cap and cortical cells appear highly vacuolized, whilst vascular cells (cc) are filled up with dense material (d).

Bar represents 1 μm.

roots. The histological organization of mycorrhizal 'hair' root tips is characterized by highly reduced (or completely absent) meristematic cells; fully differentiated, infected, epidermal cells; reduced, vacuolated and sloughing cap cells; differentiated cortical cells; and lastly, stelar cells filled with electron-dense material and showing signs of collapse.

In Rhododendron, mycorrhizal 'hair' root tips have also been found having no meristematic activity[10]; but, contrary to the present results, the number of non meristematic tips was smaller than that of the actively growing roots.

In *C. vulgaris* infected by *P. ericae*, Pearson[9] suggested structural differences between infected and uninfected root apices. Our results suggest that vigorous mycorrhizal plants, endowed with an abundant root system due to fungal growth stimulation, present, in the first phase, 'hair' root tips with a meristematic organization similar to that of uninfected tips. The presence of the fungal endophyte seems to affect meristematic activity. Initial cells no longer divide, derivates quickly differentiate and are no more replaced. These features have been observed by Burgeff[4], who considered that the fungus could become a pathogen, killing the host meristem.

In conclusion, in active and well established mycorrhizas obtained in pure culture, the appearance of the ericoid mycorrhizal tip resembles that of a senescent tip, even though the whole plant is more vigorous than uninfected one. However, a detailed comparison between tips belonging to naturally and axenically infected plants is necessary to extrapolate the present results. The symptoms of senescence in 'hair' root tips could be in fact magnified-in part at least-by the system used to grow the plants.

Our results agree with the findings concerning ectomycorrhizal fungi. The latter produce hormones, increasing root production[8] and it has been suggested that in naturally-infected, well-established ectomycorrhizas, root meristems have lower mitotic rates than without a fungal sheath[5]. However, the effect of mycorrhizal fungi on root anatomy is not fully understood, and the size, structure and longevity of root systems as well as their physiology can be affected by the presence of a fungal endophyte.

References

1 Bonfante-Fasolo P and Gianinazzi-Pearson V 1979 Ultrastructural aspects of endomycorrhizas in the Ericaceae I. Naturally infected hair roots of *Calluna vulgaris* (L.) Hull. New Phytol. 83, 739–744.

2 Bonfante-Fasolo P and Gianinazzi-Pearson V 1982 Ultrastructural aspects of endomycorrhiza in the Ericaceae III. Morphology of the dissociated symbionts and modifications occurring during their reassociations in axenic culture. New Phytol. 91, 691–704.

3 Bonfante-Fasolo P, Berta G and Gianinazzi-Pearson V 1981 Ultrastructural aspects of endomycorrhizas in the Ericaceae II. Host-endophyte relationships in *Vaccinium myrtillus* L. New Phytol 89, 219–224.

4 Burgeff H 1961 Microbiologie des Hochmoores, G. Fischer Verlag, Stuttgart pp 25–30 and 118–120.

5 Clowes F A L 1981 Cell proliferation in ectotrophic mycorrhizas of *Fagus sylvatica* L. New Phytol. 87, 547–555.

6 Clowes F A L 1981 The difference between open and closed meristems. Ann. Bot. 48, 761–767.

7 Guttenberg H von 1960 Grundzüge der Histogenese höherer Pflanzen. I. Die Angiospermen. Gebrüder Borntraeger, Berlin. 315 p.

8 Ng P P, Cole A L J, Jameson P E and McWba J A 1982 Cytokinin production by ectomycorrhizal fungi. New Phytol. 91, 57–62.

9 Pearson V 1971 The biology of the mycorrhiza in the Ericaceae. Ph.D. Thesis, University of Shieffield, 25 p.

10 Peterson T A, Mueller W C and Englander L 1980 Anatomy and ultrastructure of a Rhododendron root-fungus association. Can. J. Bot. 58, 2421–2433.

11 Read D J 1980 The role of mycorrhiza of ericaceous plants with special reference to the genus Vaccinium L. *In* Spontaneous productions. Les Colloques de l'INRA 4, 191–201.

Plant and Soil 71, 293–297 (1983). A-30

Responses of pine roots to E-strain ectendomycorrhizal fungi

H. E. WILCOX, C. S. YANG and K. F. LO-BUGLIO
State University of New York, College of Environmental Science and Forestry, Syracuse, NY 13210, USA

Key words Ectendomycorrhizae E-strain Chlamydospores Fungal characteristics Mycorrhizal synthesis *Pinus resinosa* Root morphology

Summary Cultural characteristics of 6 E-strain fungi were studied and monoxenic mycorrhizal synthesis experiments performed on *Pinus resinosa* with three isolates. Best radial growth of fungi was obtained on potato dextrose agar at pH 6, with northern isolates reaching peak growth rates at 20°C and southern at 24°C. Northern but not southern races grew at 4°C. Mycorrhizae formed on red pine by isolates from New York (NY) and West Virginia (WV) were ectendo- but those by the Mississippi (MISS) isolate were predominantly ecto-. Root systems of seedlings with the WV isolate were coarser with shorter and thicker long-root branches and fewer mycorrhizae than NY and MISS seedlings.

Introduction

Fungi responsible for the common type of ectendomycorrhizae in seedlings from northern coniferous nurseries were isolated from Finnish nurseries and designated as E-type fungi[6]. Criteria used for characterizing E-strain mycorrhizae were intracellular infection, a weakly developed mantle, and coarse hyphae in the Hartig net. The discovery of E-strain chlamydospores[7] provided an additional means of detecting the presence of these fungi in forest nurseries. The distinctiveness of these chlamydospores and their prevalence in soils from nursery beds with ectendomycorrhizal seedlings has led us to the discovery that ectendomycorrhizae are not restricted to northern nurseries. To date we have found E-strain chlamydospores in seeds beds of 17 coniferous nurseries in 11 of the United States.

The present paper compares the cultural characteristics of the following 6 E-strain fungi: Three fungi recently isolated from nurseries in Mississippi, North Carolina, and West Virginia; R-947, isolated from *Picea glauca* in Canada[2]; E-57, isolated from *Picea abies* in Finland[6]; and BDG, isolated from *Pinus resinosa* in New York[7]. For convenience of reference the three recent isolates are designated as BDG/MISS, BDG/NC, and BDG/WV. Results of a 7-month mycorrhizal synthesis experiment, with BDG, BDG/MISS, and BDG/WV are also presented.

Materials and methods

Chlamydospores of E-strain mycorrhizal fungi were isolated from nursery soils by the flotation method[1]. Spores were distinguished under the microscope by their dull and minutely roughened surface and conspicuously septate subtending hyphae. Detailed descriptions of spore morphology are presented by Wilcox *et al.*[7] and Danielson[2].

Cultures of BDG/MISS and BDG/WV were obtained by germination of chlamydospores. BDG/NC was obtained by isolation from roots of *Pinus resinosa* inoculated with chlamydospores.

Fungal isolates were compared for growth on MMN[4] agar and potato dextrose agar at 7 temperatures from 4 to 32°C. Six replicates were made of each isolate, and radial growth was measured after 30 days incubation.

Mycorrhizal synthesis experiments were performed in flask culture[5] containing MMN medium without glucose and with pH at 5.5. Inoculated seedlings were grown for 7 months in a 16°C temperature-controlled greenhouse. Numbers of the different orders of long roots were counted and measurements made of their individual and cumulative lengths. Numbers of short roots and their appearance and distribution were also recorded. Shoots were removed and their dry weights determined. Freehand transverse and longisections were prepared for study of details of fungal colonization in the various roots orders and its relationship to the different meristems.

Results

A correlation between the origin of the 6 isolates and their response to temperature is shown in Fig. 1. The three southern isolates (BDG/MISS, BDG/NC, and BDG/WV) grow best at 24°C and show no measurable growth on PDA at 4°C. Northern isolates R-947 and E-57 grow best at 20°C and show some growth at 4°C. BDG peaks at 24°C but also grows at 4°C.

Table 1 compares the radial growth of the isolates after 30 days at 16°C on three types of media. Mycelial growth is strikingly better on PDA than on either MMN or malt agar. Growth differences on MMN and PDA are inconsistent, malt agar giving significantly better growth than MMN for BDG/WV, BDG, and E-57 and poorer growth for BDG/MISS.

Results of a morphological comparison of the 7-month mycorrhizal synthesis experiment are summarized in Table 2. The seedlings inoculated with BDG and BDG/MISS are similar in size with nearly the same shoot dry weight and

Table 1. Radial growth of six ectendomycorrhizal isolates after 30 days at 16 ± 1°C on three types of media (mm)

Isolatie	PDA	MMN	2% Malt
R-947	84.17 ± .83*	34.5 ± .99	34.67 ± 2.14
BDG-NC	19.83 ± 1.49	10.17 ± .79	12.0 ± .57
BDG-MISS	15.5 ± 2.81	7.0 ± .52	1.67 ± 1.1
BDG-WV	40.83 ± 2.06	14.83 ± 1.01	22.5 ± .85
BDG	17.0 ± .45	10.67 ± .49	16.33 ± 1.38
E-57	47.5 ± 2.32	11.67 ± .80	22.83 ± 1.08

* Mycelium completely covered plate.

± Standard error.

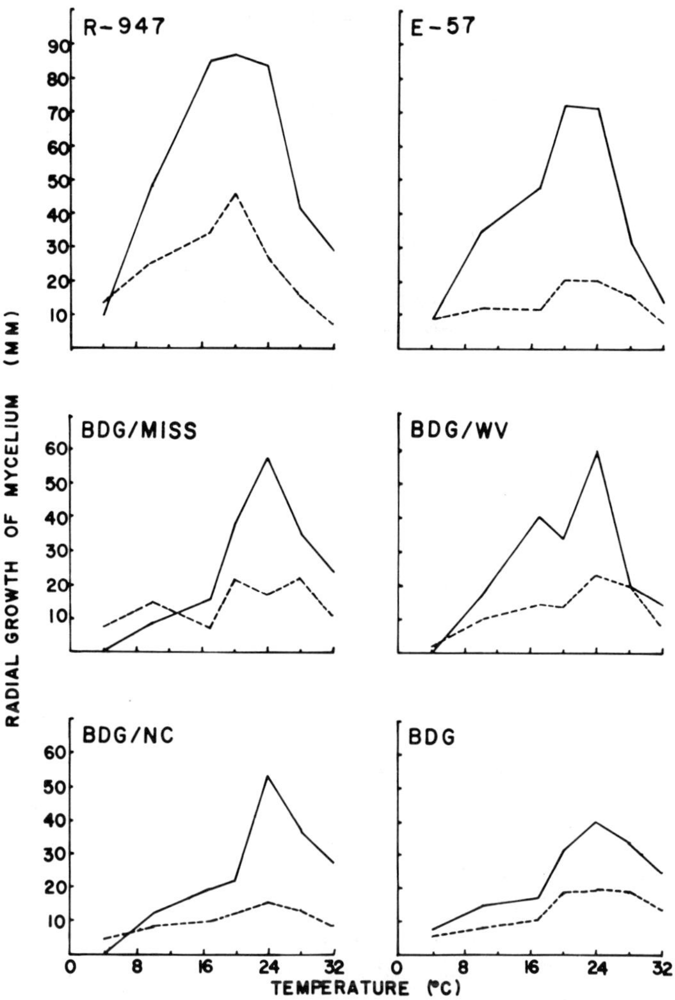

Fig. 1. Effect of temperature on the growth of the six E-strain ectendomycorrhizal fungi on PDA (———) and MMN (– – – –). Explanation in the text.

approximately the same cumulative lengths of long roots and numbers of mycorrhizal short roots. The BDG/WV seedlings have a lower average shoot dry weight and the root systems possess a significantly lower average cumulative length of long roots and fewer mycorrhizae. The difference in root morphology of the BDG/WV seedlings is readily apparent. Their root systems appear coarser with shorter and thicker long roots and fewer fine root branches than the other two treatments.

Table 2. Cumulative lengths of long roots and numbers of mycorrhizae in 7 month seedlings of *Pinus resinosa*

E-strain fungus	Average length per seedling of long roots (mm)						Average number of short roots	Ave. dry wt. of shoots (g)
	Cumu-lative	Primary root	1st lateral	2nd lateral	3rd lateral	4th lateral		
BDG (2)*	1776	80	714	831.5	225.5	5.5	461	0.45
BDG/WV (3)*	1063	83	571	443.3	48.3	0	291	0.32
BDG/MISS (4)*	2036	86.75	766.75	1140.5	128.5	0	551	0.44

* Number of seedlings sampled.

Table 3. Width of primary body between protoxylem area at 5 mm proximal to apical meristem

E-strain fungus	Width (μm)		
	1st order	2nd order	3rd order
BDG	115	63	24
BDG/WV	144	74	52
BDG/MISS	112	51	25

In BDG/MISS the infection is mostly intercellular with a well developed Hartig net in all long and short roots. A few of the larger diameter first order laterals also have intracellular infection in cells of the inner cortex. In BDG and BDG/WV infections both the intracellular infection and the Hartig net are strongly developed in roots of all orders. Although the colonization characteristics of these two fungi are similar, root infections by BDG/WV are much heavier. Fungal mantles are often more prevalent on the long roots, covering their apices. Measurement of distance between the arcs of protoxylem as viewed in transections is shown in Table 3. These data indicate that the primary body of long roots of BDG/WV seedlings is actually larger than those of seedlings infected with BDG or BDG/MISS.

Discussion

Our investigations of E-strain ectendomycorrhizal fungi show that they are cosmopolitan, and common in coniferous nurseries. Not surprisingly their growth response to temperature is correlated with the temperature of their origin. More intriguing, however, are the differences in their colonizing behavior and in their morphogenetic effect on a given host species. The apparent increase in size of the long root primary body of red pine caused by BDG/WV merits further investigation to determine the factors involved.

Laiho[3] examined various northern coniferous species and made a list of those in which the E-strain produced ectendomycorrhizae and those in which it produced ectomycorrhizae. He found almost all pine species to be ectendomycorrhizal with the E-strain. It now appears that such a generalization is not valid. The present investigation shows that some E-strains produce ectendomycorrhizal infections in a given pine while others produce ectomycorrhizal infections in the same pine. The species of pine is also important in determining whether the association will be ecto- or ectendomycorrhizal with a particular E-strain fungus. In our preliminary experiments, we have found that BDG, which was thought to be ectendomycorrhizal on all pines, is ectomycorrhizal on slash pine. The presence in the S.E. United States of E-strains such as BDG/MISS which tend towards ectomycorrhizal associations on pines, as well as pines which tend to be ectomycorrhizal with E-strain fungi, are the likely reasons why ectendomycorrhizae have not heretofore been reported in pine nurseries in that part of the country. It is thus apparent that differences between E-strains and differences between pine species are both more complex than originally thought.

We are now analyzing a larger 7-month mycorrhizal synthesis experiment with the 3 fungi used in this paper plus 6 additional isolates on *P. resinosa*, *P. elliottii*, and *P. strobus*. It is hoped that the results will clarify questions concerning the specificity of the associations between E-strain fungi and host pines in forest nurseries.

References

1 Barnard E L 1977 Mycorrhizal biology of *Liriodendron tulipifera* L. and its relationship to *Cylindrocladium* root rot. Ph.D. Thesis. Duke Univ., Durham, NC.

2 Danielson R M 1982 Taxonomic affinities and criteria for identification of the common ectendomycorrhizal symbiont of pines. Can. J. Bot. 60, 7–18.

3 Laiho O 1965 Further studies on the ectendotrophic mycorrhiza. Acta For. Fenn. 79, 1–34.

4 Marx D H 1969 The influence of ectotrophic mycorrhizal fungi on the resistance of pine roots to pathogenic infections. I. Antagonism of mycorrhizal fungi to root pathogenic fungi and soil bacteria. Phytopathology 59, 153–163.

5 Marx D H and Zak B 1965 Effect of pH on mycorrhizal formation of slash pine in aseptic culture. Forest Sci. 11, 66–75.

6 Mikola P 1965 Studies on the ectendotrophic mycorrhiza of pine. Acta For. Fenn. 75, 1–56.

7 Wilcox H E, Ganmore-Neumann R and Wang C J K 1974 Characteristics of two fungi producing ectendomycorrhizae in *Pinus resinosa*. Can. J. Bot. 52, 2279–2282.

Plant and Soil 71, 299–302 (1983). A-31

Role of *Laccaria laccata* in protecting primary roots of Douglas-fir from root rot

DAVID M. SYLVIA
Department of Plant Pathology, University of Florida, Gainesville, FL 32611, USA

Key words Antibiosis Biological control *Fusarium oxysporum* Induced resistance *Laccaria laccata* Phenolics

Summary The influence of *Laccaria laccata* on *Fusarium oxysporum*, and its ability to induce resistance to the pathogen in primary roots of Douglas-fir were investigated. Extracellular metabolites of *L. laccata* inhibited *F. oxysporum* hyphae, and delayed germination of microconidia and chlamydospores. *Laccaria laccata* or its cell-free metabolites also induced accumulation of osmiophilic materials in cortical cells of the primary root. The osmiophilic materials were primarily phenolic in nature. One-month-old primary roots, incubated with *L. laccata* or killed inoculum, were challenged with *F. oxysporum*. Rate of growth of the pathogen toward the root and intensity of surface colonization were not influenced by *L. laccata*, thus antibiosis was judged unimportant in root protection. However, cortical infection of the primary root by the pathogen was significantly less in seedlings affected by *L. laccata* than in controls. The frequency of hyphae was inversely proportional to the concentration of osmiophilic materials. Phenolics induced by *L. laccata* in the primary root are thus associated with resistance and may be the chemical basis for root protection.

Introduction

Root rot of Douglas-fir [*Pseudotsuga menziesii* (Mirb.) Franco] by *Fusarium oxysporum* Schlect. emend. Snyd. & Hans. may be lethal if the primary roots of very young seedlings become infected[2]. The ectomycorrhizal fungus *Laccaria laccata* (Scop.: Fr.) can protect young seedlings of Douglas-fir by colonizing the primary roots shortly after germination[3,11]. Of the mechanisms for root protection proposed by Marx[10], *L. laccata* may protect roots by antibiosis, induced resistance, or alteration of rhizosphere microflora.

The objectives of this research were 1) to evaluate the potential for antibiosis of *L. laccata* to *F. oxysporum*, 2) to characterize the reaction of primary roots of Douglas-fir to *L. laccata*, and 3) to determine the significance of antibiosis and induced resistance in root protection. Details of these studies have been published elsewhere[12,13].

Materials and methods

Antibiosis

Two-wk-old cultures of *L. laccata* were comminuted for 5 sec with sterile water and dispensed to one side of petri plates containing Melin-Norkrans agar medium[9] plus thiamine and biotin at various glucose concentrations and pH. After 2-wk incubation at 21°C, plugs from the margin of cultures of *F. oxysporum* were placed 40 mm from the edge of *L. laccata* colonies and growth was observed for 1 month. Fractorial experiments were conducted involving pH (4 and 6), glucose concentration (3, 10 and 50 mM), and temperature (15, 21 and 27°C). Inhibition in each treatment was evaluated by dividing the distance that *F. oxysporum* grew toward *L. laccata* by the distance it grew toward an agar slurry on a control plate.

The activity of extracellular metabolites of *L laccata* were studied by growing the fungus in Melin-Norkrans liquid medium at 3 glucose levels (3, 10 and 50 mM). One-ml amounts of slurry from stock cultures were added to 125 ml Erlenmyer flasks containing 60 ml of a test medium and incubated 14 days at room temperature on a reciprocating shaker at 100 excursions/min. The contents of each flask were poured through filter paper, and the filtrate was sterilized by aspiration through 0.2 μm Nalgene membrane filters. Microconidia or chlamydospores of *F. oxysporum* were placed in the filtrate which was then added dropwise to sterile microscope slides and incubated in moist chambers. Spore germination was monitored periodically for 24 h.

Induced host response

Seedlings were grown in a wick culture system which allowed observation of primary roots and minimized damage at harvest[12]. Roots of intact seedlings grew gnotobiotically with *L. laccata* or stem-killed inoculum in 200 × 32 mm culture tubes while shoots were exposed to ambient conditions of the growth chamber. In a modification of the wick culture system, inoculum was separated from the root by a dialysis membrane (12,000–14,000 MWCO) to allow observation of effects of extracellular metabolites. After 6 wk seedlings were removed from the wick system and observed with light microscopy when fresh or after embedment in epoxy resin. Transverse sections of fresh roots were placed in water or a histochemical reagent. The following reagents were used: for phenolics, nitrous acid[4] or saturated alcoholic vanillin-concentrated HCl[7]; for lipids, Sudan black B[1]; for lignin, phloroglucinol[8]; for proteins, naphthol blue-black[6] or acid fuchsin[5]. Autofluorescence of fresh, unstained sections was observed with incident light fluorescence microscopy. Roots embedded in plastic were fixed in 1% glutaraldehyde plus 0.5% caffeine in 0.05 M sodium cacodylate buffer, postfixed with 2% osmium tetroxide, dehydrated in an acetone series, and embedded in Spurr's epoxy resin.

Root protection

Seedlings were grown in wick culture tubes with *L. laccata* or killed inoculum for 4 wk and then aseptically transferred to microscope slides coated with 2% water agar. Surface sterilized seed of Douglas-fir, internally infested with *F. oxysporum*, were placed on the agar near the root tips. Growth of hyphae toward roots was recorded daily, and after 10 days the distal 10 mm of the root was excised and prepared for light and scanning electron microscopy. Segments for scanning electron microscopy were fixed and dehydrated as described above, subjected to critical point drying, and coated with Au-Pd before viewing.

Colonization of the root surface by *F. oxysporum* was quantified from scanning electron micrographs made at three random locations along the root. A grid of dots 5 mm apart was placed on each micrograph, and the number of dots on hyphal images was expressed as a percentage of total dots over hyphae plus root surface. A cortical infection index was determined for three roots per treatment by viewing longitudinal sections through a net reticle. The number of intersections of the grid under which hyphae were seen was expressed as a proportion of the total number of intersections over cortical tissue. The relationship between osmiophilic compounds and fungal presence in the root cortex was quantified by determining, in conjunction with the assessment of internal infection, the absorbance of light by sections stained with osmium tetroxide.

Results and discussion

Antibiosis

Substantial inhibition of hyphal elongation of *F. oxysporum* in the presence of *L. laccata* occurred only on a medium incubated at 15°C containing 10 m*M* glucose at pH 4. Under this condition, hyphal length was reduced by 70% compared to the control. Antagonism was greater at 10 m*M* than 3 m*M* glucose, but significant interactions occurred among the factors. Extracellular metabolites of *L. laccata* delayed germination of *F. oxysporum* spores. After 11.5 and 20 h incubation in filtrate of *L. laccata*, germination of chlamydospores and microconidia, respectively, were reduced 64 and 55% as compared to controls.

Induced host response

Transverse sections of 10-day-old tissue of roots grown gnotobiotically with *L. laccata* contained darkly stained osmiophilic materials in cortical cells. Similar sections produced a positive reaction with phenolic reagents throughout the cortex. Control roots lacked osmiophilic materials in the cortex and there was little reaction with phenolic reagents in this tissue. Phenolics accumulated in the cortex of roots separated from *L. laccata* by a dialysis barrier, indicating that an extracellular fungal metabolite induced this response. Induced materials failed to react with reagents for lipid, lignin and protein, and did not autofluoresce.

Root protection

Prior colonization of root by *L. laccata* or exposure to its metabolites did not reduce either the rate at which *F. oxysporum* approached the root ($\bar{x}$ = 9mm/day) or the extent of surface colonization. *Fusarium oxysporum* colonized 18% of the surface of control roots compared to 43 and 25%, respectively, for roots exposed to *L. laccata* or its metabolites. Thus we judged antibiosis unimportant in root protection although *L. laccata* had been shown capable of inhibiting *F. oxysporum* in vitro.

Cortical invasion of roots by *F. oxysporum* was reduced significantly in seedlings that had been exposed to *L. laccata* or its metabolites. *Fusarium oxysporum* occurred over 25% of the surface of sections from control roots compared to 0 and 5%, respectively, for roots exposed to *L. laccata* or its metabolites. A highly significant, inverse relationship was found between hyphal frequency and accumulation of osmiophilic material in cortical cells. This induced host response may be the chemical basis for root protection.

Acknowledgement This study was conducted at Cornell University, Department of Plant Pathology, Ithaca, NY under the direction of Wayne A. Sinclair.

References

1 Bancroft J D, Stevens A and Perse A G E 1975 Histochemical Techniques. Boston: Butterworths, 348 p.

2 Bloomberg W J 1973 Fusarium root rot of Douglas-fir seedlings. Phytopathology 63, 337–341.

3 Brown A C and Sinclair W A 1981 Colonization and infection of primary roots of Douglas-fir seedlings by the ectomycorrhizal fungus *Laccaria laccata*. For. Sci. 27, 111–124.

4 Endress A G and Thomson W W 1976 Ultrastructure and cytochemical studies on the developing adhesive disc of Boston ivy tendrils. Protoplasma 88, 315–331.

5 Feder N and O'Brian T P 1968 Plant microtechnique: some principles and new methods. Am. J. Bot. 55, 123–142.

6 Fisher D B 1968 Protein staining of ribboned epon sections for light microscopy. Histochemie 16, 92–96.

7 Gardner R O 1975 Vanillin-hydrochloric acid as a histochemical test for tannin. Stain Technol. 50, 315–317.

8 Jensen W A 1962 Botanical Histochemistry. San Francisco: W H Freeman and Co., 408 p.

9 Marx D H 1969 The influence of ectotrophic mycorrhizal fungi on the resistance of pine roots to pathogenic infections. I. Antagonism of mycorrhizal fungi to root pathogenic fungi and soil bacteria. Phytopathology 59, 153–163.

10 Marx D H 1972 Ectomycorrhizae as deterrents to pathogenic root infections. Annu. Rev. Phytopathol. 10, 429–454.

11 Sinclair W A, Sylvia D M and Larsen A O 1982 Disease suppression and growth promotion in Douglas-fir seedlings by the ectomycorrhizal fungus *Laccaria laccata*. For. Sci. 28, 191–201.

12 Sylvia D M and Sinclair W A 1983 Suppressive influence of *Laccaria laccata* on *Fusarium oxysporum* and on Douglas-fir seedlings. Phytopathology 73, *In press*.

13 Sylvia D M and Sinclair W A 1983 Phenolic compounds and resistance to fungal pathogens are induced in primary roots of Douglas-fir seedlings by the ectomycorrhizal fungus *Laccaria laccata*. Phytopathology 73, *In press*.

Plant and Soil 71, 303–308 (1983). A-32

Premiers résultats de recherches sur la compétitivité des champignons ectomycorhiziens

First results of research on the competitivity of ectomycorrhizal fungi

J. GARBAYE
C.N.R.F., Champenoux, F-54280 Seichamps, France

Key words Competitivity Ectomycorrhizae *Fagus silvatica* *Hebeloma crustuliniforme* Inoculation intensity Soil receptivity

Resumé Les premiers résultats d'un modèle expérimental visant à quantifier le comportement de différents champignons ectomycorhiziens sur différents sols sont décrits. Ils mettent en évidence que l'intensité de l'inoculation par une souche introduite est un facteur décisif pour la mycorhization par cette souche. La photopériode et le temps sont également très importants à prendre en compte pour la définition de la capacité d'accueil d'un sol. Les deux sols testés ont des comportements très différents. La microflore du sol n'est pas toujours l'obstacle principal à l'infection mycorhizienne et à son extension.

Summary The first results of an experimental model for quantifying the behaviour of different ectomycorrhizal fungi on different soils are described. They show that the inoculation intensity of the introduced strain is a decisive factor for the mycorrhization by this strain. Photoperiod and time are very important too for the definition of the soil receptivity. The two tested soils have very different behaviours. The soil microflora is not always the main obstacle to the mycorrhizal infection and to its extension.

Introduction

Les souches de champignons ectomycorhiziens utilisées pour la mycorhization contrôlée des plantations forestières doivent être sélectionnées non seulement pour leur efficacité sur la croissance des arbres, mais aussi pour leur aptitude à se maintenir sous forme de mycorhizes sur le système racinaire après la plantation et à coloniser de nouvelles racines. Ce caractère est particulièrement important car il conditionne la durée effective de l'effet bénéfique attendu de l'inoculation. Il est donc utile de disposer d'un test expérimental simple permettant de décrire le phénomène en terme quantitatif afin de comparer, pour une espèce d'arbre donnée, le comportement de différentes souches fongiques sur différents types de sol[2]. Le présent artice décrit les premiers travaux réalisés dans ce sens sur le Hêtre (*Fagus silvatica* L.) avec une souche d'*Hebeloma crustuliniforme* Pers. ex Fr. qui s'est par ailleurs avérée efficace sur la croissance du Hêtre[3].

Matériel et méthode

Les faines prégermées par stratification dans la tourbe humide à 4°C sont semées individuellement dans des pots ouvrant[5], contenant 0,41 du sol étudié. Les pots sont rangés dans des cagettes contenant 15 pots, chaque cagette constituant un traitement avec 15 répétitions. Huit semaines après le semis, les pots sont ouverts et la face visible de la motte avec les racines qui s'y sont développées reçoit 5 ml d'une suspension de mycelium d'*H. crustuliniforme* de concentration connue. Cet inoculum est préparé de la façon suivante: le champignon est cultivé 2 semaines à 25°C en milieu liquide agité (milieu de Pachlewski). Les boulettes de mycelium sont alors broyées à l'Ultra-Turax en conditions stériles. Un prélèvement de la suspension est dilué au 1/10 et 1/100 dans l'eau stérile, et chacune de ces dilutions est répandue sur 5 boîtes de Pétri (1 ml par boîte) contenant du milieu de Pachlewski gélosé et mises à incuber à 25°C, alors que la suspension mère est conservée à 4°C. Au bout d'une semaine, le comptage des colonies permet de calculer la concentration de la suspension mère en propagules vivantes. Le mycelium est alors lavé à l'eau non stérile par décantation ou filtration, et remis en suspension dans l'eau de façon à constituer l'inoculum de concentration voulue. Dès le semis, les cagettes sont placées en chambre climatisée où elles sont arrosées tous les deux jours à l'eau osmosée. L'humidité relative de l'atmosphère est maintenue à 80% et l'éclairement est de 16 000 Lux au niveau des feuilles. Les pots sont périodiquement ouverts et l'intensité de la mycorhization par *H. crustuliniforme* et par d'autres champignons est notée sur la face inoculée selon le barême suivant: 0: pas de mycorhize; 0,5: une seule mycorhize visible; 1: peu de mycorhizes, groupées en un seul point; 2: mycorhizes plus abondantes, réparties en plusieurs endroits du système racinaire visible; 3: système racinaire visible presque entièrement colonisé. La moyenne des notes des 15 plants par cagette constitute la note du traitement.

Deux expériences ont été réalisées; la première portait sur deux sols (CREDENCE: mull calcaire d'une rendzine colluviale sous hêtraie-érablière, et AMANCE: mull mésotrophe d'un sol brun lessivé à pseudogley sur limon sous hêtraie), désinfectés ou non par fumigation au bromure de méthyle, avec 4 intensités d'inoculation: 0–1,7.10^2–1,7.10^3 et 1,7.10^4 propagules vivantes par plant. L'intensité de la mycorhization a été notée à 2–4–6–8 et 10 semaines. Le climat est unique: jour long de 16 h, température de jour 25°C, température de nuit 12°C. La deuxième expérience portait sur le seul sol d'AMANCE non désinfecté, avec 5 intensités d'inoculation: 6,2–6,2.10^1–6,2.10^2–6,2.10^3 et 6,2.10^4 propagules vivantes par plant. Le jour était de 12 h, mais deux régimes thermiques ont été comparés: 25°C le jour/12°C la nuit et 20°C le jour/4°C la nuit.

Résultats

Les Figures 1 et 2 donnent les résultats de la première expérience. Sur le sol de CREDENCE (Fig.1) la mycorhization par *H. crustuliniforme* s'installe plus rapidement, et à partir d'une intensité d'inoculation plus faible, dans le sol désinfecté que dans le sol non désinfecté, et la différence s'accroît avec le temps: la mycorhization progresse sur sol désinfecté, alors qu'elle régresse sur sol non désinfecté. Sur le sol d'AMANCE (Fig. 2), la mycorhization s'installe au contraire plus rapidement sur le sol non désinfecté; par la suite il y a inversion, mais la différence reste faible. Le nombre de propagules vivantes d'*H. crustuliniforme* par plant nécessaire pour atteindre une note mycorhization moyenne de 1,5 à 10 semaines est de l'ordre de 10^3 pour le sol d'AMANCE, désinfecté ou non, de 10^2 seulement pour CREDENCE désinfecté, mais de 10^4 pour CREDENCE non désinfecté. Sur les deux sols non désinfectés, les courbes représentant les mycorhizes autres que celles d'*H. crustuliniforme* montrent que

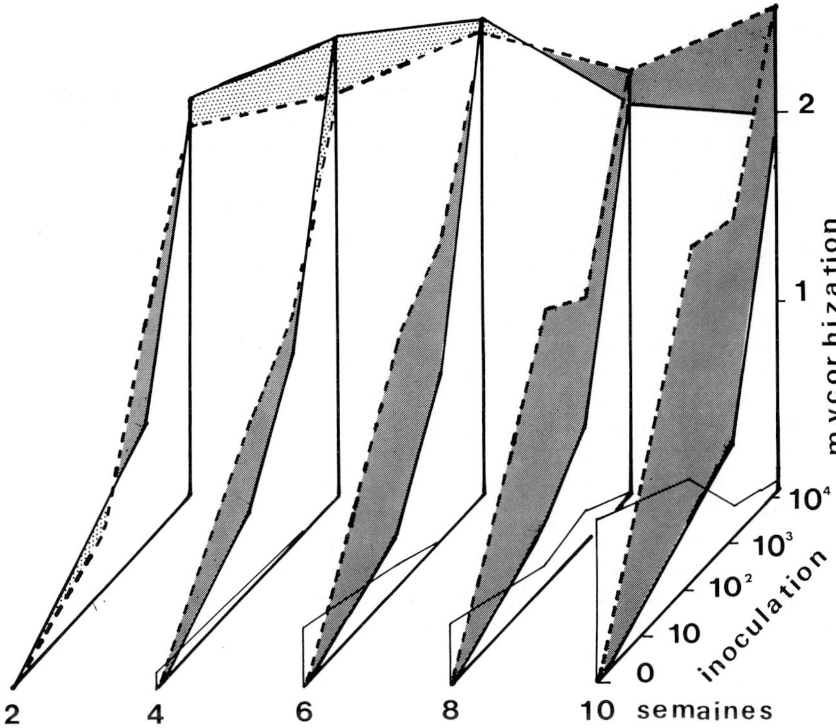

Fig. 1. Mycorhization du hêtre sur le sol de CREDENCE en fonction du temps et de l'intensité d'inoculation. L'intensité d'inoculation par *Hebeloma crustuliniforme* est exprimée en nombre de propagules vivantes par plant. La mycorhization est notée de 0 à 3 (cf. texte, § Matériel et méthode).
Trait gras continu: mycorhizes d'*H. crustuliniforme* sur sol non désinfecté;
Trait gras discontinu: mycorhizes d'*H. crustuliniforme* sur sol désinfecté;
Trait fin continu: autres mycorhizes sur sol non désinfecté.

la souche introduite est compétitive par rapport aux champignons du sol à partir d'une intensité d'inoculation comprise entre 10 et 10^2 pour AMANCE et entre 10^2 et 10^3 pour CREDENCE.

La Figure 3 donne les résultats de la deuxième expérience sur sol d'AMANCE non désinfecté, comparés à la courbe correspondante de la première expérience. On voit qu'une longue photopériode favorise fortement la mycorhization par *H. crustuliniforme*, alors que la différence de régime thermique en photopériode courte n'a qu'un effet très faible au bénéfice du climat le plus chaud.

Discussion

Dans ces deux expériences, l'inoculation a été appliquée sur des systèmes racinaires déjà développés, abondamment pourvus de racines courtes réceptives à la mycorhization. Les résultats ne tiennent donc pas compte de la phase de conservation et/ou de vie saprophytique de l'inoculum dans le sol, mais

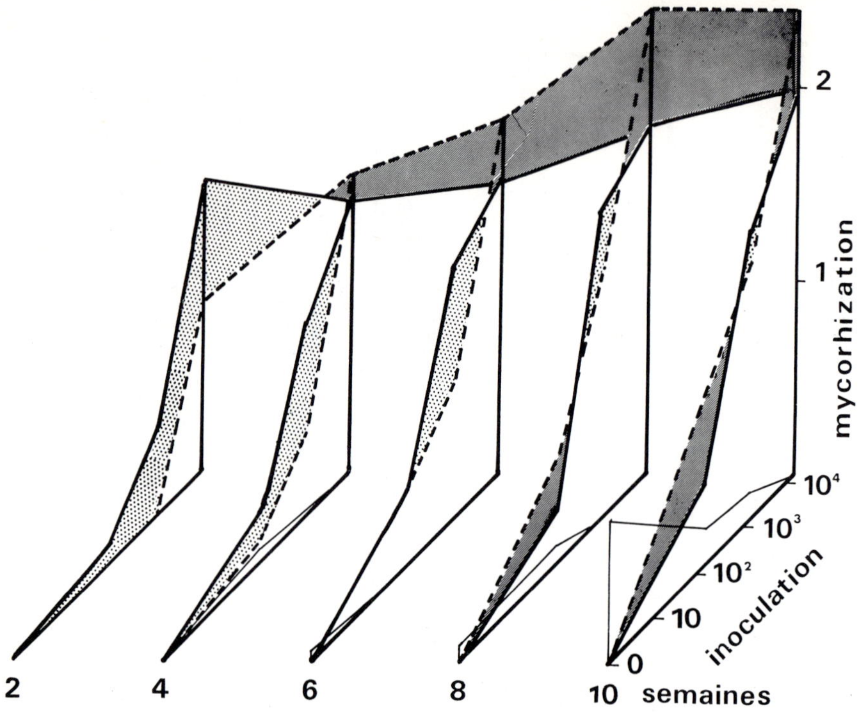

Fig. 2. Mycorhization du Hêtre sur le sol d'AMANCE en fonction du temps et de l'intensité d'inoculation. L'intensité d'inoculation par *Hebeloma crustuliniforme* est exprimée en nombre de propagules vivantes par plant. La mycorhization est notée de 0 à 3 (cf. texte, § Matériel et méthode).
Trait gras continu: mycorhizes d'*H. crustuliniforme* sur sol non désinfecté;
Trait gras discontinu: mycorhizes d'*H. crustuliniforme* sur sol désinfecté;
Trait fin continu: autres mycorhizes sur sol non désinfecté.

seulement de l'aptitude du mycelium à coloniser et infecter les racines et à propager ensuite la mycorhization sur les nouvelles racines.

Dans ces conditions qui sont aussi celles, dans la pratique, d'une inoculation directe lors du repiquage de semis en pépinière ou lors de la plantation de plants non mycorhizés en forêt[4], nos résultats permettent les conclusions suivantes: la photopériode a un effet important sur la dynamique de l'infection mycorhizienne: les jours longs favorisent la mycorhization, probablement en augmentant la photosynthèse et la disponibilité en carbone au niveau des racines.

L'intensité de l'inoculation est toujours un facteur décisif pour l'installation de la mycorhization. Il est remarquable de constater que même dans les conditions les moins favorables (sol de CREDENCE non désinfecté, ou jours courts) il est possible d'obtenir rapidement un fort taux de mycorhization par *H. crustuliniforme* à condition d'inoculer très massivement (10^4 propagules vivantes par plant). Ces résultats rejoignent ceux (non publiés) d'expériences en pépinière

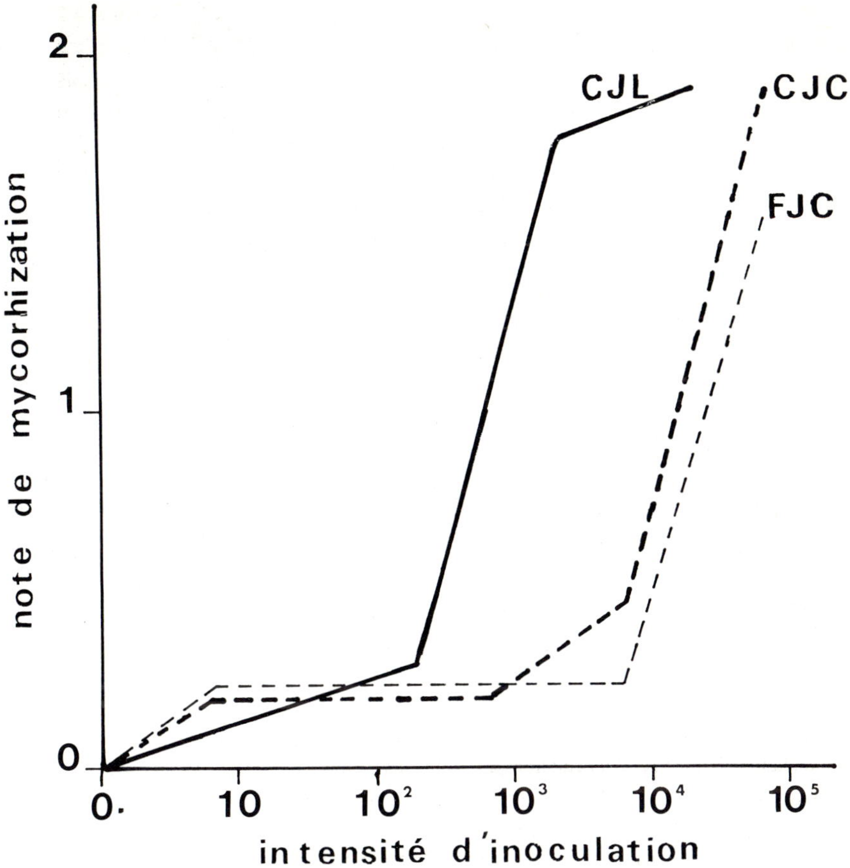

Fig. 3 Mycorhization du Hêtre par *Hebeloma crustuliniforme* sur le sol d'AMANCE non désinfecté en fonction du climat et de l'intensité d'inoculation.
L'intensité d'inoculation par *H. crustuliniforme* est exprimé en nombre de propagules vivantes par plant.
CJL: climat chaud (jour: 25°C; nuit 12°C) et jours longs (16 h).
CJC: climat chaud et jours courts (12 h).
FJC: climat froid (Jour: 20°C; nuit: 4tC) et jours courts.

où des conditions défavorables à la mycorhization par la même souche d'*H. crustuliniforme* (trop grande fertilité du sol) sont compensées par une inoculation massive.

Cette prédominance du facteur intensité d'inoculation fait qu'un bon test expérimental destiné à comparer les comportements de différents champignons sur différents sols doit être basé non pas sur la mycorhization pour une dose standard d'inoculum, mais sur la courbe de réponse en fonction de l'intensité d'inoculation, que l'on interprétera comme un bio-test, de la même façon que cela a été fait en pathologie[1] pour la mesure du potentiel infectieux des sols. La prise en compte du facteur temps semble également être discriminante. Les deux sols étudiés ont un comportement très différent vis-à-vis de l'introduction d'*H. crustuliniforme*: la capacité d'accueil du sol d'AMANCE est moyenne et dépend peu de la microflore (celle-ci a même un effet positif sur la mycorhization par *H. crustuliniforme* pendant les premières semaines), alors que cette microflore est la cause essentielle de la mauvaise capacité d'accueil du sol de CREDENCE. La composante ectomycorhizienne de la microflore n'est pas le facteur en jeu, puisque les mycorhizes spontanées disparaissent lorsque celles d'*H. crustuliniforme* sont abondantes. La compétition doit donc s'exercer essentiellement avec d'autres groupes d'organismes au niveau de la colonisation rhizosphérique. Ces dernières observations suggèrent qu'il est possible dans la pratique, sur certains sols 'réceptifs' du type de celui d'AMANCE, de procéder à des inoculations directes à la plantation sans que ni la désinfection, ni des doses d'inoculum irréalistes soient nécessaires.

Réferences bibliographiques

1 Bouhot D 1980 Le potentiel infectieux des sols. Thèse de doctorat és sciences, Université de Nancy I, France.

2 Garbaye J 1982 Quelques aspects de la compétitivité des souches ectomycorhiziennes. Colloque sur les mycorhizes, I.N.R.A. Dijon (France).

3 Garbaye J 1983 Effet du champignon mycorhizien *Hebeloma crustuliniforme* sur la croissance du Chêne et du Hêtre. R.F.F. n°1.

4 Garbaye J, Lainez J et Le Tacon F 1983 Survie, croissance et mycorhization après plantation de plants de Hêtre produits sur tourbe fertilisée. R.F.F. n°1.

5 Riedacker A 1978 Etude de la déviation des racines horizontales ou obliques issues de boutures de Peuplier qui recontrent un obstacle: applications pour la conception de conteneurs. Ann. Sci. For. 35, 1–18.

Plant and Soil 71, 309–312 (1983). A-33

Ecto- and endomycorrhizal fungus associations with *Quercus imbricaria* L.

F. M. ROTHWELL,
USDA – Forest Service, NEFES, Berea, Kentucky, USA

E. HACSKAYLO
USDA – Forest Service, NEFES, Beltsville, Maryland, USA

and D. FISHER
Kentucky Division of Forestry, Kentucky Dam State Nursery, Gilbertsville, Kentucky, USA

Key words Ectomycorrhizae Endomycorrhizae Fertilization Fumigation Glomus Shingle oak Sorghum Tree nursery *Quercus imbricaria*

Summary Ten seedlings were collected from one year old shingle oak grown in fumigated and fertilized seedbeds at a tree nursery in western Kentucky. Root material from the seedlings was washed free of soil and small segments stained for microscopic analysis of mycorrhizae development. External morphology of root segments was typically ectomycorrhizal, with characteristic hyphal mantle and Hartig net development evident in transverse sections of young ectomycorrhizae. In addition a *Glomus* species was frequently observed in the root cortex.

Introduction

Under natural growth conditions it is not uncommon for hardwoods such as Betula and Populus to form a dual symbiotic association with ectomycorrhizal and vesicular-arbuscular (VA) endomycorrhizal fungi[8,14]. Tripartite symbiotic associations have frequently been observed in Robinia and Alnus, involving the host, nitrogen-fixing bacteria, and members from one or both of the major groups of mycorrhizal fungi[2,4,7,8]. Each of the symbionts in a dual fungal association may significantly influence the physiological activity of host roots, although the ability to produce changes in root morphology apparently is not common to both. Ectomycorrhizal fungi synthesize growth regulating compounds that affect root morphogenesis in newly developing ectomycorrhizae[10], whereas VA mycorrhizal associates apparently produce no comparable substances and the external morphology of VA endomycorrhizae differ little from non-colonized roots characteristically produced by the host[1,9,13]. In contrast to the common occurrence of dual fungal associations in many hardwoods, Quercus forms symbiotic associations predominantly with ectomycorrhizal fungi[9,11], although an unusual 'beaded endotrophic mycorrhiza' has been reported for northern red oak (*Q. rubra* L.) and southern red oak (*Q. falcata* Michx.) growing in native habitats[3]. Information is presented in this paper on a dual mycorrhizal association of shingle oak (*Q. imbricaria* L.) grown under nursery conditions.

Methods

Shingle oak seedlings were grown at the John P. Rhody tree nursery near Gilbertsville, Kentucky. The soil is classified as a silty loam, with an average pH of 5.5. A sorghum hybrid was used both as a green manure cover crop during alternate years of tree growth, and to increase the inoculum density of VA endomycorrhizal propagules in the nursery soils[5].

Following incorporation of the cover crop in September 1980, seedbeds were formed and fumigated with 360 to 395 kg/ha of methyl bromide plus two percent chloropicrin. No additional fungicide, herbicide, or bird repellent seed treatments were used. Seedbeds were fertilized at the rate of 395 kg/ha with a 15–15–15 NPK formulation, and after seeding, the beds were covered with a straw mulch. During the 1981 growing season seedbeds were irrigated to maintain soil moisture equivalent to 25 mm of rain per week.

In September 1981, ten seedlings with intact rhizosphere soil clumps were collected at random from the seedbed. Soil was removed from root samples by careful washing, and small segments of young feeder roots obtained from the samples were stained for microscopic analysis of mycorrhizal associates.

Results and discussion

The fertilization and fumigation regime at the nursery seemed to have no apparent inhibitory affect on ectomycorrhizae development, or of reducing the VA mycorrhizal propagules below an effective level for endomycorrhizae formation in shingle oak seedlings. The overall appearance of root material from the seedlings was typically ectomycorrhizal, with no evidence of beaded or segmented mycorrhizae. Transverse sections of ectomycorrhizae in early stages of formation depict a relatively compact mantle about six hyphal diameters thick, with extension of the hyphae between the outer cells of the cortex parenchyma to form the Hartig net (Fig. 1). Intercellular hyphae and vesicles of what appeared to be a *Glomus* species were frequently observed in the root cortex of stained root segments (Fig. 2). Arbuscles and hyphal coils were absent, and only occasionally were intracellular vesicles noted. No external mycelium or chlamydosphores were evident when unstained root segments were viewed under low magnification. Although the sequence of events involved in the association with the host is uncertain, there was no apparent reciprocal adverse affect by the symbionts during establishment of the association. This is the first report of a VA endomycorrhizal associate in oak possessing a root morphology consistent with anatomical features of the dual mycorrhizal condition observed in other hardwood species.

Factors responsible for the endomycorrhizal association in shingle oak under these conditions are unknown, although propagule density may have influenced the association in much the same manner as suggested for the rarely mycorrhizal

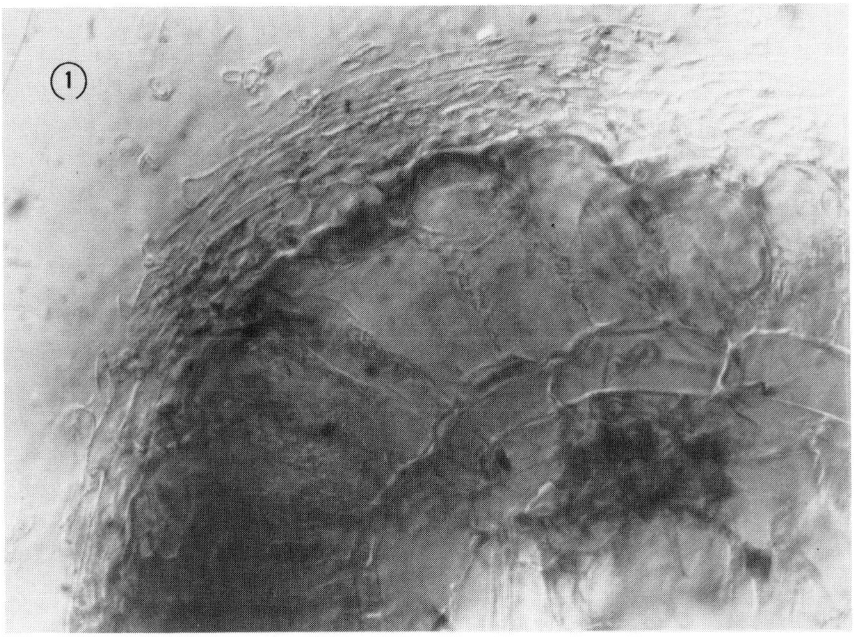

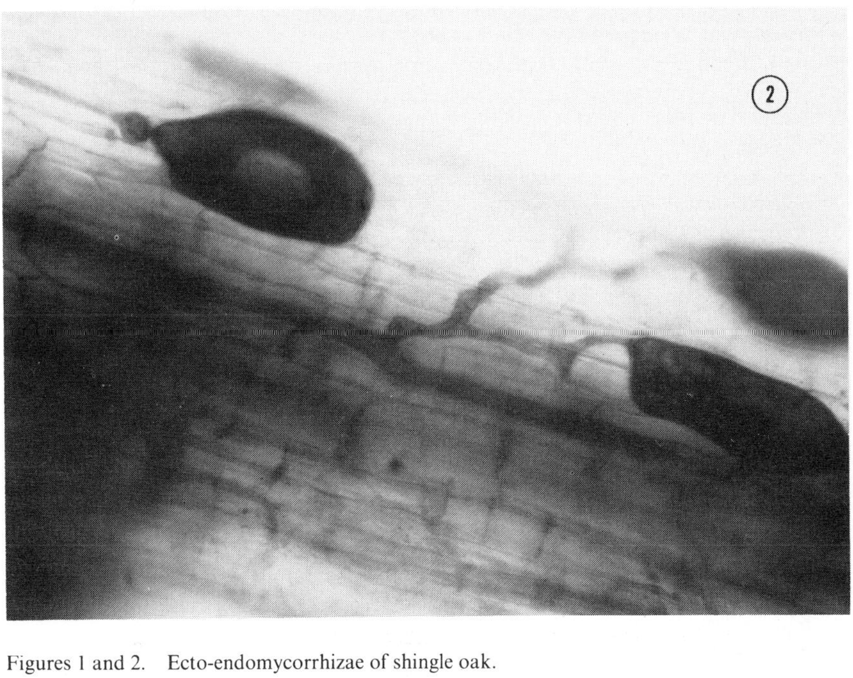

Figures 1 and 2. Ecto-endomycorrhizae of shingle oak.

Fig. 1. Transverse section of developing ectomycorrhiza. × 700.

Fig. 2. Internal hyphae and vesicles of VA symbiont. × 280.

Polygonum[8]. The results with shingle oak suggest that other oak species might also be susceptible to the dual mycorrhizal condition. However, seedlings of northern red oak that were grown contiguous to shingle oak in the seedbed, and presumably exposed to the same environmental conditions, were ectomycorrhizal only, with no evidence of early or subsequent association with the VA endomycorrhizal fungi. Although levels of propagules may have been obtained in the seedbed that exceed those encountered in nature, differences in the ability of the oak species to form an association with the VA fungi appear to be related more to host susceptibility than to increased levels of viable propagules.

Evidence presented here for the dual mycorrhizal condition in a normally ectomycorrhizal host demonstrated in addition the physiological diversity of fungi believed capable of mycorrhizae formation[12]. As previously noted, especially in those unions showing strong specificity, the means by which ectomycorrhizal hosts and fungi are able to recognize compatible partners have yet to be defined[6].

References

1 Beslow D T, Hacskaylo E and Melhuish J H Jr. 1970 Effects of environment on beaded root development of red maple. Bull. Torrey Bot. Club 97, 248.

2 Daft M J and Hacskaylo E 1976 Arbuscular mycorrhizas in the anthracite and bituminous coal wastes of Pennsylvania. J. Appl. Ecol. 13, 523–531.

3 Grand L F 1969 A beaded endotrophic mycorrhiza of northern and southern red oak. Mycologia 61, 408–409.

4 Hall R B, McNabb H S Jr., Maynard C A and Green T L 1979 Toward development of optimal *Alnus glutinosa* symbioses. Bot. Gaz. (Chicago) (Suppl.) 140, 120–126.

5 Kormanik P P, Bryan W C and Schultz R C 1980 Increasing endomycorrhizal fungus inoculum in forest nursery soil with cover crops. South. J. Appl. For. 4, 151–153.

6 Molina R 1981 Ectomycorrhizal specificity in the genus Alnus. Can. J. Bot. 59, 325–334.

7 Rose S L 1979 Mycorrhizal associations of some actinomycete nodulated nitrogen-fixing plants. Can. J. Bot. 58, 1449–1454.

8 Rothwell F M and Vogel W G 1982 Mycorrhizae of planted and volunteer vegetation on surface-mined sites. U.S. Dep. Agric. For. Serv. Gen. Tech. Rep. NE-66, 12 p. Northeastern For. Exp. Stn., Broomall, Pa.

9 Ruehle J L and Marx D H 1979 Fiber, food, fuel, and fungal symbionts. Science 206, 419–422.

10 Slankis V 1971 Formation of ectomycorrhizae of forest trees in relation to light, carbohydrates, and auxins. *In* Mycorrhizae. Ed. E Hacskaylo. U.S. Dep. Agric. For. Serv. Misc. Publ. 1189, 151–167. Washington, D. C.

11 Trappe J M 1962 Fungus associates of ectotrophic mycorrhizae. Bot. Rev. 28, 538–606.

12 Trappe J M and Fogel R D 1977 Ecosystematic functions of mycorrhizae. Col. State Univ. Range Sci. Dep. Sci. Ser. 26, 205–214.

13 Trappe J M 1979 Mycorrhiza-nodule-host interrelationships in symbiotic nitrogen fixation: a quest in need of questers. *In* Symbiotic Nitrogen Fixation in Temperate Forests. Eds. J C Gordon, C T Wheeler and D Perry. Oregon State Univ. pp 276–286. Corvallis.

14 Vozzo J A and Hacskaylo E 1974 Endo- and ectomycorrhizal associations in five populus species. Bull. Torrey Bot. Club 101, 182–186.

Plant and Soil 71, 313–317 (1983). A-34

Effect of mycorrhizal inocula on the growth of Sitka spruce seedlings in different soils

JULIA M. HOLDEN, G. W. THOMAS and R. M. JACKSON
Department of Microbiology, University of Surrey, Guildford, Surrey, GU2 5XH, UK

Key words Forest Glasshouse conditions Mycorrhizal fungi Nursery Seedling response Sitka spruce

Summary Different mycorrhizal fungi were tested for their effectiveness in promoting growth of Sitka spruce seedlings, in two contrasting soils, in a glasshouse pot experiment. In nursery soil, *Laccaria amethystina* significantly improved growth of seedlings in comparison to *L. laccata*. Seedlings inoculated with a forest isolate of *Thelephora terrestris* were significantly larger than those inoculated with a nursery isolate when grown in forest soil. The effectiveness of *Complexipes moniliformis* in forest soil was poor in comparison to other mycorrhizal fungi. Strains as well as species of mycorrhizal fungi affect seedling growth differently. These effects are further influenced by soil type.

Introduction

Artificial inoculation of nursery beds with selected mycorrhizal fungi has been recommended for producing vigorous tree seedlings for outplanting to forest sites[1,8]. Although nurseries can produce vigorous seedlings by fertilization and pathogen-control regimes, the plants may not be mycorrhizal or may possess mycorrhizas adapted to the nursery environment only. These seedlings will eventually become mycorrhizal with indigenous fungi after transplantation to forest sites, assuming infective propagules are present. But the growth and development of the transplanted seedling may be delayed until it has acquired a well-developed mycorrhizal root system.

Nursery inoculation programmes are beneficial not only because mycorrhizal plants are assured. There is much evidence that some fungi are more effective as mycorrhizal symbionts than others[3,6,9]. Inoculation with mycorrhizal fungi selected for their capacity to promote seedling growth, and their suitability for forest sites could greatly improve the establishment and growth of outplanted seedlings in the ultimate forest environment.

The benefit of a particular mycorrhizal fungus to the tree may not be constant for all environments. It is therefore important to consider the influence of soil type on the symbiotic association when screening fungal isolates for use as mycorrhizal inocula.

Materials and methods

The plant used in this study was Sitka spruce (*Picea sitchensis* (Bong.) Carr). This is at present the most important forest species grown in Great Britain. The majority of mycorrhizal fungi tested have been isolated from mycorrhizal roots of Sitka spruce trees and seedlings ranging in maturity, and growing in various nursery and forest sites around Britain.

Two soil types, both with a low nutrient status, were used in this experiment. One was a fine sand from a disused forest nursery at Ironhill, Hampshire (pH 4.1, phosphorus 5 mg/l, potassium 11 mg/l, magnesium 14 mg/l, nitrate-nitrogen 4 mg/l). In contrast, the other soil was a humus-type taken from a 45-year old Sitka spruce stand at Bramshill, Hampshire (pH 4.8, phosphorus 1 mg/l, potassium 8 mg/l, magnesium 2 mg/l, nitrate-nitrogen 5 mg/l, ammonium-nitrogen 48 ppm). The soils were sieved and partially sterilized with a chemical sterilant, dazomet, to eliminate indigenous mycorrhizal fungi. More drastic treatment is unnecessary and inadvisable as it is likely to alter the availability of plant nutrients. The soils were transferred to 7-inch diameter plastic pots and left until the phytotoxic elements could no longer be detected by a cress-seed germination test.

The soil inoculum for each isolate was prepared by culturing mycorrhizal fungi in peat: vermiculite with modified Melin-Norkrans medium[4] in 1.5 litre Kilner-jars with vented lids. These bulk cultures were incubated at 20°C for 3–4 months until the fungal mycelium had penetrated throughout the medium. The cultures were then turned out onto muslin amd rinsed thoroughly with running tap-water for several minutes in order to reduce residual nutrients in the culture medium. The inoculum matrix was drained, and 80 ml of this were mixed thoroughly into the top 5 cm of soil in each pot.

Sitka spruce seeds were surface-sterilized in hydrogen peroxide solution[7] and pregerminated on water-agar plates at 20°C under light. The germinated seeds were carefully planted in the soil contained in the pots.

The seedlings were grown in glasshouse conditions (15–25°C and 18-hour daylength) for 6–7 months, harvested and certain growth parameters measured. Shoot length (mm), root length (mm), stem diameter (mm) and shoot branching were measured directly. The shoots and roots were dried in a convection oven for 24 hours at 85°C. Shoot and root dry weights (g) were determined and from these values total dry weight and shoot-root ratio values were calculated. An estimate of percentage mycorrhizal formation was made for each root system by determining the proportion of mycorrhizal roots in a sample of 100 short roots, counted at random. Samples of root systems were checked for contaminating mycorrhizal fungi by light microscopy.

Shoot branching and root length did not appear to change consistently with fungal treatment or soil, and these data were disregarded in statistical analyses. All other data were subjected to factorial analysis of variance and a multiple range test[2].

Results and discussion

No mycorrhizal contamination was observed on the uninoculated control seedlings. Examination by light microscopy confirmed that the inoculated seedlings were mycorrhizal only with the test fungi added to the soil. Of the twelve fungal isolates used, only eight formed mycorrhizas in this experiment. *Laccaria laccata* (Scop. ex Fr.) Cooke, *L. amethystina* (Hooker) Murr., the nursery and forest isolates of *Thelephora terrestris* (Ehrh.) Fr. and an E-strain fungus all formed reasonable numbers ($> 50\%$ of short roots) of mycorrhizas in both nursery and forest soils (Table 1). An unidentified nursery isolate formed mycorrhizas on only a small number of seedlings in nursery and forest soils. A *Cenococcum geophilum* (Fr.) from Ringwood forest (Dorset) and *Paxillus involutus* (Batch ex Fr.) Fr. formed very few mycorrhizas in forest soil and none in nursery soil. A *C. geophilum* from Alice Holt forest (Hampshire), two unidentified basidiomycetes and *Pisolithus tinctorius* (Pers.) Coker and Couch formed no mycorrhizas in this experiment.

It is possible that in some cases poor mycorrhizal formation was due to a reduction in inoculum potential during the period between the inoculation of the soil with the test fungi and the formation of short roots on the seedling (about 6 weeks). Theodorou[5] has reported that the rate of mycorrhizal formation on *Pinus radiata* was higher when seedlings were inoculated at 8–13 weeks than at the sowing or postgermination stage.

From the results in Table 1 it would appear that in the nursery soil all the inoculated seedlings grew better than the controls. However, only seedlings mycorrhizal with *L. amethystina* (La) were significantly larger than the uninoculated controls (CON) (at 5% level). This fungus was significantly more effective than any of the other fungi used, including *L. laccata* (Ll). This

Table 1. Selected growth parameters of Sitka spruce seedings with different mycorrhizal inocula and soil types

Soil	Test fungus*	Mycorrhizas as % of short roots	Shoot dry weight (g)	Root dry weight (g)	Total dry weight (g)	Shoot root ratio
Nursery	CON	0	0.076	0.076	0.152	1.07
	A3	51	0.149	0.133	0.282	1.15
	L1	65	0.160	0.142	0.302	1.12
	La	81	0.380	0.421	0.801	0.95
	TtN	83	0.221	0.131	0.329	1.20
	TtF	85	0.177	0.169	0.346	1.09
	B1	48	0.583	0.675	1.259	0.85
Forest	CON	0	0.020	0.016	0.036	1.43
	A3	69	0.164	0.142	0.306	1.49
	Ll	96	0.338	0.255	0.594	1.33
	La	84	0.218	0.194	0.412	1.20
	TtN	97	0.358	0.312	0.672	1.61
	TtF	95	0.477	0.442	0.919	1.11
	Bl	39	0.057	0.045	0.102	1.32

* CON Uninoculated control
A3 E-strain fungus
Ll *Laccaria laccata*
La *L. amethystina*
TtN *Thelephora terrestris* (ex nursery)
TtF *T. terrestris* (ex forest)
B1 Unidentified nursery basidiomycete.

demonstrates that fungal species within the same genus may not necessarily exhibit the same degree of effectiveness on host plant growth. The nursery isolate of *T. terrestris* (TtN) significantly increased the shoot dry weight, but not the total dry weight, by comparison with the uninoculated control seedlings.

In the forest soil, all the fungi that formed more than 50% mycorrhizal short roots (Table 1) stimulated seedling growth above the uninoculated controls (at 5% level). In this soil the two *Laccaria* species did not differ significantly from one another. However, the forest isolate of *T. terrestris* (TtF) produced a greater seedling response than the nursery isolate of the same species. These results suggest that strains of mycorrhizal fungi may be adapted to ecologically different environments.

The E-strain fungus (A3), (also known as *Complexipes moniliformis* Walker) is a common mycorrhizal fungus in forest-tree nurseries. It is interesting to note that its effectiveness as a mycorrhizal fungus in forest soil is poor in comparison to other mycorrhizal fungi. Unfortunately the nursery basidiomycete (B_1) formed too few mycorrhizal seedlings for statistical analysis. However the results (Table 1) indicate that it is an extremely effective mycorrhizal fungus in nursery soil, but of little use in the forest-type soil. Subsequent unpublished experiments confirm this.

Seedlings inoculated with *L. laccata* and the two isolates of *T. terrestris* grew significantly better than their counterparts in the nursery soil. Whereas inoculation with *L. amethystina* proved more beneficial in nursery soil.

Significant differences shown between treatments with respect to dry weights were not necessarily reflected in the shoot length and stem diameter values. Although these are non-destructive measurements of growth they have limited value in determining seedling response. The factorial analysis of variance showed that the shoot root ratio of the seedlings was not affected significantly by soil type or mycorrhizal inoculum in this experiment.

From these results it is clear that some mycorrhizal fungi are more effective than others, with respect to stimulation of host plant growth. There is evidence here that the type of soil is also a very important factor in determining fungal effectiveness, and should undoubtedly be considered when selecting mycorrhizal fungi for artificial inoculation.

Acknowledgements The authors wish to thank the Natural Environment Research Council for financial support during this research.

References

1 Bowen G D 1965 Mycorrhiza inoculation in forestry practice. Aust. For. 29, 231–237.

2 Duncan B D 1955 Multiple range and multiple F-tests. Biometrics 11, 1–42.

3 Ekwebelam S A 1979 The effect of mycorrhizal fungi on the growth and nutrient uptake of caribbean pine seedlings. Indian For. 105, 750–757.

4 Marx D H 1969 The influence of ectotrophic mycorrhizal fungi on the resistance of pine roots to pathogenic infection. I. Antagonism of mycorrhizal fungi to root pathogenic fungi and soil bacteria. Phytopathology 59, 153–163.

5 Theodorou C 1980 The sequence of mycorrhizal infection of *Pinus radiata* D. Don. following inoculation with *Rhizopogon luteolus* Fr. and Nordh. Aust. For. Res. 10, 381–387.

6 Theodorou C and Bowen G D 1970 Mycorrhizal responses of Radiata pine in experiments with different fungi. Aust. For. 34, 183–191.

7 Trappe J M 1961 Strong hydrogen peroxide for sterilizing coats of tree seeds and stimulating germination. J. For. 59, 828–829.

8 Trappe J M 1977 Selection of fungi for ectomycorrhizal inoculation in nurseries. Annu. Rev. Phytopathol. 15, 203–222.

9 Young H E 1940 The influence of different species of mycorrhiza-forming fungi on seedling growth. J. Aust. Inst. Agric. Sci. 6, 21–25.

Plant and Soil 71, 319–323 (1983). A-35

Changes in the mycorrhizal status of Sitka spruce following outplanting

G. W. THOMAS, DOREEN ROGERS and R. M. JACKSON
Department of Microbiology, University of Surrey, Guildford, Surrey, GU2 5XH, England

Key words E-strain Forest Nursery Replacement Sitka spruce *Thelephora terrestris*

Summary Occurrence of mycorrhizal types in 4 forest sites and 4 nurseries was determined by mycorrhizal morphology. In total 25 distinct mycorrhizas were found, of which 14 were formed by identifiable fungi. The frequency of occurrence of the 'E-strain' fungus, the dominant nursery mycorrhizal fungus, decreases with the age of the outplanted seedling. Some mycorrhizal types were found in all the mature forest sites examined. The fungus *T. terrestris* was found on all age groups of Sitka spruce studied.

Introduction

Recent reviews[8,11] show the benefit, in terms of improved growth and survival in transplant sites, of inoculating seedlings with selected mycorrhizal fungi. However, it would probably be pointless to inoculate seedlings with fungi which may be adapted to growth with mature but not young trees [12] or with fungi which are rapidly replaced by the indigenous mycorrhizal fungi of the transplant site. In this study Sitka spruce nursery plants and plantation Sitka spruce outplants of different ages and in different sites have been sampled, mycorrhizas identified and isolates of mycorrhizal fungi obtained in an attempt to determine the probable changes in the mycorrhizal flora with increasing tree maturity and to find mycorrhizal fungi likely to be suitable for seedling inoculation.

Materials and methods

Over the few years 1978 to 1981 root samples and sporocarps were collected every autumn from mature Sitka spruce stands at Crychan and Glasfynydd forests Mid-Wales and at Alderholt forest Dorset (planted 1939, 1928 and 1941 respectively), and from a stand planted in 1976 in Crychan forest. Naturally regenerated seedlings, approximately 1 year-old, were collected in the autumn of 1980 from Glasfynydd forest. Nursery grown seedlings were collected in 1980 from four Forestry Commission Nurseries at Newton, Black Isle, Sowden and Ellerkers.

Root samples were removed from nursery and forest sites together with surrounding soil and taken immediately to the laboratory. After initial washing under tap water mycorrhizas were examined and characterised macroscopically. Microscopic characterisation was based on transverse sections, 6 um thick, cut on a freezing microtome and plan views of the sheath obtained from root squashes stained with lactofuchsin. Attempts were made to isolate the fungal symbiont, from mycorrhizas and sporocarps using techniques described previously.

Mycorrhizas were synthesised aseptically using the flask culture system[9] modified by using an inorganic salts medium developed for spruce[4] rather than a medium developed for the isolation of mycorrhizal fungi[7]. Flasks containing aseptically pregerminated spruce seedlings and the isolated fungi were placed in an environmental chamber and grown under fluorescent light (18 h photoperiod) at 20°C. The root systems were examined after 12 weeks and the synthesised mycorrhizas were compared structurally with the mycorrhizas from which the isolates were obtained.

Results

Macroscopic and microscopic analysis of nursery and forest root samples revealed 25 distinct types of mycorrhiza (Table 1). Fourteen of the fungi forming these mycorrhizas have been tentatively identified. Identification was based on, a) a comparison of mycorrhizal morphology with that of previously described types *e.g. C. geophilium*[3] and the 'E-strain' fungi[10]; b) comparing isolates obtained from sporocarps with those obtained from mycorrhizas *e.g. T. terrestris, L. laccata, L. rufus, L. hepaticus* and *P. involutus*; c) comparison of the morphology of mycorrhizas synthesised using isolates obtained from sporocarps with that of mycorrhizas found near the sporocarps; d) the consistent occurrence of a mycorrhizal type near to a sporocarp *e.g. R. ochroleuca, R. emetica, L. turpis, A. rubescens* and *A. muscaria*. The remaining 11 types have been characterised by their macroscopic and microscopic features and have each been assigned a type number.

The majority of seedlings in all the nurseries studied had only 'E-strain' mycorrhizas. *T. terrestris. C. geophilum* together with an unidentified basidiomycete were the only other mycorrhizal fungi found in nurseries. Examination of 2–4 years old transplants from Crychan forest revealed that although 'E-strain' mycorrhizas were present the number of different mycorrhizal fungi increases with age. The fungi which colonize the transplanted seedlings have also been found in the mature Crychan forest. In the mature sites studied 24 different mycorrhizal types were found, 16 in the Alderholt site, 13 in the Crychan and 12 in the Glasfynydd (Table 1). *T. terrestris, L. tabidus, L. rufus, R. ochroleuca* and *C. geophilum* were found in all three sites. *L. laccata, L. amethystea, P. involutus*, types 17, 18, 19 and 20 in two sites. Mycorrhizal fungi, *T. terrestris, L. rufus, P. involutus, R. ochroleuca*, types 18 and 19, found on mature trees have also been found on naturally regenerated seedlings.

Table 1 also shows the results of synthesis experiments using isolates obtained from sporocarps and mycorrhizas. Isolates were obtained from 19 of the mycorrhizal fungi, of these 16 have formed mycorrhizas.

Discussion

This study records for the first time *A. rubescens, Laccaria amethystea, Lactarius hepaticus, L. tabidus, L. turpis* and *R. ochroleuca* as mycorrhizal fungi of Sitka spruce. In addition to the 14 different mycorrhizas formed by identified fungi another 11 have been found which are formed by unidentified fungi.

Lists of mycorrhizal fungi based upon sporocarp associations with trees[15] must always be incomplete. The reasons for this are as follows; some fungi do not

Table 1. Mycorrhizal types associated with different ages of Sitka spruce on different sites, fungi isolated from sporocarps and mycorrhiza and mycorrhiza formation in synthesis tests

Mycorrhizal type	Nursery seedlings (2 year-old)				Outplanted seedlings			Mature forest			Natural	Synthesis
	Newton	Ellerkers	Sowden	Black Isle	1978	1979	1980	Alderholt	Crychan	Glasfynydd	regeneration	
Amanita muscaria (L. ex Fr.) Hooker								MS				–
A. rubescens (Pers. ex Fr.) S. F. Gray								MSC				–
Cenococcum geophilum Fr.	Mi							Mi	M	Mi		+
Laccaria amethystea (Bull ex Mérat) Murrill								MSC		MSC		+
L. laccata (Scop. ex Fr.) Cooke								MSiC	MS			+
Lactarius hepaticus Plowright apud Bund								MSiC				+
L. rufus (Scop. ex Fr.) Fr.								MSi	MSC	MSi	M	+
L. tabidus Fr.								MSC	MSC	M		+
L. turpis (Weinm.) Fr.								MSC	MSC			–
Paxillus involutus (Batsch ex Fr.) Fr.								MSiC			Mi	+
Russula emetica (Schaeff. ex Fr.) S. F. Gray								MS				
R. ochroleuca (Pers. ex Secr.) Fr.								MS	MS	MS	M	
Thelephora terrestris (Ehrh.) Fr.	M		M		Mi	Mi	Mi	MSi	M			+
'E-strain'	Mi	M	M	Mi	M	M	Mi		M			+
Type 4				Mi								+
Type 17									M	M		
Type 18									M	M	M	
Type 19							M		M	M	M	
Type 20						M	M		M	Mi		+
Type 23					M		M		M			
Type 24								Mi				+
Type 26								Mi				+
Type 28								Mi				+
Type 29										Mi		+
Type 30										Mi		+

M = identified by mycorrhizal structure
S = sporocarp present on site
i = isolate obtained from mycorrhizal
c = isolate obtained from sporocarp
+ = successful synthesis
– = unsuccessful synthesis

readily produce sporocarps, other produce inconspicuous sporocarps, some do not produce sporocarps at all or the environmental conditions at the time of sampling may be unsuitable for sporocarp production[13]. An alternative method for identifying mycorrhizal symbionts is to isolate the fungus from the mycorrhiza and identify it by comparison to known cultures[1]. However not all mycorrhizal fungi are culturable[16]. Previous studies[3,14]have shown that the morphology of a mycorrhiza is sufficiently characteristic to be used to identify the symbiotic fungus by comparison with mycorrhizas formed by known fungi. Mycorrhizas formed by unidentified fungi can also be charaterised using this method.

None of the nursery or recent transplant site mycorrhizal fungi would have been recorded by sporocarp presence. The 'E-strain' group of fungi, for example, which have no known teleomorph[2], would have been overlooked. *T. terrestris, C. geophilum*, 'E-strain' fungi and Type 4 have all been previously found with nursery grown Sitka spruce[6,14]. The predominance of the 'E-strain' fungi in forest tree nurseries[10,14] may be due in part to their production of thick walled chlamydospores that would facilitate survival in the nursery soil in which there is not a continuous presence of living roots.

The 'E-strain' fungi are frequently replaced from short roots by other fungi both in the nursery[14] and upon transplantation[10]. However an 'E-strain' fungus has been found forming mycorrhizas on mature tree roots at Crychan.

The fungi colonizing the transplanted seedlings were also found in the mature forest as were some fungi previously reported[11] as being 'pioneer' colonizers *i.e. T. terrestris* and *L. laccata*. Several fungi have only been found in mature sites, although some of these can colonize the roots of naturally regenerated seedlings e.g. *L. rufus* and *R. ochroleuca*.

The difference in the flora of young seedlings and mature trees may be due to the site environment. Disturbed nursery sites do not allow the development of a mycorrhizal community. As the transplant site matures the edaphic condition will change permitting mycorrhizal fungi adapted to these conditions to colonize short roots. No mycorrhizal fungus is dominant in the mature forest sites studied, up to 16 mycorrhizal types were found on one site.

Of the fungi found in the mature forest sites *T. terrestris, L. laccata, P. involutus, C. geophilum* and the 'E- strain' fungi have previously been found to be associated with nursery seedlings[6,14]. However only *T. terrestris* has been observed to form mycorrhizas with all tree age groups and in all outplanting sites. Because of its apparent ubiquity it may be useful for inoculation purposes.

References

1 Chu-Chou M 1979 Mycorrhizal fungi of *Pinus radiata* in New Zealand. Soil Biol. Biochem. 11, 557–562.

2 Danielson R M 1982 Taxonomic affinities and criteria from identification of the common ectendomycorrhizal symbiont of pines. Can. J. Bot. 60, 7–18.

3 Hatch A B 1934 A jet-black mycelium forming ectotrophic mycorrhizae. Svensk. Bot. Tidskr. 28, 369–383.

4 Ingestad T 1959 Studies on the nutrition of forest tree seedlings. 11 Mineral nutrition of spruce. Physiol. Plant. 12, 568–593.

5 Lamb R J 1979 Factors responsible for the distribution of mycorrhizal fungi of *Pinus* in eastern Australia. Aust. For. Res. 9, 25–34.

6 Levisohn I 1965 Mycorrhizal investigations. Forestry Commision Bulletin 37, 228–235. Her Majesty's Stationery Office London.

7 Marx D H 1969 The influence of ectotrophic mycorrhizal fungi on the resistance of pine roots to pathogenic infections. I Antagonism of mycorrhizal fungi to root pathogenic fungi and soil bacteria. Phytopathology 59, 153–163.

8 Marx D H 1980 Ectomycorrhizal fungus inoculations: a tool for improving forestation practices. In Tropical Mycorrhiza Research. Ed. P Mikola Clarendon Press Oxford. pp. 13–71.

9 Marx D H and Zak B 1965 Effect of pH on mycorrhizal formation of Slash pine in aseptic culture. Forest Sci. 11, 65–75.

10 Mikola P 1965 Studies on the ectendotrophic mycorrhiza of pine. Acta For. Fenn. 79, 1–56.

11 Mikola P 1970 Mycorrhizal inoculation in afforestation. Int. Rev. For. Res. 3, 123–196.

12 Moser M 1962 Ecological problems involved in mycorrhiza-induction. Proc. 13 Int. Congr., Internat. Union. For. Res. Org. Sect 24, 1–5.

13 Richardson M J 1970 Studies on *Russula emetica* and other agarics in a Scots pine plantation. Trans. Br. Mycol. Soc. 55, 217–229.

14 Thomas G W and Jackson R M 1979 Sheathing mycorrhizas of nursery grown *Picea sitchensis*. Trans. Br. Mycol. Soc. 73. 117–125.

15 Trappe J M 1962 Fungus associates of ectotrophic mycorrhizae. Bot. Rev. 28, 538–606.

16 Zak B 1971 Characterisation and identification of Douglas fir mycorrhizae. *In* Mycorrhizae, Ed. E Hacskylo. U.S. Dep. Agric. Misc. Publ. 1189 pp 38–53.

Plant and Soil 71, 325–329 (1983).

Inoculation of pines in a nursery with *Pisolithus tinctorius* and *Thelephora terrestris* in Korea*

K. J. LEE and C. D. KOO
Institute of Forest Genetics, Suweon, Gyeonggido 170 Korea

Key words Ectomycorrhizae *Pinus* spp. *Pisolithus tinctorius* *Thelephora terrestris*

Summary Mycelial inocula of the two ectomycorrhizal fungi, *Pisolithus tinctorius* (Pt) and *Thelephora terrestris* (Tt) were artificially introduced into fumigated nursery soil before seed sowing. At the end of the first growing season, Pt stimulated the height growth of *Pinus densiflora*, *P. thunbergii*, *P. rigida*, and *P. rigida* × *taeda* by 55, 36, 69, and 37%, respectively, compared with control seedlings (with no fumigation and no inoculation). When growth stimulation was expressed by dry weight, Pt increased dry weight of *P. densiflora* and *P. rigida* × *taeda* by 143% and 128%, respectively, over control seedlings. Total absorption (per plant top) of nitrogen and phosphorus of inoculated *P. densiflora* seedlings was also increased by 188 and 144%, respectively, even though the concentration of these two elements in the tissue was not significantly different from that of control seedlings.

Tt failed to stimulate growth of the inoculated plants possibly due to its lower level of infection (40–60% mycorrhizal tips) compared with the relatively high infection (over 80%) of Pt. *Pinus koraiensis* did not respond to inoculation during the first growing season, probably due to its inherent characteristics of producing a small number of short roots and its slow shoot growth during the first few years after germination.

Introduction

Artificial inoculation of host plants with specific mycorrhizal fungi has shown that plant growth can be significantly stimulated in conifers[11,12], broad-leaved trees[5,6], agronomic plants[2,13], and also in nitrogen-fixing plants[4]. In particular, Marx and his colleagues in U.S.A. have shown that ectomycorrhizal fungi can stimulate plant growth and enhance survival of conifer tree species. Of the fungi tested, *Pisolithus tinctorius* and *Thelephora terrestris*, appear particularly beneficial to trees planted under harsh conditions, such as eroded sites, spoiled mine areas, and reclamation sites[3].

The objectives of this study were to test the effectiveness of these two fungi and to evaluate methods of their mass culture and inoculation to nursery trees.

Materials and methods

Mass culture of ectomycorrhizal fungi

Pisolithus tinctorius #250 and *Thelephora terrestris* #223 were obtained as mycelium in tube slants from the Institute for Mycorrhizal Research and Development (USDA Forest Service) in Athens,

* This study was supported by the International Foundation for Science in Stockholm, Sweden.

Georgia, U.S.A. These fungi were first cultured in sterile Petri dishes with MMN agar medium (Modified Melin-Norkrans' medium with glucose) at 25°C for about two weeks[7].

The fungi were transferred to pre-autoclaved glass bottles (1 l in capacity) containing 770 ml of vermiculite (grade 2 or particle size larger than 2 mm), 30 ml of peat moss, and 400 ml of MMN solution (no agar added)[10]. The mouth of the bottle was plugged with cotton wool and the bottles were left at room temperature for about 3 to 4 months until the fungal mycelium permeated all the vermiculite/peat particles.

Just before inoculation of these fungi to nursery soil, the bottles were broken open and the vermiculite/peat-mycelium mixture was leached with running tap water for about 5 min to remove excess nutrients which had not been used by the fungi. After removing excess water from the vermiculite, the inoculum is ready to use.

Nursery preparation and fumigation

Fresh seeds of *Pinus densiflora*, *P. thunbergii*, *P. rigida*, and *P. rigida* × *taeda* were soaked in water for 5 days before sowing. For *P. koraiensis*, seeds were stratified for two months before sowing.

Four different treatments were tested in a randomized block design, with each block replicated four times. The treatments were: 1) Fumigation only (with no inoculation), 2) Fumigation plus inoculation with *Pisolithus tinctorius* (Pt), 3) Fumigation plus inoculation with *Thelephora terrestris* (Tt), 4) Control (with no fumigation and no inoculation).

The nursery bed was prepared as microplots described by Marx and Bryan[10]. These plots were encircled with wood panels and each plot was 90 cm × 90 cm, and 25 cm deep. Adjacent microplots were placed 1.1 m apart in each direction. The entire area of nursery bed was covered with an air-tight canvas tent and fumigated with methyl bromide (50 g/m^2 surface) for 66 hours. For control plots where no fumigation was applied, the top 30 cm of soil within the plots was removed before fumigation and put back after.

Inoculation and care of the nusery

Three days after the fumigation 1 l inoculum and 5 l fumigated pine bark (sieved with 1 cm wire net) were added to each plot and mixed thoroughly with the top soil. In inoculation plots live inoculum of Pt or Tt was used, while the fumigation only and control plots received the same amount (1 l) of autoclaved vermiculite/peat-mycelium mixture.

Seeds of the five pine species, treated as above, were sown on the following day in rows and mulched with rice straw, and bird netting was installed. No fertilizer or biocides were used during the experimental period. Mycorrhizal development was frequently examined while the seedlings were thinned to the final number of 200 trees per plot. At the end of the first growing season in mid-October, some of the seedlings were harvested to determine height, root-collar diameter, oven dry weight, shoot-root ratio, and mycorrhiza formation.

Results

Mycorrhizal development

Mycorrhizae were noticed within a month of setting up the experiment in both control and inoculated plots. Two months after seed germination, inoculated seedlings had a large number of well formed mycorrhizae although there was no difference in height between inoculated and uninoculated seedlings.

In general mycorrhizal formation by Pt was much better than by Tt in all the species. In *Pinus rigida* × *taeda* plots where we counted mycorrhizal frequency, the mycorrhizal root tips were 80–95% in Pt plots. On the other hand, seedlings inoculated with Tt had 40–60% of root tips infected. Natural mycorrhizal formation in control plots was very high resulting in about a 80% or more frequency.

Growth stimulation

The height growth of four pine species at the end of the first growing season is shown in Table 1. The height growth of *P. koraiensis* is not reported here because of the inherent limited shoot growth of this pine during the first growing season. Pt stimulated height growth of the four pine species compared with uninoculated control. The amount of growth stimulation by Pt was 55%, 36%, 69%, and 37% compared with control for *P. densiflora*, *P. thunbergii*, *P. rigida*, and *P. rigida* × *taeda*, respectively, which was statistically significant except for *P. thunbergii* at the both 5 and 1% level.

Thelephora terrestris (Tt) failed to stimulate height growth in all the pine

Table 1. Height growth (cm) of *Pinus densiflora*, *P. thunbergii*, *P. rigida* and *P. rigida* × *taeda* seedlings in the nursery inoculated with *Pisolithus tinctorius* (Pt) and *Thelephora terrestris* (Tt) at the end of the first growing season

Treatment	Species			
	P. densiflora	*P. thunbergii*	*P. rigida*	*P. rigida* × *taeda*
Fumigation	6.6	6.2	8.6	12.6
Control	5.6	5.7	7.8	11.8
Pt	8.7*	7.7	13.1*	16.2*
Tt	5.3	5.5	9.3	11.9

* Significantly different from others within the same species at 5% level.

Table 2. Oven-dry weight, top-root ratio, phosphorus and nitrogen contents (% and mg/plant top) of *Pinus densiflora* seedlings in the nursery inoculated with *Pisolithus tinctorius* (Pt) and *Thelephora terrestris* (Tt) at the end of the first growing season

Treatment	Dry weight (g)	Top/root	Phosphorus		Nitrogen	
			%·	mg/plant top	%	mg/plant top
Fumigation	0.80	1.3	0.16	0.74	1.64	7.54
Control	0.58	1.1	0.19	0.59	1.57	4.87
Pt	1.41*	1.7*	0.16	1.44*	1.56	14.04*
Tt	0.54	1.2	0.18	0.52	1.61	4.67

* Significantly different from others at 5% level

Table 3. Oven-dry weight, top-root ratio phosphorus and nitrogen contents (% and mg/plant top) of *Pinus rigida* × *taeda* seedlings in the nursery inoculated with *Pisolithus tinctorius* (Pt) and *Thelephora terrestris* (Tt) at the end of the first growing season

Treatment	Dry weight (g)	Top/root	Phosphorus		Nitrogen	
			%	mg/plant top	%	mg/plant top
Fumigation	1.63	2.42	0.19	2.20	1.51	17.52
Control	1.14	2.40	0.21	1.68	1.72	13.76
Pt	2.60*	2.51	0.18	3.35*	1.56	29.02*
Tt	1.30	2.40	0.19	1.73	1.66	15.11

* Significantly different from others at 5% level.

species except in *P. rigida* where plants were 20% taller. However, it was not significantly different from the control at 5% level.

Tables 2 and 3 show dry weight and T/R ratio (top-root ratio) of *P. densiflora* and *P. rigida* × *taeda*. Inoculation with Pt increased dry weight by 143% over control treatment in *Pinus densiflora* (1.41 g versus 0.58 g) and by 128% in *P. rigida* × *taeda* (2.60 g versus 1.14 g). In contrast, Tt had no effect. The T/R ratio in *P. densiflora* was significantly larger in the Pt treatment than in other treatments, while the T/R ratio of *P. rigida* × *taeda* was not significantly changed by these treatments.

Nitrogen and phosphorus contents of seedlings are also shown in Tables 2 and 3. In both species the total amount of N and P per plant top was significantly higher in Pt-seedlings than other seedlings. In *Pinus densiflora* total absorption (per plant top) of N and P of inoculated seedlings was increased by 188% and 144%, respectively, even though the concentration of these two elements in the tissue was not significantly different from that of control seedlings (Table 2).

Discussion

This experiment confirmed that seedling growth in the nursery can be stimulated by artificial inoculation of plants with selected mycorrhizal fungi. *Pisolithus tinctorius* is well known for its ability to improve survival and growth of pines in a variety of sites[9,10,11].

By contrast with the remarkable growth stimulation by Pt, *Thelephora terrestris* (Tt) failed to stimulate seedling growth in this experiment possibly due to the fact that Tt formed far less mycorrhizal roots (40–60%) than Pt (80–95%). The poor performance of Tt has also been observed in other situations[8]. Marx and Artman[9] postulated that Pt, which is ecologically adapted to soils of low fertility, was 'probably more efficient than Tt in maximizing nutrient absorption from soil'. In the present experiment, no fertilizer was applied during the experimental period and this fact could have enhanced the effect of Pt on seedling growth. However, Pt has also been shown to stimulate plant growth in well-fertilized nursery soil[8].

Pisolithus tinctorius has been reported to occur in Korea[1]. The authors found large population of this fungus on sandy soil in Chilbo Mountain in Gyeonggi Province where soil erosion has been a serious problem for reforestation. The Pt strain #250 used in this experiment was originally isolated from eroded sites with sandy soil in U.S.A. When Pt #250 was cultures on the artificial medium (MMN), it grew much faster than Pt isolated from Chilbo Mt. in Korea, indicating that the Pt introduced from U.S.A. might be better source of inoculum than Korean Pt. An experiment is under way to test the potential of Korean Pt as a nursery inoculum.

Literature cited

1 Anonymous 1978 Suggestions on standard Korean names of mushrooms in Korea. Korean J. Mycology 6, 43–55.

2 Bagyaraj D J and Manjunath A 1980 Response of crop plants to VA mycorrhizal inoculation in an unsterile Indian soil. New Phytol. 85, 33–36.

3 Berry C R and Marx D H 1978 Effects of *Pisolithus tinctorius* ectomycorrhizae on growth of loblolly and Virginia pines in the Tennessee copper basin. USDA For. Serv. Res. Note SE-264, 6 p.

4 Islam R and Ayanaba A 1981 Effect of seed inoculation and preinfecting cowpea (*Vigna unguiculata*) with *Glomus mosseae* on growth and seed yield of the plants under field conditions. Plant and Soil 61, 341–350.

5 Kormanik P P, Bryan W C and Schultz R C 1977 Influence of endomycorrhizae on growth of sweetgum seedlings from eight mother trees. For. Sci. 23, 500–505.

6 Kormanik P P, Bryan W C and Schultz R C 1981 Effects of three vesicular-arbuscular mycorrhizal fungi on sweetgum seedlings from nine mother trees. For. Sci. 27, 327–335.

7 Marx D H 1969 The influence of ectotrophic mycorrhizal fungi on the resistance of pine roots to pathogenic infections. I. Antagonism of mycorrhizal fungi to root pathogenic fungi and soil bacteria. Phytopathology 59, 153–163.

8 Marx D H 1980 Ectomycorrhizal fungus inoculations: a tool for improving forestation practices. *In* Tropical Mycorrhiza Research Ed. P Mikola. pp 13–71. Clarendon Press, Oxford.

9 Marx D H and Artman J D 1978 Growth and ectomycorrhizal development of loblolly pine seedlings in nursery soil infested with *Pisolithus tinctorius* and *Thelephora terrestris* in Virginia. USDA For. Serv. Res. Note SE-256.

10 Marx D H and Bryan W C 1975 Growth and ectomycorrhizal development of loblolly pine seedlings in fumigated soil infested with the fungal symbiont *Pisolithus tinctorius*. For. Sci. 21, 245–254.

11 Marx D H, Bryan W C and Cordell C E 1976 Growth and ectomycorrhizal development of pine seedlings in nursery soils infested with the fungal symbiont *Pisolithus tinctorius*. For. Sci. 22, 91–100.

12 Marx D H, Morris W G and Mexal J G 1978 Growth and ectomycorrhizal development of loblolly pine seedlings in fumigated and nonfumigated nursery soil infested with different fungal symbionts. For. Sci. 24, 193–203.

13 Owusu-Bennoah E and Mosse B 1979 Plant growth responses to vesicular-arbuscular mycorrhiza. New Phytol. 83, 671–679.

Plant and Soil 71, 331–337 (1983). A-37

The influence of mixed inocula on endomycorrhizal development

MELVIN J. DAFT
Department of Biological Sciences, University of Dundee, Dundee DD1 4HN, Scotland

Key words Endomycorrhiza *Glomus* spp. Mixed inocula Spores

Summary Multiple mycorrhizal infections are common in nature, both in nodulated and non-nodulated herbaceous and woody plant species. Woodlands and bulb fields contained more species than coal tips in Scotland or in the United States. In mixed inocula *Glomus clarum* was a good fungal competitor whereas *G. caledonium* was not. Sporulation in *G. clarum* was reduced when phosphate was applied but much less so in *G. caledonium*.

Introduction

Woodland plants are invariably infected with endomycorrhizal fungi[6,11,12,14] and their contribution to the recycling of nutrients in such permanent communities may be significant. However, little or no data are available on the influence of endomycorrhizal infections on plant growth in natural woodlands. Woodland plants are often infected with several different endophytes and it has been suggested that each endophyte may be important at different times during the growing season[6]. Several nodulated tree species have been found to be infected with mycorrhizas and hence benefit from the fixation of atmospheric nitrogen and an increased uptake of phosphorus. Crop plants can benefit from inoculation with selected endophytes in the field[10]. Not all endophytes are equally effective[16] and some produce different growth rates and phosphorus contents of the hosts[1]. Field inoculations with endomycorrhizas poses certain practical problems. Selection for the particular host, effectiveness, strategies, survival and competitive abilities with respect to the indigenous endophyte flora must be taken into consideration. It has been suggested that inocula containing several endophytes with various strategies might prove more beneficial from one growing season to the next[7].

This paper describes the occurrence of mixed infections in field grown plants and growth effects in plants infected experimentally.

Materials and methods

Field survey

Soil samples from the woodland site at the Den of Fowlis (National Grid reference NO 328328) and from the bulb fields at Drumforber (NGR NO 723622) were collected monthly over a complete growing season. The coal tip at Alloa (NGR NN 957897) was sampled 3 times and the anthracite waste at Tamaqua (Pennsylvania, USA) only once. Plants and soil samples, at least 6, from each site were bulked, wet sieved[9] and if clean enough the spores were counted and identified under a stereoscopic zoom microscope. When the samples contained large amounts of organic debris the spores were subjected to a modified sucrose density centrifugation[2] before identification.

Root infections

Fresh samples (1 g) from 3–4 plants were mixed, cleared, stained and then estimated for infection levels as decribed previously[3].

Glasshouse experiments

Single onion sets (mean wt. 3.8 g), variety Stuttgart, were grown in sterile, washed seashore sand, pH 7.5 with $Ca_3\ (PO_4)_2$ added at the rate of 3.2 g to 40 kg sand, in 5" plastic pots. The plants were grown for 18 weeks and fed twice weekly with 60 ml Hoaglands nutrient solution minus the phosphate ion. In some experiments an aqueous solution of potassium hydrogen phosphate (2 ml of 0.4 g in 100 ml H_2O) was injected into the tubular onion leaves twice weekly.

At the time of planting the onion sets were inoculated with the mycorrhizal fungi by pipetting a total of 60 spores into the hole made for the bulb. The inocula comprised of spores only. Four endophytes were used as inocula, either singly or all four: *Glomus caledonium*, *G. clarum*, *G. geosporum* and *G. mosseae*. These spores were selected because they can be recognised on the basis of size, colour and hyphal attachments. At harvest, the roots were washed and the spores from the roots and sand media collected by wet sieving.[9]. After suitable dilution the spores were counted and identified from each treatment. Plant growth was estimated by measuring the heights of the plants and then drying the shoots and roots at 80°C to constant weight.

Root segments with attached nodules were fixed in Craft's solutions under vacuum, dehydrated in tertiary butyl alcohol then embedded in paraffin. Sections were stained in safranin and light green and mounted in DPX.

Results and discussion

Field survey of multiple infected plants

Under natural conditions, plant roots often contain more than one type of mycorrhizal endophyte[4]. Apparently an exclusion mechanism does not appear to operate amongst differing endophytes. Spore types identified from 4 differing habitats and infecting a range of host plants are given in Table 1. The woodland soil and the bluebells, at the time of sampling contained 5 different endophytes. It is not understood why sporulating types are so common in such a permanent community. Actively growing roots are present throughout most of the year. The ground flora is dominated in the early Spring by blue-bells (*Hyacinthoides non-scripta*) and this plant produces roots in the Autumn and at the beginning of the year[6]. In a previous survey of a bulb field[3] four endophyte species were recognised from their spores. Most roots of the daffodils die off during the summer months and the new roots are produced in two periods, Autumn and Spring. Coal tips being very harsh environments contained fewer spore types, but those present may be tolerant varieties[4,5]. It has been suggested that different endophytes may have varying efficiencies throughout the growing season[6]. Different hosts produce varying responses to different endophytes[7]. On the Scottisch coal tip *G. tenuis* predominated in the colonising plants and a mixed

Table 1. Species of mycorrhizal fungi collected and identified from 4 different habitats and the major host plants sampled at each site

		Coal tips*	
Woodland – Den of Fowlis	Bulb field – Drumforber	Scotland – Alloa	USA – Tamaqua
Hyacinthoides non-scripta	*Narcissus pseudonarcissus*	*Agrostis tenuis* & *Chamaenerion angustifolium*	*Coronilla varia* & *Agropyron repens*
Acaulospora laevis	*Gigaspora calospora*	*Glomus calosporum*	*Glomus fasciculatum*
Glomus caledonium	*Glomus caledonium*	*Glomus fasciculatum*	*Gigaspora gigantea*
Glomus fasciculatum	*Glomus fasciculatum*	*Glomus tenuis*	
Glomus tenuis	*Glomus geosporum*		
Glomus mosseae			

fungal population was found in the established ground cover growing on the stable slopes.

Dually infected plants

Enhanced phosphate uptake and fixation of atmospheric nitrogen by herbaceous plants has been the subject of a great deal of research recently[17]. Tree species such as *Alnus glutinosa*, *Robinia hispida*, *R. pseudoacacia*, *Eleagnus umbellata*, *Hippophaea rhamoides*, *Acacia sp.* and *Casuarina sp*[5,8] are both

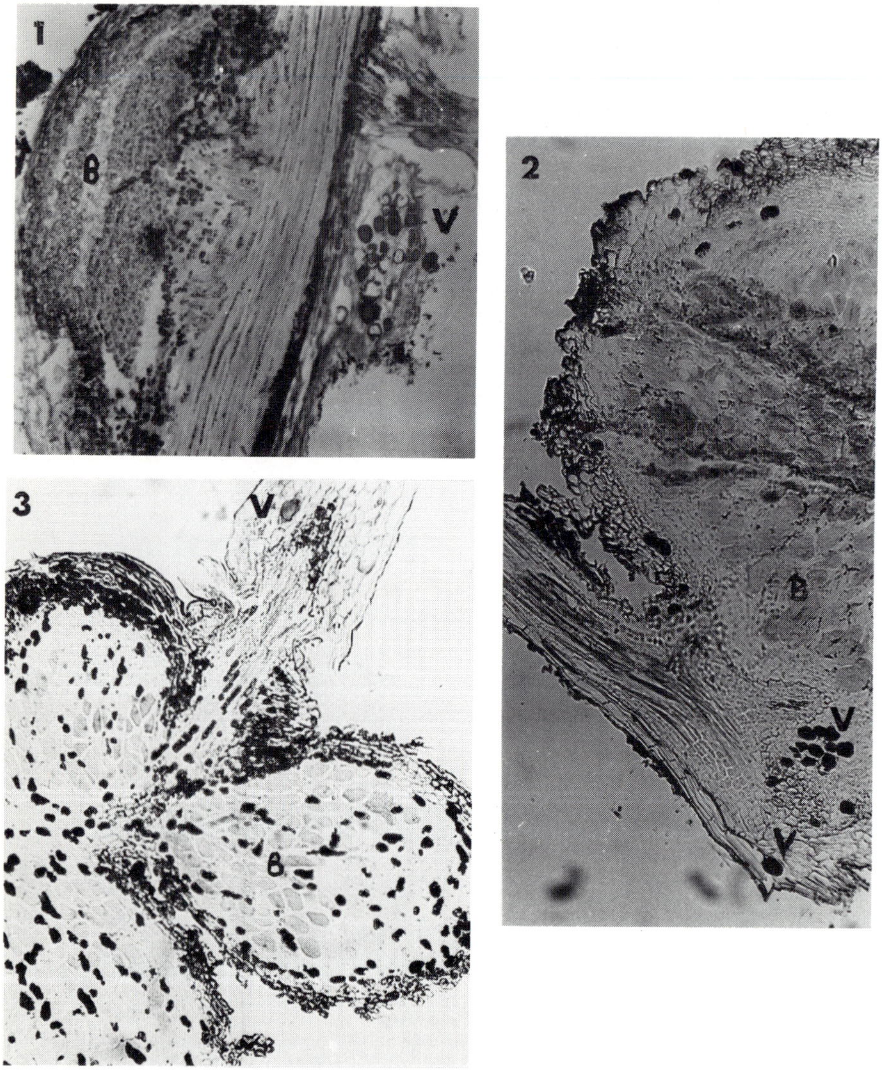

Fig. 1. Sections of nodulated and mycorrhizal roots showing fungal vesicles (V) and bacteroids (B). 1. *Alnus glutinosa;* 2. *Robinia hispida* and 3. *Coronilla varia*.

nodulated and endomycorrhizal. As these are long lived hosts the benefits derived from these multiple endophytic associations are probably considerable. Sections of roots containing nodules have shown that the fungal endophyte does not enter the zone of the nodule containing the bacteroids but may be in the outer cortex of the nodule (Fig. 1) or adjoining root.

Experimental effects of multiple inocula

Onion plants inoculated singly with *G. caledonium*, *G. clarum*, *G. geosporum* and *G. mosseae* at the rate of 60 spores per plant were larger than the uninoculated control plants after 126 days growth (Table 2). Only two species, *G. caledonium* and *G. clarum* significantly increased the dry weight of the plants. Spore numbers for these two species, produced after 18 weeks, were also greater than those recovered from plants infected with either *G. geosporum* or *G. mosseae*. All single inocula gave similar final % infection levels. The quadruple inoculum, containing 15 spores of each endophyte, produced plants of intermediate height and dry weight. Total yield of spores of (29.5×10^3) was made up of 400 *G. caledonium*, 25×10^3 *G. clarum*, 3×10^3 *G. geosporum* and 1×10^3 *G. mosseae*. It would appear that *G. clarum* is a good fungal competitor whereas *G. caledonium* is not when onion is used as the host plant.

The effect of soluble phosphate on onions infected with 2 endophytes is given in Table 3. Soluble phosphate was injected through the tubular leaves, a technique used previously by Sanders[15]. This method of applying soluble phosphorus ensures that its influence on the fungi is *via* the plant. The total dry

Table 2. Plant heights, dry weights, spore yields and percentage infection of onion plants inoculated with single (60 spores plant $^{-1}$) and quadruple (15 + 15 + 15 + 15 spores plant^{-1}) inocula

Inocula	Height (mm)	Dry weight (g)	Spore numbers ($\times 10^3$)	Infection (%)
Single				
G. caledonium	600	13.0*	77	79
G. clarum	640	14.1*	51	87
G. geosporum	440	8.0	12	75
G. mosseae	500	9.2	2	76
Quadruple				
G. caledonium				
G. clarum				
G. geosporum				
G. mosseae	550	12.2*	29	90
Control	400	7.3	–	–

* Significantly different from controls at 1% level.

Table 3. Effect of phosphate application and type of inoculum on growth, spore production and infection levels in onions

Phosphate applied	Inoculum	Total dry weight(g)	Spore number	Percentage infection
($0.4\,g\ KH_2PO_4$	*G. caledonium*	11.6	12,918	35
$100\,ml^{-1}$)	*G. clarum*	11.6	2,018	39
	G. caledonium +	11.7*	3,564	31
	G. clarum		2,510	
	Control	9.8	–	0
0	*G. caledonium*	4.9*	17,820	56
	G. clarum	4.8*	13,336	63
	G. caledonium +	5.9*	22,960	64
	G. clarum		15,650	
	Control	3.8	–	8

* Significantly different from control at 1% level.

weight of the onions given soluble P was greater than the untreated plants. Only the double inoculum plus soluble P treatment was significantly different from the corresponding control whereas all three singly inoculated treatments were different from the controls at the 1% level of probability. Spore production was only slightly reduced with *G. caledonium* plus P whereas the two other treatments were affected markedly. Without P sporulation was high for all infected treatments. Percentage infection levels were similar within each main treatment but approximately 50% lower in the plus P than in the no P treatments.

Considerable variation can occur in the growth of plants inoculated in the field[13]. The introduced endophytes have to compete with the indigenous flora. By using a mixed inoculum, containing endophytes with differing strategies, might reduce variation and give more consistent benefits to the host plants. An ideal endophyte would need to possess many properties. For example, the ability to infect plants early in their growth period, efficiently exploit the soil, transfer materials readily to the host, spread and multiply, compete effectively and to infect a range of crop plants under a range of environmental conditions. All these properties may not be found in one single endophyte and so a multiple inoculum would perhaps give an effective compromise.

Acknowledgements The work was supported by the Agricultural Rerearch Council and the Natural Environmental Research Council.

References

1 Abbott L K and Robson A D 1979 A quantitative study of the spores and anatomy of mycorrhizas formed by a species of *Glomus* with reference to its taxonomy. Aust. J. Bot. 27, 363–375

2 Chilvers M T 1981 Studies on mycorrhizas of the Liliiflorae with special reference to *Hyacinthoides non-scripta*, *Narcissus*, *Tulipa*, *Crocus* and *Allium cepa*. PhD Thesis, University of Dundee.

3 Chilvers M T and Daft M J 1981 Mycorrhizas of the Lilliiflorae. II. Mycorrhiza formation and incidence of root hairs in field grown *Narcissus* L., *Tulipa* L. and *Crocus* L. cultivars. New Phytol. 89, 247–261.

4 Daft M J and Nicolson T H 1974 Arbuscular mycorrhizas in plants colonizing coal wastes in Scotland. New Phytol. 73, 1129–1138.

5 Daft M J and Hacskaylo E 1976 Arbuscular mycorrhizas in the anthracite and bituminous coal wastes of Pennsylvania. J. Appl. Ecol. 13, 523–531.

6 Daft M J, Chilvers M T and Nicolson T H 1981 Mycorrhizas of the Liliiflorae. I. Morphogenesis of *Endymion non-scriptus* (L) Garcke and its mycorrhizas in nature. New Phytol. 85, 181–189.

7 Daft M J and Hogarth B C 1982 Competitive interactions amongst four species of Glomus on maize and onion. Trans. Br. Mycol. Soc. *In press*.

8 El-Giahmi A A, Nicolson T H and Daft M J 1976 Endomycorrhizal fungi from Libyan soils. Trans. Br. Mycol. Soc. 67, 164–170.

9 Gerdemann J W and Nicolson T H 1963 Spores of mycorrhizal *Endogene* species extracted from soil by wet-sieving and decanting. Trans. Br. Mycol. Soc. 46, 235–244.

10 Hayman D S 1982 Practical aspects of vesicular-arbuscular mycorrhiza. *In* Advances in Agricultural Microbiology. Ed. N S Subba Rao. pp 325–373. Oxford & IBH Publishing Co, New Delhi.

11 Johnson P N 1977 Mycorrhizal Endogonaceae in a New Zealand Forest. New Phytol. 78, 161–170.

12 Mosse B and Hayman D S 1973 Effects of mycorrhiza on phosphate uptake of some plant species from Meathop wood soil. Bull. Br. Ecol. Soc. 4, 6.

13 Rangeley A, Daft M J and Newbould P 1982 The inoculation of white clover with mycorrhizal fungi in unsterile hill soils. New Phytol. 92, 89–102.

14 Read D J, Koucheki H K and Hodgson J 1976 Vesicular arbuscular mycorrhiza in natural vegetation systems. I. The occurrence of the infection. New Phytol. 77, 641–653.

15 Sanders F E T 1975 The effect of phosphorus on the development and extent of endomycorrhizal infection of onion roots. *In* Endomycorrhizas. Eds. F E T Sanders, B Mosse and P B H Tinker, London, Academic Press.

16 Sanders F E, Tinker P B, Black R T B and Palmerly S M 1977 The development of endomycorrhizal root systems. I. Spread of infection and growth promoting effects with four species of vesicular-arbuscular endophyte. New Phytol. 78, 257–268.

17 Smith F A and Smith S E 1981 Mycorrhizal infections and growth of *Trifolium subterraneum*: use of sterilized soil as a control treatment. New Phytol. 88, 299–310.

Plant and Soil 71, 339–344 (1983). A-38

Growth and survival of container-grown *Pinus caribaea* infected with various ectomycorrhizal fungi

M. H. IVORY
Commonwealth Forestry Institute, Oxford OX1 3RB, UK

and F. M. MUNGA
Kenya Agricultural Research Institute, P.O. Box 30148, Nairobi, Kenya

Key words Ectomycorrhizas Establishment Inoculation Kenya *Pinus caribaea*

Summary Young pine seedlings were inoculated with mycelial cultures of 8 fungi, and basidiospore cultures of 4 fungi. Four months later distinctive mycorrhizas had been formed by all 4 spore cultures and by 5 of the mycelial cultures.

At 10 months one further mycelial culture had given rise to distinctive mycorrhizas. However, by that time contamination between treatments, or from outside, was at a higher level.

Survival of outplanted seedlings after 6 months was generally poor, but was slightly greater in some treatments.

Introduction

A few ectomycorrhizal trees are native to southern and central Africa, including Kenya[9], but none are conifers. Nevertheless, when pines were first introduced to Kenya they grew well. Later, plantings grown from imported seed encountered deficiency problems[1], which were only overcome after the significance of mycorrhizas had been demonstrated in forest nurseries[2,6]. The pine mycorrhizal fungi presently in Kenya were probably introduced from Europe *via* parts of southern Africa[7]. They appear to be well-suited to the cool upland forests, but may be less suitable in lowland forest areas where conditions are much hotter and drier.

The Unit of Tropical Silviculture of the Commonwealth Forestry Institute (CFI) has been involved in establishing provenance trials of pines throughout the Tropics. However, little attention was paid to their fungal symbionts until it was demonstrated that mycorrhizal fungi varied considerably in their ability to benefit the host under particular site conditions[3,8]. Consequently an investigation of the mycorrhizas of some tropical pines in their native habitats, and in exotic plantations, was initiated in 1976.

As part of this scheme, a joint project with the Kenya Forestry Department was started in 1980 to study the effects of several ectomycorrhizal fungi on the field performance of *Pinus caribaea* Mor. var. *hondurensis* Barr. & Golf.

Material and methods

Seed of *P. caribaea* ex Kwale, Kenya (provenance origin MPR, Belize) was sown at Muguga in November, 1980 in trays of sterile sand.

Isolates of 8 fungi were grown in peat and vermiculite mixture at CFI and taken to Kenya in January, 1981. These cultures were washed, mixed with vermiculite, and put into Spencer Lemaire 'Five' containers. Washed seedlings were then transplanted into each container. The fungi were (a) *Pisolithus tinctorius* (Pers.) Coker and Couch, *Rhizopogon nigrescens* Coker & Couch, *Scleroderma texense* Berk., *Thelephora terrestris* Fr., *Phylloporus rhodoxanthus* (Schw.) Bres. ex Central America; (b) *Suillus granulatus* (L. ex Fr.) Kuntze ex Zambia; (c) *Scleroderma bovista* Fr. ex Tanzania; and (d) *Rhizopogon luteolus* Fr. & Nordh. ex Ghana. Control treatments consisted of plants inoculated with autoclaved cultures or with nursery soil.

At the same time, further seedlings were inoculated with spores of 4 fungi by adding dry spores, or spores in water suspension, to small quantities of vermiculite, and then applying these to the immediate rooting zone of the seedlings in Spencer Lemaire 'Ferdinand' containers. Further vermiculite was used for the remainder of the rooting medium. Spore cultures of *P. tinctorius*, *S. texense* and *R. nigrescens* ex Belize (col. October 1980 and stored at 5°C) were used. Insufficient seedlings were available at this time to set up a control treatment in these small containers.

All seedlings were kept at Muguga inside a polythene house to reduce contamination and to raise the temperature. They were sprayed with $FeSO_4$ in February, in addition to regular fertilization with N.P. and K.

Table 1. The effect of inoculation on mycorrhiza formation and plant height after 4 months in the nursery

	% Plants infected		Mycorrhizal development* Type				Plant height (cm)
Treatment	Total	Inoculated fungus	1	2	3	4	
P. rhodoxanthus	4	0	–	–	0.1	–	16.0
P. tinctorius	100	100	–	*4.0*	–	–	14.4
P. tinctorius (spores)	98	93	–	*3.1*	–	0.3	13.7
R. luteolus	18	12	–	–	0.1	0.2	13.8
R. nigrescens (534)	28	28	–	–	–	*0.7*	14.4
R. nigrescens (534 spores)	93	93	–	–	–	*2.5*	12.3
R. nigrescens (553 spores)	100	100	–	–	–	*2.7*	11.8
S. bovista	74	74	*2.2*	–	–	–	14.7
S. texense	58	58	*1.6*	–	–	–	17.2
S. texense (spores)	100	100	*4.0*	–	–	–	12.7
S. granulatus	30	0	–	–	–	0.7	15.6
T. terrestris	96	96	–	–	*2.7*	–	15.6
Nursery soil control	84	–	–	–	0.7	1.2	17.5
Control	2	–	–	–	–	Tr	16.4

* Grade 0 = nil; 1 = few mycorrhizas; 2 = $<\frac{1}{3}$ roots mycorrhizal, 3 = $\frac{1}{3}$ to $\frac{2}{3}$ roots mycorrhizal; 4 = $>\frac{2}{3}$ roots mycorrhizal.

In April some plants were transferred to Gedi F.S. nursery one month before outplanting there in May. They were planted as line plots of 10 trees, replicated 5 times in a randomized block design (4 replications for spore inoculum treatments), in an area with deep sandy soil of pH 7.7. Each plant was measured for size, assessed for mycorrhizal infection and its position recorded. After 6 months in the field this trial was assessed for survival and the survivors measured for height increment.

The remaining seedlings were kept at Muguga until November 1981, most having been transplanted into larger containers with additional vermiculite in August. Fifty in each treatment were then assessed for mycorrhizal infection, and measured for height and stem diameter, and outplanted at Kibwezi F.S.

Results

The results of assessments made in May at Gedi (4 months after inoculation) and in November at Kibwezi (10 months after inoculation) are given in Tables 1 and 2 respectively. At 4 months, inoculation with basidiospores of *P. tinctorius*, *R. nigrescens* and *S. texense*, and with mycelial cultures of *P. tinctorius*, *R. nigrescens*, *S. bovista*, *S. texense* and *T. terrestris*, led to the formation of abundant distinctive mycorrhizas on many of the plants, whereas the control

Table 2. The effect of inoculation on mycorrhiza formation and plant height after 10 months in the nursery

	% Plants infected		Mycorrhizal development*					Plant height (cm)
			Type					
Treatment	Total	Inoculated fungus	1	2	3	4	5	
P. rhodoxanthus	70	0	–	–	1.2	0.6	–	24.2
P. tinctorius	100	90	–	*2.9*	1.4	Tr	Tr	23.0
P. tinctorius (spores)	100	100	–	*3.2*	0.2	1.4	Tr	16.9
R. luteolus	94	30	–	–	2.1	*0.8*	–	22.1
R. nigrescens (534 spores)	84	40	–	–	1.4	*1.3*	–	20.7
R. nigrescens (534 spores)	100	100	–	–	–	*3.3*	–	15.6
R. nigrescens (553 spores)	100	100	–	–	–	*3.0*	–	14.0
S. bovista	92	42	*1.1*	–	1.7	0.3	–	23.3
S. texense	88	34	*0.8*	–	1.5	–	–	26.9
S. texense (spores)**	100	96	*3.2*	–	Tr	0.1	–	16.5
S. granulatus	86	0	–	–	1.3	0.8	Tr	22.7
T. terrestris	80	78	–	–	*1.6*	0.1	0.1	23.7
Nursery soil control	100	–	–	–	0.8	2.4	–	29.7
Control	72	–	–	–	0.9	1.1	–	26.5

* Grade 0 = nil; 1 = few mycorrhizas; 2 = $<\frac{1}{3}$ roots mycorrhizal; 3 = $\frac{1}{3}$ to $\frac{2}{3}$ roots mycorrhizal; 4 = $>\frac{2}{3}$ roots mycorrhizal.

** 25 plants only

plants were virtually uninfected. A few mycorrhizas of type 3 and 4 were formed on plants inoculated with mycelial cultures of *P. rhodoxanthus*, *R. luteolus* and *S. granulatus* but most probably resulted from contamination with inoculum from the *T. terrestris* and *R. nigrescens* treatments. Plants inoculated with nursery soil developed mycorrhizas of 2 types, thought to be associations with *T. terrestris* and *Rhizopogon rubescens* (Tul. & Tul.) Tul. & Tul., both of which are known to occur in Muguga nursery.

At 10 months results were similar except that some plants inoculated with mycelial culture of *R. luteolus* had also developed abundant type 4 mycorrhizas. The specific identity of the fungus associate of this type could not be confirmed as mycorrhizas of the 3 *Rhizopogon* spp. present at this site are very similar in appearance. Contamination in most treatments with type 3 and 4 mycorrhizas was much more prevalent at this time, probably because of the profuse presence of *T. terrestris* sporocarps on treated plants and the ease with which waterborne *Rhizopogon* spp. spores are distributed. Attempts to isolate the fungal symbionts from all mycorrhiza types were unsuccessful.

Plant height varied greatly within each treatment after 4 and 10 months, but differences between treatments were insignificant. However, it was noted that seedlings at the edge of the trays, and seedlings in the smaller containers tended to be smaller. After 6 months at Gedi, during which time only 320 mm of rain fell, survival was poorest for the plants grown in 'Ferdinand' containers and for those

Table 3. Survival of outplanted seedlings after 6 months, and height increment of survivors at Gedi F.S.

Treatment	Height at time of planting (cm)	Mycorrhizal* development at time of planting	Survival %	Height increments
P. tinctorius	14.5	3.9	52	7.1 cm
S. texense (spores)	12.2	3.9	22	6.1 cm
P. tinctorius (spores)	12.6	3.3	32	5.7 cm
S. bovista	14.6	2.2	56	5.2 cm
R. nigrescens (534 spores)	11.4	3.3	14	4.3 cm
R. nigrescens (553 spores)	11.2	3.3	16	4.2 cm
Nursery soil	17.4	1.8	44	3.9 cm
S. texense	15.7	1.2	42	3.7 cm
S. granulatus	15.6	0.9	50	2.8 cm
R. luteolus	13.2	0.7	36	2.7 cm
P. rhodoxanthus	16.6	0.0	28	2.5 cm
R. nigrescens	14.9	0.8	24	2.4 cm
Uninoculated	15.3	0.1	30	2.2 cm
T. terrestris	16.4	3.0	18	– 0.4 cm

* Grade 0 = nil; 1 = few mycorrhizas; 2 = $<\frac{1}{3}$ roots mycorrhizal; 3 = $\frac{1}{3}$ to $\frac{2}{3}$ roots mycorrhizal; 4 = $>\frac{2}{3}$ roots mycorrhizal.

with poor mycorrhizal development, with the exception of those inoculated with *T. terrestris* (Table 3). Height increment of the survivors similarly was related to the initial mycorrhizal status of the outplanted seedlings.

Discussion

Robertson[10] showed that basidiospores could be used in some instances as effective mycorrhizal inocula, and the results obtained in Kenya with spores of *P. tinctorius*, *R. nigrescens* and *S. texense* confirm this. The results also show that spore inocula are as good, or better than, mycelial inocula for the three species tested in both forms, which is contrary to the assertions by Marx[5]. In addition infection in the spore treatments must have been quite rapid as it had reached high levels within 4 months.

A standard inoculation procedure was used for all mycelial inocula, based on that used for *P. tinctorius*[5]. However, several authors have indicated that other fungi may require different handling procedures for infection to occur[4,5]. The variation in infection levels attained (0–100% plants infected) may therefore be related to the technique used rather than the infectivity of the fungi under test.

Growth effects in the nursery were minimal, probably because growth was limited by the small volumes of growing medium employed[4].

Effects on survival and growth of the trees after outplanting have been demonstrated by many authors[5]. Preliminary data from this trial suggest that similar results can be achieved in Kenya also, but that larger more uniform seedlings should be used on lowland sites.

Acknowledgements We thank the Kenya Forestry Department for their generous technical support and financial assistance towards this project in Kenya; and the Overseas Development Administration for funding the mycorrhiza research schemes at CFI, of which the project in Kenya is an essential part.

References

1 Gibson I A S 1963 Eine Mitteilung über die Kiefernmykorrhiza in den Wäldern Kenias. *In* Mycorrhiza. Eds. W Rowald and H Lyr. pp 49–51. F Fischer, Jena.

2 Kessel S H 1927 Soil organisms: the dependence of certain pine species on a biological soil factor. Emp. For. J. 6, 70–74.

3 Lamb R J and Richards B N 1971 Effect of mycorrhizal fungi on the growth and nutrient status of slash and radiata pine seedlings. Aust. For. 35, 1–7.

4 Le Tacon F and Valdenaire J M 1980 La mycorrhization controlee en pepiniere. Premiers resultats obtenus a la pepiniere du fonds forestier national de Peyrat-le-Chateau (Haut-Vienne) sur epicea (*Picea excelsa*) et douglas (*Pseudotsuga douglasii*). Rev. For. Francaise 32, 281–293.

5 Marx D H 1980 Ectomycorrhizal fungus inoculations: a tool for improving forestation practices. *In* Tropical Mycorrhiza Research. Ed. P Mikola. Oxford Univ. Press, pp. 13–71.

6 Melin E 1923 Experimentelle Untersuchungen über die Birken- und Espen-Mykorrhizen und ihre Pilzsymbioten. Sven. Bot. Tidskr. 17, 479–520.

7 Mikola P 1970 Mycorrhizal inoculation in afforestation. Int. Rev. For. Res. 3, 123–196.

8 Theodorou C and Bowen G D 1970 Mycorrhizal responses of radiata pine in experiments with different fungi. Aust. For. 34, 183–191.

9 Redhead J F 1980 Mycorrhiza in natural tropical forests. *In* Tropical Mycorrhiza Research. Ed. P Mikola. Oxford Univ. Press. pp. 127–142.

10 Robertson N F 1954 Studies on the mycorrhiza of *Pinus sylvestris* 1. The pattern of development of mycorrhizal roots and its significance for experimental studies. New Phytol. 53, 253–283.

Plant and Soil 71, 345–351 (1983). A-39

Influence of ectomycorrhizae on infectivity of Pythium-infested soils and substrates

R. PERRIN
INRA, Station de Recherches sur la Flore Pathogène dans le Sol 17, Rue Sully, F-21034 Dijon, CEDEX, France

and J. GARBAYE
CNRF Champenoux, Unité de Microbiologie des Sols Forestiers F-54280 Seichamps, France

Key words Greenhouse *Hebeloma crustuliniforme* Microbial nursery protection *Paxillus involutus* *Pythium* sp. Soil infectivity

Summary Greenhouse experiment and nursery investigations were conducted to determine the influence of ectomycorrhizae on the infectivity of conducive soil or substrates artificially infested or colonized by *Pythium* spp.

A standard bioassay method was used to estimate soil infectivity (S.I.) and to follow changes with time under various conditions. S.I. decreased in substrates in which Beech seedlings infected with *Hebeloma crustuliniforme* had grown. Principal component analysis of the data, and *in vitro* confrontation between *H. crustuliniforme* and *Pythium ultimum* suggest a direct effect of the mycorrhizal species against Pythium. Further investigations are necessary to specify and determine this phenomenon. Such methods may permit the better selection of mycorrhizal strains for protecting tree seedlings against root pathogens.

Introduction

Pythium are redoutable participants in keen competition to conquest plant roots. Numerous publications have shown that mycorrhizae can protect plants successfully against fungal pathogens. Recently Marx[3] published a review in which he described and postulated several mechanisms by which mycorrhizal fungi may afford disease protection against root pathogens. Despite its fundamental advantage in biological control one of these mechanisms – the protective role of the microbes in the 'mycorrhizosphere' – has received very little attention.

Now, the concept of soil infectivity and the methods developed for measuring it[1] allow the effect of the mycorrhizosphere on plant pathogens to be studied more closely. Soil infectivity has been defined[1] as the ability of soil carrying or containing a pathogen to transfer it to a host plant and to cause infection. Infection is determined by pathogenic, physiochemical and microbiological components in the infested soil. The experiments described here concern the influence of ectomycorrhizae on Pythium infested soil infectivity.

Methods

Soil infectivity

It may be estimated by a standard bioassay, using cucumber seedlings as plant baits for Pythium infested soils. Quantification of this bioassay is obtained by diluting a soil to be analysed with a standard disinfected soil. Resolution of the linear part of the regression between percentage of dead seedlings and the soil concentration for obtaining 50% mortality gives the value of SI 50 which is used as a basis for comparison of infectivity of different soils, or change in infectivity of soil with time. Joannes and Bouhot[1] using principal components analysis (PCA) interpretated their results on the basis of competition for space.

Greenhouse experiment

Mycorrhizal and non-mycorrhizal beech and oak seedlings, were produced on peat as previously described[2] then, transplanted into pots containing a conducive soil infested by *Pythium* spp. Infestation was realised by adding naturally infested garden soil in order to obtain approximatively 0, 1 USI 50/g. Pots were placed in a greenhouse, watered twice a week; the temperature ranged between 20 and 35°C. Soil samples of the different treatments (Table 1) were taken from the region of the rhizosphere after 2, 4 and 7 months respectively and their SI measured using the standard bioassay.

Nursery experiment

Lainez and Garbaye[2] have conducted investigations with mycorrhizal Beech and Oak seedlings in a nursery in the North East of France. The growing medium were fertilized peat, desinfected with different ectomycorrhizal fungi. Soil infectivity of peat samples taken at the end of the growing season (oct. 81) indicated the influence of ectomycorrhizae on recolonization of the peat by Pythium from the underlying soil. Soil infectivity was measured for all the treatments recorded in the Table 2.

In vitro confrontation

A strain of *Pythium ultimum* isolated from the soil (used for infestation) was selected for the *in vitro* confrontation with *Hebeloma crustuliniforme* on M.M.N. medium (Melin, Norkrans medium as modified by Marx) in petri dishes. Because of its relative slow growth *H. crustuliniforme* was transfered on M.M.N. 15 to 18 days prior to the Pythium strain. The growth rate of Pythium was assessed every hour by mesuring the radial length of each fungal colony along two axis. The second one at an axis of 45° to the first.

Table 1. Nature of the mycorrhizal fungi and the tree seedlings involved in various treatments of the greenhouse experiment

Code	Soils samples taken from rhizosphere of:
{ T_1* / T_1 bis	Non inoculated oak, or naturally inoculated with *Telephora terrestris*
T_2	Bare soil
T_3	Non inoculated beech
M_1	Beech inoculated with *Hebeloma crustuliniforme* (H.c.) Strain Sth
M_2	Oak inoculated with *Hebeloma crustuliniforme* (H.c.) Strain Sth

* five replications – each one is one plant in 15 liter pot.

Table 2. Nature of the mycorrhizal fungi inoculated in the growing medium of beech seedlings in nursery

Non inoculated beech
Beech inoculated with *Hebeloma crustuliniforme* Strains Siv, Lit 2, Sth
Beech inoculated with *Paxillus involutus* Strain Sth

Results

Greenhouse experiment

Changes in soil infectivity with time among the different treatments is represented in Fig. 1. A significant decrease of SI occurred in treatment M_1 (Beech infected with *H. crustuliniforme*) after four months growth. In spite of a low level of mycorrhizal roots ($\simeq 10\%$) there was a strong SI reduction in the rhizosphere but only for a limited period.

The relationships between cucumber seedlings death and concentration of soil samples taken in four treatments can be seen on Fig. 2. This relationship is linear overall, except for beech infected with *H. crustuliniforme*. The different shape of the curve, can be interpreted as indicating a microbiological influence[1].

Nursery experiment

The results were analysed by PCA (Fig. 3) where each point represents a regression curve. According to Joannes and Bouhot[1] the points moving along vertical axis (axis 1) from a lower to a higher position indicate a change from a low level to a high level of seedling disease closely correlated with SI; therefore soil infectivity of substrates where *H. crustuliniforme* was the mycorrhizal fungus associated with beech, is very low in comparison with the non mycorrhizal treatment. In contrast the SI of samples taken from the rhizosphere of beech infected with *Paxillus involutus* is higher (but not significantly different) than the control. Movement along the horizontal axis (axis 2) from left to the right indicates a change in the shape of the regression line, from linearity to nonlinearity. The more right position of the *H. crustuliniforme* points signifies (1) a microbiological influence on disease severity.

These results indicate that all the strains of *H. crustuliniforme* act as a brake on the recolonization of the peat by *Pythium* sp. and substantiate the results of greenhouse experiment. In contrast *P. involutus* does not appear to have the same ability to reduce soil infectivity under conditions of this experiment.

In vitro study

The figures for the growth of *Pythium ultimum* (Py 18 strains) in the presence and absence of *H. crustuliniforme* are presented in Table 3.

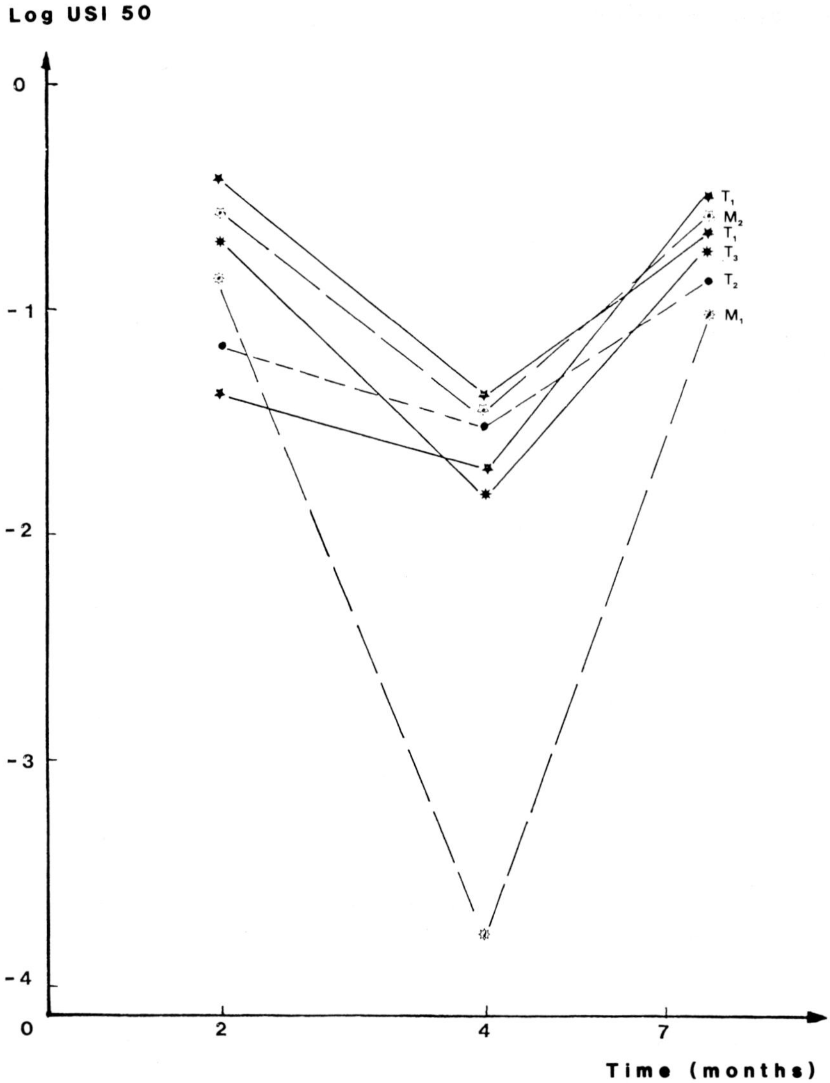

Fig. 1. Time evolution of the soil infectivity of samples taken from the rhizosphere of inoculated and noninoculated tree seedlings.
Key: T_1 Non inoculated oak
T_2 Bare soil
T_3 Non inoculated beech
M_1 Beech inoculated with *Hebeloma crustuliniforme*
M_2 Oak inoculated with *Hebeloma crustuliniforme*

On the one hand, growth rate of *Pythium ultimum* was strongly suppressed on the medium were *H. crustuliniforme* has previously grown within two weeks (radial length < 10 mm). These observations suggest an effect of a diffusible or volatile inhibitory substances produced by *H. crustuliniforme*.

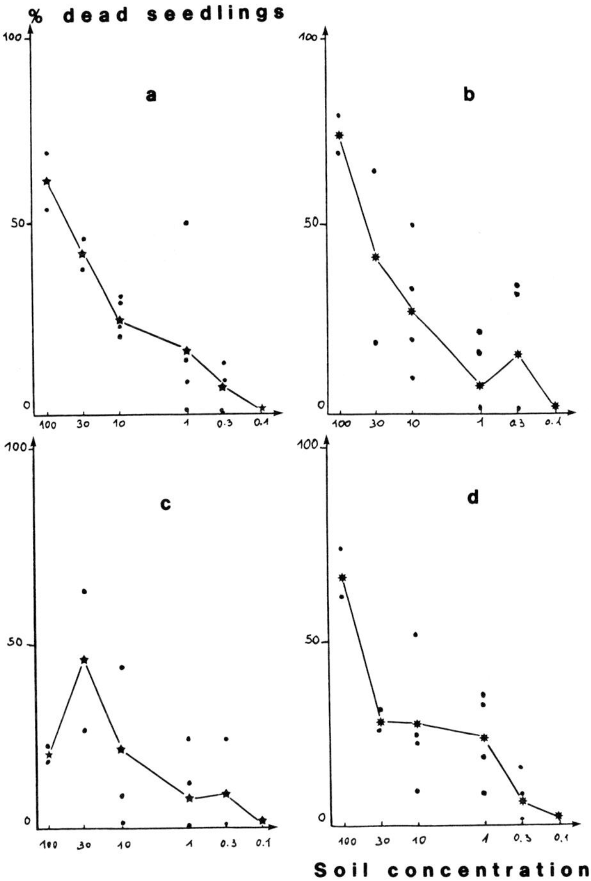

Fig. 2. Regression curves between percentage of dead cucumber seedlings and soil concentrations of samples taken after 4 months growth of oak and beech seedlings in greenhouse.

a: Oak inoculated with *H. crustuliniforme*

b: Bare Pythium infested soil

c: Beech inoculated with *H. crustuliniforme*

d: Non inoculated beech

In addition, the growth of *Pythium ultimum* was very similar along both axes, indicating a lack of direct antagonism between both the species. However, a loose hyphal covering of Pythium appearing as a halo on the petri dishes could be observed in the close vicinity of *H. crustuliniforme*.

Discussion

The preliminary results reported here provide evidence that some ectomycorrhizal fungi like *Hebeloma crustuliniforme* may decrease the infectivity of Pythium infested soil to such a level that it protect tree seedlings from disease under field conditions. Larges differences observed between mycorrhizal fungi,

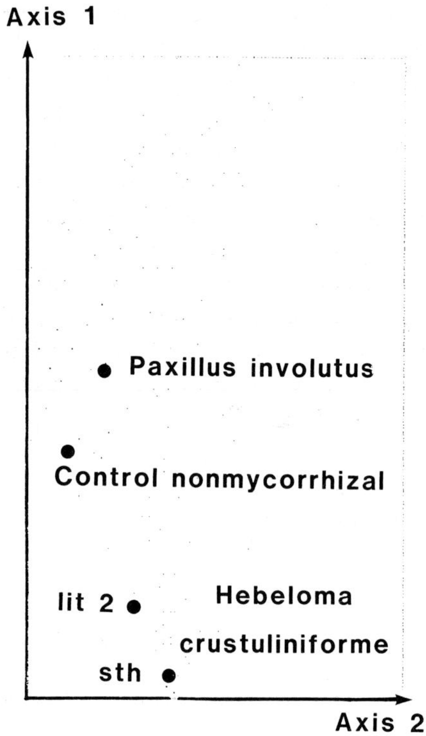

Fig. 3. Principal components analysis of bioassay data related to four samples taken in nursery at the end of the growing season.

Table 3. *In vitro* confrontation of *H. crustuliniforme* with *Pythium ultimum.* Mean growth rate of Pythium (mm/hour) on MMN medium at 20°C

Fungi	Pythium growth along two axes: opposite to *H. crustuliniforme* colony	45° to first axis
Pythium alone	2.39	2.39
H. crustuliniforme prior to		
Pythium strain 1	1.32	1.34
Pythium strain 2	1.28	1.27

may be of sufficient ecological significance in influencing damping off disease or root rot disease.

The few observations made here suggest that the decrease in soil infectivity is the result of a direct effect of *H. crustuliniforme* against the population of *Pythium* sp. *In vitro* studies need to be supported by confrontation studies in the soil.

Measurement of the saprophytic colonization ability of *Pythium* sp. in a soil containing or not containing mycorrhizal fungi is currently in progress.

The concept of soil infectivity and the methods used to measure it (available for *Pythium* sp., *Rhizoctonia solani* and *Fusarium oxysporum* damping off and root rot diseases) enable one to demonstrate microbiological protection in the rhizosphere, to study the mechanism and its expression in the nursery or forest. We can, now, add a new criteria for the selection of mycorrhizal fungi: protection against root pathogens, that is even the principal interest of mycorrhizae in high level nutrients soils. The extent of the decrease in soil infectivity observed in this study indicates that mycorrhizae may provide the most promising way of biological control in forest nursery, in forests, as well as in gardens and fields crops. To achieve this aim we have to make ecological selection of effective strains from natural large supplies.

Acknowledgements The authors wish to thank Miss Nadine Gautheron and M. G. Courrier for technical assistance and R. Sampangi for improving the english.

References

1 Joannes et Bouhot D 1980 Le potentiel infectieux des sols. Thèse de Doctorat d'Etat. Université de Nancy I, 151 p.

2 Lainez J et Garbaye J 1981 La mycorrhization du Hêtre et du Chêne en Lorraine. Rapport de stage E.N.I.T.E.F., 79 p.

3 Marx D J 1973 Mycorrhizae and feeder root disease. *In* Ectomycorrhizae. Eds. G C Marks and T T Kozlowski. Academic Press, New York, pp 351–389.

Plant and Soil 71, 353–361 (1983). A-40

Effect of pesticides on ectomycorrhizae of *Pinus sylvestris* seedlings

P. CUDLÍN, V. MEJSTŘÍK and J. SKOUPÝ*
Institute of Landscape Ecology, Czechoslovak Academy of Sciences, CS-252 43 Průhonice, Czechoslovakia

Key words Artificial inoculation Ectomycorrhizae Fungicide Dithane M-45 Herbicide Gramoxone *Pinus sylvestris*

Summary Knowledge of the effect of pesticides on the formation of forest tree mycorrhizae is important as pesticides are nowadays used in forestry. The effect of the fungicide Dithane M-45** and the herbicide Gramoxone** on the growth of *Pinus sylvestris* L. seedlings and on the development of their mycorrhizae was studied. Investigations involved seedlings inoculated with pure cultures of mycorrhizal fungi in flasks with perlite under aseptic conditions, in Mitscherlich pots filled with perlite under semi-aseptic conditions, and on peat substrate in outdoor beds. No change in seedling growth and the mycorrhiza formation occurred when water suspension of the fungicide Dithane M-45 was used at the recommended dose. The highest rates of this fungicide had no phytotoxic effect although the growth of treated seedlings was reduced due to complete or partial inhibition of mycorrhizal formation. In contrast, even low doses of Gramoxone reduced the growth of the inoculated and non-inoculated seedlings which were more sensitive than their mycorrhizal fungi. The soil sterilization of outdoor beds with an application of a water suspension of Dithane M-45 at recommended doses reduced mycorrhizal development and seedling growth. Seedlings inoculated simultaneously with pure cultures of *Suillus granulatus* showed a slightly better growth than untreated controls.

Introduction

Effects resulting from the use of pesticides in forestry is of general interest. Their effects on pure cultures of mycorrhizal fungi have been studied and results reviewed[2]. Experiments on the effects of pesticides on the development of mycorrhizae have been reported for seedlings cultivated in nurseries[4,5,6,9], although controlled experiments using inoculations of pure cultures of mycorrhizal fungi under aseptic or semi-aseptic conditions are limited. The results of this type of experiments are reported in this paper.

Materials and methods

Experiments in erlenmeyer flasks with perlite under aseptic conditions

Seedlings of *Pinus sylvestris* were grown in 500-ml erlenmeyer flasks specially adapted for this purpose. The flasks were filled with perlite, saturated with distilled water and Koniferin II solution

* Institute of Applied Ecology and Ecotechnics, University of Agriculture, CS-281 63 Kostelec nad Černými lesy, Czechoslovakia

** Dithane M-45 or mankozeb; Gramoxone or Paraquat

with the addition of 2.5 g of glucose per 1 liter of the solution[3]. Inoculation was carried out using a pure culture of *Pisolithus arrhizus* V-5 strain grown on agar-medium KHO[8] for one month after sowing. The fungicide Dithane M-45 (Rohm and Haas: 80% of mankozeb) was applied in 50, 500, and 5,000-μg doses and the herbicide Gramoxone (Spolana Neratovice: 20% of paraquat) in 25, 50, and 250-μg doses per flask in 10-ml aliquots of distilled water. The pesticide suspensions were applied at three different times: (i) directly before sowing seed, (ii) with the artificial inoculation, and (iii) three months after the inoculation. The seedlings were grown in a growth chamber providing a day temperature of 22°C, day humidity of 70%, illumination of 13,000-lx and a night temperature of 16°C, night humidity of 80% and a 16 h photoperiod. Two blocks containing 12 seedlings of each treatment were set up. The seedlings were harvested and evaluated after six months.

Eleven parameters were measured *i.e.* root collar diameter, shoot height, shoot dry weight, root dry weight, shoot to root dry weight ratio, total lateral root length, total number of short roots, number of short roots per 10 mm of total lateral root length, total number of mycorrhizae, number of mycorrhizae per 10 mm of total lateral root length, mycorrhizae as a percentage of total number of short roots. The results were analysed using an IBM 370 computer.

Experiments in Mitscherlich pots with perlite under semi-aseptic condition

The seedlings were grown in 5-l Mitscherlich pots situated in an unheated greenhouse. Pots were filled with perlite saturated with Koniferin solution. Two months after sowing, inoculation was made using pure culture fragments of *Pisolithus arrhizus* V-2 strain grown on agar-medium KHO.

The fungicide Dithane M-45 was applied in two concentrations (5 and 500 μg/ml water suspension): (i) directly before sowing, 70 ml of water suspension per pot; (ii) 4 days before artificial inoculation, 85 ml of water suspension sprayed per pot; (iii) 2 months after inoculation, 100 ml of water suspension sprayed per pot; (iv) applications comprising i, ii, and iii methods, plus an application 5 weeks after inoculation as 100 ml of water suspension sprayed per pot; (v) applications as in iv, but without i; (c) untreated control. The herbicide Gramoxone was applied at periods corresponding to application i, ii, and iii of Dithane M-45 as a water suspension, two concentrations (2.5 and 25.0-μg/ml) representing the doses of 175 and 1,750-μg of herbicide per pot. In addition to the inoculated and non-inoculated seedlings without pesticide treatment, non-inoculated seedlings were treated with higher concentrations of both pesticides at times ii and iii.

Measurements and statistical tests were made in the same way as that mentioned earlier. In addition, the surface of the root system was investigated by a modified gravimetric method[1].

Experiments on outdoor beds with peat substrate

Four outdoor beds, 100 × 40 cm were prepared and filled with a substrate composed of 4 parts of peat to 1 part of perlite plus 0.5 kg of Cererit fertilizer per 1 m^3 of substrate. All beds were partly sterilized with the fungicide Dithane M-45 owing to the big quantity of fungal spores in substrate; 4 l of 0.4% water suspension was applied to each square metre. After one week the beds were sown. Three months after sowing the beds were split in half and one half of each plot was inoculated with *Suillus granulatus* IV-3 strain grown in submerged culture. The beds were treated again with Dithane M-45 as: (i) pre-emergent application 2 weeks after sowing with 1,500 ml of a 0.5% water suspension per plot, plus three sprays at monthly intervals (80 ml, 150 ml, and 150 ml of a 0.5% water suspension per plot); (ii) same as (i) but no pre-emergent application; (iii) only the last final spray; (c) untreated control. Ten seedlings were taken from each bed for evaluation after a period of 7 months. Measurements and statistical analysis were as for the Mitscherlich pots.

Results

Experiments in erlenmeyer flasks with perlite under aseptic conditions

The results presented in this paper provide evidence of differences between the two pesticides with respect to their effect on seedling growth and mycorrhiza formation. The fungicide Dithane M-45 was non-phytotoxic at rates up to 5,000-μg per flask. The herbicide Gramoxone decreased all measured values compared with the control even at concentrations of 25 μg per flask only. This was true for applications given simultaneously with artificial inoculations. In the first and second application periods the highest doses, resulted in a reduced growth rate of the inoculated seedlings compared with non-inoculated controls. Direct phytotoxic effect of Gramoxone is responsible. In contrast, the fungicide Dithane M-45 affected seedling growth in several ways; it partly or completely inhibited mycorrhizal formation and decreased the size of root system. Phytotoxic effects were evident only for the fungicide application before sowing.

Both pesticides differed in the way they reduced mycorrhizal formation. Dithane M-45 inhibited fungal growth in the first and second application periods at the highest application rate. In all other cases mycorrhizae were formed as long as the mycelium had grown at least to a small extent. Gramoxone, however, inhibited mycelial growth although successful fungal growth did not always result in mycorrhizal formation. Gramoxone may inhibit mycorrhizal formation by affecting the root system.

At four months after sowing applications of neither pesticides affected growth (Table 1 and 2).

Experiments in Mitscherlich pots with perlite under semi-aseptic conditions

Shoot dry weight was significantly reduced by treatment with the fungicide Dithane M-45 immediatly prior to inoculation (application method ii). The root system including mycorrhizae developed poorly in all pesticide treatments with the exception of application before sowing (application method i). The mycorrhiza formation was strongly suppressed with repeated pesticide applications (application methods iv and v); (Table 3).

The differences in the growth observed in inoculated and non-inoculated seedlings appeared to be statistically significant in the shoot to root dry weight ratio only. The herbicide application before sowing reduced seedling growth and the higher doses also reduced mycorrhiza formation compared to the other application methods.

Experiments on outdoor beds with peat substrate

Preemergent application by the fungicide (application method i) resulted in a remarkable decrease in shoot growth and in a smaller decrease in root growth. The application methods i and ii decreased the number of mycorrhizae, compared to the control. Inoculation stimulated root development and

Table 1. Effect of three different doses (μg per flask) of the fungicide Dithane M-45 applied in three different periods on mycorrhiza formation and seedling growth of inoculated and non-inoculated *Pinus sylvestris* seedlings. The seedlings were cultivated in Erlenmeyer flasks with perlite under aseptic conditions

Investigated feature*	Before sowing			Simultaneously with inoculation			3 months after inoculation			Control
	50	500	5000	50	500	5000	50	500	5000	0
Inoculated seedlings										
RCD (mm)	1,3	1,2	1,0b	1,2	1,2	1,3	1,2	1,3	1,2	1,2b
AH (mm)	95	110i	80bi	87c	104	87d	110	97	97	122bcdr
ADW (g)	0,096h	0,077i	0,046bhi	0,071	0,084	0,067	0,065	0,069	0,059	0,097b
UDW (g)	0,050	0,042	0,019bx	0,051	0,039	0,037	0,045	0,067	0,052x	0,055b
ADW/UDW	2,24	2,13j	2,75kx	1,52	2,37	1,82	1,73	1,063fj	1,37x	2,04f
LRL (mm)	919h	755	341bhx	968	825	817	1166	821	949x	991b
NSR (piece)	484	409	105bx	595	456	357	633	499	653x	506b
NSR/10 mm (piece)	5,24h	5,33i	2,47bhix	5,40m	5,41n	3,61mn	4,87	5,40	6,57x	4,97b
NM (piece)	20	109i	0bix	27c	61	1d	34p	53	123xp	120bcd
NM/10 mm (piece)	0,53	1,19i	0bix	0,32c	0,081	0,01d	0,38ep	1,03	1,54xp	1,42bcde
% M	7,3	23,9i	0bix	7,2	15,3	0d	5,2	19,9	19,8x	24,5bd
Non-inoculated seedlings										
RCD (mm)				1,1	1,3	1,2				1,0
AH (mm)				83	94	95				90r
ADW (g)				0,072	0,076	0,072				0,050
UDW (g)				0,052	0,033	0,036				0,031
ADW/UDW				1,39	2,09	2,69				1,87
LRL (mm)				1496	706	1255				726
NSR (piece)				755	477	585				402
NSR/10 mm (piece)				4,81	6,18a	4,58				4,64a
NM (piece)				0	0	0				0
NM/10 mm (piece)				0	0	0				0
% M				0	0	0				0

* RCD = root collar diameter, AH = above-ground height, ADW = above-ground dry weight, UDW = under-ground dry weight, ADW/UDW = above- to under-ground dry weight ratio, LRL = total lateral root length, NSR = total number of short roots, NSR/10 mm = number of short roots per 10 mm of total lateral root length, NM = total number of mycorrhizae, NM/10 mm = number of mycorrhizae per 10 mm of total lateral root length, % M = percentage of mycorrhizae from total number of short roots.

Treatments with difference significant in 1% level are marked with identical indexes for every feature.

Table 2. Effect of three different doses (µg per flask) of the herbicide Gramoxone applied in three different periods on mycorrhiza formation and seedling growth of inoculated and non-inoculated *Pinus sylvestris* seedlings. The seedlings were cultivated in Erlenmeyer flasks with perlite under aseptic conditions

Investigated feature*	Before sowing		Simultaneously with inoculation			3 months after inoculation		Control
	25	50	25	50	250	25	250	0
Inoculated seedlings								
RCD (mm)	1,3	1,1n	1,2p	1,2	1,0r	1,1	1,1	1,2s
AH (mm)	89a	81b	83c	85	78e	106	93g	122abcegs
ADW (g)	0,775	0,054	0,063c	0,042d	0,057e	0,067f	0,062g	0,097cdefgs
UDW (g)	0,032a	0,027b	0,027	0,028d	0,037em	0,037	0,044m	0,055abdes
ADW/UDW	2,43	2,34	1,48k	1,83	2,42km	1,89	1,48m	2,04
LRL (mm)	745	534	303ct	546	287em	909t	996m	991ce
NSR (piece)	320	276	255	379	219m	458	562m	506
NSR/10 mm (piece)	3,77	4,39	3,63	5,00	4,33	5,18	4,89	4,97
NM (piece)	8a	5b	39c	0d	1e	39f	26g	120abcdefg
NM/10 mm (piece)	0,11ai	0,11b	0,41c	0d	0,05er	0,64fi	0,28g	1,42abcdefg
% M	3,5a	3,3h	12,0cj	0dj	0,08e	10,2f	4,7g	24,5abcdefg
Non-inoculated seedlings								
RCD (mm)			1,0np		0,8hnr			1,0hs
AH (mm)			69		69			90s
ADW (g)			0,041		0,041			0,050s
UDW (g)			0,018		0,016			0,031s
ADW/UDW			2,47		3,06h			1,87h
LRL (mm)			436		241			726
NSR (piece)			215		108h			402h
NSR/10 mm (piece)			4,67		3,14h			4,64h
NM (piece)			0		0			0
NM/10 mm (piece)			0		0			0
% M			0		0			0

* See Table 1.

Table 3. Effect of five different application methods of the fungicide Dithane M-45 in two concentrations (μg/ml) of water suspension on mycorrhiza formation and seedling growth of inoculated and non-inoculated *Pinus sylvestris* seedlings. The seedlings were cultivated in Mitscherlich pots with perlite under semi-aseptie conditions

Investigated feature*	Application i Before sowing		Application ii 4 days before inoculation		Application iii 2 months after inoculation		Application iv i + ii + iii + 5 weeks after inoculation		Application v ii + iii + iv		Control
	5	50	5	50	5	50	5	50	5	50	0
Inoculated seedlings											
RCD (mm)	1,5	1,5	1,3	1,4	1,3	1,4	1,3	1,5	1,4	1,5	1,5
AH (mm)	105ef	95	87e	93	88f	95	98	103	100	107	100
ADW (g)	0,324efgh	0,248	0,158e	0,171k	0,209f	0,246k	0,212g	0,252	0,212h	0,220	0,250
UDW (g)	0,162g	0,132	0,109	0,120	0,110	0,132	0,053bgn	0,142n	0,114	0,113	0,174b
ADW/UDW	2,08g	2,04	1,66i	1,48r	1,97j	1,99	2,94bginp	1,87n	1,98p	1,98	1,50b
LRL (mm)	3488g	3002	2274	2650	2559	2553	834bg	2698	2202	2490	3996b
RSS (g)	0,52egh	0,49	0,31ae	0,40	0,46j	0,46m	0,23bgj	0,36	0,37h	0,28dm	0,53abd
NSR (piece)	948g	726	631	725	664	639	212gn	789n	505c	758	1037c
NSR/10 mm (piece)	2,79	2,59	2,99	3,17	2,59	2,73	2,55	3,03	2,31	3,08	2,94
NM (piece)	60gh	37	19	23	19	17	3g	0	0h	10	55s
NM/10 mm (piece)	0,27h	0,11	0,08	0,10	0,08	0,08	0,06	0	0h	0,03	0,21
% M	8,5h	4,3	2,7	2,9	3,0	2,8	2,5	0	0h	0,9	6,2
Non-inoculated seedlings											
RCD (mm)				1,3		1,3					1,5
AH (mm)				94		99					96
ADW (g)				0,216		0,260					0,251
UDW (g)				0,086		0,142					0,135
ADW/UDW				2,96r		1,91					2,19
LRL (mm)				1910		2954					3206
RSS (g)				0,30		0,44					0,41
NSR (piece)				580		656					826
NSR/10 mm (piece)				3,35		2,14					2,83
NM (piece)				0		4					2s
NM/10 mm (piece)				0		0,01					0,01
% M				0		0,4					0,4

Table 4. Effect of two doses (µg per pot) of the herbicide Gramoxone applied in three different periods on mycorrhiza formation and seedling growth of inoculated and non-inoculated seedlings. The seedlings were cultivated in Mitscherlich pots with perlite under semi-aseptic conditions

Investigated feature*	Before sowing		Simultaneously with inoculation		2 month after inoculation		Control
	175	1750	175	1750	175	1750	0
Inoculated seedlings							
RCD (mm)	1,2	1,0b	1,5	1,4	1,2	1,4	1,5b
AH (mm)	73c	50bde	100c	100df	82	97e	100b
ADW (g)	0,174	0,071	0,267	0,216	0,216	0,214e	0,250b
UDW (g)	0,087	0,039b	0,119	0,122	0,087	0,121	0,174b
ADW/UDW	5,17a	8,22bde	2,49	1,99	2,58	2,02e	1,50ab
LRL (mm)	1762	1046b	2651	3070	2296	3016	3996b
RSS (g)	0,26	0,08be	0,46	0,34	0,34	0,43e	0,53b
NSR (piece)	444	241b	613	905	597	741	1037b
NSR/10 mm (piece)	1,51a	0,46bde	2,36	2,82d	2,71	2,55e	2,94ab
NM (piece)	15a	3b	16	11	13	45	55abg
NM/10 mm (piece)	0,08	0,01	0,08	0,04	0,05	0,20	0,21
% M	3,6	0,2	3,1	1,4	2,5	7,6	6,2
Non-inoculated seedlings							
RCD (mm)				1,3		1,3	1,5
AH (mm)				84f		91	96
ADW (g)				0,208		0,184	0,251
UDW (g)				0,113		0,104	0,135
ADW/UDW				1,87		1,85	2,19
LRL (mm)				2490		2589	3205
RSS (g)				0,43		0,29	0,41
NSR (piece)				626		670	826
NSR/10 mm (piece)				2,51		2,71	2,83
NM (piece)				4		6	2g
NM/10 mm (piece)				0,02		0,05	0,01
% M				0,8		1,6	0,4

* See Table 1. RSS = root system surface.

Table 5. Effect of three different application methods of the fungicide Dithane M-45 on mycorrhiza formation and seedling growth of inoculated and non-inoculated seedlings. The seedlings were cultivated on outdoor beds with peat substrate

Investigated feature*	Application method			
	Sterilization and three sprayings	Three sprayings	Once spraying	Control
Inoculated seedlings				
RCD (mm)	1,2d	1,2	1,2	1,1
AH (mm)	72	70	72g	72
ADW (g)	0,200	0,200	0,196	0,202
UDW (g)	0,095	0,057	0,080	0,059
ADW/UDW	3,20	3,60	3,93	3,25
LRL (mm)	1790	948	1160	1288
RSS (g)	0,42hi	0,29h	0,21gi	0,27
NSR (piece)	528	344	443	411
NSR/10 mm (piece)	3,20	3,61	3,93	3,25
NM (piece)	395	323	347	301
NM/10 mm (piece)	2,45	2,70f	3,00	2,40
% M	74,9	74,9f	78,9	73,7
Non-inoculated seedlings				
RCD (mm)	1,1d	1,2	1,4d	1,3
AH (mm)	61acd	82c	85dg	76a
ADW (g)	0,204d	0,214e	0,358bde	0,216b
UDW (g)	0,078	0,084	0,089	0,069
ADW/UDW	2,99	2,99	4,09	3,55
LRL (mm)	1170	1638	1334	1320
RSS (g)	0,27d	0,37	0,41dg	0,27
NSR (piece)	351	487	455	451
NSR/10 mm (piece)	3,10	2,78	3,53	3,50
NM (piece)	206	230	307	319
NM/10 mm (piece)	1,82	1,43f	2,43	2,53
% M	59,6	48,8f	67,4	71,3

* See Table 1. RSS = root system surface.

mycorrhizal formation but only in the treatment with pre-emergent application when compared with other pesticide treatments (Table 5).

Discussion

These experiments have led to an improved understanding of the action of the fungicide Dithane M-45 and the herbicide Gramoxone on the growth of seedlings of *Pinus sylvestris* and on the development of their mycorrhizae. However, the results from the experiments carried out under aseptic conditions must be interpreted with care especially as the uneven concentration of pesticide in each flasks as well as the non-symetrical position of both the seedlings and fungal inoculum probably led to uneven mycorrhiza formation.

As *Pinus sylvestris* seedlings are so susceptible to Gramoxone the screening of the effects of a pesticide on seedlings grown in outdoor beds were assessed only for Dithane M-45 only. The results obtained demonstrate that single or repeated sprayings only slightly affect the microflora soil components and thus seedling growth and mycorrhiza formation.

Partial soil sterilization decreased the number of mycorrhizae produced on non-inoculated seedlings and affected growth. Inoculated seedlings showed an increase in both mycorrhizal numbers and root growth. Dithane M-45 limits the viability of mycorrhizal fungi in soil and thus the number of mycorrhizae produced by the natural fungal flora. It may therefore give better conditions for effective inoculation stimulating mycorrhiza formation and root system development.

References

1 Carley H L and Watson R D 1966 A new gravimetric method for estimating rootsurface areas. Soil Sci. 102, 289–291.

2 Cudlín P, Mejstřík V and Šašek V 1980 The effect of the fungicide Dithane M-45 and herbicide Gramoxone on the growth of mycorrhizal fungi *in vitro*. Čes. Mykol. 34, 191–198.

3 Hacskaylo E 1953 Pure culture syntheses of pine mycorrhizae in terralite. Mycologia 45, 971–975.

4 Hacskaylo E and Palmer J G 1957 Effects of several biocides on growth and incidence of mycorrhizae in field plots. Plant Dis. Rep. 41, 354–358.

5 Iyer J G and Trautman W L 1967 Effect of DMTT on the growth of Monterey pine at different contents of moisture. Weed 15, 282–284.

6 Laiho O and Mikola P 1964 Studies on the effect of some eradicants on mycorrhizal development in forest nurseries. Acta Forest. Fenn. 106, 1–72.

7 Marx D H and Bryan W C 1975 Growth and ectomycorrhizal development of loblolly pine seedlings in fumigated soil infested with the fungal symbiont *Pisolithus tinctorius*. Forest Sci. 21, 245–254.

8 Šašek V and Musílek V 1967 Cultivation and antibiotic activity of mycorrhizal basidiomycetes. Folia Microbiol. 12, 515–523.

9 Wilde S A and Persidsky D J 1966 Effect of biocides on the development of ectotrophic mycorrhizae in Monterey pine seedlings. Soil Sci. Soc. Am. Proc. 20, 107–110.

Plant and Soil 71, 363–366 (1983). A-41

Mycorrhiza in some plant desert species in Algeria

V. K. MEJSTŘÍK and P. CUDLÍN
Environmental Ecology Institute of the Czechoslovak Academy of Sciences, CS-25243 Průhonice, Czechoslovakia

Key words Algeria Casuarina Cedrus Desert plants Ectomycorrhiza Eriobotria Eucalyptus Globularia Helianthemum Herniaria Noaea Peganum *P. halepensis* Pistacia *P. maritima* Quercus Rosmarinus Stipa Tamarix V-A mycorrhiza

Summary Twenty plant species from four environmentally harsh sites in Algeria which were associated with seven ectomycorrhizal and nine endomycorrhizal species. Ectomycorrhizal associations were found on *Quercus illex*, *Pinus halepensis*, *Pinus maritima*, *Juniperus phoenica*, *Cedrus atlantica*, *Peganum harmala* and *Globularia alypum*.

Endomycorrhizal associations were with *Stipa tenacissima*, *Rosmarinus Tournefortii*, *Helianthemum Lippii*, *Tamarix* sp., *Eryobotria japonica*, *Casuarina equisetifolia*, *Herniaria Fontanesii*, *Noaea mucronata* and *Pistacia lentiscus*.

Both ecto- and endomycorrhizal associations occurred on *Eucalyptus globosa* and *Eucalyptus cupressus*. Ectomycorrhizas varied from well-defined to loose with or without a Hartig net.

Introduction

There is not much information about mycorrhizas in desert plant species[3,4,5,6,8]. It is well known that ecto- and endomycorrhizal associations are more widespread both taxonomically and geographically[2]. The stresses associated with desert ecosystems are related to both nutrient deficiences and water stress. The growth advantages attributed to mycorrhizas are believed due to an increase in nutrional status of the plant due to an increased phosphorus uptake and water transport[2]. Mycorrhizas are likely to be important to desert species. This paper reports on the status of a range of arid land species from Algeria, where relatively few plants are known to be mycorrhizal

Materials and methods

Four sites were surveyed and evaluated for the presence of mycorrhizas. 1) near El Harrach. A site 2 km from the sea coast, at altitude 28 m, with an average annual temperature 17.5°C and average precipitations of 647 mm. This site is used for intensive horticulture.

2) Chrea south of Alger, in Tell Atlas. Samples were from natural high mountain forest, 30 km from the sea coast at an altitude of 1,550 m with an average temperature of 11.1°C and annual precipitations of 1,400 mm.

3) About 2 km south of the city Aïn Oussera between Tell and Sahara Atlas. The natural arid desert steppe communities have considerably deteriorated. The characteristic plant species are *Stipa tenacissima* and *Artemisia herba alba*: soil pH is 7.8, the altitude 750 m, average temperature is 17°C and annual precipitations 250 mm.

4) The Forest Reservation at Senalba Chergui near Djelfa in the Atlas Saharien 260 km south-east of Alger. Here the altitude is 1,160 m, average temperature 13.1°C and annual precipitation is 308 mm. The major plant community is *Pinetum halepensis* with semi-arid vegetation dominated by *Quercus illex, Stipa tenacissima, Artemisia herba alba.* The soil pH is 4.0 at 0–3 cm and 7.0–8.0 at 3–10 cm.

Root samples from plants occurring on each site were collected during June, fixed in chinosol, cut into 1 cm segments, cleared with KOH, and stained with trypan blue in lactophenol. A Hartig net and hyphal mantle was scored as a positive ectomycorrhizal infection. The presence of hyphae, arbuscules, vesicules or any combination of the three indicated the presence of a va mycorrhiza.

Results and discussion

The occurrence of mycorrhizae of the two types at the four sites is shown in Table 1.

The ectomycorrhizae of *P. halepensis* and *P. maritima* were similar to those on *P. sylvestris*[7]. Long brown roots may provide at least three orders of branches, any of which may become mycorrhizal while unbranched or simple mycorrhizae are found in the nursery on small seedlings. The colour of mycorrhizal roots ranges from white to yellow, orange, brown or black colour. Especially at site 2), young mycorrhizae are intimately associated with dense mycelia and there are clear strands and rhizomorphs at sites 3) and 4).

Ectomycorrhizae are found mostly in the humus-rich upper soil layer from 0–15 cm. The highest mycorrhizal frequency was at site 2) and the lowest at site 3). At site 3) and 4) a high percentage of mycorrhizal roots were dead.

The fungal mantle around the roots was 2–35 μm thick, on pine roots from site 4) there were two layers of fungal mantles in the outer pseudoparenchymatous and the inner synenchymatous. Mycorrhizae of *Cedrus atlantica* were similar to those of Pinus.

Quercus illex had the best ectomycorrhizal associations at site 2) and the poorest at 3). The mycorrhizal roots were white, ochre or brown colour and the mantle 2–15 μm thick.

In *Juniperus phoenica* the ectomycorrhizae possessed a thin (15 μm) synenchimatous sheath and brown taninoforous cells in the cells of epidermis to which the Hartig net was restricted.

In the roots of *Globularia alypum* mycorrhizae were formed in the whole root system. Mycorrhizal roots were simple or dichotomously bifurcated, slightly thickened, brown colour with rhizomorphs on the surface. The mantle was 20–40 μm thick, strawy or brown colour with the Hartig net well developed and sometimes reaching the endodermis.

Ectomycorrhizae in *Peganum harmala* occurred only in the upper portion of the root system in simple mycorrhizal roots, dark brown in colour. The mantle was 0–10 μm thick, colourles, built of loosely knitted hyphae, with the Hartig net usually restricted to one layer of cortical cells.

The ectomycorrhizae of *Eucalyptus cupressus* and *E. globosa* were similar to those of *E. regnans*[1]. Ectendotrophic mycorrhizae were common especially in the

Table 1. The presence of different types of mycorrhizae on plants in different sites

Plant species	Site	Mycorrhiza	
		EM	VAM
Pinus halepensis	Senalba Forest	+	–
Juniperus phoenica		+	–
Quercus illex		+	–
Globularia alypum		+	–
Rosmarinus Tournefortii		–	+
Stipa tenacissima		–	+
Centaurea incana		–	–
Artemisia herba alba		–	–
Pinus halepensis	Aïn Oussera	+	–
Eucalyptus cupressus		+	+
Eucalyptus globosa		+	+
Quercus illex		+	–
Pistacia lentiscus		–	+
Tamarix sp.		–	+
Herniaria Fontanesii		–	+
Noaea mucronata		–	+
Helianthemum Lippii		–	+
Rosmarinus Tournefortii		–	+
Stipa tenacissima		–	+
Artemisia herba alba		–	–
Scabiosa stellata		–	–
Pinus halepensis	Chrea	+	–
Cedrus atlantica		+	–
Quercus illex		+	–
Peganum harmala		+	–
Pinus halepensis	El Harrach	+	–
Pinus maritima		+	–
Eucalyptus cupressus		+	–
Eriobotria japonica		–	+
Casuarina equisetifolia		–	+

EM – ectomycorrhiza, VAM – vesiculo-abuscular mycorrhiza.

stand at Aïn Oussera (site 3). The fungal mantle was relatively thin and the Hartig net was associated with intracellular hyphae in the cells of the epidermis and outer cortex.

Endotrophic mycorrhizae were found in thirteen species (Table 1). Cells of the epidermis and cortex were infected with different amounts of hyphae. Mycorrhizae were usually abundant except in *Pistacia lentiscus* and *Herniaria*

Fontanesii. Vesicular development consisted of both inter and intracellular vesicles.

No endomycorrhizae were found in *Centaurea incana*, *Scabiosa stellata* and *Artemisia herba alba*.

The relationship between the mycorrhizal fungi and the plant species investigated is complex and depends on the environmental response of both partners and seems necessary for optimum growth and possibly even survival.

References

1 Ashton D H 1976 Studies on the mycorrhizas of *Eucalyptus regnans* F. Muell. Aust. J. Bot. 24, 723–741.

2 Harley J L 1969 The biology of mycorrhiza. Plant Sci. Monogr. 2nd Ed. Leonard Hill London, 334 p.

3 Kadic B 1978 Influence de la mycorrhization sur la croissance de jeunes plants forestiers (*P. halepensis*). Proc. Symp. Root Physiol. and Symb. Nancy 11–15 September 1978, 444–448.

4 Khan A G 1974 The occurrence of mycorrhizas in halophytes, hydrophytes and xerophytes, and of Endogene spores in adjacent soils. J. Gen. Microbiol. 81, 7–14.

5 Lebtähi-Akermi M 1981 Contribution a l'etude des mycorrhizas du pin d'Alep. *P. halepensis*. Thes. Univ. Sci. et Technol. d'Alger, 65 p.

6 Miller R M 1979 Some occurrence of vesicular-arbuscular mycorrhiza in natural and disturbed ecosystems of the Red Desert. Can. J. Bot. 57, 619–623.

7 Mejstřík V 1969 Anatomicko-morfologické sledování mykorrhizy k účelům rekultivačním. Sbor. Věd. Úst. VŠZ v Praze, 12, 221–249.

8 Souidi C 1976 Analyse de la mycorrhization de juenes d'Alep dans une station de l'Algérios. Rapport interve BV/76-01, 13 p.

Plant and Soil 71, 367–370 (1983). A-42

Environmental factors determining the distribution of *Suillus luteus* fructifications in *Pinus radiata* grazing-forest plantations*

H. PEREDO, M. OLIVA** and A. HUBER
Instituto de Silvicultura and Instituto de Geociencias respectively, Universidad Austral de Chile, Casilla 567, Valdivia, Chile

Key words Environmental factors Mycorrhizal fruitbody productivity Pine *Pinus radiata* Silvopastoral system *Suillus luteus*

Summary The environmental factors influencing the production of *Suillus luteus* in a *Pinus radiata* – Sheep sylvopastoral system, were measured in La Unión, Chile during the first two weeks of April 1981.

The most important factors were humidity and soil temperature at 1 cm depth.

Introduction

The important increase in agroforestry systems in the underdeveloped countries in the past decade, has allowed a better use of land[4]. One of these, the sylvopastoral system *Pinus radiata* – Sheep has been stablished in the X Región of Chile as means of improving the use of cattle-forestry lands of low quality which belong to small owners[5]. This allows the farmers to obtain benefits almost from the second year after the establishment of the system by means of wool and meat from sheep. In addition and because *P. radiata* is an ectomycorrhizal tree species, there is the possibility of benefitting from the system by collecting the *Suillus luteus* fructification harvest, which can be consumed by the land owners as being sold commercially as well.

The environmental and soil factors which affect fruitbody production in forest plantations are relatively well known[2,6,8]. However, information about these factors in a sylvopastoral system as their effect on fruitbody productivity as well is still insufficient.

Phenological studies carried out in chilean *P. radiata* plantations show a least fruitbody productivity period between Sept-Oct and a maximum one between March-May each year[3]. This study was carried out at the beginning of the last mentioned period.

* Dirección de Investigación Universidad Austral de Chile. Project IC–82–8: 'Uso Silvopastoral en terrenos forestales ganaderos' (Sylvopastoral use of cattle-forestry lands) financed by ODEPA/UACH Agreement.

** Senior student. Facultad Ciencias Forestales, Universidad Austral de Chile, Casilla 853, Valdivia, Chile.

Material and methods

A five year old plantation of *P. radiata*, located in La Unión, Chile (40° 15' lat S, 73°15' long W), was selected. It had a 2,5 × 7 m spacing with its rows orientated East-West.

The variables temperature and soil humidity at 1 cm depth, relative humidity and temperature at 5 cm above-ground and luminosity were measured with an aspiration electrical psychrometer, electrical geothermometer and luxmeter Metrux K respectively. The measurement were taken in each of the rectangle formed in a 20 × 16 m grid, having its length perpendicular to the rows and its width divided in 4 m lines with each one of them marked every metre.

From the data taken every 4 h during 48 h period (27–28 March), a daily average was calculated and plotted as isolines on the grid. The number and location of the mycorrhizal fructifications collected inside the grid during the first two weeks of April were superimposed onto the distributions of temperature, humidity and illumination.

Results and discussion

The zone showing the highest mycorrhizal fruitbody concentration was located in the central third of the interrowspace (Fig. 1), where the optimal soil temperature and humidity for mycorrhizal fructification were recorded (Figs. 2, 3).

The fruitbody distribution as shown in Fig. 3 was confined to the narrow soil temperature range 13–15°C with the fall of in production more noticiable at the lower temperatures than the higher. The optimal temperature were however lower than those found by Theodorou and Bowen[7].

The soil humidity results, (Fig. 3) show that fruitbody production occured from 25–38% with optimum productivity in the 32–33% range.

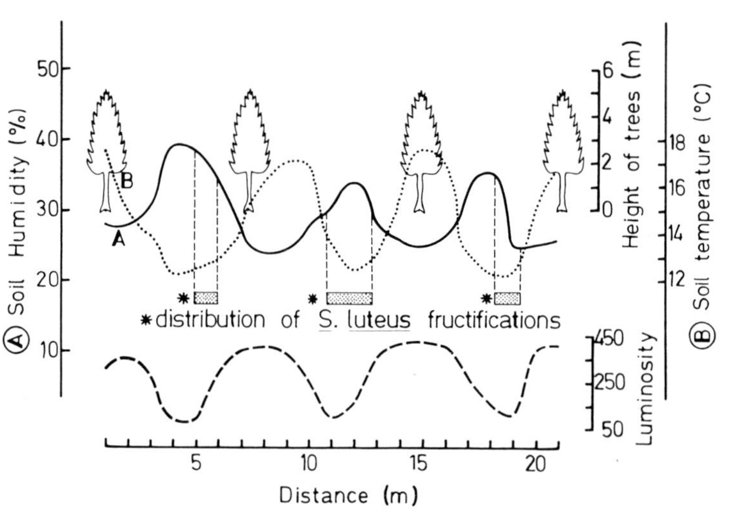

Fig. 1. North-South schematic representation of the temperature and humidity of the soil, the luminosity at 5 cm above the ground level and the distribution of *S. luteus* fructifications.

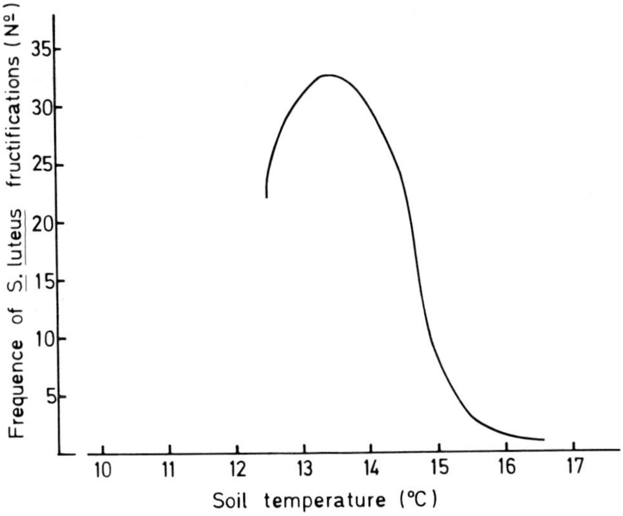

Fig. 2. The influence of temperature on the appearance of *S. luteus* fructifications.

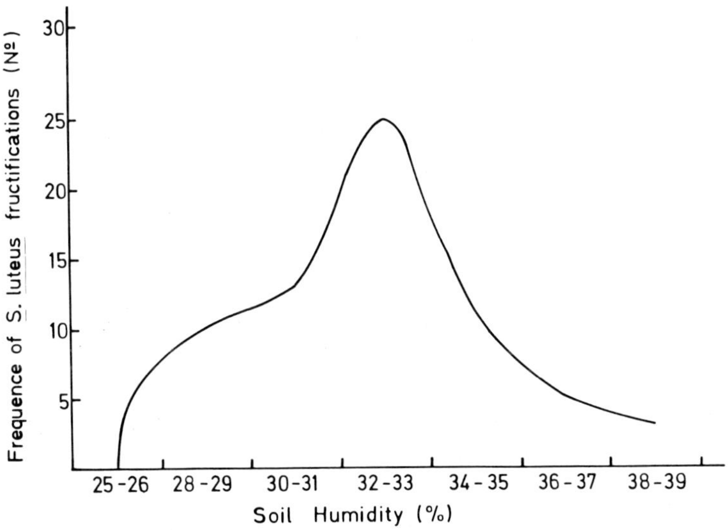

Fig. 3. The influence of soil humidity on the appearance of *S. luteus* fructifications.

No correlation was found between air temperature, relative air humidity above ground level and fruitbody distribution.

Of the factors measured, soil temperature and humidity appeared to be the most important for fruitbody production. However solar radiation may also be important in fruitbody productivity by not only affecting carbohydrate accumulation in roots[1] but also by regulating soil temperatures.

Acknowledgement The authors thank to the Project Leader Dr. Rubén Pañaloza for the support during the trials and further presentation of the results. The former author gratefully acknowledges to The Royal Society, CONICYT (Chile), and Universidad Austral de Chile, sponsors of a Study Visit in England where this work was presented at the Tree Root Systems and their Mycorrhizas Symposium-Edinburgh and Canterbury.

References

1 Azcon C 1980 Investigación y Ciencia 47, 8–16.
2 Bakshi B K 1966 World Symposium on man-made forests and their industrial importance, Canberra 14–25 April 1967. pp 1037–1042.
3 Garrido N 1981 Tesis Licenciado en Biología, Universidad de Concepción. 113 p.
4 King K F S 1979 Agroforestry. Department of Agricultural Research Royal Tropical Institute Amsterdam. Bulletin 303, 1–10.
5 Peñaloza R, Hervé M and Sobarzo L 1982 Universidad Austral de Chile Facultad de Ciencias Forestales. Serie Técnica. Informe de Convenio N° 48. 57 p.
6 Slankis V 1974 Annu. Rev. Phytopathol. 12, 437–457.
7 Theodorou C and Bowen G D 1971 Aust. J. Bot. 19, 13–20.
8 Westerlund I and Ingeloeg T 1981 For. Ecol. Manage. 3, 269–294.

Plant and Soil 71, 371–380 (1983). A-43

Nutrient inflows into apple roots

K. K. S. BHAT
East Malling Research Station, East Malling, Maidstone, Kent ME19 6BJ, UK

Key words Apple Calcium Copper Iron Irradiance Magnesium Manganese Nitrogen Nutrient uptake rate Phosphorus Potassium Scion effects Season Source of nitrogen Temperature Tissue nutrient level Zinc

Summary The rates of uptake of nutrients from solution by apple roots were measured (a) in a root laboratory, using intact roots of mature trees growing under field conditions and (b) in controlled environment using young trees. Maximum nitrate inflows into Discovery/M.9 roots under field conditions were only slightly lower than those into roots of the same genotype in controlled environment, but up to 80 times lower than those into roots of Worcester Pearmain seedlings. At any given external P concentration, P inflows into roots of field-grown trees were about 2.5-times lower than those into the roots of young trees in controlled environment.

Nitrate inflows were constant above a solution concentration of 20 mmol m^{-3} in both field-grown and small trees. In both cases, phosphate inflows increased linearly with solution concentration up to 10 mmol m^{-3}.

Among the various plant and environmental factors influencing nutrient uptake characteristics of apple roots were: the scion genotype, tissue nutrient levels, root origin, the form in which N is supplied, level of irradiance of the shoot, root temperature and the season of the year. The effects of these factors are illustrated with examples.

Introduction

Although the measurement of the rates of uptake of nutrients by excised or intact roots of herbaceous species under laboratory or glasshouse conditions has been the subject of innumerable investigations, very limited information is available in the literature on (a) the kinetics of nutrient uptake by roots of perennial trees[1] and (b) nutrient inflows into plant roots in general under field conditions[6]. This paper considers the rates of nutrient uptake by intact apple (*Malus domestica* Borkhausen) roots, measured (a) in controlled environment room experiments using young trees, and (b) in root laboratory experiments on mature trees growing under field conditions. Wherever possible, the results of the two sets of experiments are compared. The effects of some of the environmental and plant factors are also discussed.

Determination of inflows

Complete details of most of the experiments referred to in this paper are given elsewhere[2,3,4]. Briefly, the rate of uptake from solution of a given nutrient in the presence of adequate levels of all other nutrients was measured using either entire root systems (controlled environment room experiments with young trees) or a part of the root system (root laboratory experiments with mature trees). In all experiments mentioned here except that concerning the effect of the form of nitrogen supply, uptake rate per unit root was calculated from the rate of removal of the nutrient from solution and the corresponding root dimensions, measured by destructive or non-destructive methods. The relationship used throughout to describe the phosphate depletion curves was[2]:

$$C = a + \frac{b + ct}{1 + dt + et^2} \tag{1}$$

where C is the P concentration in mmol m^{-3}, t the time in seconds and a, b, c, d and e are constants and the rate of solution depletion

$$-\frac{dC}{dt} = \frac{(c - bd) - 2bet - cet^2}{(1 + dt + et^2)^2} \tag{2}$$

In the case of nitrate, at solution concentrations above 20 mmol m^{-3}, the depletions were adequately described by straight lines and the uptake rates were calculated from the coefficients of the linear regression of concentration on time.

In the experiment on the effect of the form of N supply on nutrient uptake rates, sequentially harvested plants were analysed for the various elements mentioned and at each harvest, the root lengths were also measured. Mean inflows (mol m^{-1} s^{-1}) between harvests were calculated as:

$$\frac{U_2 - U_1}{t_2 - t_1} \cdot \frac{\ln L_2 - \ln L_1}{L_2 - L_1} \tag{3}$$

where U_1 and U_2 are the total contents of the element (mol per plant) and L_1 and L_2 are the root lengths (m) after times t_1 and t_2 (s) respectively.

Inflow *versus* solution concentration

The relationships between the rates of uptake of nitrate and orthophosphate per unit root and the concentrations of these ions in solution at the root surface were studied both in controlled environment using young trees and in a root laboratory using intact roots of mature trees growing under field conditions[2,3,4].

Phosphate

In young M.9 rootstocks and Worcester Pearmain seedlings, the apparent minimum P concentration in solution below which P uptake is negligible was of the order of 0.25–0.50 mmol m^{-3}. Both in young trees and in mature trees under field conditions, P inflows increased approximately linearly with P concentration in external solution at least up to 10 mmol m^{-3}. However, at any given external

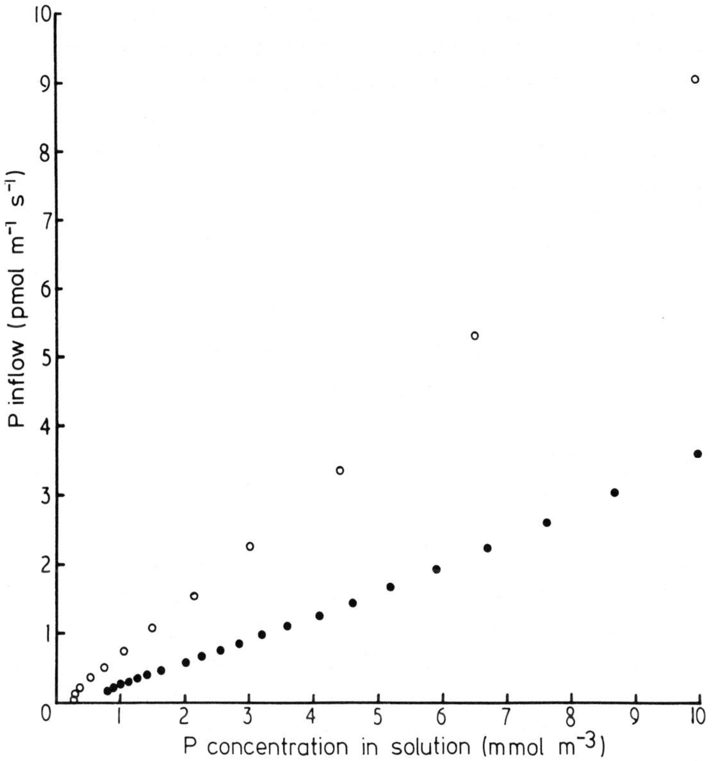

Fig. 1. Relationship between P uptake rate per unit root length and P concentration in solution: ○, young M.9 rootstocks in controlled environment; ●, mature Discovery/M.9 tree under field conditions.

concentration, the rate of P uptake per unit root of a 3.5-year-old, field-grown Discovery/M.9 tree measured in the root laboratory in September was about 2.5 times lower than that in a young M.9 rootstock carrying its own shoot in controlled environment, although the age of the roots used in these measurements were similar (Fig. 1).

Comparing the inflow-solution concentration relationship for apple with that for other species reported in the literature, it would appear that the P concentration in solution above which inflow does not increase is higher for apple than for most other species.

Nitrate

In roots of 4.5-year-old field-grown trees, the minimum nitrate concentration in solution below which net nitrate uptake is reduced to zero was of the order of 1 mmol m^{-3} [4]. Inflows did not respond to increasing solution concentration above 20 mmol m^{-3} so that the concentration dependence of uptake rate was restricted to the narrow concentration range of 1–20 mmol m^{-3}. Therefore, in most field situations, where it would be difficult to maintain the nitrate concentration in soil solution at the root surface within this range (unless the soil nitrification rate were very high or slow-release N fertilizers were applied), the concentration dependence of nitrate inflow would be only of limited significance.

Maximum nitrate inflows into roots of 5-year-old Discovery/M.9 roots under field conditions were only slightly lower than those into the roots of young trees of M.9 rootstocks, but up to 80 times lower than those into the roots of Worcester Pearmain seedlings grown in controlled environment (Bhat, unpublished).

Effects of scion genotype

Experiments conducted in the root laboratory using roots growing at two different depths in the soil profile, of two 4.5-year-old trees (Discovery/M.9 and Worcester Pearmain/M.9) revealed the following effects of scion genotype on the nitrate uptake characteristics of the M.9 rootstock[4]:

1) Presence or absence of a 'lag' phase in uptake: In nitrate depletion experiments designed to measure inflows, marked differences were observed in the nature of the depletion curves obtained for the roots of the two trees. With Worcester Pearmain as the scion variety, an initial phase of slow uptake lasting up to 5 h, similar to the 'lag' or 'induction' phase frequently observed in various herbaceous species (especially when they are N-deficient) in glasshouse or controlled environment experiments, was noticed at both depths in the soil profile. This lag phase in nitrate uptake by M.9 roots was absent in all cases when Discovery was the scion. The previous histories of the two trees including the pre-treatment conditions to which the roots used in measuring the uptake rates were subjected being identical, it appears that the presence or absence of a lag phase in nitrate uptake by apple roots is, at least in part, a shoot-dependent phenomenon.

2) Inflows after the lag phase: Nitrate inflows after the lag phase into Worcester Pearmain/M.9 roots were greater than those into Discovery/M.9 roots irrespective of the position of the roots in the soil profile.

3) Distribution of uptake efficiency in the soil profile: With Discovery as the scion variety, roots of M.9 situated at 46 cm depth in the soil profile were much more efficient in absorbing nitrate than those at 110 cm. In contrast, with Worcester Pearmain as the scion, inflows into roots at 104 cm depth were higher than those at 46 cm.

Plant genotype and root origin

Nitrate and phosphate inflows into roots of young, ungrafted apple trees of different genotypes were measured in controlled environment using micropropagules of rootstocks and scions and seedlings, grown for several weeks in solution culture. Under identical conditions of culture, at any given external P concentration, P inflows into the (adventitious) roots of M.9 rootstock were higher than those into the (seminal) roots of Worcester Pearmain seedlings[2].

Maximum nitrate inflows into the (adventitious) roots of M.9 and M.26 rootstocks carrying their own shoots and Bramley scions on their own roots varied within the range 30–60 pmol m^{-1} s^{-1}, whereas those into the (seminal) roots of Worcester Pearmain seedlings were of the order of 350 pmol m^{-1} s^{-1} (Bhat, unpublished).

Plant nutrient status

In young trees of M.9 rootstocks and Worcester Pearmain seedlings, phosphate uptake rate per unit root increased with decreasing internal P levels[2]. The effect of tissue P levels on inflows was more marked in M.9 than in Worcester. Inflows appeared to be more closely related to the concentration of P in the shoot than in the root.

In controlled environment room experiments on the effect of tissue N levels on nitrate inflows, N-starved plants of M.9 and M.26 rootstocks showed a 'lag' phase in nitrate uptake, lasting up to 5 hours (Bhat, unpublished). This initial phase of slow uptake was not noticed in the corresponding plants well supplied for N. In the N-starved M.26, inflows after the lag phase were similar to those in the healthy plants, whereas in N-deficient M.9, inflows after the lag phase were greater than those in corresponding healthy plants.

Form of N supply

T. Fyfield working in our laboratory, studied the effect of supplying N (at a concentration of 2 mmol m^{-3}) entirely in the form of NH_4^+ or in the form of NO_3^- on the maximum mean inflows of 9 different elements into the roots of Worcester Pearmain seedlings at different stages of growth in solution culture in controlled environment. The results (Table 1) show that the inflows of P, K, Ca, Mg, Mn and Fe were lower when N was supplied as NH_4^+ whereas those for N and Zn were similar in the two treatments at most stages of growth. Cu inflows were somewhat higher in the NH_4^+ treatment than in NO_3^-.

In addition to the direct effects of N source on the inflows of various nutrients, two other indirect effects likely to modify the nutrient uptake characteristics of apple roots growing in soil were observed in these experiments. The first of these was the well-known effect on the pH of the medium, which increased when NO_3^- was the only source of N and decreased when N was supplied solely as NH_4^+. Secondly, within a week of applying the treatments, marked differences were noticed in root morphology. In the nitrate treatment the roots were thin and long, whereas with the ammonium source they were thick and short with characteristic swollen regions behind the tip (Fig. 2a). The NO_3^--fed roots were almost devoid of root hairs whereas those fed with NH_4^+ were covered with dense, long root hairs over their entire lengths (Fig. 2b). The diameter of the root hair cylinder in the NH_4^+ treatment was about three times that of the root cylinder in the NO_3^- treatment. Therefore, NH_4^+-fed roots growing in soil may be much more effective sinks for slow-moving solutes[5].

Table 1. Mean inflows of 9 elements into roots of Worcester Pearmain seedlings (pmol $m^{-1} s^{-1}$) over different stages of growth in the two N treatments

Element	Source of N	Growth interval (days)			
		0–6	6–13	13–20	20–27
N	NO_3	258.35	348.16	308.45	279.40
	NH_4	232.98	348.75	367.80	144.31
P	NO_3	13.92	17.86	14.72	13.55
	NH_4	12.10	17.20	14.89	7.92
K	NO_3	85.54	93.49	68.55	57.83
	NH_4	52.30	66.39	64.51	36.86
Ca	NO_3	26.32	31.91	21.27	27.83
	NH_4	13.80	19.63	18.86	8.51
Mg	NO_3	9.74	8.97	9.51	5.87
	NH_4	4.83	8.35	9.04	4.75
Mn	NO_3	0.86	0.71	1.01	0.46
	NH_4	0.22	0.55	0.51	0.27
Zn	NO_3	0.22	0.09	0.12	0.08
	NH_4	0.25	0.13	0.13	0.05
Cu	NO_3	0.028	0.029	0.030	0.025
	NH_4	0.044	0.043	0.028	0.020
Fe	NO_3	1.07	0.50	0.52	0.76
	NH_4	0.64	0.53	0.52	0.24

After two weeks, large differences were also observed in the branching patterns of roots in the two N treatments (Fig. 2c). In NO_3^- the laterals were well spread out, whereas in NH_4^+ they grew in clusters. Therefore, in the case of a plant growing in soil, the form in which N is presented to the roots may influence the volume of soil exploited for various elements indirectly through its its effect on root morphology.

Effect of the level of irradiance of the shoot

The effect of the level of irradiance to which the aerial parts of the plant are subjected on the relationship between phosphate uptake rate per unit root and P concentration in the external solution was studied using M.9 rootstocks grown in

Fig. 2. Effect of the form of N supply (left, NH_4^+; right, NO_3^-) on root morphology: **a**. after one week, showing difference in root diameters; **b**. after one week, showing differences in root thickness and root hair development; **c**. after two weeks, showing difference in branching patterns.

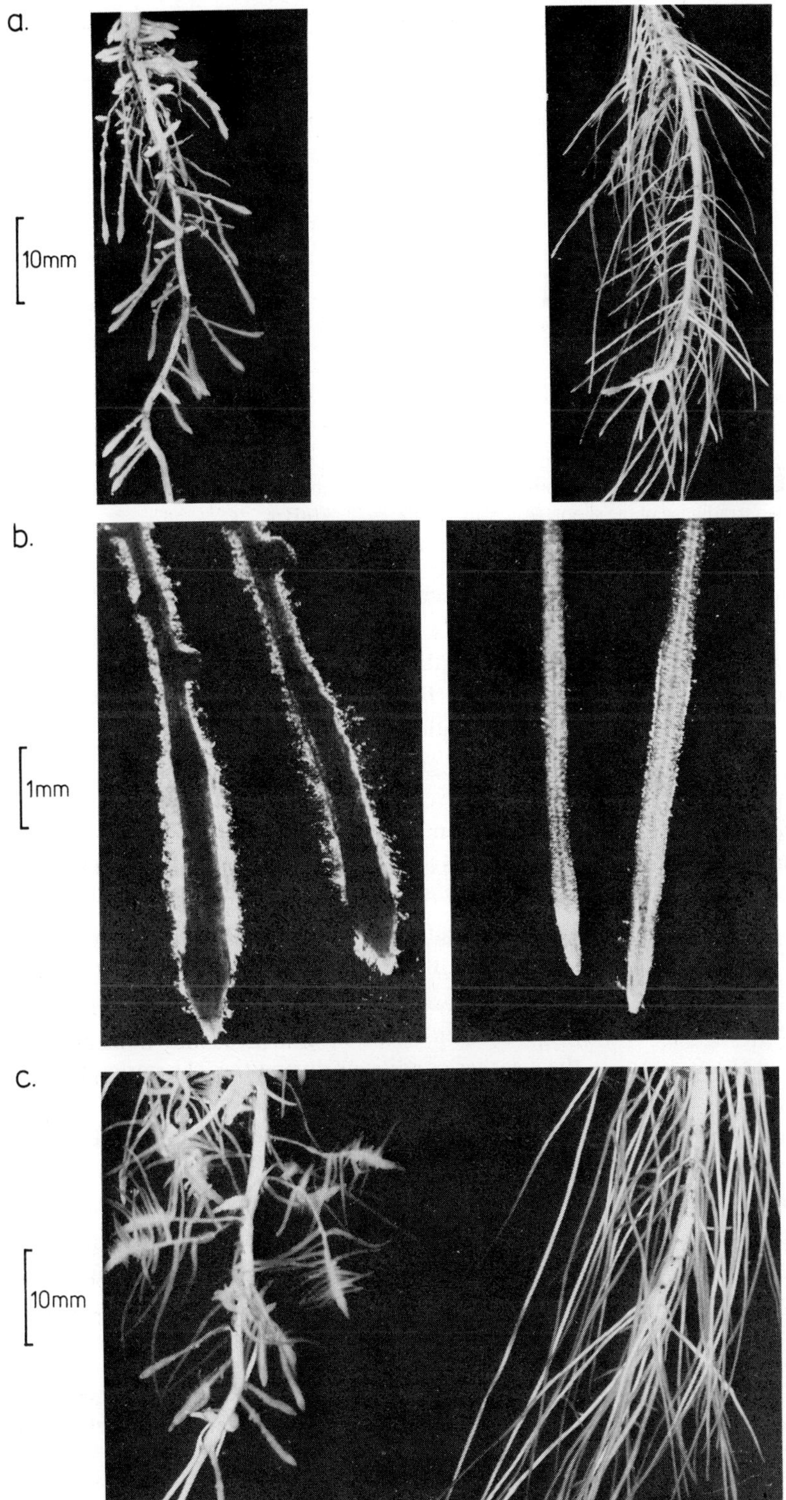
a.
10mm
b.
1mm
c.
10mm

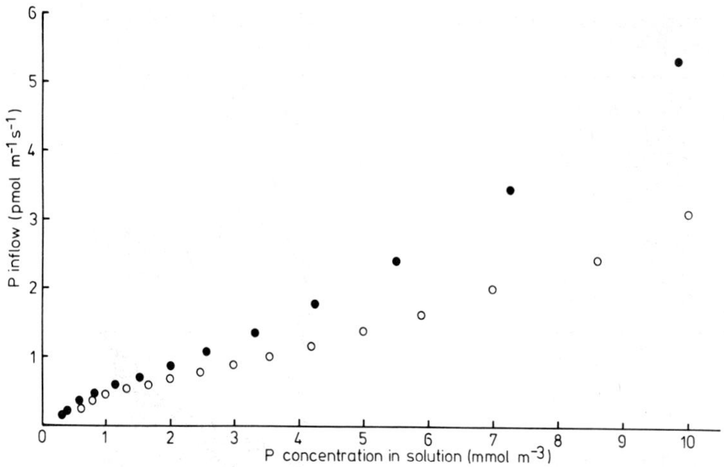

Fig. 3. Relationship between P uptake rate per unit root length and P concentration in solution at two levels of irradiance of the shoot: ●, 290 μE m^{-2} s^{-1}; ○, 80 μE m^{-2} s^{-1}.

solution culture in controlled environment. After growing the plants for 8 weeks at a constant flux density of 290 μE m^{-2} s^{-1}, P depletion curves were determined at this level of irradiance as well as at 80 μE m^{-2} s^{-1}, after allowing an adaptation period of 12 hours in the latter case. The inflow-concentration relationships calculated from the slopes of the depletion curves and the corresponding root lengths are presented in Fig. 3. At any given external concentration, P inflows were greater at the higher irradiance level. The difference between the two light levels increased with increasing solution concentration.

In controlled environment room experiments with 3-month-old trees of Greensleeves/M.9 (micropropagules of the scion and the rootstock grown and grafted in solution culture), similar effects of the level of irradiance of the shoot were also observed on nitrate inflows (Bhat, unpublished).

However, in root laboratory experiments on roots of a 4.5-year-old Discovery/M.9 tree, diurnal variations in nitrate inflows were found to be small[4], despite large variations in the weather conditions during the 24 h period within which these measurements were made. Inflows measured between 2–5 p.m. on a dry, warm and sunny day were very nearly the same as those between the following midnight-4 a.m. period and between 7–11 a.m. the following morning when the weather was cold, cloudy, windy and rainy throughout. It would appear therefore, that in mature trees growing under field conditions, the variations in root activity in response to variations in the aerial environment are much less pronounced (or, perhaps, much slower to appear) than in the case of the young plants used in the controlled environment room experiments.

Table 2. Effect of root temperature on nitrate inflow into roots of a 4.5-year-old Discovery/M.9 tree under field conditions[4]

Temperature °C	Nitrate inflow pmol m^{-1} s^{-1}
5	3.68
10	8.43
15	9.30
20	12.13
25	12.66

Root temperature

The effects of local variations in root temperature on nitrate inflows were studied in the root laboratory using the roots of a Discovery/M.9 tree growing at 46 cm depth, over the range of 5°–25°C in steps of 5°C[4]. The results are given in Table 2. The largest increase in nitrate inflow for a 5°C rise in root temperature occurred over the 5°–10°C range, the value at 10°C being 2.3 times that at 5°C. With further successive 5°C increments in temperature up to 25°C, nitrate inflow increased by 10, 30 and 5% respectively. Arrhenius plots of the same data showed a major change in the activation energy for nitrate uptake by apple roots occurring at a transition temperature located between 5°C and 10°C[4].

Seasonal variations

For following seasonal variations in nitrate inflows under field conditions, determinations of inflows were made in the root laboratory at monthly intervals, over a period of twelve months (Bhat, unpublished). The roots used were those on the Discovery/M.9 tree growing at depths of 46 cm and 110 cm, mentioned previously. Very significant nitrate uptake occurred throughout winter, especially at the 46 cm depth, including days on which the air temperature was as low as −15.5°C and the temperature of the surface soil remained below 0°C throughout the week immediately preceding the measurement. The largest rise in nitrate uptake rate per unit root was observed between mid-April and mid-May at both depths and the largest decrease towards the end of September. Maximum inflows occurred during May, June and August, with a decrease in July which was accompanied by a decrease in new root growth during this month. Relative inflows into roots at the two depths in the soil profile varied significantly at different times of the year.

References

1 Atkinson D 1980 The distribution and effectiveness of the roots of tree crops. Hortic. Rev. 2, 424–490.

2 Bhat K K S 1981 Nutrient inflows into apple roots. I. ^{32}P-orthophosphate uptake from solution by M.9 rootstocks and Worcester Pearmain seedlings. Plant Cell Environ. 4, 297–302.

3 Bhat K K S 1982 Determination of the relationship between nutrient uptake rate and solution concentration at the root surface under field conditions: ^{32}P-orthophosphate uptake by apple roots. J. Exp. Bot. 33, 190–194.

4 Bhat K K S 1982 Nutrient inflows into apple roots. II. Nitrate uptake rates measured on intact roots of mature trees under field conditions. Plant Cell Environ. 5, 461–469.

5 Bhat K K S 1982 Some factors affecting phosphate uptake efficiency of apple roots. J. Sci. Food Agric. 33, 734–735.

6 Mengel D B and Barber S A 1974 Rate of nutrient uptake per unit of corn root under field conditions. Agron. J. 66, 399–402.

Plant and Soil 71, 381–385 (1983). A-44

Phosphorus absorption (^{32}P) by apple trees under drip irrigation as influenced by the physical properties of the soil

R. HABIB
Station d'Agronomie, INRA, Domaine Saint Paul, F-84140 Montfavet, France

Key words Apple tree Drip irrigation Phosphorus Soil physical properties

Summary P absorption by apple tree roots (Golden delicious/M2) is studied using carrier-free ^{32}P. A qualitative model of the influence of some physical properties of the soil is proposed combining individual tree responses to ^{32}P injection.

Introduction

Some authors have worked with ^{32}P on fruit trees[1,2,7] but not within the frame-work of the influence of physical properties of soil on P uptake. Studies in this area by the author[5] led to qualitative modelling of the effects of interactions of some soil parameters on P absorption by apple trees. It was found that no P absorption occured with high soil density or at water saturation, but absorption did occur over a large range of soil moisture and aeration *between the trunk and the dripper*. This paper describes the results of a second year's work and a new model which uses both years' results and assesses year to year variability.

Materials and methods

The two experiments were carried out on adult Golden delicious trees/M2 4 × 4.5 m spaced (30 trees in 1979 and 27 in 1980) in the beginning of July 1979 and mid-August 1980 respectively. Trickle irrigation was applied with one dripper per tree 1 m from the trunk along the tree row. Carrier free ^{32}P was supplied on the same day to all trees included in each experiment in one injection hole per tree at 0.35 m depth and at a randomly chosen position.

Leaves were sampled, and activity measured, two hours after ^{32}P injection into the soil. Soil bulk density[8] ρ_d, gravimetric moisture content (%) and distance from trunk and dripper D_T amd d_D (m) of ^{32}P injection location were recorded. Laboratory measurements were made to estimate water potential ψ (bars) and structural porosity[9] n_s (%). Volumetric water content θ and soil water saturation θ_s were calculated from H and ρ_d.

Two sets of date were combined: (i) The soil measurements and ^{32}P location coordinates. (ii) The qualitative tests on the trees' responses to ^{32}P injection into the soil, *i.e.*, testing leaf samples for the presence or absence of radioactivity, A *Principal component analysis* (P.C.A.) was done with the first set of data from the matrix of correlations. The purpose of this analysis is to summarize N p-variate observations in some biologically or physically meaningful manner by N two (or more) -variate observations. The *Principal components*, (P.C.s), are linear combination of the ρ variates. They are orthogonal to each other. Each P.C. accounts for a computed percentage of the total variance of the standardized N p-variate data. The sum of two of these percentages is the effectiveness of the corresponding plan to summarize observations. Trees' responses are 'projected' on the *first principal*

Table 1. Correlation between variables and principal axes 1 and 2*

Variables	M	θ	ψ	θ_s	d_T	d_D	ρ_d	n_s
Axis 1	0,78	0.96	–0,69	0.84	–0.54	–0.70	0.35	–0.61
Axis 2	–0.46	0.08	0.43	0.52	0.46	0.53	0.88	–0.74

* For the explanation of symbols, see the text

plan (F.P.P.). The location of P absorption on the F.P.P. was studied combining individual trees' responses.

Results and discussion

Table 1 presents the correlations between the variables for the data of 1979 and 1980 experiments and the first two P.C.s. It can be seen that P.C . 1 is mainly explained by 'hydric' variables (H, θ, ψ, θ_s) and P.C. 2 by 'physical status' variables (ρ_d, n_s). Distances from trunk and dripper are less correlated with both components. P.C. 1 accounts for 50% and P.C. 2 for 30 % of the variance of data respectively. The F.P.P. thus accounts for 80% of the initial variation. In most cases, therefore, two points close to each other on F.P.P. will have close values of 'hydric' and 'physical status' variables. It becomes possible to divide the F.P.P. into homogeneous areas, six are shown in Fig. 1 derived as follow:

–A: defined by $\rho_d > 1.6$

–B: defined by θ_s close to 100 and $\rho_d < 1.6$

–C: defined by $70 < \theta_s < 90$ and $\rho_d < 1.6$

–D: defined by $50 < \theta_s < 70$ and $\rho_d < 1.6$

–E: defined by $\theta_s < 50$ and $\rho_d < 1.6$

–F: defined by $\theta_s < 70$, $\rho_d < 1.6$ and $d_T > 1$ m for trees of the first year's experiment. F is included in D and E.

Absorption zone

P absorption has been observed ('P' on the F.P.P. in Fig. 1) 23 times out of 39 for $\theta_s < 90\%$ (areas C, D and E); ten out of 13 trees which did not absorb the labelled P ('O' in Fig. 1) in low θ_s areas (D and E) had ^{32}P injection location at $d_T > 1$ m. In area C (higher θ_s) P absorption was observed at both positions where $d_T > 1$ m. When d_T is > 1 m such high θ_s values are less likely because d_D tends to be large and water content low. Nevertheless it seems that being far from the trunk decreased the probability of P absorption at low θ_s. P absorption still occurred at $d_T > 1$ m, and was noted five times in 1980 but only once in 1979 (see 'P' inside the shaded circle in Fig. 1). Rainfall was higher in spring 1980 than in spring 1979, and was the main climatic difference between the two years (Table 2). Consequently, the mean soil moisture content outside the dripper zone ($d_D >$ 1 m) was 18.4% in 1980 and 14.6% in 1979. Furthermore in the areas studied d_T

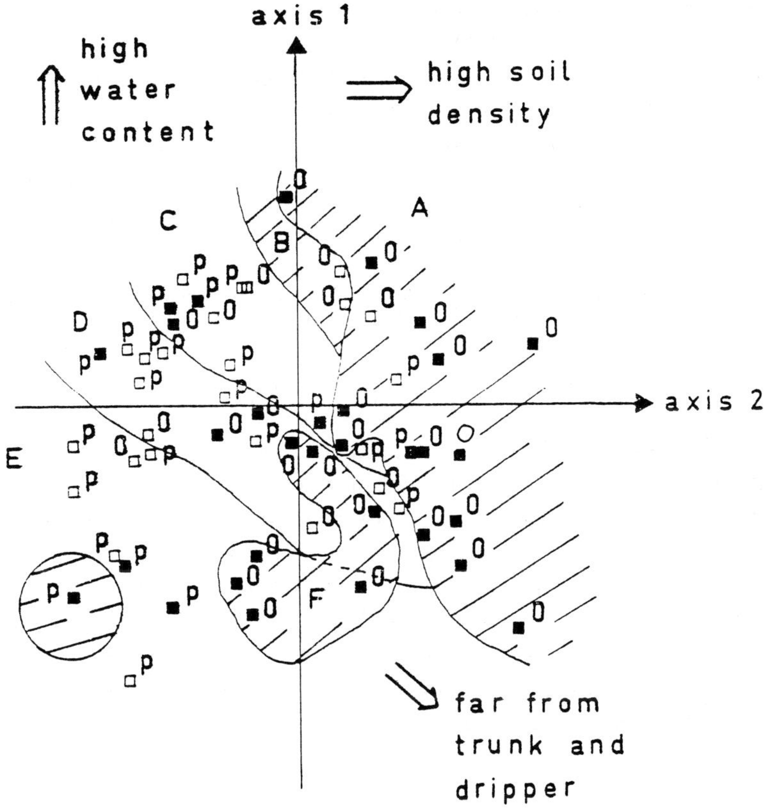

Fig. 1. First principal plan.
P: ^{32}P absorption by roots
O: no absorption
■: 1979 experiment
□: 1980 experiment

Table 2. Rainfall (mm) during springs 1979 and 1980

Months	March	April	May	June	July
1979	69	28	24	15	45
1980	63	61	66	29	15

is likely to be negatively correlated with root density at 0,35 m depth[3]. In the dry year root density was probably the limiting factor for P absorption far from the trunk (area F).

Non absorption zone

The first non absorption zone (shaded in Fig. 1) is area F plus shaded circle. P absorption was uncommon at high soil density (area A), and was observed only twice out of 16 measurements. Structural porosity was low at high density ($n_s < 7\%$ at $\rho_d > 1.6$). Possibly there were few pores for rapid air movement. Moreover at low n_s root growth is less likely to occur than in well structured soil (high n_s). High ρ_d values were much more frequent in the dry year. Only two trees '0' are on area B (θ_s close to 100%), hence it is difficult to draw conclusions. But it has been demonstrated that water saturation under the dripper system prevents P absorption[4], probably because of oxygen deficiency[6].

Qualitative modelling

From the combination of individual tree responses a qualitative model is proposed. It states that:

* No P absorption occurs at high soil density or at water saturation,
* P absorption can occur over a large range of soil humidity and aeration between trunk and dripper ($d_T < 1$ m with dripper at 1 m from the trunk),
* P absorption > 1 m from the trunk depends on climatic conditions during the root growth season. It can occur in a rainy year and is improbable in a dry one but can some times be observed because of local variations in root density.

Conclusion

Previous conclusions[5] on the influences of soil water content, aeration and bulk density are confirmed by the combined results of both years' experiments. Climatic conditions during spring have to be taken into account. For the absence of absorption the model gives good results for about 90% of the observations, and for absorption, about 70%. But absorption depending on actual root density is much more difficult to predict.

References

1 Atkinson D 1974 Some observations on the distribution of root activity in apple trees. Plant and Soil 40, 333–342.

2 Broeshart H and Nethsing D A 1972 Studies on the pattern of root activity of tree crops using isotopes techniques. I.A.E.A., Isotopes and Radiation in Soil-Plant Relationships including Forestry pp 465–479.

3 Cabibel B 1978 Influence du mode d'irrigation sur la structure et le système racinaire de jeunes pommiers C.R. Acad. Agric. France 64, 616–624.

4 Guennelon R and Habib R 1979 Conditions d'absorption du phosphore en irrigation localisée. C.R. Acad. Agric. France 65, 1453–1459.

5 Habib R 1981 Etude, pour des végétaux pérennes (Pommiers) soumis à une irrigation localisée, des conditions d'absorption racinaire du phosphore. Thèse 3ème Cycle – U.S.T.L.-E.N.S.A.M.

6 Lawton K 1961 Plant nutrition and the utilization of fertilizer phosphorus. *In* Phosphorus and its Compounds. Interscience Publishers, IN,C., New-York v. 2, pp 1461–1545.

7 Singh M M, Pushparajah E, Soong N and Talibudeen O 1972 Radiotracer studies on P uptake by Haevea brasiliensis from malayan soils for determining active root distribution. I.A.E.A., Isotopes and Radiations in Soil-Plant Relationships including Forestry pp 465–479.

8 Soane B D, Campbell D J and Herkes S M 1971 Hand-held gamma-ray transmission equipment for the measurement of bulk density of field soils. J. Agric. Eng. Res. 16, 146–156.

9 Stengel P 1983 Etude expérimentale du retrait de mélanges argile-limon. *Accepted for publication in* Agronomie.

Plant and Soil 71, 387–394 (1983). A-45

Responses of loblolly pine (mesophyte) and swamp tupelo (hydrophyte) seedlings to soil flooding and phosphorus*

D. D. HOOK,
Clemson University, Southeastern Forest Experiment Station, Charleston, SC

D. S. DEBELL,
USFS, Pacific Northwest Forest Experiment Station, Olympia, WA

W. H. McKEE JR.,
USFS, Southeastern Forest Experiment Station, Charleston, SC

and J. L. ASKEW
Clemson University, Georgetown, SC

Key words Anaerobiosis Ion imbalance Leaky roots Loblolly pine *Nyssa silvatica* var. *biflora* *Pinus taeda* L. Swamp tupelo

Summary In a pot study, flooding reduced growth of loblolly pine (LP) seedlings and caused an imbalance of nutrients, particularly Fe, in the tops. Swamp tupelo (ST) suffered none of these problems. Also under anaerobiosis, LP roots leaked 85 percent of total ethanol production into solution whereas ST leaked only 55 percent. Application of phosphorus increased LP growth under flooded and drained conditions but ST only under drained conditions. The differential response of the two species to flooding suggests that poor energetics and the leaky nature of LP roots under anaerobiosis restricts its growth in such habitats. Application of phosphorus (P) appears to amend these conditions in LP by reducing moisture stress, reducing nutrient imbalances, and improving energetics in older seedlings. Our results confirm that ST, the hydrophyte, avoids the problems of soil flooding via adaptations in anaerobic metabolism and oxidation of the rhizosphere.

Introduction

Application of phosphorus fertilizer to poorly drained soils has been reported to increase growth of pine trees[13] and 2 year-old loblolly pine (*Pinus taeda* L.) seedlings[11] and increase the capacity of excised roots of loblolly pine to respire anaerobically[2]. Also, there are phosphorus-zinc interactions in a number of plant species but the nature of the interactions varies among species and site conditions[6]. In lowland rice, applied P reduced Zn uptake but it and applied Zn, individually and in combination, increased dry weight production in waterlogged soils[6].

For further insight into why loblolly pine responds to phosphorus under poor soil aeration, seedlings of loblolly pine and swamp tupelo (*Nyssa sylvatica* var.

* Supported in part by a Cooperative Agreement with the USDA, Forest Service, Southeastern Forest Experiment Station, Asheville, NC.

biflora (Walt.) Sarg.), were grown in flooded and drained pots with and without applied P in a growth chamber. Growth and physiological responses were compared.

Methods

Loblolly pine* and swamp tupelo seedlings from coastal South Carolina sources were grown separately in 4 l stone crocks filled with a 1 : 1 volume of sand and finely ground peat. The pots, 12 for LP and 8 for ST, were arranged in a growth chamber in a complete randomized design as a 2 × 2 × 2 factorial experiment. They were grown under well drained conditions for 48 days and one-half were flooded (0–4 cm above the soil surface) until day 118.

Hoagland's nutrient solution was modified by substituting NH_4NO_3 in place of $NH_4H_2PO_4$ for the non-phosphorus treatment by doubling the standard amount of NH_4NO_3 in the final solution. In the phosphorus treatment the amount of $NH_4H_2PO_4$ was doubled so the amount of nitrogen in each treatment was 14 g. A total of 740 ml of one strength with and without applied P was added to each pot on four separate occasions at 3-to-4 week intervals; this amounted to 200 mg P per treated pot. The photoperiod was 16 h ($20 \pm 1 \times 10^3$ lux at 18″ from lights) and temperatures were 30°C ± 2.5 (light) and 20°C ± 1.5 (dark).

Height growth was measured periodically and at harvest. Soil redox potential was sampled 4 times (6 cm depth) only in the flooded pots during the final 30 days of the experiment. The stem of one of the four seedlings from each pot was clipped at harvest, at the beginning of the light cycle, and water potential was measured with a pressure bomb.

Soil solutions were sampled from the bottom of drained and top of flooded pots and stabilized in dilute HNO_3. Soil samples were taken from the pots as seedlings were washed out and seedling parameters were measured. CO_2 production was measured on excised root tips for 4 h in N_2. After incubation, root tips and reaction solutions were analyzed separately for ethanol[7] and malate[5].

Plant nitrogen was determined with a modified micro Kjeldahl procedure[12]. Phosphorus was determined by the molybdovanadate reduction[9] and K, Ca, Cu, Mn, Fe, Mg, and Zn by atomic absorption.

Soil exchangeable bases were determined by atomic absorption on extracts made with 1 *N* NH_4OAC[9] and available phosphorus by extracting 2.5 g of soil with 20 ml of Bray P2 solution[1]. Soil pH was measured with a glass electrode. Total soil nitrogen was determined by micro Kjeldahl digestion[12]. Soil solution samples were analyzed for Ca, Mg, K, Na, Mn, Cu, Fe, and Zn by atomic absorption. Phosphorus was determined by molybdate reduction[9] and nitrogen was determined by Kjeldahl digestion[12].

Results and discussion

Soil

Soil redox potential in the flooded pots averaged +24.7 mv and ranged from +84.7 mv to −13.7 mv which indicated that reducing conditions in the flooded soil were sufficient to reduce O_2, NO_3^-, Mn^{+4} and Fe^{+3} (ref.[4]). Such conditions probably reduced the extra nitrate added in the nutrient solution. Soil pH for all treatments at harvest averaged 4.65 and ranged from 4.10 to 5.52.

* LP seeds were donated by the Westvaco Nursery at Ravenel, South Carolina.

Table 1. Growth, and water potential of loblolly pine and swamp tupelo seedlings and nutrient concentration of the foliage at harvest

Variable	Species	Treatment*			
		Drained		Flooded	
		P_0	P_{200}	P_0	P_{200}
(a) Height (cm)	LP	16.0a	18.2a	7.9b	9.4b
	ST	45.9c	53.0d	55.3d	46.7c
Root length (cm)	LP	37.5a	37.1a	11.4b	15.4b
	ST	30.0a	28.9a	15.0b	12.3b
Dry weight (g)					
Foliage	LP	3.02a	4.94b	1.22c	1.76c
	ST	1.90d	4.29b	1.91d	1.97dc
Stem	LP	0.59a	1.15b	0.31a	0.30a
	ST	3.13c	4.44d	4.45d	4.37d
Root	LP	1.30a	1.59b	0.22a	0.28a
	ST	3.80b	4.28c	2.14d	2.11d
(b) Water potential (MPa)	LP	−1.05a	−1.35a	−1.40a	−1.15a
	ST	−0.93b	−0.85b	−0.73b	−0.77b
(c) Major element (mg/g)**					
N	LP	15.6	14.5	14.2	8.7
	ST	15.4	17.5	9.9	11.4
P	LP	1.7	2.3	0.9	1.2
	ST	1.0	1.6	1.0	1.7
K	LP	12.8	12.4	5.9	5.9
	ST	10.3	9.7	12.4	12.9
Ca	LP	2.3	2.0	0.9	0.9
	ST	5.4	5.1	6.5	5.3
Mg	LP	1.3	1.5	0.6	0.7
	ST	3.6	3.6	3.3	2.8
Minor elements (mg/g $\times$ 10^{-2})					
Zn	LP	5.5	6.0	4.1	3.8
	ST	2.1	3.5	3.1	2.2
Mn	LP	25.6	23.6	9.9	8.9
	ST	121.5	50.7	37.4	46.3
Fe	LP	4.9	5.3	17.9	16.9
	ST	4.2	4.9	6.2	4.8

Table 1 (continued)

(d) *Ratio* – Top/root of concentrations (F/D)

Na	LP	.34	.23	.87	1.01	3.2
	ST	.25	.11	.24	.22	1.3
Mn	LP	2.52	1.12	3.05	4.00	1.9
	ST	13.94	10.07	13.03	13.95	1.1
Zn	LP	.57	.59	1.30	1.00	2.0
	ST	.49	.46	.40	.37	0.8
Fe	LP	.14	.18	.34	.28	1.9
	ST	.31	.30	.07	.07	0.2

* The data was analyzed as a completely randomized design with a 2 × 2 × 2 factorial experiment. Mean values are statistically significant at 0.05 level only if they do not have a letter in common.
** Because of the small amount of pine material in the flooded treatments all plant material for each treatment was composited by component (foliage, stem and root) for nutrient analyses, so it was not possible to statistically analyze the data.

The concentration of most nutrients in the soil solution was much lower than in the soil. There was essentially no nitrogen in solution in any treatment and iron concentration in solution from the drained soil was only about one-half that in the soil. However, in the flooded pots, the iron concentration in solution averaged 2882 ppm as compared to 290 ppm in solution in the drained pots.

Nitrogen, sodium, and iron concentrations in the soil were not influenced by treatments but as expected phosphorus concentration was highest in the soil to which P was applied. There was less P, K, and Mg in the soil in which ST grew but this was attributed to greater uptake by the larger ST seedlings. Calcium and potassium concentrations were lower in soils from flooded than drained pots. Applied P resulted in lower concentrations of Mn in flooded soils whereas it tended to be higher in flooded soils without P.

Growth

Average height growth of ST and LP was 0.43 cm d^{-1} and 0.15 cm d^{-1} respectively (Table 1a). This rapid initial height growth of ST is apparently an inherent adaptative survival characteristic of hydrophytic tree species[3], which enables them to extend their foliage as far above the soil surface of the swamp as possible during dry periods. Seeds of swamp tree species do not germinate under water[3].

Flooding reduced the root length of both species but it had a more pronounced effect on LP roots (Table 1a). Water potential (Table 1b) was lower in ST than LP seedlings and applied P tended to decrease it in LP and increase it in ST in the flooded pots but had the reverse effect in drained pots (3 way interaction was significant). The lower stress for LP correlates with increased growth in the flooded treatment.

Applied P increased dry weight of both species in drained pots but only of LP in the flooded pots (Table 1a). LP weight was primarily in foliage whereas ST had a high percentage of weight in the stems. Even though P increased total dry weight of ST in the drained pots it had no more stem weight than in flooded pots. LP total dry weight responded slightly more to applied P in the drained than the flooded pots.

Anaerobic metabolism of roots

Excised roots of swamp tupelo produced more CO_2 (3.53 μl/mg ODW roots/h) than LP (2.14 l/mg ODW roots/h) and higher concentrations of ethanol and malate than LP from both drained and flooded pots (Fig. 1). ST roots from flooded pots produced slightly more ethanol and much more malate than roots from drained pots, but LP roots from flooded pots produced no detectable

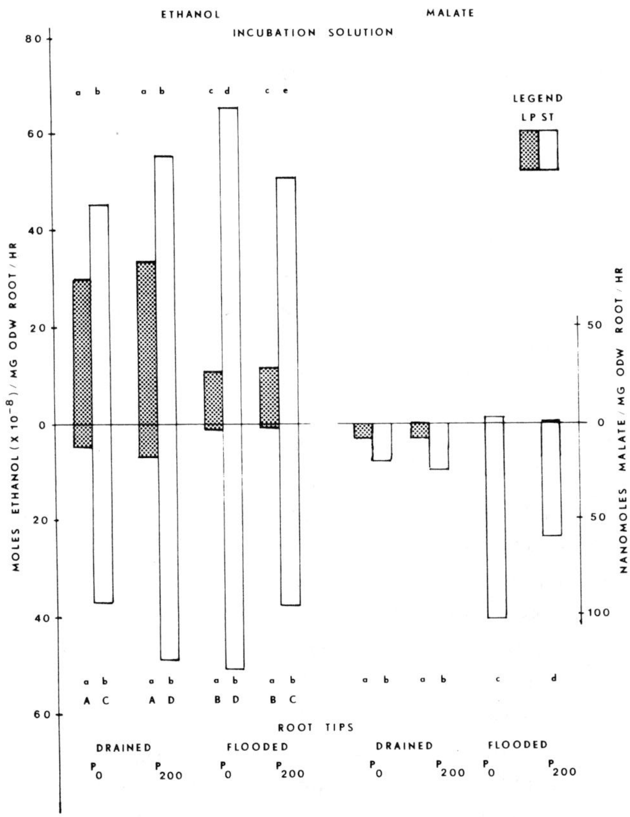

Fig. 1. Ethanol and malate concentrations in excised roots and incubation solution in relation to water and phosphorus treatments. Bars with similar letters were not significantly different at the 0.10 level; small letters refer to incubation solution and root tips; capital letters refer to total production.

malate. LP roots leaked about 85 percent of the total ethanol produced into solution whereas ST leaked only about 55 percent. Essentially no malate leaked into solution from either species. The relatively small increase of ethanol production in ST roots from flooded compared with drained pots and lack of malate production in LP roots from flooded treatment is attributed to age of the seedlings. It has been reported that between one month and one year of age, ethanol production in flooded ST roots decreased significantly and oxygen diffusion rate increased[10]. Based on the data presented here and in a companion paper[2], accumulation of end products of anaerobic metabolism by LP roots also appears to change with age. Roots incubated in the same manner as above from two year-old LP seedlings which were grown in flooded tanks (tanks were located outside hence plants and soil were exposed to natural environmental factors) and harvested in early spring (April 15), contained much higher concentrations of ethanol and malate than from well drained tanks under the same conditions. Also, under these semicontrolled conditions applied phosphorus was associated with increased ethanol and malate content in the incubated roots from flooded tanks. However, no influence of applied P on anaerobic metabolism was found in the experiment reported here with seedlings of approximately 4 months age, even though roots from applied P treatments had three times higher P concentrations than non-phosphorus treatments.

Nutrient status

The concentration of P, K, Ca, and Mg in LP foliage in the flooded pots was only about one-half that in the drained pots (Table 1c) whereas ST foliage contained much higher concentrations of Ca, Mg, and Mn than LP in all treatments. However, in comparing the relative concentration within seedlings (concentration top/root), it can be seen that LP appears to distribute relatively more Na, Mn, Zn, and Fe to the tops under soil flooding (cf F/D in Table 1d) than in drained soils. The LP also accumulated higher concentrations of Fe in the shoot under waterlogged than under drained conditions in contrast to ST (Fig. 2). ST oxidizes its rhizosphere in waterlogged soils[8] and presumably it oxidized the iron in the 'free space' of its roots, thereby preventing excessive uptake of this element, even though Fe concentration in the soil solution was 10 times higher in flooded than in drained pots. In the flooded treatments, the nutrients in ST probably entered the symplast and were transported to the top via active processes. However, in LP in the flooded pots several elements appear to be distributed more to the tops, *i.e.*, Na, Mn, Zn, and Fe (Table 1c), hence it appears that under anaerobiosis LP roots not only leak ethanol, a non ionized compound, to the external environment, but may also lose their discriminatory capacity to uptake and transfer to the transpirational stream. Roots of beans and peas (mesophytes) appear to lose their discriminatory capacity under anaerobiosis and lose both inorganic and organic ions[14,15].

The increased concentrations of several elements in the shoot of LP relative to

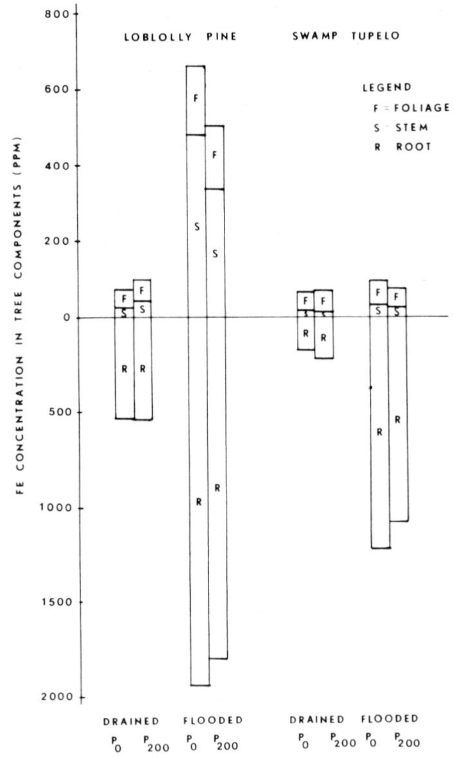

Fig. 2. Iron concentration in foliage (F), stems (S), and roots (R) of loblolly pine and swamp tupelo seedlings in relation to water and phosphorus treatments.

root in the waterlogged soils and not in ST indicates that the mesophyte under anaerobiosis stress may have incurred an ion imbalance between shoot and root and/or even toxic levels of some elements in the shoot.

Since application of P appeared to reduce the concentration of N, Zn, Mn, and Fe in the foliage of LP under waterlogging (Table 1b), its beneficial aspects may be related to lessening the apparent ion imbalance. Considering the improved nutrient relationships in the younger seedlings with applied P and the enhanced anaerobic metabolism of older seedlings[3], the response of LP to applied P may be related to improved energetics which may help maintain the normal structure of the root plasma membrane and thereby stabilize nutrient uptake.

In summary, application of P under drained conditions influenced both species positively but under flooding only LP benefited. It appears from this and companion research that P facilitates the adaption of loblolly to wet sites *via* improving root penetration, lowering water stress, stabilizing nutrient uptake, and improving energetics in older seedlings. Applied P had no noticeable beneficial effect on ST under flooding and may even have been detrimental. This is in contrast to the response reported for lowland rice mentioned previously[6].

References

1 Bray R H and Kurtz L T 1945 Determination of total, organic, and available forms of phosphorus in soils. Soil Sci. 59, 39–45.

2 DeBell D S, Hook D D, McKee W H Jr, and Askew J L 1983 Water table level and applied phosphorus affect morphology and physiology of loblolly pine root systems. *In preparation.*

3 DeBell D S and Naylor A W 1972 Some factors affecting germination of swamp tupelo seed. Ecology 53, 504–506.

4 Gambrell R P and Patrick W H Jr 1978 Chemical and microbiological properties of anaerobic soils and sediments. *In* Plant Life in Anaerobic Environments. Eds. D D Hook and R M M Crawford. Ann. Arbor. Science, Ann Arbor, MI. 564 p.

5 Gutman I and Wahlefeld A W 1974 L-(–)-Malate determination with malate dehydrogenase and NAD. *In* Methods in Enzymatic Analysis. Ed. H V Bergmeyer. Academic Press, New York. pp 1585–1589.

6 Haldar M and Mandal L N 1981 Effects of phosphorus and zinc on the growth and phosphorus, zinc, copper, iron, and manganese nutrition of rice. Plant and Soil 59, 415–425.

7 Hook D D and Brown C L 1973 Root adaptations and relative flood tolerance of five hardwood species. For. Sci. 19, 225–229.

8 Hook D D and Brown C L 1971 Inductive flood tolerance in swamp tupelo (*Nyssa sylvatica* var. *biflora* (Walt.) Sarg.). J. Exp. Bot. 22, 78–89.

9 Jackson M L 1958 Soil chemical analysis. Prentice-Hall, Inc. 498 p.

10 Keeley J E 1979 Population differentiation along a flood frequency gradient: Physiological adaptations in *Nyssa sylvatica.* Ecol. Monogr. 49, 89–108.

11 McKee W H Jr, Hook D D, DeBell D S, and Askew J L 1983 Effects of controlled water depth and phosphorus on growth and nutrient uptake of two-year-old loblolly pine. *In preparation.*

12 Nelson D W and Sommers L E 1973 Determination of total nitrogen in plant material. Agron. J. 65, 109–112.

13 Pritchett W L 1979 Properties and Management of Forest Soils. John Wiley and Sons, New York. 500 p.

14 Shadan M 1980 Fixation, translocation and root exudation of $^{14}CO_2$ by *Phaeseolus vulgaris* L. subjected to root anoxia. M.S. Thesis, Michigan State University.

15 Smuckers A J M and Erickson A E 1982 Unpublished data on anaerobiosis as it affects the exudation, growth and infection of pea roots.

Plant and Soil 71, 395–399 (1983). A-46

Root surface/leaf area ratios in arctic dwarf shrubs

J. KUMMEROW
Botany Department and Systems Ecology Research Group, San Diego State University, San Diego, CA 92182, USA

Key words *Betula nana* Fine roots Leaf area *Ledum palustre* Root surface area Tundra dwarf shrubs Tussock *Vaccinium uliginosum*

Summary Individual shoots of the shrubs *Ledum palustre* L., *Vaccinium uliginosum* L., and *Betula nana* L. were severed from their parent plants beneath the moss surface in an Alaskan tussock tundra. These shoots remained one year in their original position in peat moss cushions without further disturbance. After this period fine root dry weight, fine root surface area, leaf dry weight, and leaf area of these shoots were measured and compared with equivalent values from unsevered control shoots. Dry weight ratios of fine roots/leaves were similar in cut and control shoots, with the exception of *B. nana*. The fine root surface/leaf area ratios showed significant differences between control and cut shoots except in *L. palustre*. Without tedious rootlet extractions it should be possible to estimate fine root surface area from leaf area of *L. palustre*.

Introduction

Below-ground/above-ground biomass ratios ranging from 1 to 4 are common in arctic shrubs[2,5,8]. However these high ratios do not quantitatively describe the balance between the root and shoot system, as is the case with most temperate climate zone shrubs and trees, but rather reflect the large amount of below-ground stem biomass. A characteristic of arctic dwarf shrubs is the abundant formation of stemborn roots in the upper 20 cm of the organic substrate, probably stimulated by the relatively high temperatures and aeration in this close to surface layer[6].

When investigating root/shoot ratios, *e.g. Betula nana*, it is possible that a given shoot belongs to a branch system of an established dwarf shrub as such branches can reach the substrate surface more than a meter apart from each other. Are these individual shoots functionally independent? If so the amount of fine roots attached to such a shoot will allow an estimate of the absorbing root surface which is required to sustain the development of the respective shoot and the quantity of rootlets in balance with the leaf area of the respective shoot. Knowledge regarding this balance would have considerable theoretical interest. Further it would become possible to predict the absorbing root surface from estimates of the leaf area.

The present work has the specific purpose to explore the degree of physiological independence of dwarf shrub shoots detached from their parent plant and to estimate their midsummer fine root surface and leaf area. This is viewed as a first step towards a quantitative assessment of the seasonal fine root and leaf area growth dynamics in the tussock tundra.

Material and methods

Studies were conducted in the Eagle Creek research area, located about 140 km northeast of Fairbanks, Alaska, close to the Steese Highway (65° 25′N, 145° 30′W). The specific tussock site is situated at 750 m elevation on a gentle northwest facing slope with enough drainage to prevent water logging. The floristic composition of the area was described earlier[4,8]. The 30–40 cm tall tussocks are dominated by *Eriophorum vaginatum L.* Other important species of vascular plants are *Carex bigelowii Torr.* and dwarf shrubs such as *Ledum palustre L.*, *Vaccinium uliginosum L.*, *V. vitis-idaea L.*, *Rubus chamaemorus L.*, *Betula nana L.*, and the less frequent, *Salix pulchra Cham.* Peat moss cushions fill the intertussock area in many instances.

The shrub species *Ledum palustre*, *Vaccinium uliginosum*, and *Betula nana* were chosen for the experimental work. In June 1979, 13 shoots from each of these species were marked. Criteria for their selection were that they had to originate from different individual shrubs and that each of the shoots had a length of 10–20 cm above the moss surface. The connections with the next order branch for five shoots per species, were severed with a long knife 10–20 cm below the peat moss surface. After severing the shoots remained for one year in their original location without further disturbance.

In July 1980, the branches of the cut and uncut (= control) shoots, (the latter ones severed before harvest), were carefully lifted out of the peat moss retaining as many of their fine roots as possible. Each shoot was divided into leaves, stems, fine roots (diam. < 1 mm) and larger roots. Total leaf area, and dry weight per shoot were recorded. The length of representative portions of fine roots was measured by using a modified intersect method[7]. The dry weight of the measured rootlets was recorded and the length/dry weight ratio used to calculate the total fine root length per shoot. Fine root diameters were measured microscopically. Total length and mean diameter were used to calculate the fine root surface area per shoot[1]. Due to the nature of the extraction procedure, the rootlet quantities obtained are considered minimum values because up to 25% of the fine roots may have been lost.

Results and discussion

About eight weeks after the separation of the test branches from their parent plants (early August 1979) some detrimental effects became apparent. Leaves were smaller than with control branches, some were slightly chlorotic while most flowers and fruits aborted. One year later, in June 1980, these differences had disappeared.

Analysis of the harvested branches showed that the dry weights of leaves, stems, and roots were quite similar in treated and control branches (Table 1). In *B. nana* the root/shoot ratio of the control branches was significantly larger than in the cut ones. The low root/shoot ratio for the experimental species resulted from few roots with diameters > 1 mm being attached to the upper part of the stems although larger roots with few rootlets occurred in deeper layers of the substrate. Other studies on the same species also stress the importance of the upper substrate layers for fine root formation[4,5].

The measurements of fine root diameter gave values ranging from 0.03 to

Table 1. Dry weight of not severed (control) shoots, severed shoots, and root/shoot ratios of three shrub species. Values are means of 8 shoots for uncut controls and 5 for cut shoots. Standard errors in parentheses. Only the root/shoot ratio in *B. nana* showed a significant ($\alpha = 0.05$, asterisk) difference between control and treatment. Statistical analysis by ANOVA and Newman-Keuls

	Leaf dry wt. g		Stem dry wt. g		Root dry wt. g		Root/shoot ratio	
	Contr.	Cut	Contr.	Cut	Contr.	Cut	Contr.	Cut
Ledum	0.56	0.43	1.79	1.90	0.26	0.24	0.11	0.10
palustre	(0.067)	(0.064)	(0.336)	(0.551)	(0.036)	(0.061)	(0.014)	(0.031)
V. uligi-	0.27	0.12	2.25	2.83	0.24	0.36	0.10	0.12
nosum	(0.054)	(0.032)	(0.595)	(0.459)	(0.049)	(0.072)	(0.015)	(0.012)
Betula	0.81	0.58	9.30	12.32	2.88	1.93	0.28	0.15*
nana	(0.174)	(0.174)	(1.847)	(2.859)	(0.645)	(0.761)	(0.043)	(0.045)

Table 2. Diameter (mm) and extension (m/g) of fine roots of three experimental shrubs. Standard errors in parentheses. For diameter measurements n = 100, for extension estimates n = 5

	Diam.(mm)	Range	Extension (m/g)
L. palustre	0.123 (0.010)	0.030–0.573	67 (5.98)
V. uliginosum	0.124 (0.015)	0.050–0.788	43 (2.42)
B. nana	0.212 (0.017)	0.045–0.895	22 (1.12)

Table 3. Fine root/leaf dry weight and fine root surface/leaf area ratios. Values are means of 8 shoots for uncut controls and 5 for severed shoots. Standard errors are in parentheses. The asterisks indicate significance of differences between control and severed shoots ($\alpha = 0.05$). Statistical analysis by ANOVA and Newman-Keuls

	Weight ratios		Surface area ratios	
	Control	Cut	Control	Cut
L. palustre	0.31	0.40	2.03	2.23
	(0.039)	(0.058)	(0.265)	(0.3222)
V. uliginosum	0.77	1.92*	1.41	10.55*
	(0.175)	(0.379)	(0.32)	(1.164)
B. nana	1.27	1.21	1.08	2.44*
	(0.163)	(0.046)	(0.138)	(0.092)

0.89 mm. *Vaccinium uliginosum* had the thinnest and *B. nana* the thickest rootlets although the difference between *L. palustre* and *V. uliginosum* was not significant (Table 2). Length per unit weight varied between the three species and followed the same pattern as the diameters (Table 2). These values are in agreement with rootlet measurements of the same species collected at a different site[4].

The ratio of fine root dry weight/leaf dry weight was significantly higher for the cut-off shoots in *V. uliginosum* (Table 3). Only with *L. palustre* was the fine root/leaf area ratio not significantly affected by the treatment. In *V. uliginosum* and *B. nana* the cutting may have stimulated fine root formation but with insufficient time between treatment and harvest to reach equilibrium between leaves and rootlets. The ratio leaf area to rootlet surface area is relatively high in these species. Given more time rootlet surface area relative to leaf area may increase. Alternatively the higher photosynthetic efficiency of the leaves of the deciduous species (*B. nana* and *V. uliginosum*) relative to evergreen (*L. palustre*) foliage[3], may provide more carbohydrate for fine root growth in deciduous shrubs allowing a smaller leaf area to maintain a larger fine root system.

The treatments had an initial effect on the amount of foliage per branch. Experimental branches were not completely independent of their parents although they adjusted during the growing season. Fine roots and leaves did not appear to be completely balanced although their amounts seemed to be adequate to produce mutual support. The leaf area of shoots of *L. palustre* may therefore give a coarse measure of fine root surface area. This is not yet possible for *V. uliginosum* and *B. nana* and is valid only for midseason. Observations to assess seasonal dynamics of fine root and leaf growth as well as information on fine root turnover rates are needed to draw a more complete picture. A major unknown factor is the extension of the hyphal system of the abundant mycorrhizal fungi which would substantially alter the nutrient absorbing surface of these tundra shrubs.

Acknowledgements Michael Russel's and Ann Jackson's help with the field work is gratefully acknowledged. Dr. David Rayle assisted with the preparation of the manuscript. The study was supported by the U.S. Department of Energy, Grant No. AS03–77 EVD 1525.

References

1 Böhm W 1979 Methods of studying root systems. Ecological Studies 33, Springer-Verlag, Berlin.

2 Dennis J G and Johnson P L 1970 Shoot and rhizome-root standing crops at Barrow, Alaska. Arct. Alp. Res. 2, 243–266.

3 Limbach W E, Oechel W C and Lowell W 1982 Photosynthetic and respiratory responses to temperature and light of three Alaskan tundra growth forms. Holarct. Ecol. 5, 150–157.

4 Miller P C, Mangan R and Kummerow J 1982 Vertical distribution of organic matter in eight vegetation types near Eagle summit, Alaska. Holarct. Ecol. 5, 117–124.

5 Shaver G R and Cutler J C 1979 The vertical distribution of live vascular phytomass in cottongrass tussock tundra. Arct. Alp. Res. 11, 335–342.

6 Stuart L and Miller P C 1982 Soil oxygen flux measured polarographically in an Alaskan tussock tundra. Holarct. Ecol. 5, 139–144.

7 Tennant D 1975 A test of a modified line intersect method of estimating root length. J. Ecol. 63, 995–1001.

8 Wein R W and Bliss L C 1974 Primary production in arctic cotton grass tussock tundra communities. Arct. Alp. Res. 6, 261–274.

Plant and Soil 71, 401–413 (1983).

A-47

Nutrient reserves in roots of fruit trees, in particular carbohydrates and nitrogen

J. TROMP
Research Station for Fruit Growing, 4475 AN Wilhelminadorp, The Netherlands

Key words Carbohydrates Cultural factors Environment Fruit trees Nitrogen Reserves Seasonal pattern

Summary In trees, nutrient reserves built up in the previous year are of primary importance for early spring growth. Despite the relatively great importance of roots for nutrient storage, the root system should not be regarded as a special storage organ. Quantitatively, carbohydrates predominate in these reserves, but qualitatively N and other minerals are of more than minor significance. In roots carbohydrates are usually stored in insoluble form, mainly as starch; sorbitol is the predominant soluble compound in apple and peach. For nitrogen reserves, the soluble form predominates in roots, especially arginine in apple and peach, followed by asparagine. The level of reserves usually becomes maximal early in the winter. During leafing-out the reserves are drawn on until, later in the season, the supply of newly produced or absorbed nutrients exceeds the demand and replenishment occurs. The initial carbohydrate reserves do not determine the amount of new growth, whereas reserve nitrogen is of decisive importance for shoot growth vigour. Environmental factors such as light intensity and temperature affect the level of carbohydrates in roots; the concentration can be reduced by defoliation and summer pruning and increased by ample supply of nitrogen fertilizer in the autumn. The main cultural factors that influence nitrogen reserves are the amount and the time of nitrogen fertilization.

Introduction

The importance of reserves for survival of the tree during the winter and for the resumption of growth activity in the spring can hardly be overestimated. The dry weight of expanding buds increases before any photosynthesizing parts are present. Similarly, the amount of nitrogen increases appreciably in developing buds and young shoots. As substrate, reserve carbohydrates maintain respiration and other processes during the leafless period. By reserves is understood here 'those food substances that are not used directly in assimilation and respiration but are stored somewhere in the tree until required'[7]. This excludes short-term diurnal storage of assimilates in leaves when phloem export lags during the day.

Carbohydrates make up the greater part of reserve material in the tree, but qualitatively, nitrogenous reserves are equally important and there is no reason to assume that the other minerals are of only minor importance as reserve nutrients. However, so little information is available that in the present paper the discussion of these nutrients must be restricted to a few instances for K, an example of a very mobile element, and the almost immobile element Ca. A review of the literature on carbohydrate resources in fruit trees has been published by Priestley[27] and for nitrogen storage and mobilization by Taylor[42].

Two other points should be made. The first concerns the more general question of how to express the results of chemical analyses of plant material. Dry weight is usually taken as the basis of expression but it is evident that especially when a rapid short-term change in dry weight occurs, conclusions drawn from data expressed in terms of dry weight may be unreliable. Therefore, residue dry weight was recommended instead, residue being defined as structural material that cannot be metabolized further[31]. Expression of the results as total values for each plant or plant part avoids most of the problems and provides useful quantitative data on distribution within the tree, but for practical reasons the method is only applicable to small trees.

The second point concerns the distinction between reserves and current supply of newly produced carbohydrates or freshly taken up N and minerals. Of course, in this respect carbohydrates do not form a problem during the leafless period; otherwise, the use of ^{14}C has been shown to be very helpful[12,17,18]. For nitrogen and the other minerals, this difficulty is not easy to eliminate. Unlike the situation in most herbacious plants, in fruit trees – as in many other tree species – nitrate absorbed from the soil is completely reduced in the roots[3,13,25]. Both freshly taken up and reserve N are present in organic from and consequently are not easy to distinguish from each other. Useful information has been obtained in experiments where uptake of N was mainly restricted to certain periods[14,16,40,44,47,51]. In a single experiment, ^{15}N was used to make that distinction[15].

The root as storage organ

The emphasis placed on the root as storage organ in the present paper is not meant to imply that in trees the root system has a specialized function in this respect[17,28]. Although the relative importance of root reserves can be great[12,19,32], there is no difference in potential value between reserves from different regions of the tree, and all parts contribute to the new growth activity early in the spring, as was found for carbohydrates[17,28] as well as for nitrogenous compounds[15,43,44,47].

Nevertheless, the situation in which roots function differs from that of the above-ground tissues, mainly because of the environmental conditions they are exposed to. Roots are rarely exposed to very low temperatures that can affect various biochemical processes. For example, in certain tree species that store carbohydrates in the form of fat, these fat reserves are only found in roots when in extremely cold winters the soil temperature remains below 0°C for many months[57].

With respect to the location of reserves within the various tissues, relevant information for fruit trees is lacking, and we have to rely on what is known about forest trees[57]. Storage of reserves is a function of the living cell and occurs in parenchymatous cells. Specialized cells such as sieve-tube cells do not store reserve nutrients, which would disturb their physiological function in transport.

Since it is essential for the storage parenchyma in the phloem and the xylem to be in direct contact with vascular tissue, a special function in the transport of reserve compounds is attributed to radially-running ray tissue by which the storage cells are connected with the main conducting elements in the phloem and xylem[37,57,59]. It is clear from all this that the anatomy of woody tissues is strongly related to the need for efficiency in the supply and distribution of their reserves[5].

One difficulty in the study of reserves in roots is that the longevity of at least part of the root system is limited. Because the roots lose appreciable quantities of organic material to the soil[35,36], the reliability of data on reserves is not always certain. The problem may be less serious in the usually very young trees used in almost all experimental work on reserves in fruit trees, but this last point raises the question of the extent to which findings obtained in young trees in experiments done under artificial growing conditions are valid for full-grown trees under orchard conditions.

Nature of reserves

Reserve carbohydrates are the predominant components of the total reserves distributed over all parts of the tree. For instance, as Table 1 shows for 18-year-old apple trees, the roots may contain more than 45% carbohydrates (on a dry weight basis) as against about 30% for the above-ground parts taken as a whole[22]. Complete exhaustion usually will not occur, but it has been shown that of the roughly 25% carbohydrates present in the small roots of nine-year-old pecan trees in November-December, half was used in the spring[55]. Total N concentration in winter, in contrast, was about 1.5% for roots of well-fertilized young apple trees, and about 0.8% was mobilized during the spring (Fig. 1). A study on the distribution of minerals in five-year-old apple trees showed that in the root bark the K concentration of 0.5% in the winter decreased to 0.3% in early May[21].

Carbohydrates are present in soluble and insoluble forms, the solubility being defined by the extraction solvent, usually methanol. For most trees starch is considered to be the main form of carbohydrate reserve in all woody tissues, especially roots. In Murneek's[22] apple trees the ratio between starch and the soluble fraction was between 3 and 4 for roots of different ages (Table 1) and a little above unity for the above-ground parts. Other studies have confirmed these findings[12,26]. Hemicelluloses form an appreciable proportion of the carbohydrates present in the root and other regions (Table 1). There is some controversy about the extent to which these compounds have a function as reserve food besides being cell-wall constituents[7,20,26]. It is not necessary to give too much attention to such questions as which component of the total reserves is the most important. Dependent on environmental conditions or metabolic activity, a ready interconversion of the various components occurs.

Table 1. Quantitative distribution of carbohydrates in 18-year-old apple trees cv. Jonathan in Mid-October (From 22)

	Dry weight	Percentage of dry weight			
	(kg)	Total carbo-hydrates	Starch	Hemi-cellulose	Sugars
Total above ground	203.7	30.8	4.4	23.1	3.4
Root stump	21.9	37.0	11.0	23.0	2.7
18–14 year roots	28.2	39.0	11.4	24.7	2.9
13–7 year roots	16.3	43.8	12.3	26.4	5.1
6–1 year roots	2.6	42.0	11.5	25.0	4.7
Total below ground	69.0	39.6	11.5	24.6	3.5

There is ample evidence that in apple, within the soluble-carbohydrate fraction, the sugar alcohol sorbitol is the main component in all tissues, although sucrose, glucose, and fructose may be found in fair amounts[4, 12, 33]. The occurrence of sorbitol is widespread in the Rosaceae, including commercial fruit species such as pear and cherry[53]. Sorbitol is supposed to be the main end-product in photosynthesis in apricot leaves[2], and conversion from sorbitol into glucose is known to occur in leaf tissue of apple[56]. Whether glucosides have a function in carbohydrate storage in the sense of 'food' is not certain as yet. A special glucoside seems to be characteristic for a certain genus, *i.e.*, phloridzin for apple (high in roots), arbutin for pear, and prunasin for plum[27]. Although phloridzin varied markedly throughout the year, the fluctuation differed from that of the other carbohydrate components[26].

Like carbohydrates, storage N is present in alcohol-soluble and alcohol-insoluble forms, mainly protein. There is some controversy as to whether protein really functions as storage nitrogen. Some data indicate that in apple and peach N is stored mainly in soluble form[23,44], but other findings for similar material show that not only soluble N but also proteins are very important as storage N[14,38,47,49,50]. It has even been reported that apple trees possess a special storage protein characterized by a high arginine content[50]. For roots, however, proteins can be much less important for N storage (Fig. 1)[47,51]. Arginine is also the main component of the soluble N fraction. It may reach values of 60–70% in roots of young apple trees (Figs. 2 and 3), which fits in well with other data for apple[14,23] and peach[38,43]. The amide asparagine is always a fairly important constituent. A number of other amides and amino acids occur only in minor quantities (Fig. 3).

It is unlikely that the very mobile element K is present in a special storage form. For Ca the situation is quite different in that it is especially concentrated in the bark of stem and roots as oxalate precipitate and is therefore hardly available

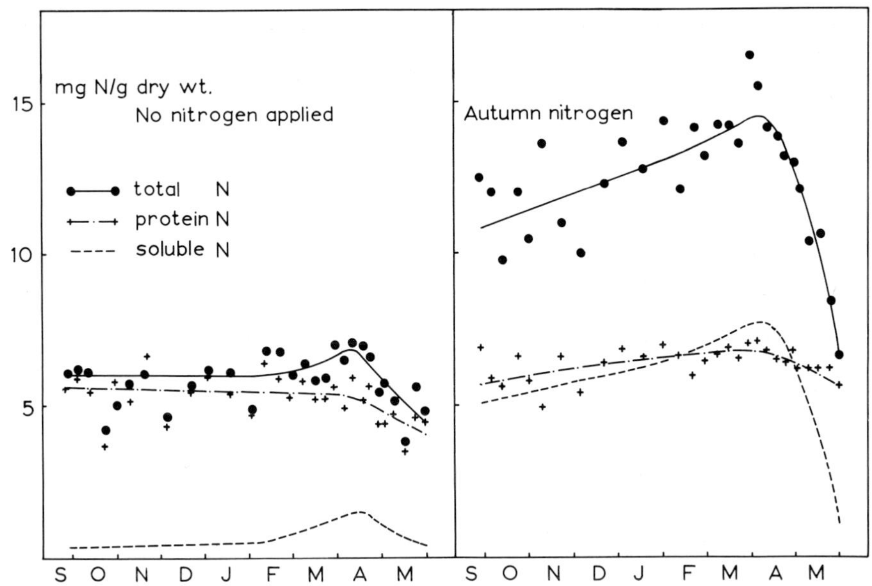

Fig. 1. The total-, protein-, and soluble N levels in roots of one-year-old M.7 apple rootstocks throughout autumn, winter and spring. *Left:* unfertilized, *right:* N applied at the end of September. The difference between total and protein N is considered as soluble N.

for remobilization. However, reversible retention in other forms, *e.g.* 'storage' at adsorption sites in wood, might be possible[6]. In any case, redistribution of Ca from the roots has been shown to occur in apple after the resumption of growth in spring[46, 54].

Seasonal pattern

Broadly speaking, the level of the various reserve substances shows the same rhythm during the course of the year in all perennial parts of the tree. In early spring, when the swelling and leafing-out buds start to draw nutrients, a decrease is found until sometime in late spring or early summer, when the supply of newly produced or absorbed material exceeds demand and replenishment occurs. However, detailed scrutiny reveals differences. In various apple tissues, including roots, carbohydrates were readily depleted but quickly replenished before the actual shoot extension started; a second minimum could occur in July when growth was most vigorous. From other work on apple it was concluded that the amount of new growth was not determined by the level of the initial reserves[29]. From studies wit ^{14}C it was calculated that only the development of the first 5 to 6 leaves was dependent on reserves; after that, the shoots became self-sufficient by photosynthesis[11, 12]. For N in apple, to the contrary, the depletion of reserves occurred independent of the current year's N supply provided the storage level was high[23, 47].

Although there can be little doubt that the roots supply carbohydrates to the above-ground parts early in the spring[12,34], it is a matter of conjecture whether this contribution is significant quantitatively. When root activity intensifies in this early period, appreciable amounts of carbohydrates are lost by respiration, bound in the structure of new cells, and incorporated into amino compounds. It seems quite likely that the greater part of the carbohydrate reserves of roots is used up during metabolism. In the N economy of the above-ground tree, however, the root reserves are an important factor. This is clear from the considerable drop in total N in apple roots occurring during April and May (Fig. 1). A good illustration is provided by data obtained in an experiment with apple, where the use of ^{15}N permitted distinction between reserve and newly absorbed N[15]. Nitrogen applied in October was mainly stored in the roots and the rootstock and at least half of the total N present in the new leaves the following May came from root reserves.

For most trees the insoluble fraction of storage carbohydrates (mainly starch) in the above-ground parts shows two peaks, one early in the spring and the other at the beginning of winter when the temperature drops, the subsequent winter minimum being caused by the conversion of starch into sugars[58]. For roots of fruit trees this pattern is less clear or absent, possibly due to the fact that in the winter soil temperatures usually are higher than air temperatures; moreover, the data show little consistency[12,18,39,55]. It is supposed that – as found in apple seedlings – starch is synthesized directly (or perhaps indirectly via glucose) from sorbitol, the main soluble storage carbohydrate in apple[4], but in similar material others recently found a rather low level of conversion of sorbitol in non-photosynthesizing tissues except root apices, where it was a ready substrate in respiration[8]. Other experiments done in apple showed that over the period under study (November-May) the relative importance of the various soluble carbohydrates in the roots was not constant, but a clear pattern did not emerge[12].

An example of the seasonal course of N reserves of apple roots is given in Fig. 1. The variation mainly concerned soluble N; the protein concentration was apart from a small decrease after the end of March, almost constant throughout the experimental period. Due to uptake from the soil and to translocation from the leaves during senescense[24], total and soluble N increased during the autumn and winter, *i.e.*, until a substantial drop started in April. The level of arginine, the main component of soluble N, rose throughout the autumn and early winter, its share in the soluble N fraction increasing from 35–40% in October to 60–70% in February. After that there was little change in this proportion in spite of a dramatic decrease of the concentration (Fig. 2). A similar accumulation of arginine in the autumn and winter has also been reported for peach roots[38,43]. It was argued that arginine is synthesized at the expense of asparagine and glutamine[43], which were shown to accumulate in the preceding period. In the spring, the reverse occurred: arginine decreased and the amides increased, which

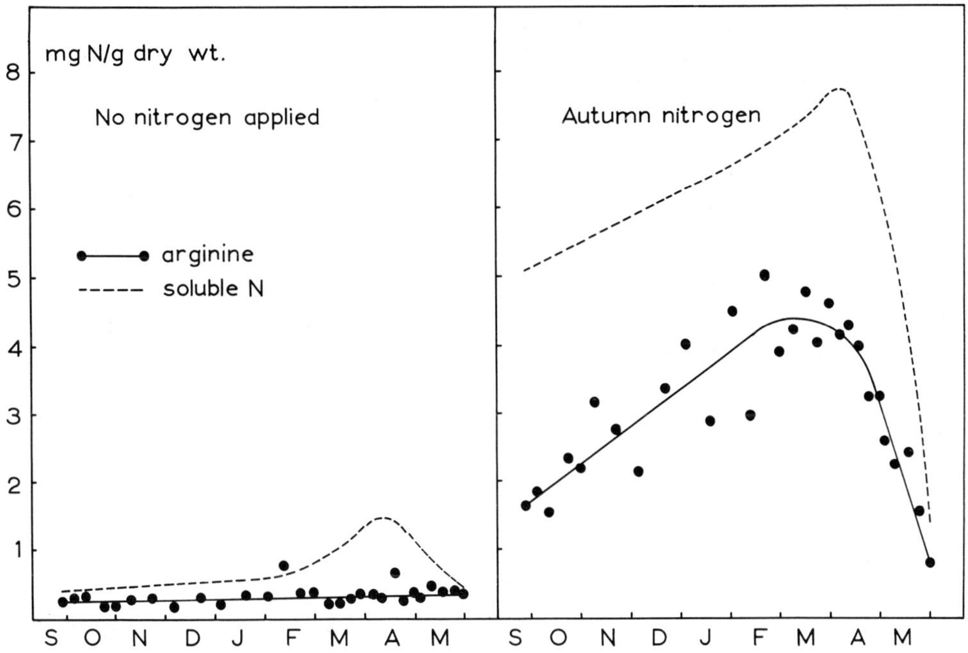

Fig. 2. The arginine level and the soluble N level (redrawn from Fig. 1) in roots of one-year-old M.7 apple rootstocks throughout autumn, winter and spring. *Left*: unfertilized, *right*: N applied at the end of September.

might indicate hydrolysis of arginine followed by incorporation of the released ammonia into the amides. At first sight this does not seem to hold for apple-tree roots since the composition of the amino N fraction remained constant while total N decreased markedly (Fig. 3). However, a surplus of amides would not be noticed if, for example, there were a preferential export of amides to the above-ground parts, as is suggested by the observation that in the xylem sap sampled from the same trees asparagine + aspartic acid predominated and arginine was clearly less important (Fig. 3).

The seasonal variation of the reserves can also be discussed with more emphasis on function, *i.e.*, in terms of the sink-source concept. According to this concept, in the spring the activated growth centres in the swelling buds use their capacity to draw nutrients from sources, (*i.e.*, the storage tissues) elsewhere in the tree. Later, however, the new shoots become self-sufficient, at least for carbohydrates, and sink location shifts to other parts of the tree that are more active in attracting nutrients. In this system of competing sinks, roots initially seem to have a low competitive ability. Although lower leaves of actively growing shoots have been shown to export assimilates to the roots[10,34], root growth is usually reduced when shoot growth is most active[1]. Later in the season, when shoot growth slows down, nutrient accumulation in the perennial parts

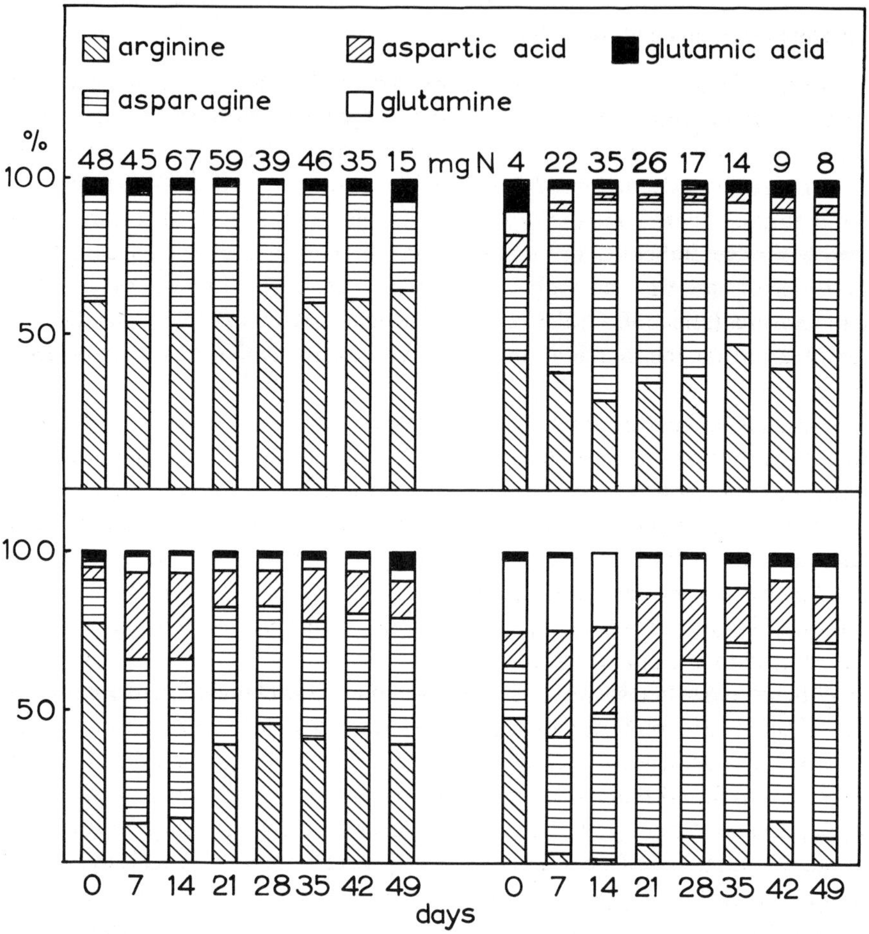

Fig. 3. Percentage composition of the main amino compounds in the soluble N fraction of the roots (*top*) and the xylem sap of the stem (*bottom*) of one-year-old M.7 rootstocks at various intervals starting at bud-break. Nitrogen applied either in September (*left*) or in the spring shortly before day 0 (*right*). The values at the top of the bars denote total amino N in the roots (mg N/tree).

increases in importance and predominates in the period before leaf fall. It is remarkable that in the autumn a preferential accumulation in the roots occurs, as various ^{14}C-studies with apple have shown[10,18,19,34]. This preference may be related to the ability of roots to maintain activity at lower temperatures than the above-ground plant can, and also to the relatively high soil temperatures in the autumn. Of course, this activity may not concern root growth, since growth means nutrient consumption rather than accumulation. For accumulation, sink strength must be maintained by compartmentalization or by conversion of the moving in substances into an insoluble fraction as starch or proteins.

Environmental and culture factors

The knowledge that reserves are of primary importance for tree life and that ambient factors control plant behaviour to a high degree raises the question of the relationship between reserve 'physiology' and the environment. Unfortunately, little information is available, certainly for roots. With respect to temperature, apple rootstocks kept at an elevated root temperature from October to December lost more dry weight,especially of starch and sugars, than those kept at the normal winter temperature[32]. Supra-optimal root temperatures (above 30°C) also decreased root dry matter in apple[9]. The effect of temperature on the degradation and resynthesis of starch has already been mentioned[37].

It should be kept in mind that with respect to root reserves, not only soil factors that affect root metabolism directly, but also all factors that influence above-ground growth determine the extent to which the reserves of the various tree parts, including the roots, will be exhausted. For example, shading reduced root growth and the amount of root carbohydrates in young apple trees[30]. In a similar experiment depletion of reserves was the same at all light intensities although initial growth varied greatly, which indicated that growth was not determined by the amount of reserves[29].

With respect to N reserves it was supposed that the accumulation of arginine in the autumn is a response to, *e.g.*, short days and low temperature[44], but experimental evidence is lacking. For a few rootstock-scion combinations of apple it was found that the accumulation of Ca relative to K occurring in the roots, decreased clearly with rising root temperatures in the range 6–30°C[48].

In addition to the effect of shading already mentioned, another cultural factor influencing carbohydrate reserves is defoliation. An experiment with pecan trees showed that early defoliation only had a transitory effect on the carbohydrate level in all tree parts, probably due to refoliation. For the roots, the most pronounced and lasting effects were found when defoliation occurred in September and October[55]. Removal of part of the leaf area by summer pruning in apple led to a marked reduction of dry weight of roots and the basal stem section[41]. In young peach trees ample supply of N, especially when given in the autumn, increased markedly the level of starch in late autumn, particularly in the roots, as compared with trees that were hardly fertilized or that received N only in the summer[40]. However, this difference had disappeared by late spring, when the starch level was minimal. The build-up of starch later in the season again led to higher values for the autumn-fed trees, probably due to advanced leafing-out (longer season).

The main cultural factors that affect the N reserves in the fruit tree are the amount and the time of N dressing. For all perennial tissues of apple and peach, including the roots, total N increased with increasing supply of fertilizer N; especially the response of soluble N and in particular arginine was considerable[23,40,44,45,47]. The time of N application may have a pronounced effect on the distribution of the N reserves over the tree[14,15,16,23,40] and consequently can influence, *e.g.*, growth and fruit set the following year. As shown for apple, a large proportion of the N given in the spring was used up for new growth; summer N came too late to have much effect on growth but was rapidly distributed over all parts of the tree. In contrast, N given in the autumn (end of October) mainly remained in the root and rootstock, in particular as insoluble N and asparagine; later in the winter arginine predominated[14,16]. The effect of the time of N application on the soluble N concentration of roots of young apple trees and of xylem sap sampled from the stem during an eight-week period after bud-burst is shown in Fig. 3. Nitrogen was given either in the preceding autumn or in the spring around bud-burst. In the root of the autumn-fertilized trees arginine predominated, followed by asparagine, but in the spring-treated trees asparagine predominated over arginine and the share taken by the remaining compounds was larger. The main difference in xylem-sap composition concerned the minor role of arginine and the abundance of asparagine in the spring-fertilized trees, whereas in the autumn-fed trees arginine usually took a much greater share at the expense of asparagine. Other work on apple showed that besides the amount and time of N application, the type of N given (nitrate or ammonium) may affect the composition of reserve N in roots[51].

References

1 Atkinson D and Wilson S A 1980 The growth and distribution of fruit tree roots: some consequences for nutrient uptake. *In* Mineral Nutrition of Fruit Trees. Eds. D Atkinson, J E Jackson, R O Sharples and W M Waller. Butterworth, London-Boston, pp 137–150.

2 Bieleski R L and Redgwell R J 1977 Synthesis of sorbitol in apricot leaves. Aust. J. Plant Physiol. 4, 1–10.

3 Bollard E G 1957 Composition of the nitrogen fraction of apple tracheal sap. Aust. J. Biol. Sci. 10, 279–287.

4 Chong C 1971 Study of the seasonal and daily distribution of sorbitol and related carbohydrates within apple seedlings by analysis of selected tissues and organs. Can. J. Plant Sci. 51, 519–525.

5 Crafts A S and Crisp C E 1971 Phloem Transport in Plants. Freeman and Co, San Francisco, 481 p.

6 Ferguson I B 1980 The uptake and transport of calcium in the fruit tree. *In* Mineral Nutrition of Fruit Trees. Eds. D Atkinson, J E Jackson, R O Sharples and W M Waller. Butterworth, London-Boston, pp 183–192.

7 Glerum C 1980 Food sinks and food reserves of trees in temperate climates. N. Z. J. For. Sci. 10, 176–185.

8 Grant C R and ap Rees T 1981 Sorbitol metabolism by apple seedlings. Phytochemistry 20, 1505–1511.

9 Gur A, Hepner J and Mizradi Y 1976 The influence of root temperature on apple trees. I Growth responses related to the application of potassium fertilizer. J. Hortic. Sci. 51, 181–193.

10 Hansen P 1967 ^{14}C-Studies on apple trees. III The influence of season on storage and mobilization of labelled compounds. Physiol. Plant. 20, 1103–1111.

11 Hansen P 1971 ^{14}C-Studies on apple trees. VII The early seasonal growth in leaves, flowers and shoots as dependent upon current photosynthates and existing reserves. Physiol. Plant. 25, 469–473.

12 Hansen P and Grauslund J 1973 ^{14}C-Studies on apple trees. VIII The seasonal variation and nature of reserves. Physiol. Plant. 28, 24–32.

13 Hill-Cottingham D G and Bollard E G 1965 Chemical changes in apple tree tissues following applications of fertilizer nitrogen. N. Z. J. Agric. Res. 8, 778–787.

14 Hill-Cottingham D G and Cooper D R 1970 Effect of time of application of fertilizer nitrogen on the distribution and identity of the nitrogenous constituents of young apple trees. J. Sci. Food Agric. 21, 172–177.

15 Hill-Cottingham D G and Lloyd-Jones C P 1975 Nitrogen-15 in apple nutrition investigations. J. Food Sci. 26, 165–173.

16 Hill-Cottingham D G and Williams R R 1967 Effect of time of application of fertilizer nitrogen on the growth, flower development and fruit set of maiden apple trees, var. Lord Lambourne, and on the distribution of total nitrogen within the tree. J. Hortic. Sci. 42, 319–338.

17 Kandiah S 1979 Turnover of carbohydrates in relation to growth in apple trees. I Seasonal variation of growth and carbohydrate reserves. Ann. Bot. 44, 175–183.

18 Kandiah S 1979 Turnover of carbohydrates in relation to growth in apple trees. II Distribution of ^{14}C assimilates labelled in autumn, spring and winter. Ann. Bot. 44, 185–195.

19 Katzfuss M 1979 ^{14}C-Verteilung im Herbst und -Mobilisierung in jungen Apfelbäumen. Arch. Gartenbau 27, 119–123.

20 Kramer P J and Kozlowski T T 1979 Physiology of Woody Plants. Academic Press, New York-San Francisco, 811 p.

21 Mason A C and Whitfield A B 1960 Seasonal changes in the uptake and distribution of mineral elements in apple trees. J. Hortic. Sci. 35, 34–55.

22 Murneek A E 1942 Quantitative distribution of nitrogen and carbohydrates in apple trees. Res. Bull. Mo. Agric. Exp. Stn. 348, 28 p.

23 Oland K 1959 Nitrogenous reserves of apple trees. Physiol. Plant. 12, 594–648.

24 Oland K 1963 Changes in the content of dry matter and major nutrient elements of apple foliage during senescence and abscission. Physiol. Plant. 16, 682–694.

25 Pate J S 1973 Uptake, assimilation and transport of nitrogen compounds by plants. Soil Biol. Biochem. 5, 109–119.

26 Priestley C A 1960 Seasonal changes in the carbohydrate resources of some six-year-old apple trees. Annu. Rep. East Malling Res. Stn for 1959, 70–77.

27 Priestley C A 1962 Carbohydráte Resources within the Perennial Plant. Commonwealth Bureau of Hort. and Plantn. Crops. Tech. Comm. 27, 116 p.

28 Priestley C A 1964 The location of carbohydrate resources within the apple tree. Proc. XIV th. Int. Hortic. Congr, 1962, 319–327

29 Priestley C A 1963 The carbohydrate resources of young apple trees under four levels of illumination. Ann. Bot. 27, 435–446.

30 Priestley C A 1969 Some aspects of the physiology of apple rootstock varieties under reduced illumination. Ann. Bot. 33, 967–980.

31 Priestley C A 1973 Bases for the expression of the results of chemical analysis of plant tissue. Ann. Bot. 37, 943–953.

32 Priestley C A 1981 Perennation in woody fruit plants and its relationship to carbohydrate turnover. Ann. Appl. Biol. 98, 548–552.

33 Priestley C A, Catlin P B and Olsson E A 1976 The distribution of ^{14}C-labelled assimilates in young apple trees as influenced by doses of supplementary nitrogen. II Soluble carbohydrates and amino acids. Ann. Bot. 40, 1171–1176.

34 Quinlan J D 1969 Mobilization of ^{14}C in the spring following autumn assimilation of $^{14}CO_2$ by an apple rootstock. J. Hortic. Sci. 44, 107–110.

35 Rogers W S 1968 Amount of cortical and epidermal tissue shed from roots of apple. J. Hortic. Sci. 43, 527–528.

36 Rogers W S and Head G S 1969 Factors affecting the distribution and growth of roots of perennial woody species. *In* Root Growth, Ed. W J Whittington. Butterworth, London, pp 280–295.

37 Sauter J J 1966 Untersuchungen zur Physiologie der Pappelholzstrahlen. I Jahresperiodischer Verlauf der Stärkespeicherung im Holzstrahlparenchym. Z. Pflanzenphysiol. 55, 246–258.

38 Stassen P J C, Stindt H W, Strydom D K and Terblanche J H 1981 Seasonal changes in nitrogen fractions of young Kakamas peach trees. Agroplantae 13, 63–72.

39 Stassen P J C, Strydom D K and Stindt H W 1981 Seasonal changes in carbohydrate fractions of young Kakamas peach trees. Agroplantae 13, 47–54.

40 Stassen P J C, Terblanche J H and Strydom D K 1981 The effect of time and rate of nitrogen application on development and composition of peach trees. Agroplantae 13, 55–61.

41 Taylor B H and Ferree D C 1981 The influence of summer pruning on photosynthesis, transpiration, leaf abscission, and dry weight accumulation of young apple trees. J. Am. Soc. Hortic. Sci. 106, 389–393.

42 Taylor B K 1967 Storage and mobilization of nitrogen in fruit trees: a review. J. Aust. Inst. Agric. Sci. 33,23–29.

43 Taylor B K 1967 The nitrogen nutrition of the peach tree. I Seasonal changes in nitrogenous constituents in mature trees. Aust. J. Biol. Sci. 30, 379–387.

44 Taylor B K and May L H 1967 The nitrogen nutrition of the peach tree. II Storage and mobilization of nitrogen in young trees. Aust. J. Biol. Sci. 20, 389–411.

45 Taylor B K and Van den Ende B 1969 The nitrogen nutrition of the peach tree. IV Storage and mobilization of nitrogen in mature trees. Aust. J. Agric. Res. 20, 869–881.

46 Terblanche J H, Wooldridge L G, Hesebeck I and Joubert M 1979 The redistribution and immobilization of calcium in apple trees with special reference to bitter pit. Commun. Soil Sci. Plant Anal. 10, 195–215.

47 Tromp J 1970 Storage and mobilization of nitrogenous compounds in apple trees with special reference to arginine. *In* Physiology of Tree Crops. Eds. L C Luckwill and C V Cutting. Academic Press, London-New York, pp 143–159.

48 Tromp J 1978 The effect of root temperature on the absorption and distribution of K, Ca and Mg in three rootstock clones of apple budded with Cox's Orange Pippin. Gartenbauwissenschaft 43, 49–54.

49 Tromp J and Ovaa J C 1971 Spring mobilization of storage nitrogen in isolated shoot sections of apple. Physiol. Plant. 25, 16–22.

50 Tromp J and Ovaa J C 1973 Spring mobilization of protein nitrogen in apple bark. Physiol. Plant. 29, 1–5.

51 Tromp J and Ovaa J C 1976 Effect of time of nitrogen application on amino-nitrogen composition of roots and xylem sap of apple. Physiol. Plant. 37, 29–34.

52 Tromp J and Ovaa J C 1979 Uptake and distribution of nitrogen in young apple trees after application of nitrate or ammonium, with special reference to asparagine and arginine. Physiol. Plant. 45, 23–28.

53 Wallaart R A M 1980 Distribution of sorbitol in Rosaceae. Phytochemistry 19, 2603–2610.

54 Wieneke J und Führ F 1975 Untersuchungen zur Translokation von ^{45}Ca im Apfelbaum. IV Sekundäre Ca-Verlagerung nach der Ruheperiode. Gartenbauwissenschaft 40, 106–112.

55 Worley R E 1979 Fall defoliation date and seasonal carbohydrate concentration of pecan wood tissue. J. Am. Soc. Hortic. Sci. 104, 195–199.

56 Yamaki S 1980 A sorbitol oxidase that converts sorbitol to glucose in apple leaf. Plant Cell Physiol. 21, 591–599.

57 Ziegler H 1964 Storage, mobilization and distribution of reserve material in trees. *In* The Formation of Wood in Forest Trees. Ed. M H Zimmermann. Academic Press, New York, pp 303–320.

58 Zimmerman M H and Brown C L 1971 Trees, Structure and Function. Springer Verlag, Berlin-Heidelberg-New York, 336 p.

Plant and Soil 71, 415–432 (1983). A-48

Nitrogen nutrition, photosynthesis and carbon allocation in ectomycorrhizal pine

C. P. P. REID, F. A. KIDD* and S. A. EKWEBELAM**
Department of Forest and Wood Sciences, Colorado State University, Fort Collins, Co 80523, USA

Key words Carbon allocation Light Mycorrhizas Nitrogen Phosphorus Photosynthesis *Pinus contorta* *Pinus taeda* *Pisolithus tinctorius* Respiration Source-sink *Suillus granulatus*

Summary Studies examined net photosynthesis (Pn) and dry matter production of mycorrhizal and nonmycorrhizal *Pinus taeda* at 6 intervals over a 10-month period. Pn rates of mycorrhizal plants were consistently greater than nonmycorrhizal plants, and at 10 months were 2.1-fold greater. Partitioning of current photosynthate was examined by pulse-labelling with $^{14}CO_2$ at each of the six time intervals. Mycorrhizal plants assimilated more $^{14}CO_2$, allocated a greater percentage of assimilated ^{14}C to the root systems, and lost a greater percentage of ^{14}C by root respiration than did nonmycorrhizal plants. At 10 months, the quantity of $^{14}CO_2$ respired by roots per unit root weight was 3.6-fold greater by mycorrhizal than nonmycorrhizal plants. Although the stimulation of photosynthesis and translocation of current photosynthate to the root system by mycorrhiza formation was consistent with the source-sink concept of sink demand, foliar N and P concentrations were also greater in mycorrhizal plants.

Further studies examined Pn and dry matter production of *Pinus contorta* in response to various combinations of N fertilization (3, 62, 248 ppm), irradiance and mycorrhizal fungi inoculation. At 16 weeks of age, 6 weeks following inoculation with either *Pisolithus tinctorius* or *Suillus granulatus*, Pn rates and biomass were significantly greater in mycorrhizal than nonmycorrhizal plants. Mycorrhizal plants had significantly greater foliar %P, but not %N, than did nonmycorrhizal plants. Fertilization with 62 ppm N resulted in greater mycorrhiza formation than either 3 or 248 ppm. Increased irradiance resulted in increased mycorrhiza formation.

Introduction

A number of papers have suggested or implied that the processes involved in the development of mycorrhizas on the root systems of woody plants should lead to increased rates of photosynthesis and translocation[12,13,37]. Harley and Lewis[13] have suggested that irreversible diversion of carbohydrates to the fungus, as shown for beech mycorrhizas, should have repercussions on patterns of translocation within the host plant. They further suggest that the fungal sink might indirectly increase the rate of photosynthesis, an idea consistent with the concept that photosynthesis is under some control by sink demand[42]. Several studies have indirectly indicated that the mycorrhizas of conifers cause increased translocation of photosynthate to roots and increased

* Present address: Potlatch Corporation. Lewiston, ID 83501, USA.
** Principal Research Officer on Study leave from the Savanna Forestry Resarch Station, P.M.B. 1039, Samaru-Zaria, Nigeria.

photosynthesis[19,27,33,35]. Recent studies with nonwoody plants forming vesicular-arbuscular mycorrhizas have also indicated stimulation of photosynthesis by mycorrhiza formation[1,21,30]. Although mycorrhizas might establish an assimilation sink by utilization of carbohydrate for fungal biomass, by conversion of carbohydrates to storage products, or by conversion of carbohydrates into energy for maintenance metabolism, other interacting factors may be involved in stimulation of photosynthesis and translocation. Improved nutrient status of the plant or changes in hormone balances might be a very important result of mycorrhiza formation[36].

The following studies were therefore initiated to determine if net photosynthesis (Pn) is enhanced by mycorrhiza formation and if such enhancement is consistent with the classical 'source-sink' hypothesis. Two series of experiments were conducted in order to examine 1) the response in net photosynthesis to mycorrhiza formation as related to carbon allocation in *Pinus taeda*, and 2) the interactions between light, N fertilization, mycorrhiza formation, and net photosynthesis of *P. contorta*.

Materials and methods

Photosynthesis and carbon allocation in Pinus taeda

Seedling establishment Full-sib seed of loblolly pine (*Pinus taeda* L.) were surface sterilized with dilute sodium hypochlorite solution and then planted into Ray-Leach pots containing 150 cm^3 autoclaved peat-vermiculite (1 : 2.5 v/v) mix. Seedlings were grown in a HEPA-filtered fibreglass chamber within a standard greenhouse to minimize airborn contamination by fungal spores. Day/night temperatures were 26/22 $\pm$ 2°C. Plants were fertilized weekly with Hocking's nutrient solution[16]. Four weeks following germination, seedlings other than controls were inoculated with a slurry of mycelium of *Pisolithus tinctorius* (Pers.) Coker and Couch prepared from rinsed mycelial mats previously grown in liquid nutrient media.

At 1, 2, 3, 5, 7 and 9 months following fungal inoculation, 4 to 6 plants of the control (noninoculated) and fungal-inoculated treatments were removed from the greenhouse. The root systems were gently washed with distilled water to remove the potting mix and clinging debris was removed by forceps. Plants were then transplanted into the root chamber with half-strength Hoagland's nutrient solution[15] containing 25 ppm gentamicin to inhibit bacterial growth and aerated with millipore filtered air. Previous studies had indicated little apparent effect of gentamicin on mycorrhizas metabolism or growth of *P. tinctorius* in pure culture[17]. After transfer, each seedling was placed in a controlled-environment growth chambre with a 16-h photoperiod, 20 $\pm$ 1°C day/night temperature, and fluorescent/incandescent lighting about 350 μEinstein (E)/m^2/s at seedling shoot height.

Net photosynthesis, respiration, and transpiration measurements After 24 hours in the growth chamber, the shoot was sealed in a plexiglas, 700 cm^3 volume assimilation cuvette, isolated from the root system by Terostat sealant (Terosan GmbH Heidelberg), and the seedling was placed under a bank of seven, 300W incandescent flood lamps with a waterbath filter. Quantum flux density within the shoot cuvette was 1200 μE/m^2/s and the temperature was maintained at 25 $\pm$ 2°C. After 30 min of acclimation, Pn rates were determined by CO_2 analysis using a Beckman Model 315A infrared gas analyzer (IRGA) in an open system. Ambient air was pumped through a 210-l mixing reservoir, dehumidified with anhydrous $CaSO_4$ and divided into sample and reference lines. The length and volume of the pathway was identical for both sample and reference lines. Flow rates (*ca.* 1.4 l/min) were adjusted to maintain CO_2 concentration in the sample cuvette within ± 10% of ambient. The CO_2 concentration differential between the reference and sample lines was measured following calibration of the IRGA for a full scale deflection of 50 ppm using CO_2 standard gases.

Dark respiration (Rs) was determined by placing the seedling immediately in the dark and measuring CO_2 exchange under similar temperature conditions as above.

Calculation of CO_2 exchange rates (mg CO_2/g/h) for Pn and dark Rs was on the basis of dry weight of green shoot. Transpiration (Tr) was measured under the same conditions by measuring the dewpoint temperature differential between inflow and outflow air of the cuvette with a Cambridge Model 880 dewpoint hygrometer. Transpiration was calculated as g H_2O/g/h on the basis of the dry weight of green shoot.

Pulse labelling with $^{14}CO_2$ Following measurements of Pn, dark Rs and Tr, seedlings were aseptically transferred to cuvettes designed for $^{14}CO_2$ uptake and subsequent determination of shoot and root respiration. Roots were placed in fresh Hoagland's nutrient solution containing gentamicin. The shoot and root system were separately sealed in individual cuvettes, and maintained in the same growth chamber conditions as before except with a day/night temperature of 30/20 ± 1°C. Following a 1–h equilibration period, 25 μCi of $NaH^{14}CO_3$ were converted to $^{14}CO_2$ by excess 85% lactic acid in a side-arm receptacle of the shoot cuvette. A continuous flow of air was maintained from the $^{14}CO_2$ source to the seedling cuvette by a closed circulation system. After a 5-h assimilation period, residual $^{14}CO_2$ was removed from the system by scrubbing the air with 10 *N* NaOH until CO_2 concentration was negligible (as determined by the IRGA). A flow of millipore filtered air sufficient to cause vigorous bubbling in 20 cm³ of 4*N* NaOH was then maintained through the cuvette to trap CO_2 during a subsequent 48-h period. This was done similarly for the root system cuvette. Correction for trapping of ambient CO_2 was made.

Determination of respired CO_2 Total carbon respired ($^{12}CO_2 + {}^{14}CO_2$) was determined by reacting aliquots of the NaOH (diluted to 2*N*) with 1 *M* $BaCl_2$ and titrating against 2*N* HCL to pH 9.3 (thymolphthalein blue indicator). Radioactive $^{14}CO_2$ was measured by diluting aliquots of the NaOH to 2*N* and counting by liquid scintillation counting (LSC) in a solution of toluene-cellosolve[31].

Determination of ^{14}C *in plant tissues* Fifty-three hours after initial introduction of $^{14}CO_2$, the plants were harvested and divided into components of green foliage, stem, woody roots, mycorrhizal short roots, and nonmycorrhizal short roots. Nonwoody apices which were not short roots were included in the woody fraction. Mycorrhizas per cent (number of mycorrhizal short root tips/total number of short root tips) was determined visually with a dissecting scope. Mycorrhizal infection was verified by randomly selecting root tips for sectioning and staining with phloxine-B[22]. All separated plant components were freeze-dried and weighed. Each component was combusted in a Packard Model 306 Sample Oxidizer and the resulting $^{14}CO_2$ from each sample was trapped in Permafluor IV-Carbosorb II (Packard Inst. Co.) (12:7 v/v) LSC solution. Specific activity of plant tissue was expressed as disintegrations per min (dpm)/mg dry weight after correction of LSC counts for background and counting efficiency. The quantity of ^{14}C considered to be exuded by the root was determined by freeze-drying the root-bathing solution and rewetting with distilled water back to 5 ml. Aliquots were then counted in Ista-gel (Packard Inst. Co.) LSC solution. Activity was expressed on the basis of root dry weight (dpm/mg).

Foliar analysis The green shoot material from 4 to 6 plants treated similary to those used above for Pn measurements were pooled together and freezedried. Foliar N concentration was determined by the ammonia electrode method[9] following Kjeldahl digestion. Total P was determined by the molybdate blue method[24] after ashing of tissue with $Mg(NO_3)_2$[2].

Nitrogen and light effects on growth and photosynthesis of Pinus contorta

The overall objective was to examine the interaction between N, light intensity, and mycorrhizal fungi inoculation on growth and photosynthesis. To accomplish this, seedlings were grown for 10 weeks under differing light and N regimes, then inoculated with mycorrhizal fungi and grown for an additional 6 weeks. This particular design was partly dictated by our need to examine root carbohydrate levels immediately prior to fungal inoculation and shortly after infection took place. Carbohydrate data will not be reported here.

Seedling establishment Seed of lodgepole pine (*Pinus contorta* L.) were surface-sterilized in dilute hypochlorite solution and direct-sown into autoclaved vermiculite-peat (5:2 v/v) potting mix contained in Ray-Leach pots (150 cm^3). All seedlings were grown in a HEPA-filtered fibreglass chamber as before. Day/night temperatures were $30/26 \pm 2°C$, and a 16-h photoperiod was maintained by supplemental lighting.

Nitrogen fertilization and light regimes Although ambient sunlight in the greenhouse varied during the 16-week growing period, three relative levels of irradiance were established using commercial shadecloth netting (nominal rating of 55% shade). Placing two, one or no layer(s) of shadecloth above the seedlings gave about 100, 210, and 470 $\mu E/m^2/s$ quantum flux density, respectively, at midday during the part of the growing season with the lowest ambient sunlight. Maximum values reached without shadecloth were near 610 $\mu E/m^2/s$. Seedlings were fertilized with Hocking's nutrient solutions[16] with the following modifications: N was added as NH_4NO_3, to give levels of 3, 62, and 248 ppm, $CaCl_2$ was substituted for $Ca(NO_3)_2$, and S was changed from 150 ppm to 64 ppm. To maintain container concentrations of N near the applied levels, nutrient solution was added daily in excess to allow flushing of the potting mix. No other irrigation was necessary. Nine groups of 45 seedlings each were randomly assigned to the light and N treatments. These groups were then rerandomized on the greenhouse bench weekly. At 5 and 10 weeks, 5 seedlings were randomly selected from each treatment combination for determination of Pn, dark Rs, growth parameters and foliar N and P.

Mycorrhizal fungi inoculation At 10 weeks, mycorrhizal treatments were superimposed on the light and N treatments. Fungal inocula of *P. tinctorius* and *Suillus granulatus* (L. ex Fr.) O. Kuntze were prepared in vermiculite-peat[23] and stored at 4°C until used. Thirty seedlings from each N and light treatment combination were removed from their containers and their roots rinsed with sterile tap water. Seedlings were then replanted into a mix of vermiculite-peat and inoculum (3:1 v/v); 10 seedlings from each fungal species, and 10 seedlings as controls (noninoculated). The replanted seedlings were then placed in the same N and light treatment groups and rerandomized on the bench. After 6 weeks, 5 seedlings from each treatment combination were randomly selected for determination of Pn, dark Rs, growth parameters, mycorrhiza development and foliar N and P. The remaining 5 seedlings in each treatment combination were utilized in carbohydrate analyses (not reported here).

Photosynthesis and respiration measurement Net photosynthesis and dark Rs were determined as for *P. taeda* above with the following exceptions: Seedlings remained in the vermiculite-peat mix and were well-watered prior to measurement. At 10 and 16 weeks, Pn rates were determined at 100, 210, and 470 $\mu E/m^2/sec$ by use of shadecloth for each appropriate light treatment; at 5 weeks, all seedlings were measured at 470 $\mu E/m^2/sec$. The sequence of selecting seedlings for Pn measurements from each treatment was completely random. The Pn and dark Rs rates were determined at $31 \pm 2°C$ in a 700 cm^3 volume cuvette continuously stirred by a small fan to minimize boundary air resistances. Individual needle length and widths were measured and surface area calculated by modification of the method of Bingham[5]. Rates of Pn and Rs were expressed as mg $CO_2/dm^2/h$.

Dry weight and foliar analyses Root and shoot weights were determined after oven-drying at 68°C for 48 hours. Total N was determined by the salicylate-isocyanurate method[4] and P by the molybdate blue method[24] after digestion of needle tissue[40].

Statistical analysis

Measured parameters were examined by analysis of variance[38]. When significant differences were found, treatment means were compared by Tukey's test for Honestly Significant Difference (HSD) or by a two-tailed Student's 't' test[38].

To simplify data presentation and interpretation, in some cases treatments were pooled if a significant interaction did not occur in the analysis of variance.

Results

Photosynthesis and carbon allocation in Pinus taeda

Photosynthesis and growth response to mycorrhizal fungi inoculation The increase in mycorrhiza formation with age is illustrated in Fig. 1. Four weeks following inoculation with *P. tinctorius*, 2-month-old seedlings had an average of 9% mycorrhizas and at 10 months of age, 73%. Noninoculated seedlings remained free of mycorrhizas. The increase of Pn with age of mycorrhizal seedlings (Fig. 2) corresponded with the increase in mycorrhizas per cent. The Pn rates of mycorrhizal seedlings were significantly greater than nonmycorrhizal seedlings at all ages except 3 months, and at 10 months, the average Pn rate of mycorrhizal seedlings was 2.1 times greater than nonmycorrhizal seedlings. The positive effect of inoculation on Pn was also reflected in total plant biomass (Fig. 3). At 4 months of age, where a large increase in mycorrhiza development had occurred (Fig. 1), dry weight of mycorrhizal seedlings (Fig. 3) was significantly

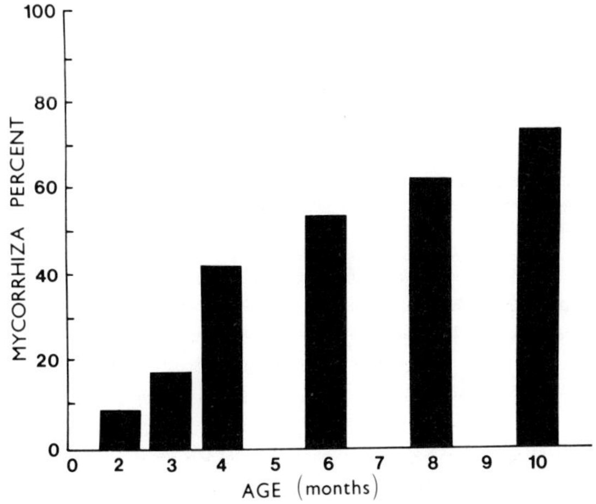

Fig. 1. Mycorrhiza formation in *Pinus taeda* at different ages following inoculation with *Pisolithus tinctorius* at one month of age. All comparisons of mycorrhizas per cent are significantly different ($p = 0.05$) from one another except for 6 and 8 months.

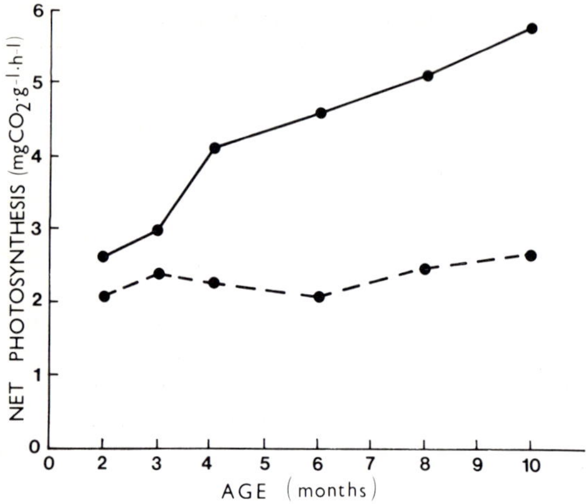

Fig. 2. Net photosynthesis (Pn) of mycorrhizal (●—●) and nonmycorrhizal (●- -●) *Pinus taeda* seedlings at different ages. Pn of mycorrhizal plants are significantly different ($p = 0.05$) from nonmycorrhizal plants at each age except 3 months.

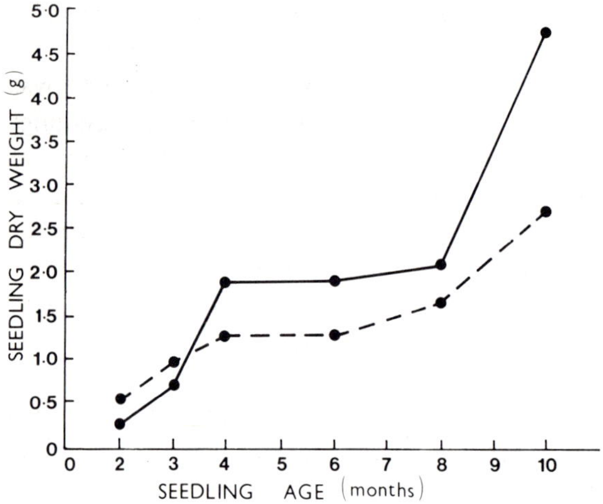

Fig. 3. Total plant dryweight of mycorrhizal (●—●) and nonmycorrhizal (●- -●) *Pinus taeda* seedlings at different ages. Dry weight of mycorrhizal seedlings is significantly different ($p = 0.05$) from nonmycorrhizal seedlings at each age except 3 months.

greater than nonmycorrhizal seedlings and continued to be so up to 10 months. However, dry weight of mycorrhizal seedlings was significantly lower at 2 months. A comparison of transpiration rates of mycorrhizal and nonmycorrhizal seedlings indicated no significant differences ($p = 0.05$) at any age, suggesting that stomatal conductance was not an explanation for differences

Table 1. Foliar nutrient concentration of N and P of mycorrhizal and nonmycorrhizal *P. taeda*

Seedling age	Nitrogen (%)		Phosphorus (%)	
in months	Nonmycorrhizal	Mycorrhizal	Nonmycorrhizal	Mycorrhizal
2	0.8	1.1	0.08	0.09
3	0.9	1.3	0.07	0.10
4	1.1	1.7	0.09	0.12
6	1.4	2.1	0.10	0.16
8	1.5	2.4	0.07	0.15
10	1.2	2.2	0.08	0.18

Foliar concentrations represent nonreplicated analyses of the pooled shoots of 4–6 seedlings at each age.

in Pn rates. The Tr rates increased slightly with age and averaged 0.75 and 2.1 g H_2O/g/h at 2 and 10 months of age, respectively.

The foliar contents of N and P were consistently higher in mycorrhizal than nonmycorrhizal seedlings at all ages (Table 1). However, it isn't possible to determine if the differences are statistically significant, since needles of seedling replicates were combined in order to have large enough samples for analysis.

Carbon allocation after $^{14}CO_2$ pulse labelling The quantity of ^{14}C was determined in tissue components of mycorrhizal and nonmycorrhizal seedlings 53 hours after introduction of $^{14}CO_2$. However, the quantity of ^{14}C attributed to shoot and root respiration represents that trapped in alkali over 48 hours since a 5-h assimilation period occurred before removal of residual $^{14}CO_2$ from the shoot cuvette. Since the actual quantity of $^{14}CO_2$ assimilated by each seedling over the 5-h period increased with age, distribution of ^{14}C was 'normalized' by expressing ^{14}C in various components as a percentage of total assimilated ^{14}C (Table 2), *i.e.*, total recovered in tissues and as respiration by alkali trapping. The total ^{14}C assimilated, also presented in Table 2, indicates that greater amounts were fixed at older ages. Mycorrhizal plants generally fixed greater amounts than nonmycorrhizal plants and this is consistent with their larger size and greater Pn rates. Analysis of variance of data represented in Table 2 (excluding the mycorrhizal short root fraction) indicated that age significantly ($p = 0.05$) influenced the ^{14}C distribution in all measured components, while mycorrhizal fungi inoculation significantly ($p = 0.05$) influenced distribution in all components except needles.

The loss of ^{14}C to both root and shoot respiration was significantly greater in the mycorrhizal plants. For example, the percentage of ^{14}C allocated to root respiration of mycorrhizal plants was 5, 4 and 2 times greater than nonmycorrhizal plants at 4, 6, and 10 months, respectively. Measurement of total carbon ($^{14}C + {}^{12}C$) respired by the roots during the same period (Fig. 4)

Table 2. Distribution of ^{14}C (as per cent of total fixed) into various components of mycorrhizal (MYC) and nonmycorrhizal (NM) seedling at various ages

Age (months)		Shoot				Roots						Total ^{14}C fixed (10^3 dpm)
		Needles	Stem	Respiration	Total	Woody	MYC short	NM short	Respiration	Exudation	Total	
2	MYC	28.62	27.30**	5.52†	61.4	32.24**	0.10	0.18	6.02*	0.12	38.7	77
	NM	26.94	53.02	3.80	83.8	13.28	—	0.12	2.84	0.10	16.3	270
3	MYC	27.58	47.02	3.16	77.8	16.90	0.06	0.08	5.00†	0.20	22.2	315
	NM	28.96	53.04	3.44	85.4	11.04	—	0.10	3.30	0.12	14.4	513
4	MYC	36.80	30.10*	6.30	73.2	14.67	0.10	0.10	11.82**	0.17	26.9	856
	NM	32.47	48.05	4.80	85.3	12.00	—	0.10	2.42	0.25	14.7	587
6	MYC	15.73*	13.92**	5.70	35.4	33.52**	0.10	0.10†	30.42**	0.43	64.6	1528
	NM	23.87	42.55	5.22	71.6	20.40	—	0.17	7.60	0.20	28.4	698
8	MYC	19.32	26.47	7.32†	53.1	24.62	0.10	0.12	21.65	0.37	46.9	1363
	NM	20.57	35.40	3.52	59.5	21.47	—	0.20	18.42	0.22	40.3	1076
10	MYC	29.26	25.40*	6.96	61.6	14.36	0.10	0.02*	23.50*	0.40	38.4	4222
	NM	28.76	38.66	6.14	73.6	14.30	—	0.16	11.66	0.32	26.4	1701

**, *, † indicates that the component mean of the mycorrhizal plants is significantly different from nonmycorrhizal plants at each age by Student's t-test at $p = 0.01$, 0.05, and 0.10, respectively.

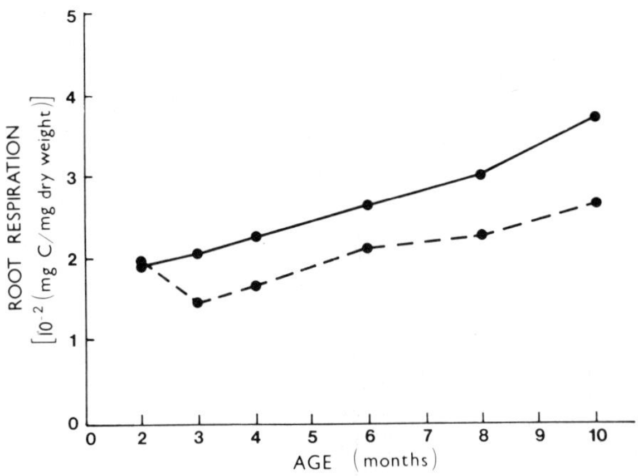

Fig. 4. Total root respiration ($^{12}CO_2$ + $^{14}CO_2$) of mycorrhizal (●——●) and nonmycorrhizal (●– –●) *Pinus taeda* seedlings over a 48-h period. Total carbon respired by mycorrhizal seedlings is significantly different ($p = 0.05$) from nonmycorrhizal seedlings at each age except 2 months.

also showed significantly greater carbon loss by mycorrhizal plants. The ratio of carbon respired by mycorrhizal to that by nonmycorrhizal plants (MYC Rs/NM Rs), after 2 months, did not change appreciably with age (ratio $\simeq$ 1.3). If the total carbon respired at 10 months is calculated (mg C/mg dry weight × mg total root weight), the MYC Rs/NM Rs ratio is 1.95, similar to the ratio of %^{14}C loss at 10 months (23.5%/11.7% = 2.01, Table 2).

Apparent effects on translocation by mycorrhizal development are indicated by the significantly greater ^{14}C distribution into the woody root component and lesser distribution in the stem component of mycorrhizal plants (Table 2). Specific activity of tissue components of 10-month-old plants is presented in Table 3. The specific activities in the stem (422 dpm/mg) and woody roots (508 dpm/mg) are unexpectedly high relative to mycorrhizal (242 dpm/mg) and nonmycorrhizal short roots (160 dpm/mg), but may indicate considerable levels

Table 3. Specific activity of ^{14}C (dpm/g dry weight) of tissue components of mycorrhizal and nonmycorrhizal seedlings of *P. taeda* pulse-labelled at 10 months of age

	Tissue component				
	Needles	Stem	Woody root	Mycorrhizal short roots	Nonmycorrhizal short roots
Mycorrhizal seedlings	1050 ± 46.3[1]	442 ± 79.5	508 ± 40.1	242 ± 18.2	160 ± 22.1
Nonmycorrhizal seedlings	846 ± 196	526 ± 105	286 ± 61.7	–	162 ± 44.1

[1] Mean ± standard error.

of current photosynthate in the phloem not yet transported to ultimate sinks. Although mycorrhizal short roots had higher specific activities than uninfected short roots, this alone is not an indication of sink strength since Table 2 and Fig. 4 indicate a considerable loss of current photosynthate as respiration from mycorrhizal rootsystems on a per unit dry weight basis.

Nitrogen and light effects on growth and photosynthesis of Pinus contorta

Photosynthesis and growth response of seedlings to light and N Analysis of variance indicated a significant ($p = 0.05$) effect of N and light treatments on Pn. Since there were no significant interactions between the N, light or mycorrhizal fungi inoculation treatments at the 16-week harvest, all observations on Pn were pooled within N or light treatments (Table 4). Increases in both N and light resulted in increased Pn rates. Similarly, both light and N treatments generally resulted in significant increases in seedling biomass. Since there was a significant interaction between the two treatments, treatment means are presented in Figure 5 for both light and N relative to age of noninoculated seedlings. At 5 and 10 weeks, the effect of N treatment on root biomass was greater as light intensity increased. However, at 16 weeks, root dry weight doubled in response to an increase of N from 3 ppm to 248 ppm at low light, even though overall growth was small compared to high light. Although shoot dry weight increased with increasing light level at any given age, N had little influence on dry weight except at 248 ppm. The increase in root-to-shoot ratio with age is well illustrated at all three light treatments.

Table 4. Effect of N and light levels on net photosynthesis (mg $CO_2/dm^2/h$) of *P. contorta* measured at 16 weeks[1]

	Nitrogen (ppm)			Relative quantum flux density ($\mu E/m^2/s$)		
	3	62	248	100	210	470
Net Pn	1.52^{a}[2]	1.64^{a}	1.99^{b}	1.45^{a}	1.64^{a}	2.05^{b}

[1] Determined 6 weeks after inoculation.
[2] Mean values followed by similar letters are not significantly different ($p = 0.05$) by HSD within either nitrogen or light treatments.

Mycorrhiza formation as affected by light and N After only 6 weeks following inoculation, both light and N treatments significantly ($p = 0.05$) influenced mycorrhiza formation, but without a significant interaction. A level of 62 ppm N gave a significantly higher mycorrhizas per cent (7.04) than either 3 or 248 ppm N (Table 5). Mycorrhizas per cent increased with increasing light intensity and was more than double at 470 $\mu E/m^2/s$ (6.40%) as compared to 100 $\mu E/m^2/s$ (2.74%).

Table 5. Effect of N and light levels on mycorrhizas per cent of 16-week-old *P. contorta* at 6 weeks following inoculation

	Nitrogen (ppm)			Relative quantum flux density (μE/m²/s)		
	3	62	248	100	210	470
MYC %	4.41^{ab1}	7.04^{a}	2.37^{b}	2.74^{a1}	4.68^{ab}	6.40^{b}

[1] Mean values followed by similar letters are not significantly different ($p = 0.05$) by HSD.

Effect of inoculation on photosynthesis, biomass, and foliar N and P Table 6 presents the mean values for Pn and biomass of *P. contorta* seedlings at 16 weeks of age, 6 weeks after inoculation. Since there was no significant interaction in analysis of variance between inoculation, N and light treatments at this harvest date, N and light treatment observations were pooled within inoculation treatments. Even though the average mycorrhizas per cent in the *P. tinctorius* and *S. granulatus* treatments were 4.4 (range 0.8 to 10.5) and 4.8 (range 1.2 to 12.9), respectively, they significantly influenced Pn, and shoot and root dry weight. Although Pn and shoot biomass means of the *P. tinctorius* treatment were not significant at $p = 0.05$, they were at $p = 0.10$.

Examination of foliar contents of N and P (Figs. 6 and 7) suggests that P, but not N, might be responsible for the marked response to mycorrhizal fungi inoculation. The effects of inoculation on %N and %P are presented for each N and light treatment because of a significant interaction between the two treatments. It is obvious that inoculation with either *P. tinctorius* or *S. granulatus* significantly increased %P over noninoculated controls at all three light treatments, even though the shoots of the inoculated plants had greater dry weights (an expected dilution of P). This increase in %P is accentuated by the fact that %P decreased with increased age in noninoculated seedlings (data not

Table 6. Effect of mycorrhizal inoculation on net photosynthesis and biomass of 16-week-old *P. contorta*[1]

Inoculation treatment	Net photosynthesis ($mg\,CO_2/dm^2/h$)	Biomass (mg) Shoot	Biomass (mg) Root
Control	1.41^{a}	62.2^{a}	139.4^{a}
P. tinctorius	1.87^{b}	70.1^{ab}	163.8^{b}
S. granulatus	1.85^{b}	71.6^{b}	155.2^{ab}

[1] Determined 6 weeks after inoculation.
[2] Column mean values followed by similar letters are not significantly different ($p = 0.05$) by HSD.

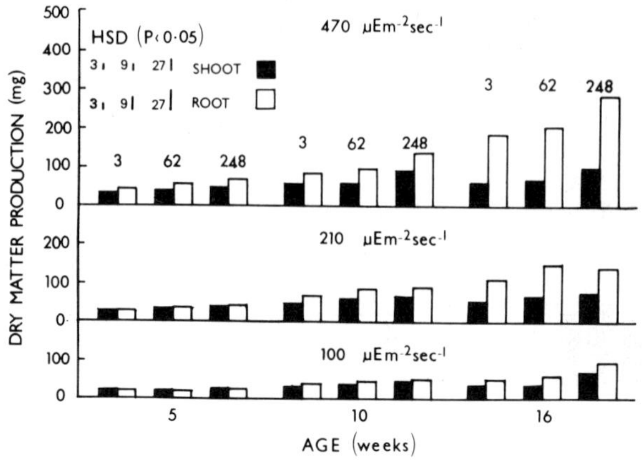

Fig. 5. Shoot and root biomass of nonmycorrhizal seedlings of *Pinus contorta* at 3 ages in response to 3 levels of irradiance (100, 210, 470 μE/m²/s) and 3 levels of N fertilization (3, 62, 248 ppm). Vertical bars labelled 3, 9 and 27 respresent HSD ($p \leqq 0.05$) for a) comparison of 3 means within any one treatment, b) comparison of 9 means within any two treatments, and c) comparison of 27 means among all treatments, respectively.

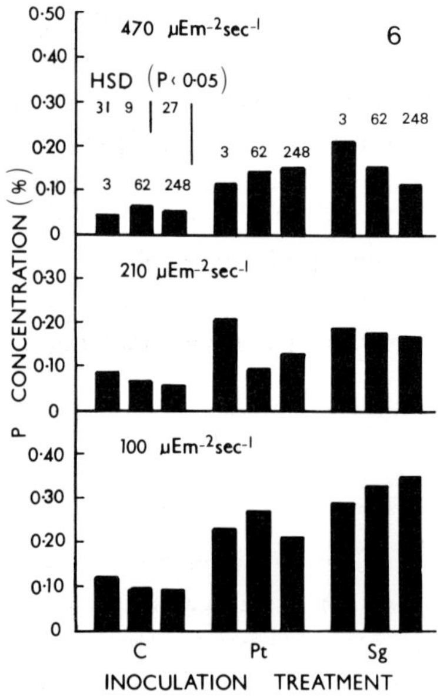

Fig. 6. Foliar concentration of P in 16-week-old *Pinus contorta* seedlings in response to mycorrhizal fungi inoculation at 10 weeks of age, 3 levels of irradiance (100, 210, 470 μE/m²/s), and 3 levels of N fertilization (3, 62, 248 ppm). C = noninoculated, Pt = inoculated with *Pisolithus tinctorius*, Sg = inoculated with *Suillus granulatus*. Vertical bars for HSD are as in Figure 5.

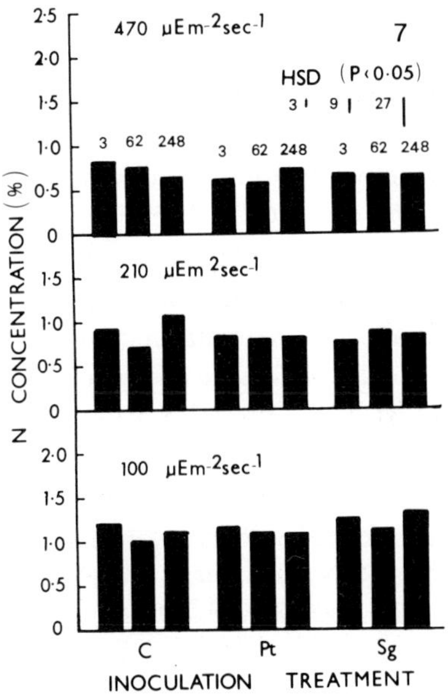

Fig. 7. Foliar concentration of N in 16-week-old *Pinus contorta* seedlings in response to mycorrhizal fungi inoculation, irradiance and N fertilization. See Figure 6 for details.

presented). Although mycorrhizas per cent was lowest at the lowest light level (Table 5), inoculation resulted in a substantial increase in P concentration (Fig. 6). Lower concentrations of P at high light and high N treatments are likely to be a dilution effect caused by greater dry matter production (*e.g.*, Fig. 5) at those combinations of light and N. The concentration of N was little affected by inoculation treatment (Fig. 7). Even though shoot biomass was increased by inoculation, analysis of variance indicated that total foliar N was not significantly changed. However, some statistically significant differences were observed between light and N treatments within particular inoculation treatments.

Discussion

Unquestionably mycorrhiza formation resulted in increased Pn rates and growth in both *P. taeda* and *P. contorta*. The causes for such increases, however, are equivocal. It is possible that increased photosynthetic rates and translocation rates, as observed in *P. taeda*, are the result of a greater sink by mycorrhiza formation since respiration of mycorrhizal root systems was significantly greater. The proportion of ^{14}C distributed to the roots of mycorrhizal plants was always greater than that distributed to roots of nonmycorrhizal plants of the same age (Table 2). Of the ^{14}C allocated to roots, an average of 39.5% and

28.8% was released as respiration by mycorrhizal and nonmycorrhizal root systems, respectively. Since ^{14}C distribution values represent a percentage of the total fixed, a quantity that varied with age as the shoots became larger, it is of interest to note that mycorrhizal plants also fixed more $^{14}CO_2$ than nonmycorrhizal plants when expressed as total activity per unit shoot weight or as per unit plant weight. For example, at 10 months, the ^{14}C fixed per unit shoot weight by mycorrhizal seedlings (73% infection) was 1.3-fold greater than by nonmycorrhizal seedlings. This would seem consistent with the 2.1-fold greater Pn rates (5.8 *vs* 2.7 mg CO_2/g/h) and larger size of the mycorrhizal plants.

It is important to emphasize that a much greater difference in respiration between mycorrhizal and nonmycorrhizal short roots probably exist than indicated by the %^{14}C distribution values. For example, the dry weight of short roots of 10-month-old seedlings (both mycorrhizal and nonmycorrhizal) was less than 2% of the total root dry weight. The mean dry weight values at 10 months for total roots, mycorrhizal short roots, and nonmycorrhizal short roots, were 1184, 15.2, and 4.6 mg, respectively, for mycorrhizal plants; and, 866, 0, and 14.8 mg, respectively, for nonmycorrhizal plants. A shift in 77% (15.2 mg/19.8 × 100) of short root biomass to mycorrhizas, representing 1.3% of the total root biomass, resulted in a 2-fold change in total root respiration based on the per cent of ^{14}C assimilated. The actual quantity of ^{14}C respired by root per unit total root weight was 838 dpm/mg for mycorrhizal plants and 230 dpm/mg for nonmycorrhizal plants at 10 months of age, a 3.6-fold difference. Although it is impossible to determine the actual respiration rates of the various root components from these studies, there are reports of mycorrhizas respiration exceeding noninfected root respiration by 2 to 4-fold[3,32,33].

In addition to the direct contribution by mycorrhizal short roots to respiration, overall root respiration might be stimulated by improved mineral nutrition, increased translocation of carbohydrate to the root system as a whole or changes in hormonal balances.

The consistently larger percentage of ^{14}C allocated to the root system and smaller percentage in the stem of mycorrhizal seedlings (Table 2) suggest that the rate of translocation from shoot to root was also enhanced by mycorrhiza formation. This might be the result of greater utilization of current photosynthate by the mycorrhizas, especially in respiration. Greater utilization of photosynthate by sinks has been shown to increase translocation rates[8,41].

However, in the studies with *P. taeda*, the observed increases in foliar N and P concentration of mycorrhizal plants could have influenced both Pn and translocation. Both N and P content have been implicated, either directly or indirectly, in the photosynthesis of a number of plants[20,28,39]. Increases in Pn following increases in either N or P can result from a number of factors, including increased chlorophyll concentration[10], decreased resistances (both stomatal and mesophyll) to CO_2 diffusion[20], increased activity of carboxylating enzymes[25,29], and improved utilization of assimilates by sinks[42]. In Douglas-fir,

Brix[10] found that the rate of Pn increased with increases in foliar N concentration up to an optimum of 1.74%. The foliar N of mycorrhizal *P. taeda* in our studies was well above 1.7% after 4 months, whereas nonmycorrhizal seedlings' foliar concentration was between 1.1 and 1.5%.

In our studies with *P. contorta*, mycorrhiza formation enhanced Pn as did both increases in irradiance and N fertilization. However, the response of Pn to mycorrhizal fungi inoculation was apparently not directly related to N since inoculation had little effect on foliar N concentration (Fig. 7). In contrast, the influence of inoculation on foliar P concentration was dramatic. Foliar %P of mycorrhizal plants increased to more than 3-fold over controls in some treatments (Fig. 6). Plants were not intentionally placed under P stress during the growing period. In fact, it was believed that daily application of 31 ppm of P would be adequate for such small seedlings. Although the *P. taeda* seedlings were not grown under as closely controlled fertilization application as *P. contorta*, N and P limitations also were not intentionally imposed. It is interesting that a substantial difference in Pn and biomass between mycorrhizal and nonmycorrhizal *P. taeda* seedlings still occurred at 10 months when the small rooting volume (150 cm^3) of the containers was likely to have been well-occupied by roots, and the increased surface area expected to occur with mycorrhizal root systems might have been less significant in nutrient uptake. Perhaps this indicates that the enhanced growth response to mycorrhizal development cannot be entirely attributed to nutrient uptake, a view long maintained by Slankis[36].

In the case of *P. contorta*, it seems remarkable that such low mycorrhizas per cent, and after only 6 weeks, had such a marked influence on Pn, biomass and nutrient content. Although utilization of carbohydrate by the fungal symbiont may have established an assimilate sink and thereby stimulated translocation and photosynthesis, fungal-produced hormones or changes in root production of hormones could have influenced source-sink relationships or photosynthesis directly[18,26,42].

Not unexpectedly, the greatest mycorrhizal formation of *P. contorta* occurred at the highest light level and intermediate N level. Although the highest N fertilization level almost always resulted in greater plant growth, regardless of age or light treatments, the inhibition of mycorrhiza formation at the highest N level should be considered in selecting fertilization regimes for containerized seedlings. The level of N in proportion to P may be critical in affecting mycorrhiza development[6,7,11,14,34]. In our study, the best formation at 62 ppm N represented a N/P ratio of 2 : 1. The N/P ratios at the fertilization levels which gave the poorest formation (248 ppm N) and intermediate formation (3 ppm N) were 8 : 1 and 1 : 10, respectively.

Acknowledgements This research was supported in part by NSF Grand DEB 76-00344 and Cooperative States Research Service Project M/S 5312. We thank R Kent Reid for assistance in statistical analyses and John Rousseau for helpful discussions.

References

1 Allen M F, Smith W K, Moore T S and Christensen M 1981 Comparative water relations and photosynthesis of mycorrhizal and nonmycorrhizal *Bouteloua gracilis* H.B.K. Lag ex Steud. New Phytol. 88, 683–693.

2 Association of Official Agricultural Chemists 1965 Chapter 2. Fertilizers. Ed. W Horwitz. pp. 11–12. AOAC, Washington, D.C.

3 Barnard E L and Jorgensen J R 1977 Respiration of field-grown loblolly pine roots as influenced by temperature and root type. Can. J. Bot. 55, 740–743.

4 Bigelow D S, Scott G R and Adamsen F J 1982 Automated methods for ammonium, nitrate and nitrite in 2M KCL-PMA soil extracts. (*In prep.*).

5 Bingham G E 1980 Leaf area measurement of pine needles. *In* Instruction Manual. LI-1600 steady state porometer. LI-COR, Inc./LI-COR, Ltd., Lincoln, Nebraska, USA.

6 Bjorkman E 1942 Uber die Bedingungen der Mykorrhizabildung bei Kiefer und Fichte. Symb. Bot. Upsal. 6, 1–191.

7 Bjorkman E 1970 Mycorrhiza and tree nutrition in poor forest soils. Stud. Forstl. Suec. 83, 1–24.

8 Borchers-Zampini C, Glamm A B, Hoddinott J and Swanson C A 1980 Alterations in source-sink patterns by modifications of source strength. Plant Physiol. 65, 1116–1120.

9 Bremner J M and Tabatabai MA 1972 Use of an ammonia electrode for determination of ammonium in Kjeldahl. Commun. Soil Sci. Plant Anal. 3, 159–165.

10 Brix H 1981 Effects of nitrogen fertilizer source and application rates on foliar nitrogen concentration, photosynthesis, and growth of Douglas-fir. Can. J. For. Res. 11, 775–780.

11 Hacskaylo E 1957 Mycorrhiza of trees with special emphasis on physiology of ectotrophic types. Ohio J. Sci. 57, 350–357.

12 Hacskaylo E 1973 Carbohydrate physiology of ectomycorrhizae. *In* Ectomycorrhizae: Their Ecology and Physiology. Eds. G C Marks and T T Kozlowski. pp. 207–230. Academic press, New York.

13 Harley J L and Lewis D H 1969 The physiology of ectotrophic mycorrhizas. *In* Advances in Microbial Physiology, Vol. 3. Eds. A H Rose and J F Wilkinson. pp 53–80. Academic press, New York.

14 Hatch A B 1937 The physical basis of mycotrophy in the genus Pinus. Black Rock For. Bull. 6. 188 p.

15 Hoagland D R and Arnon D I 1950 The water-culture method for growing plants without soil. Calif. Agric. Exp. Stn. Circ. 347. (Revised edition).

16 Hocking D 1971 Preparation and use of a nutrient solution for culturing seedlings of lodgepole pine and white spruce, with selected bibliography. North. For. Res. Cent. Inf. Rep. Nor-X-1, 14 p. Can. For. Serv. Dep. Environ., Edmonton, Alberta, Canada.

17 Kidd F A, Wullschleger S D, Dawley K and Reid C P P 1982 Use of gentamicin in axenic culturing of ectomycorrhizal plants. Appl. Environ. Microbiol. 44, 506–508.

18 Ledig F T 1976 Physiological genetics, photosynthesis and growth models. *In* Tree Physiology and Yield Improvement. Eds. M G Cannell and F T Last. pp 21–54. Academic press, New York.

19 Lister G R, Slankis V, Krotkov G and Nelson C D 1968 The growth and physiology of *Pinus strobus* L. seedlings as affected by various nutritional levels of nitrogen and phosphorus. Ann. Bot. (N.S.) 32, 33–43.

20 Longstreth D J and Nobel P S 1980 Nutrient influences on leaf photosynthesis. Effects of nitrogen, phosphorus, and potassium for *Gossypium hirsutum* L. Plant Physiol. 65, 541–543.

21 Lösel D M and Cooper K M 1979 Incorporation of ^{14}C-labelled substrates by uninfected and VA mycorrhizal roots of onion. New Phytol. 83, 415–426.

22 Marx D H and Bryan W C 1971 Formation of ectomycorrhizae on half-sib progenies of slash pine in aseptic culture. For. Sci. 17, 488–492.

23 Marx D H and Bryan W C 1975 Growth and ectomycorrhizal development of loblolly pine seedlings in fumigated soils infested with the fungal symbiont *Pisolithus tinctorius*. For. Sci. 21, 245–254.

24 Murphy J and Riley J P 1962 A modified single solution method for determination of phosphate in natural waters. Anal. Chem. Acta. 27, 31–36.

25 Nátr L 1975 Influence of mineral nutrition on photosynthesis and the use of assimilates. *In* Photosynthesis and Productivity in Different Environments. Ed. J P Cooper. pp 537–555. IBP Programme 3, Cambridge University press, Cambridge.

26 Neales T F, Treharne K J and Wareing P F 1971 A relationship between net photosynthesis, diffusive resistance, and carboxylating enzyme activity in bean leaves. *In* Photosynthesis and Respiration. Eds. M D Hatch, C B Osmond and R O Slatyer. pp 89–96. Wiley-Inter-science, New York.

27 Nelson C D 1964 The production and translocation of photosynthate C^{14} in conifers. *In* The Formation of Wood in Forest Trees. Ed. M H Zimmermann. pp 243–257. Academic press, New York.

28 Osman A M, Goodman P J and Cooper J P 1977 The effects of nitrogen, phosphorus and potassium on rates of growth and photosynthesis of wheat. Photosynthetica 11, 66–75.

29 Osman A M and Milthorpe F L 1971 Photosynthesis of wheat leaves in relation to age, illumination and nutrient supply. II. Results. Photosynthetica 5, 61–70.

30 Paul E A and Kucey R M N 1981 Carbon flow in plant microbial associations. Science 213, 473–474.

31 Reid C P P and Woods F W 1969 Translocation of ^{14}C-labeled compounds in mycorrhizae and its implications in interplant nutrient cycling. Ecology 50, 179–187.

32 Routien J B and Dawson R F 1943 Some interrelations of growth, salt absorption, respiration and mycorrhizal development in *Pinus echinata*. Am. J. Bot. 30, 440–451.

33 Schweers W and Meyer F H 1970 Einfluss der Mykorrhiza auf den Transport von Assimilaten in die Wurzel. Ber. Dtsch. Bot. Ges. 83, 109–119.

34 Shemakhonova N M 1962 Mycotrophy in woody plants. Akad. Nauk. SSSR. Inst. Mikrobiol. Moskva (US Dep. Commerce Transl. 1967).

35 Shiroya T, Lister G R, Slankis V, Krotkov G and Nelson C D 1962 Translocation of the products of photosynthesis to the roots of pine seedlings. Can. J. Bot. 40, 1125–1135.

36 Slankis V 1973 Hormonal relationships in mycorrhizal development. *In* Ectomycorrhizae: Their Ecology and Physiology. Eds. G C Marks and T T Kozlowski. pp 231–298. Academic press, New York.

37 Smith D, Muscatine L and Lewis D 1969 Carbohydrate movement from autotrophs to heterotrophs in parasitic and mutualistic symbiosis. Biol. Rev. 44, 17–90.

38 Steel R G D and Torrie J H 1980 Principles and procedures of statistics. A biometric approach. McGraw-Hill Book Co., New York. 633 p.

39 Swada S, Igarashi T and Miyachi S 1982 Effects of nutritional levels of phosphate on photosynthesis and growth studied with single, rooted leaf of dwarf bean. Plant Cell Physiol. 23, 27–33.

40 Thomas R L, Sheard R W and Moyer J R 1967 Comparison of conventional and automated procedures for nitrogen, phosphorus, and potassium analysis of plant material using a single digestion. Agron. J. 59, 240–243.

41 Thorne J H and Koller H R 1974 Influence of assimilate demand on photosynthesis, diffusive resistances, translocation, and carbohydrate levels of soybean leaves. Plant Physiol. 54, 201–207.

42 Wareing P F and Patrick J 1975 Source-sink relations and the partition of assimilates in the plant. *In* Photosynthesis and Productivity in Different Environments. Ed. J P Cooper. pp 481–499. IBP Programme 3, Cambridge University press, Cambridge.

Plant and Soil 71, 433–443 (1983). A-49

The structure and function of mycelial systems of ectomycorrhizal roots with special reference to their role in forming inter-plant connections and providing pathways for assimilate and water transport

C. BROWNLEE, J. A. DUDDRIDGE, A. MALIBARI and D. J. READ
Botany Department, The University of Sheffield, Sheffield S10 2TN, UK

Key words Assimilate transfer Ectomycorrhiza Mycelial strands Pine *Pinus sylvestris Suillus bovinus* Vessel hyphae Water transport

Summary In the past function of mycorrhizas has been examined at the level either of the isolated individual root or of the individual entire plant. It is shown here that ectomycorrhizal mycelial strands can extend from plant to plant, thus initiating infection in seedlings, and that the resulting fungal interconnections provide functional pathways for the transfer of labelled assimilate between individuals. Mycelial strands also provide a pathway for the transport of physiologically significant quantities of water. Strand functions are examined in relation to structure, and their role as morphological and physiological extensions of the root system is emphasized. The significance of the experimental observations is discussed in relation to nutrient cycling processes in natural ecosystems.

Introduction

Most studies of the function of mycorrhizas have been carried out at the level of the isolated individual plant or root and their possible role at the community and ecosystem level has been largely neglected. In nature, however, mycorrhizas do not occur as isolated individual roots neither do plants normally occur as isolated organisms but rather as communities in which roots of individuals of a number of species grow in close juxtaposition. We know, also, that in contrast to the situation in most pathogenic host-fungus relationships, the mycorrhizal association shows little specificity. Thus some basisdiomycete fungi are capable of forming ectomycorrhizal associations with a wide range of forest trees[11] and a similar situation is found in relationships between the fungi and hosts of both VA[6] and ericoid mycorrhiza[8]. This general lack of specificity may be a significant factor in the natural ecosystem since seedling radicles growing into soil in established swards of vegetation will come rapidly into contact with roots supporting compatible mycorrhizas at a very early stage in their development. The nature of the infection process is potentially of profound importance since in the case where infection arises from contact with mycelium of an established plant the newly infected individual becomes, for some time at least, incorporated into a mycelial network through which it is connected to other plants. In contrast, infection of a seedling root by a spore or independent vegetative propagule leaves the plant as an individual entity. Some evidence for the existence of functional interconnections between ectomycorrhizal plants of pine was presented recently[2]. We now investigate the capacity of mycorrhizal mycelial strands of *Suillus bovinus* to initiate infection in developing pine seedlings and to act as functional pathways for the transport of current assimilate between plants, and of water to individual plants. These functions are assessed in relation to an analysis of the ultrastructure of the strands.

Materials and methods

Synthesis of mycorrhizas

Ectomycorrhizas were synthesized under aseptic conditions in 250 ml Erlenmeyer flasks containing a peat-vermiculite (1:4 v/v) mixture moistened with modified Melin-Norkrans solution (MMN). The medium was first inoculated with a disc of the appropriate mycorrhizal fungus. After the fungus had become established an aseptically germinated seedling of Pine was transferred to the flask which was then maintained in a growth chamber where it was exposed to a 16 hour day with day temperature of 15°C and night temperature of 10°C and irradiance of 38 Wm^{-2}.

After mycorrhiza formation had occurred the plants were transferred to chambers of different designs, each one appropriate for a particular type of experiment.

Experiments on initiations of infection

Perspex root chambers were employed in attempts to simulate the field circumstance in which an uninfected seedling root grows into a soil already occupied by mycorrhizal roots of established plants with their associated mycelial strand systems. Plants of *P. sylvestris* with mycorrhizas synthesized in Erlenmeyer flasks as described above were transferred to fresh non-sterile milled peat in Perspex chambers (10 × 10 × 1 cm) with detachable sides and base. The infected plant was placed on the surface of the peat so that most of its mycorrhizal roots were visible when the transparent wall of the root chamber was replaced. After approximately 21 days when the surface of the peat could be seen to be colonized by mycelium growing from the established plant, a germinating seedling of the same or of a related Pine was introduced into the chamber at about 5 cm from the established plant. In other chambers pine plants with established mycorrhizas were placed together with 5 day old uninfected seedlings of the same species on the surface of peat with approximately 5 cms between then. The spread of mycelium through the peat and the pattern of infection of the young seedlings was followed over the subsequent period of five weeks.

Transfer of carbon from plants with established mycrorrhizas to plants with developing mycorrhizas

Using associations of the type described above, the capacity of the established mycelial strands to act as pathways for the transport of carbon between plants was investigated. The shoot of the older plant was enclosed in a perspex box and exposed to $^{14}CO_2$ for 24 hours. 50 μCi of $^{14}CO_2$ was released from $NaH^{14}CO_3$ by addition of lactic acid and the entire system was maintained in the illuminated growth chamber. After 24 hours the fed 'donor' shoot was removed and the complete chamber was taken to air dryness at 40°C. The distribution of label was examined by autoradiography.

Transfer of water through rhizomorphs to seedlings

Seedlings of *Pinus sylvestris* with mycorrhizas formed by *Suillus bovinus* in Erlenmeyer flasks were transferred to divided Petri dishes containing unsterile milled moss peat moistened to 80% water content with one-fifth strength MMN but lacking glucose and malt extract. The seedling was arranged so that its roots were confined to the upper section of the petri-dish. In addition to the central plastic barrier in the dish, a peat free gap was maintained alongside the barrier in the lower half of the dish. Dishes were wrapped in aluminium foil so that the shoot alone was exposed to the atmosphere. The dishes were stacked in groups of five in plastic trays which were covered by clear perspex domes (Stewart Plant Propagators), the vents of which were loosely plugged with cotton wool. They were maintained in the same growth chamber environment as that described above, but the humidities around the shoots were higher (95% ± 5%) because of the presence of the propagator lids.

Over the subsequent period of 5 weeks during which time additional moisture was applied to the peat, strands spread over the surface of the peat in both sections of the dishes thus crossing the barriers formed by the central wall and the gap (Plates 2a, b). By this time the water content of the peat in the upper sections of the dishes had been reduced from approximately 80% to less than 60% in all dishes as a result of absorption of water by the mycorrhizal roots concentrated in that part of the system. No further water additions were made to the upper sections but water was added to peat in the lower section to maintain a supply of moisture to the mycelial strands growing in that section. The capacity of strands to maintain a water supply to the shoot was tested by measurement of the change in needle water potential before and after cutting of the mycorrhizal strands. Water potentials were measured using a Scholander pressure bomb.

Ultrastructure of mycelial strands

Mycelial strands of *Suillus bovinus* were dissected away from the peat, immediately fixed using 2.5% gluteraldehyde in 0.1 *M* cacodylate buffer, post-fixed in 1% osmium tetroxide and dehydrated in a graded ethanol series. Specimens were infiltrated in Spurr's low viscosity resin. Sections were cut on a Reichert OMU 3 ultramicrotome and viewed on a Jeol Jem T7 electron microscope at 60 Kw.

Results

The spread of infection from established mycorrhizas to uninfected seedlings. Rates of development of mycelial systems and of mycorrhizal infection differ from chamber to chamber depending upon peat consistency and moisture level, upon distances between infected and uninfected seedling roots and direction and speed of root development. In a representative circumstance fans of mycelium can be seen emerging from established mycorrhizal roots within 10 days of transferring the infected plant to fresh peat. (Plate 1a). The first contact with the expanding root system of the uninfected plant was made within two weeks in this chamber. Mycelial strands and individual hyphae contacting the already suberized outer surface of the primary root of this plant form a loose weft of hyphae on the surface but do not form a Hartig net. A major proliferation of hyphae takes place at the point of emergence of the short roots on which a sheath and Hartig net are quickly formed. Within five weeks from planting (Plate 1b, c) the inoculated seedling is heavily mycorrhizal and interconnections between the two seedlings can clearly be seen. These physical connections are maintained for several months after the initiation of the infection.

The function of inter-plant mycelial strand connections

Autoradiographs of the plants shown in Plate 2 taken 24 hours after the administration of $^{14}CO_2$ to the more mature plant show clearly that labelled assimilate has been distributed from the 'donor' plant to the young seedling. (Plate 1d). Particularly high levels of activity are seen in the recently formed mycorrhizal roots of the younger seedling. Low levels of activity were also detected in the shoot of the unfed plant.

This observation, which has been repeated on several occasions using the same experimental design, confirms that the mycorrhizal mycelial strands act not only as infective propagules but also as pathways for the transport of assimilate from one plant to another.

Water transport through mycorrhizal mycelial strands

Both experimental and observational analysis suggest that strands are capable of the absorption and transport of water in sufficient quantities to satisfy the requirements of the shoots. Determination of needle water potential before and after severing mycelial strands connecting plants to a water source (Table 1) shows that decreased water potentials are experienced in the needles very soon after a cutting. This suggests that water is absorbed along the length of that part of the strand associated with moist peat and then transported to the roots and shoots in the dry areas of the system across the various barriers. The addition of an anhydrous lanolin layer to the central barrier does not affect the capacity of strands to act as pathways for water movement which indicates that water transport occurs internally rather than by way of external capillary channels.

Further indication that strands play a significant role in water transport is provided by studies of whole plant responses to the presence or absence of connecting strands. Seedlings remained in an apparently healthy condition for periods in excess of ten weeks when the only source of water was mycelial strands growing through moist peat. In contrast, seedlings lost colour and died within one week of severing the mycelial strand.

Table 1. Water potential (ψ – MPa) in detached needles of *Pinus sylvestris* after excision from plants growing with or without roots or rhizomorphs in contact with moist peat

1. Needles from plants with roots and rhizomorphs penetrating most peat (controls).
   0.4, 0.6, 0.6, 0.8, 0.5, 0.6.

2. Needles from plants with all connections with moist peat severed.
   ψ measured after 1 and 2 hours.

1 hour	0.9	2.6	1.4
2 hours	4.9	7.3	3.8

3. Needles from plants with rhizomorphs only in most peat. Mean ψ of 2 needles per plant.
   0.5, 1.0, 0.7, 0.6

4. Needles from same plants as (3) but with rhizomorphs cut. ψ measured after 1 and 2 hours.

1 hour	1.7	1.9	2.6	2.8
2 hours	4.2	3.8	6.7	7.8

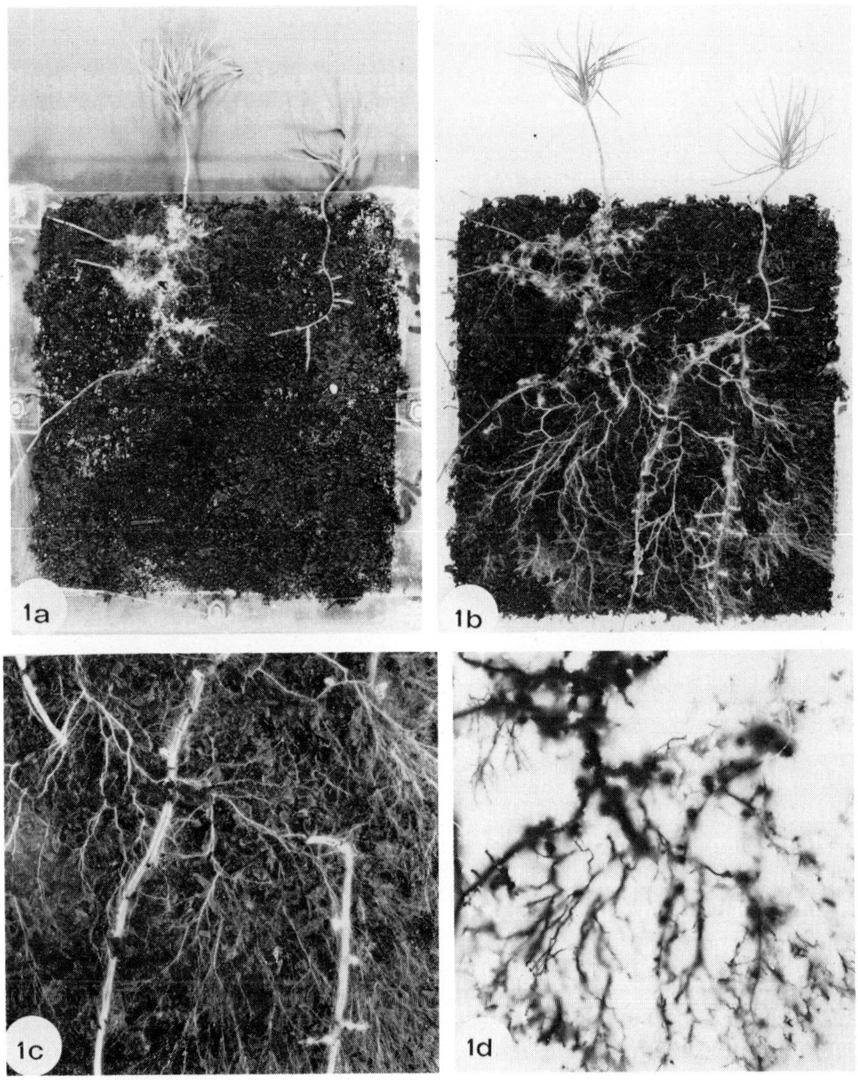

Plate 1.

a. Perspex root chamber containing infected and uninfected and seedlings of *Pinus sylvestris*, ten days after transplantation of seedlings to the peat. Fans of mycelium of *Suillus bovinus* which already show some differentiation into strands can be observed.

b. The same chamber as in la but after 5 weeks. Extensive infection of the younger seedling is now evident and the chamber is almost completely exploited by mycelial strands and hyphae.

c. More detailed view of part of the chamber at 5 weeks showing surface colonisation of the primary root of the young seedlings and conversion of first order laterals to mycorrhizas. The very intensive exploitation of the peat is also evident.

d. Autoradiograph of chamber shown in Plate 1b taken 24 hours after exposure of the older seedling shoot to $^{14}CO_2$. Extensive transfer of labelled materials to mycelial strands and to mycorrhizas of the newly infected seedlings has occurred.

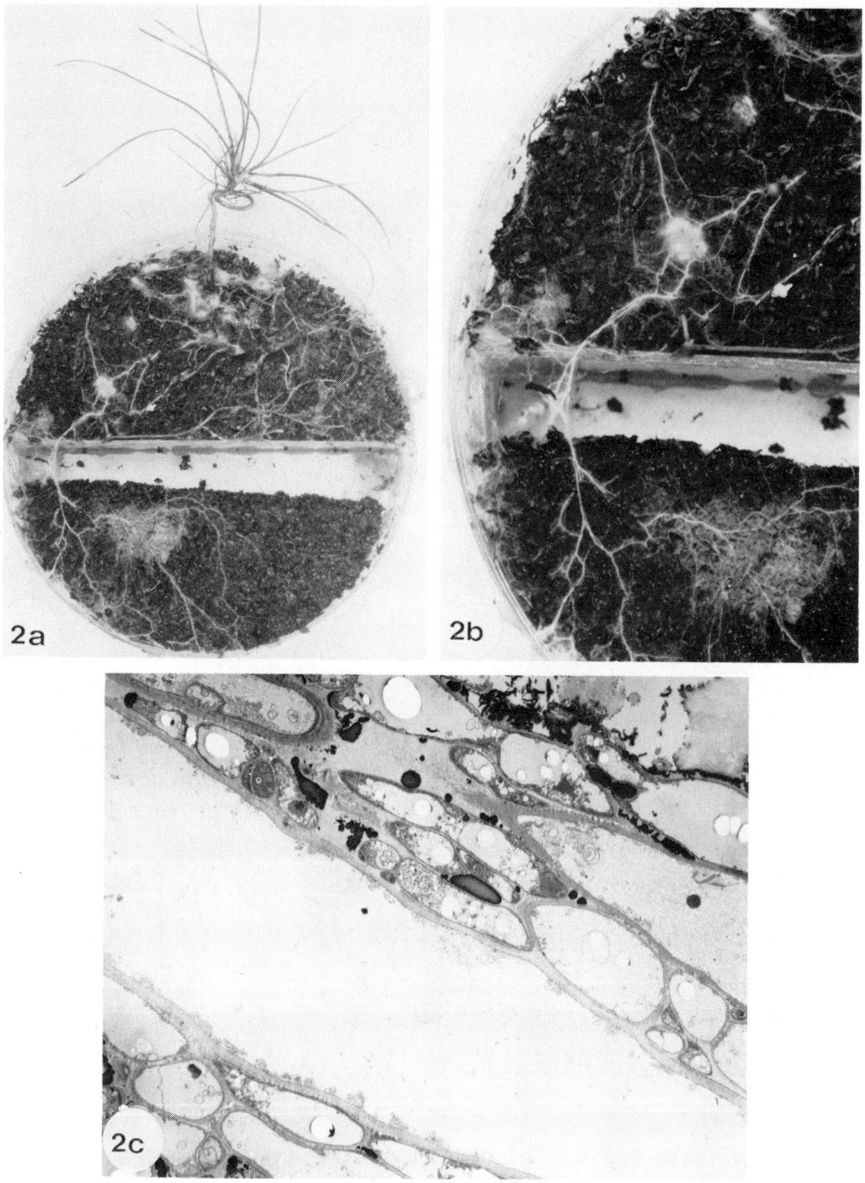

Plate 2.
a. Divided petri-dish system for analysis of water transfer in mycelial strands.
b. Detail of the petri-dish system showing the two translocation barriers over which the strands grow to exploit moist peat at the base of the dish.
c. TEM of longitudinal section of mature mycelial strand of *Suillus bovinus* showing fully differentiated central vessel hypha, lacking cytoplasmic content, surrounded by a sheath of hyphae with cytoplasmic contents × 8000.

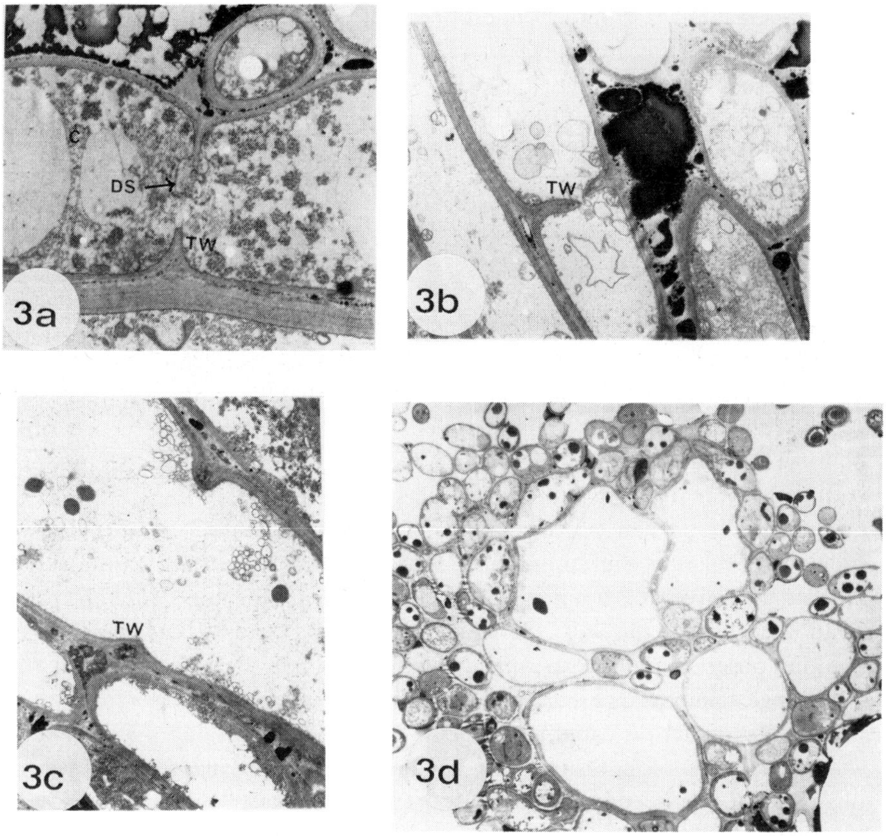

Plate 3.

a. Longitidinal section through young differentiating vessel hypha showing breakdown of the transverse wall, remnants of the dolipore septum (DS), and cytoplasm (C). × 7300.

b. Later stage in differentiation of vessel with little cytoplasm remaining but transverse wall still in evidence. × 4000.

c. Breakdown of transverse wall almost complete. Cytoplasm absent from the cell. × 3800.

d. TEM of transverse section through mycelial strands of *S. bovinus* showing central cluster of vessel hyphae and sheath of cytoplasmic hyphae. × 1800.

Structure of mycelial strands

TEM analysis of longitudinal sections through mycelial strands of *Suillus bovinus* reveal that the structure is differentiated into a core region of elongate cells with little or no cytoplasmic content and an outer 'sheath' of less elongate cells some with dense cytoplasmic contents. Sections also reveal various stages in the differentiation of the central hyphae from the cytoplasmic young condition towards the mature condition in which they resemble the 'vessel' hyphae seen in *Serpula lacrymans.*[3]. Stages of breakdown of the transverse walls of the central hyphae are shown from a condition in which the dolipore septum and cytoplasmic contents are visible (Plate 3a) to those in which only vestiges of the original cross wall are visible (Plates 3b, c). Transverse sections of mature *Suillus bovinus* strands viewed by TEM show an irregular distribution of vessel hyphae of large diameter and lacking cytoplasm, and a sheath of hyphae of smaller diameter most of which are densley cytoplasmic (Plate 3d).

Discussion

The experiments reported in this paper clearly demonstrate not only that mycelial strands growing from mycorrhizal roots can colonize and produce mycorrhizal infection in young seedling roots, but that they provide a functional pathway through which assimilates can be transported from a 'source' to a 'sink' plant. The autoradiographic evidence indicates not only that the mycelial strands provide a direct pathway to the seedling roots but also that activity accumulates in the newly formed mycorrhizas in those roots. It has been shown previously[9] that mycelial strands have the capacity to take up and transport ^{14}C labelled carbohydrates but no evidence of movement of isotope from one plant to another through mutually shared mycelium could be obtained. Transport of the kind observed in our experiments would be of considerable ecological and physiological importance in nature where the germination of seeds of forest trees often occurs under the canopy of overstorey trees which produces considerable shade. The present results suggest that under these circumstances the root system of the developing seedling would become rapidly infected by fungi already in mycorrhizal association with mature trees and that the shaded seedling would act as a sink to which assimilates, originally synthesized in the fully illuminated crown of the overstorey trees, would be transported.

While the presence of low levels of activity in the shoots of the sink plant suggest that some transfer of assimilates has occured beyond the roots, it is clear that the major accumulation of label occurs in the mycorrhizal roots. The sheaths of ectomycorrhizal roots are known to be major sinks for assimilate, and conservative estimates suggest that 10% of current photosynthate may be required to maintain these structures[4]. Such a drain on host carbon resources might pose a threat to the survival of seedlings growing in poorly illuminated environments. The fact that some or all of the assimilate requirement can be met by supplies from neighbouring trees could therefore be of great ecological significance. The seedling is able to develop, with little energy expenditure, a mycorrhizal system which will lead to improved exploitation of soil nutrient resources in the intensely competitive surface layers of the forest floor. It is possible also, that in addition to inter-plant transfer of assimilates, nutrients such as phosphorus may be supplied directly from well established neighbouring plants through these connections. Such transfer has been shown to occur in plants connected by VA mycorrhizal hyphae[1,12].

Several factors may be involved in determining the polarity of movement of assimilates in mycelial strands. The individual mycorrhizal strand is dependent upon the host root for a supply of assimilates which probably move towards the advancing front of the structure along a gradient of decreasing concentration. In the case where the leading hyphae of the strands reach an uninfected lateral root they will be stimulated to greater growth by exudates of that root. The extent to which the growth enhancement is supported by carbon in the exudate is not known, but will presumably be dependent upon the carbon balance of the plant contacted. If, as would be expected in the case of a poorly illuminated seedling, the carbon balance is weak, the contact will increase rather than decrease the concentration gradient from original 'parent' root to developing mycorrhiza and thus increase the magnitude of the sink effect. This situation would explain the enhancement of transfer of carbon to the newly infected seedling observed in our experiments. It will be interesting in future to investigate the extent to which polarity can be reversed by shading older seedlings while illuminating the newly infected younger plants to which they are connected.

Previous studies demonstrated that mycelial strands provide a pathway for water transport along the ψ gradient from moist soil to transpiring seedlings[2]. It is now clear that under some circumstances at least, this supply can be adequate to satisfy seedling requirements. The present experiments were carried out in conditions conductive to low rates of transpiration since atmospheric water potential was high, and low air speeds in the growth chamber produced high boundary layer resistances. However, in some cases seedlings were connected to the moist peat by only one mycelial strand so that efficiency of exploitation of the medium was in turn much less than would be experienced in nature. Further experiments using a range of atmospheric water potentials, air flow rates and fungal associates are now being undertaken to provide a more detailed assessement of the role of mycelial strands in host water relations.

Ultrastructural analyses show that the differentiated mycelial strand is well adapted to function in a pathway for such movement. Large vessel hyphae with relatively few cross walls and no cytoplasmic contents all provide an efficient system for long distance transfer. The hagen-Poiseuille equation indiciates that the volume flow of water in each conduit will be dependent on the fourth power of the conduit radius, which means that in a mycelial strand with a range of conduit dimensions a small proportion of large bore vessel hyphae will carry the bulk of water flow. The vessel hyphae do not pass through the sheath and the precise pathway of water in this region is not known. It has been demonstrated however, in the case of roots of *Pinus taeda* infected with *Pisolithus tinctorius*, that the presence of a mycorrhizal sheath had no effect on hydraulic conductivity of the root[10]. From this it may be concluded that the major pathway of water movement is from vessel hyphae through the apoplast of the sheath and then to the cortical cell walls of roots and that this zone does not present a significant resistance to flow.

In the circumstances of the individual root with its attached mycorrhizal mycelial strand it is interesting to consider the pathway of movement of assimilates and water in the system. Water may be assumed to move from the tip of the strand backwards through vessel hyphae towards the seedling, while assimilates move in the opposite direction from the seedling, presumably through the outer cytoplasmic hyphae, to the growing area. Such a situation is analogous to that seen in xylem and phloem in the higher plant, and the mycorrhizal strand system can therefore be regarded as being both structurally and functionally analogous to the plant root system. It has long been known that ectomycorrhizal woody species have low root densities, values of LA (root length cm/cm^2 of ground surface) being as low as 5 in pine[5] compared with values of 1000–4000 in grasses and 50–300 in herbs[7]. The possibility emerges that in plants such as pine, carbon is invested in the formation of mycorrhizal mycelial strands which fulfil the function of roots and that the effective absorptive surface is therefore as high as or higher than those of plants which rely upon root hairs or VA mycorrhizal hyphae for soil exploitation.

In addition, since mycelial strands provide for inter-plant distribution of nutrient and water it is necessary to reasses the pattern of nutrient acquisition and cycling in forest ecosystems. Clearly plants in such systems do not occur as individuals but as part of an interconnected system through which nutrients flow along concentration gradients both at the intra and interspecific levels. The significance of these interconnections for seedling survival and plant productivity is currently being examined.

References

1 Chiariello N, Hickman J C and Moony H A 1982 Endomycorrhizal role for interspecific transfer of phosphorus in a community of annual plants. Science 217, 941–943.

2 Duddridge J, Malibari A and Read D J 1980 Structure and function of mycelial rhizomorphs with special reference to their role in water transport. Nature London 287, 834–836.

3 Falck R 1912 Die Merulius–Fäule des Bauholzes. Hausschwammforschungen 6, 1–405.

4 Harley J L 1975 Problems of Mycotrophy. *In* Endomycorrhizas. Eds. F E Sanders, B Mosse and P Tinker. Academic Press, London.

5 Kramer P J and Bullock H C 1966 Seasonal variations in the proportions of suberized and unsuberized roots of trees in relation to the absorption of water. Am. J. Bot. 53, 200–204.

6 Mosse, B 1975 Specificity in VA mycorrhizas. *In* Endomycorrhizas. Eds. F E Sanders, B Mosse and P Tinker. Academic Press London.

7 Newman E I 1969 Resistance to water flow in soil and plant. I Soil resistance in relation to amounts of root: theoretical estimates. J. Appl. Ecol. 6, 1–12.

8 Pearson V and Read D J 1973 The biology of mycorrhiza in the Ericaceae. I. The isolation of the endophyte and synthesis of mycorrhizas in aseptic cultures. New Phytol. 72, 371–379.

9 Reid C P P and Woods F W 1969 Translocation of C^{14}-labelled compounds in mycorrhizal and its implications in interplant nutrient cycling. Ecology 50, 179–187.

10 Sands R, Fiscus E L and Reid C P P 1982 Hydraulic properties of pine and bean roots with varying degrees of suberization, vascular differentiation, and mycorrhizal infection. Aust. J. Plant Physiol. (In press).

11 Trappe J 1962 Fungus associates of ectotrophic mycorrhizas. Bot. Rev. 28, 538.
12 Whittingham J and Read D J 1982 Vesicular-arbuscular mycorrhiza in natural vegetation systems. III. Nutrient transfer between plants with mycorrhizal interconnections. New Phytol. 90, 277–284.

Plant and Soil 71, 445–454 (1983). A-50

Effects of ammonium and nitrate on growth and nitrogen uptake by mycorrhizal Douglas-fir seedlings

CAROLINE S. BLEDSOE and R. J. ZASOSKI
College of Forest Resources AR-10, University of Washington, Seattle, WA 98195, USA

Key words Clay type CEC Douglas-fir *Hebeloma crustuliniforme* Mycorrhizas Nitrogen Phosphorus *Pseudotsuga menziesii* Rhizosphere pH Root-soil interface

Summary In a greenhouse pot study, mycorrhizal and non-mycorrhizal Douglas-fir seedling were grown in a forest soil amended with clays and ammonium or nitrate fertilizer. Mycorrhizal seedlings grew taller, increased root and shoot dry weight more, and accumulated more nitrogen and phosphorus than non-mycorrhizal seedlings. Mycorrhizal effects were most pronounced in the nitrate treatment. Ammonium enhancement of phosphorus uptake was demonstrated for both mycorrhizal and control seedlings. Rhizosphere pH was altered by nitrogen source and the presence of mycorrhizas. Ammonium fertilization decreased pH, while nitrate fertilization increased pH.

Introduction

Mycorrhizal associations of forest trees can dramatically increase the ability of the host to accumulate phosphorus from soils with low available phosphorus levels[2]. However effects of mycorrhizas on nitrogen acquisition have received less attention[3]. Nitrogen supply is a major limitation to increased growth of Pacific Northwest conifers and responses to nitrogen fertilization are well documented[5,15]. The role that mycorrhizas play in the nitrogen nutrition of NW conifers is less well understood.

Mycorrhizas can affect nitrogen uptake both by increased uptake at the root surface and by increased hyphal growth and soil exploration[3]. In forest soils where ammonium is the major inorganic nitrogen source, both processes may be important. Since ammonium is a poorly mobile ion, fungal growth through the soil should facilitate ammonium uptake. Bowen[2] and others have demonstrated that mycorrhizal roots can grow rapidly through the soil and remain active for longer periods of time than uninfected roots. Mycorrhizal Douglas-fir roots in solution culture had faster ammonium uptake rates than non-mycorrhizal roots[9]. Mycorrhizas may also enhance uptake of more mobile elements, such as nitrate, especially when soil-nitrate levels are low, as is the case in Pacific NW forests. This study reports on how the interacting factors of mycorrhizas, nitrogen fertilization, and clay type affect growth and nutrition of Douglas-fir, a prominent NW coniferous species.

Methods and materials

Experimental design

In this soil pot study, a sandy forest subsoil was modified by addition of different clays and ammonium or nitrate fertilizers at two application rates. Combination of these treatments with mycorrhizal and nonmycorrhizal trees resulted in 96 pots arrayed in the greenhouse in four blocks.

Soil preparation

Soil and clays The soil treatments were designed to modify cation exchange capacity (CEC) while maintaining a constant clay addition. Ten percent (weight basis) Bentonite, Kaolinite, or a 1 : 1 mixture were added to air-dried soil. The soil, collected from the B horizon of an Indianola loamy sand (mixed mesic Dystric Xeropsamment) was sterilized by autoclaving. Soil characteristics were: pH 5.6, CEC 12 meq/100 g, 4 ppm KCl extractable ammonium, 0.6 ppm available nitrate, 15 ppm available phosphorus, and 2.2% organic matter.

Large batches of the clays were leached with chloride salts of calcium, magnesium, and potassium (20 : 4 : 1) to replace exchangeable ions and to produce similar exchangeable cation complements on the clays. Excess salts were leached with distilled water. The clays were dried and ground. The clay CEC's originally were 10 and 62 meq/100 g for Kaolinite and Bentonite, respectively. Silt in the clays lowered the CEC slightly. In the clay-soil mixes, the resulting CEC's were 11, 14, and 17 meq/100 g for the Kaolinite, 1 : 1 mixture, and Bentonite clay additions. For simplicity, these soil treatments are referred to as the Bentonite, mix and Kaolinite treatments.

Nitrogen as ammonium chloride or sodium nitrate was supplied in solution at rates of 50 and 100 ppm. Fertilized soils (1500 g) with added clay were placed in plastic-lined pots to prevent leaching losses. During the study, seedlings in pots were weighed, then watered to known weight equal to less than field capacity, thus preventing anaerobic soil conditions.

Seedlings Douglas-fir (*Pseudotsuga menziesii* [Mirb.] Franco) seedlings were grown in peat-vermiculite in plastic tubes (4 × 21 cm) in a greenhouse for 6 months. Then half the seedlings were inoculated with the mycorrhizal fungus *Hebeloma crustuliniforme* (Bull. ex St. Am.) Quel, using a cellulose thimble technique[10]. This fungus was chosen because it is a demonstrated mycorrhizal former with Douglas-fir, grows rapidly in culture, and is common in coniferous seedling nurseries. The fungal isolate was obtained from J Trappe (U.S. Forest Service, Corvallis, Oregon). A control group of seedlings was treated identically except that no fungal inoculum was added. Seedlings were grown for an additional 6 months in the greenhouse to allow mycorrhizal development.

At the time of the experiment, seedlings were 10 cm tall. Mycorrhizal seedling dry weight was slightly lower than that of control seedlings (0.37 g *vs* 0.46 g). The shoot/root ratio of mycorrhizal plants was slightly higher (1.4 *vs* 1.2). Two one-year-old seedlings were planted in each pot and grown for 16 weeks. Height measurements were taken monthly. At harvest, seedling shoots and roots were separated at the root collar. Tops were dried, weighed, and ground. Roots from half of the seedlings were gently removed from the pots, dried, and rhizosphere soil samples collected[17]. The remaining two replicate root samples were examined for the presence of mycorrhizas, washed, dried, and weighed. Bulk soil samples were collected and dried.

Analyses

Soil and clays Bulk and rhizosphere pH were measured potentiometrically using a combination glass electrode on a 2 : 5 soil: water mixture (rhizosphere soil) or saturated paste (bulk soil). The clay CEC's were determined using unbuffered ammonium chloride and an ammonium acetate procedure. Soil ammonium and nitrate were determined by extraction with 2*N* potassium chloride[4]. Available phosphorus was estimated by extraction with dilute hydrochloric acid and ammonium fluoride[12].

Seedlings Foliar nitrogen and phosphorus content were determined colorimetrically (Technician Industrial Methods 108–71W and 155–81W, respectively) after micro-Kjeldahl digestion[13]. Data were analyzed using analysis of variance routines and significance among means was tested using Duncan's new multiple range test[19].

Results and discussion

Soil chemical properties

Field capacity was altered by clay additions: 24% (Bentonite), 21% (mix), and 19% (Kaolinite). Initial available phosphorus levels were 18 ppm. Following phosphorus fertilization (150 ppm) and 16 weeks of tree growth, the average extractable phosphorus level was 7 ppm. This substantial reversion of added phosphorus, resulted in available phosphorus levels which were low by agronomic standards[12], but comparable to levels in local forest soils[6]. Soil solution electrical conductivity values were 1.3, 1.8, and 2.3 μSeimens/cm respectively for the Bentonite, mix, and Kaolinite treatments at the end of the experiment.

Bulk and rhizosphere pH

Bulk pH Bulk pH of the original soil was 5.6. Addition of clays altered the pH (Kaolinite, 5,3; mixed clays, 5.8; Bentonite, 6.2). Both nitrogen source and mycorrhizas significantly affected bulk pH (Table 1). Nitrogen source effects were particularly evident in the mixed clay treatment, where bulk pH after

Table 1. Bulk soil pH and rhizosphere pH after 16 weeks growth of mycorrhizal(+) and non-mycorrhizal (−) Douglas-fir seedlings. Values are means of four replications with 50 and 100 ppm data pooled. Values in columns for bulk soil and rhizosphere pH followed by the same letter are not significantly different ($p \leq 0.05$, Duncan's new multiple range test). Asterisks indicate significance of differences between mycorrhizal and non-mycorrhizal values at $p \leq 0.05$* and $p \leq 0.01$**

			Soil treatment		
Soil source	Fertilizer	Mycorrhizas	Kaolinite	Mix	Bentonite
Bulk soil	Ammonium	+	5.4a	5.8a	6.1a
		−	5.6b	5.8a	6.2a
		△	−0.2**	0	−0.1
	Nitrate	+	5.6b	6.0b	6.2a
		−	5.6b	6.0b	6.4b
		△	0	0	−0.2*
Rhizosphere	Ammonium	+	5.2a	5.7ab	5.5a
		−	−	5.5a	5.6a
		△	−	+0.2	−0.1
	Nitrate	+	5.8b	6.6c	6.6c
		−	5.6b	5.9b	6.3b
		△	+0.2	+0.7**	+0.3*

ammonium fertilization was less than the bulk pH after nitrate fertilization. Similar trends were noted for the other two clay treatments. The pH decrease due to ammonium fertilization and pH increase due to nitrate fertilization are well-recognized phenomena for both agricultural species[7,14,16,18] and coniferous species[1,8,17,20]. Mycorrhizas also significantly affected bulk pH in the ammonium + Kaolinite treatment (Table 1). However, in this treatment, survival was low; mycorrhizal seedlings apparently modified the pH initially before survival decreased. Mycorrhizal effects on bulk pH were not noted for nitrate treatments.

Rhizosphere pH Nitrogen source significantly affected rhizosphere pH in all clay treatments, mirroring effects noted for bulk soil, although effects were more pronounced. For example, in the mix treatment, the mean difference between ammonium and nitrate treatments was only 0.2 pH units for bulk soil, but was 0.6 pH units for rhizosphere soil. These significant rhizosphere pH changes can be important, as Nye stated, 'Clearly pH changes... may have far reaching implications for a host of rhizosphere processes'[11].

Mycorrhizas significantly affected rhizosphere pH in the nitrate treatments. Rhizosphere pH was higher in the presence of mycorrhizas for all trees clay types. Since mycorrhizal seedlings grew more and took up more nitrate (see Figs. 2,4), more prominent increases in pH were expected. In the ammonium treatments, although mycorrhizal plants took up more ammonium, rhizosphere pH was altered very little. No substantial decrease in pH was observed. Since some nitrification occured in spite of the presence of N-Serve, the net effect of accumulating both nitrogen forms could negate rhizosphere pH changes. In addition, mycorrhizas may have altered cation/anion balance during ammonium uptake, since related solution culture work (P. Rygiewicz, pers. comm.) demonstrated that mycorrhizal conifer roots took up ammonium with less acidification of the medium (per ammonium absorbed) than did non-mycorrhizal roots.

Mycorrhiza

Seedling roots were visually inspected for the presence of mycorrhizas. Initially, uninoculated seedlings were non-mycorrhizal. For seedlings inoculated with *H. crustuliniforme*, more than half of the root system was infected. After 16 weeks growth, mycorrhizal infection of uninoculated seedlings was less than 5%, due to the presence of *Thelephora terrestris*. Inoculated seedlings remained mycorrhizal with Hebeloma and with a small amount of Thelephora. There was 50–75% infection in the 50 ppm nitrate treatment and 15–30% in the 50 ppm ammonium treatment. Mycorrhizal infection decreased slightly at the 100 ppm nitrogen rates. There was no detectable effect of clay type on mycorrhizal infection.

Seedling survival and growth

Survival Survival of Douglas-fir seedlings was markedly affected by clay type,

Table 2. Percent survival of mycorrhizal (+) and non-mycorrhizal (−) Douglas-fir seedlings grown in clay-amended soils fertilized with ammonium or nitrate at 50 or 100 ppm

Fertilizer			Soil treatment		
N source	Rate	Mycorrhizas	Kaolinite	Mix	Bentonite
Ammonium	50	+	26	63	100
		−	13	50	88
	100	+	13	50	38
		−	0	38	88
Nitrate	50	+	100	100	100
		+	63	100	100
	100	+	63	100	100
		−	88	88	75

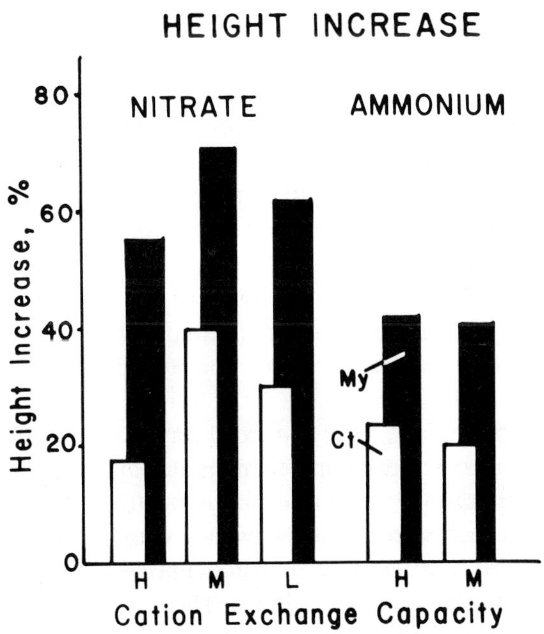

Fig. 1. Effects of mycorrhizas on height growth of Douglas-fir seedlings grown in fertilized, clay-amended soil. Soil CEC's are ranked as H (high, bentonite), M (medium, mix) and L (low, kaolinite). Nitrate or ammonium fertilization is noted. Seedlings were either mycorrhizal (My) or uninoculated controls (Ct). Values are means of 16 samples. Data from 50 and 100 ppm nitrogen application rates were pooled.

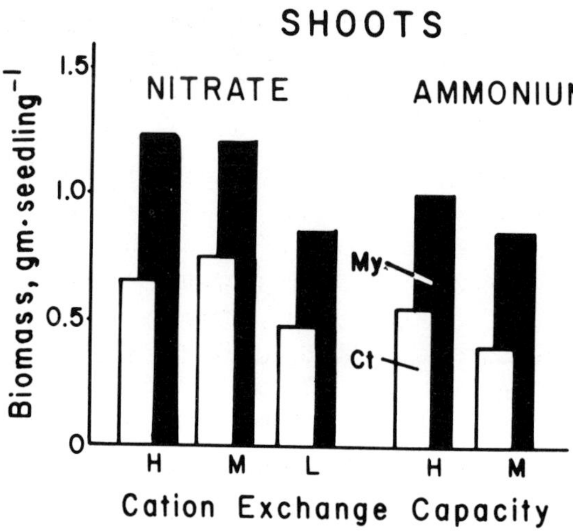

Fig. 2. Effects of mycorrhizas on shoot dry weight of Douglas-fir seedlings. Values are means of 8 samples; all other conditions and symbols as in Fig. 1.

nitrogen source and rate, and mycorrhizal presence (Table 2).Average survival declined from 86% to 74% to 46% in the Bentonite, mix, and Kaolinite treatments respectively. This decreased survival was particularly evident in the ammonium treatments, where average survival declined from 90% (nitrate) to 48% (ammonium). Ammonium toxicity apparently decreased survival. In the 100 ppm ammonium treatment, survival was 38%, while at 50 ppm ammonium, survival was increased to 57%. Mycorrhizal seedlings survived better than uninoculated seedlings, particularly in the ammonium treatments where survival of the mycorrhizal seedlings was 57%, compared to 46% for the non-mycorrhizal treatment.

Height Mycorrhizal seedlings grew significantly taller than control seedlings in all clay and fertilizer treatments (Fig. 1). Height data for the 50 and 100 ppm fertilizer rates were averaged since the differences were not significant. The mycorrhizal effect was highly significant ($p \leq 0.001$). Mycorrhizas had a greater effect on height growth in the nitrate treatments than in the ammonium treatments (63% height increase, nitrate, 43%, ammonium; $p \leq 0.025$).

In the control group of seedlings, there were no significant differences in height growth except in the nitrate + mixed clay treatment. Seedlings in this treatment had significantly greater growth than did seedlings in the higher CEC, Bentonite treatment (46% mixed clay vs. 19% Bentonite $p \leq 0.025$).

Shoot dry weight Shoot dry weight of mycorrhizal seedlings was greater than that of nonmycorrhizal seedlings in all treatments ($p \leq 0.10$, Fig. 2). This mycorrhizal enhancement was particularly evident in the nitrate treatment (average mycorrhizal shoot dry weight, 1.1 g/seedling compared to 0.64 g/seedling for non-mycorrhizal seedlings; $p \leq 0.025$). Since very few seedlings survived in the Kaolinite treatment, dry weight data were not presented for this treatment.

Mycorrhizal enhancement of shoot growth was similar within each clay treatments, suggesting that mycorrhizae did not provide any additional benefits in soils with higher CEC's. Shoot dry weight of seedlings in the Kaolinite treatment was significantly lower than in the other two treatments, suggesting growth reduction due to nutrient limitations or toxicity. Differences in shoot dry weight due to nitrogen source and clay type were significant for mycorrhizal seedlings, but not for control seedlings.

Root dry weight Root dry weight data were more variable than shoot dry weight data, due in part to the difficulties of removing soil from the fine roots. Mycorrhizas significantly enhanced root growth ($p \leq 0.001$) in all clay and nitrogen source treatments, except the ammonium + Kaolinite treatment where there were insufficient data due to low survival. The mycorrhizal effect was most pronounced in the nitrate + Bentonite treatment where mycorrhizal root dry weights averaged 2.7 times greater than non-mycorrhizal roots ($p \leq 0.05$). Root dry weight increased significantly for both mycorrhizal and non-mycorrhizal seedlings in the nitrate treatment with increasing soil CEC ($p \leq 0.05$). While a

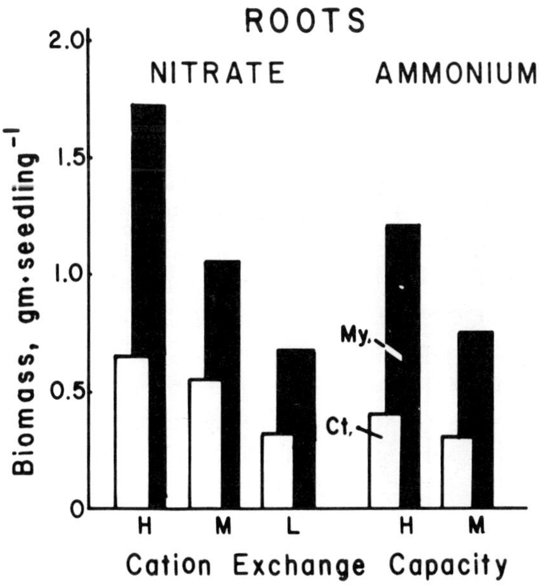

Fig. 3. Effects of mycorrhizas on root dry weight of Douglas-fir seedlings. All other conditions, symbols and replications as in Fig. 2.

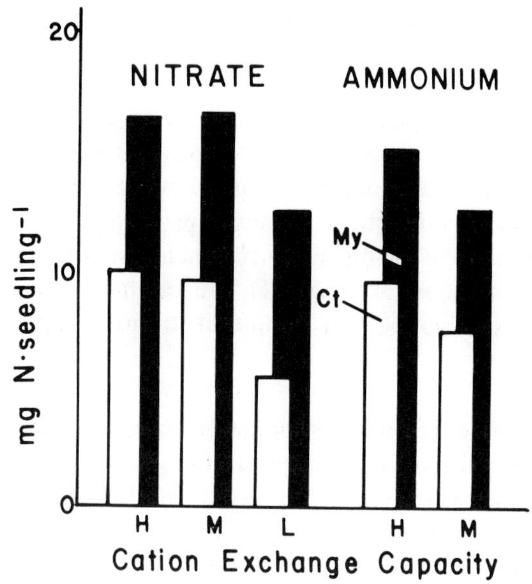

Fig. 4. Effects of mycorrhizas on nitrogen content of shoots of Douglas-fir seedlings. All other conditions, symbols and replications as in Fig. 2.

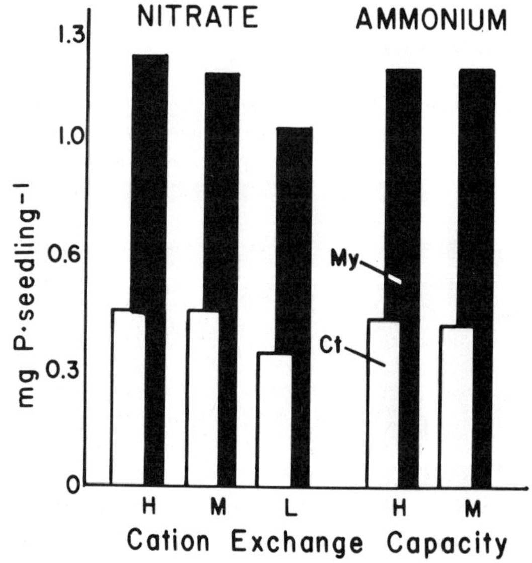

Fig. 5. Effects of mycorrhizas on phosphorus content shoots of Douglas-fir seedlings. All other conditions, symbols and replications as in Fig. 2.

similar trend was noted with ammonium fertilization, results were not statistically significant.

Nitrogen Nitrogen content of Douglas-fir shoots (stems + needles) was calculated and used for treatment comparisons, rather than % nitrogen. Nitrogen source did not significantly affect nitrogen uptake, since nitrogen contents of nitrate-grown and ammonium-grown seedling tops were similar. Mycorrhizas did significantly affect nitrogen uptake in all three clay treatments particularly for nitrogen supplied as nitrate. Since mycorrhizal seedlings were taller and grew more than non-mycorrhizal seedlings, increased nitrogen uptake was expected.

Phosphorus Even though phosphate fertilizer was added to all treatments, these data show a dramatic phosphorus response by mycorrhizas in both ammonium and nitrate treatments for all clay types. Mycorrhizal effects were all highly significant ($p \leq 0.01$) for all treatments.

Ammonium fertilization has been shown to increase phosphate availability due to lower rhizosphere pH and increased phosphate availibility[16,18]. A similar effect was demonstrated for Douglas fir seedlings (Fig. 2 and 5). Although shoot dry weight of ammonium-grown seedlings was less than that of nitrate-grown plants, phosphorus contents were equal (Fig. 5). The ammonium enhancement of phosphate uptake was evident for both control and mycorrhizal seedlings.

References

1 Bigg W S and T W Daniel 1978 Effects of nitrate, ammonium and pH on the growth of conifer seedlings and their production of nitrate reductase. Plant and Soil 50, 371–385.

2 Bowen G D 1973 Mineral nutrition of ectomycorrhizae. *In* Ectomycorrhizae: Their Ecology and Physiology. Eds. G C Marks and T T Kozlowski. Academic Press, NY, pp 151–205.

3 Bowen G D and S E Smith 1981 The effects of mycorrhizas on nitrogen uptake by plants. In Terrestrial Nitrogen Cycles. Eds. F E Clark and T Rosswall. Ecol. Bull (Stockholm) 33, 219–232.

4 Bremner J M 1965 Total nitrogen, Inorganic forms of nitrogen, Organic forms of nitrogen. *In* Methods of Soil Analysis. Ed. C A Black. Am. Soc. Agron., Madison, WI, pp. 1035–1049.

5 Gessel S P, T N Stoate and K J Turnbull 1969 The growth and behavior of Douglas-fir with nitrogenous fertilizer in western Washington. Inst. of Forest Prod. Contrib No. 7, Univ. of Washington, Seattle WA, 119 p.

6 Heilman P E, H W Anderson and D M Braumgartner 1979 Forest Soils of the Douglas-fir Region. Washington State University Cooperative Extension Service, Pullman, WA, 298 p.

7 Kirkby E A 1981 Plant growth in relation to nitrogen supply. *In* Terrestrial Nitrogen Cycles. Eds. F E Clark and T Rosswall. Ecol. Bull (Stockholm) 33, 249–267.

8 Krajina V J, S Madoc-Jones and G Mellor 1973 Ammonium and nitrate in the nitrogen economy of some conifers growing in Douglas-fir communities of the Pacific Northwest of America. Soil Biol. Biochem. 5, 143–147.

9 Littke W R 1982 Nitrogen Uptake by Mycorrhizal Fungi and Mycorrhizal Douglas-fir. Ph. D. Diss. University of Washington, Seattle WA, 181 p.

10 Littke W R, C S Bledsoe, N Nadkarni and R L Edmonds 1980 A technique for rapid mycorrhizal colonization of container-grown Douglas-fir by *Hebeloma crustuliniforme*. Soil Biol. Biochem. 12, 575–578.

11 Nye P H 1981 Changes of pH across the rhizosphere induced by roots. Plant and Soil 61, 7–26.

12 Olsen S R and L A Dean 1965 Phosphorus. *In* Methods of Soil Analysis, Ed. C A Black. Am. Soc. Agron., Madison WI, 1035–1049.

13 Parkinson J R and S E Allen 1975 A wet oxidation procedure suitable for the determination of nitrogen and mineral nutrients in biological materials. Comm. Soil Sci. Plant Anal. 6, 1–11.

14 Raven J A and F A Smith 1976 Nitrogen assimilation and transport in vascular land plants in relation to intracellular pH regulation. New Phytol. 76, 415–431.

15 Regional Forest Nutrition Research Project 1976 Inst. of Forest Prod. Contrib. No. 25, University of Washington, Seattle WA. 67 p.

16 Riley D and S A Barber 1971 Effect of ammonium and nitrate fertilization on phosphorus uptake as related to root-induced pH changes at the root-soil interface. Soil Sci. Soc. Am. Proc. 35, 301–306.

17 Rollwagen B A 1981 Effects of ammonium and nitrate applications on rhizosphere pH, growth, and nutrient uptake by Douglas-fir, Sitka spruce and Western hemlock. M.S. Thesis, Unviversity of Washington, Seattle WA.

18 Smiley R W 1974 Rhizosphere pH as influenced by plants, soils and nitrogen fertilizers. Soil Sci. Soc. Am. Proc. 38, 795–799.

19 Steel R G D and J H Torrie 1960 Principles and Procedures of Stastistics, McGraw-Hill, NY. 481 p.

20 Van den Driessche R 1978 Response of Douglas-fir seedlings to nitrate and ammonium nitrogen sources at different levels of pH and iron supply. Plant and Soil 49, 607–623.

Plant and Soil 71, 455–462 (1983). A-51

Phosphatase production by mycorrhizal fungi

J. DIGHTON
Institute of Terrestrial Ecology, Merlewood Research Station, Grange-over-Sands, Cumbria, UK

Key words Birch Mycorrhizas Phosphorus Pine Saprophytes enzymes

Summary Sheathing mycorrhizal fungi have been shown to possess phosphatase enzymes which can hydrolyse inositol hexaphosphate. In a range of mycorrhizal fungi, this activity was often greater than in two common decomposer basidiomycetes. Mycorrhizal birch and pine roots both produce phosphatases. In birch production is inversely related to the inorganic phosphorus concentration in the growth medium. Mycorrhizas reduce phosphatase activity compared with non-mycorrhizal plants.

Phosphatase production by basidiomycete fungi in liquid culture is independent of P in the medium. Saprophytic basidiomycetes tend to incorporate hydrolysed phosphate into their biomass. In contrast mycorrhizal fungi release more into solution than they absorb. The significance of this difference in relation to the supply of nutrients to plants is discussed.

Introduction

The role of mycorrhizas in decomposition is not clear. Although mycorrhizas produce phosphatase and phytase enzymes[1,2,13,22,23], mycorrhizal roots have been suggested as inhibiting litter decomposition[7,8]. Relative phosphatase production by mycorrhizal and decomposer basidiomycetes has not been studied. In the evolution of the mycorrhizal habit physiological changes in the fungus may have occurred which may affect the degree of obligatoriness of the association. Giltrap[9] alludes to this idea in his study of the decomposer acitivity of a range of mycorrhizal fungi and suggests that *Lactarius* spp. behave as saprophytes so may have a facultative mycorrhizal association.

This study aims to compare the phosphatase activity of mycorrhizal and saprophytic basidiomycetes and to relate their roles in supplying nutrients to host plants and in nutrient mineralisation.

Materials and methods

Mycorrhizal root systems

Mycorrhizal fungi were inoculated onto 10 week old lodgepole pine (*Pinus contorta* Dougl.) and birch (*Betula pubescens* Ehrh.) seedlings in sterile vermiculite/peat culture[18], hardened off and planted into perlite in 12.5 cm dia. plastic plant pots with five plants per pot. Treatments with birch were control (non-mycorrhizal), *Hebeloma sacchariolens* Quél, *Lactarius pubescens* (Fr. ex Krombh.) Fr. and *Amanita muscaria* (L. ex Fr.) Hooker and of pine, non mycorrhizal, *Hebeloma crustuliniforme* (Bull. ex St. Amans) Quél, *Suillus luteus* (Fr.) S. F. Gray and *Amanita muscaria*. Nutrient levels were maintained by flushing the pots with 300 ml Hewitts solution[12] containing 1, 2, 4, 8, 12, 25, 50 or 100 ppm P as orthophosphate. After six weeks growth 40 mg root samples were assayed for acid phosphatase production at 20°C and pH 5.0 using the method of Hoffman[10,14] which measures phenol release from disodium phenyl phosphate.

Liquid culture of mycorrhizal fungi

Fungi were isolated from fruit bodies and grown on petri dishes of Hagem's agar. Three 5 mm dia. discs of agar were taken from the edge of actively growing colonies and inoculated into 250 ml Erlenmeyer flasks containing 100 ml Hagem's medium modified to contain 10, 50 or 100 ppm P as orthophosphate-P or inositol hexaphosphate-P (IHP-P).

Experiment 1: Flasks with 10 ppm P were incubated at 20°C for 42 days in the dark. Fungi were filtered through 0.45 grade Oxoid membrane filters, for *Mycena galopus* (Pers. ex Fr.) Kummer and *Marasmius androsaceus* (L. ex Fr.) Fr. 54 grade cellulose filters. In each case fungi were rinsed with 10 ml sterile distilled water which was added to the filtrate. Phosphatase determinations were carried out as above on mycelium and filtrate but at pH 4.5 for 3 h.

Experiment 2: Fungi were grown with 10, 50 or 100 ppm P as PO_4- or IHP-P in Hagem's medium at 20°C. The incubation period was varied for each fungus in an attempt to achieve similar fungal biomass at the end of the incubation period. Fungi were filtered through 54 grade cellulose filters and phosphatase assayed as above.

Experiment 3: Culture was as in experiment 2 but using inositol hexaphosphate only. Following filtration through 54 grade cellulose filters orthophosphate-P released by the fungal activity was determined in the filtrate and perchloric/nitric acid digests of the fungal mycelium using an arsenomolybdate colorimetric determination[15].

Results and discussion

Phosphatase production by roots

Roots can hydrolyse organic and insoluble phosphate in soil by the production of phosphatases[20, 24]. Acid phosphatases are produced by mycorrhizal roots of beech[2] and Sitka spruce[1] from the field although phosphatase production may be negatively related to extractable (inorganic) phosphorus in the rooting zone[1]. A significant negative correlation between phosphatase production and PO_4-P concentration was found in birch (Fig. 1). This substantiates the findings of Alexander and Hardy[1], who suggested a negative feedback mechanism as controlling acid phosphatase production. There was no correlation between phosphatase production and PO_4-P concentration with pine (Table 1). In birch the presence of mycorrhizas tended to suppress phosphatase production (Fig. 1). In pine only Hebeloma significantly reduced phosphatase production compared to non-mycorrhizal roots (Table 1). This broadly agrees with results quoted by Morrison[19] who suggested that the poorer P-solubilization by mycorrhizal roots of *Pinus radiata* associated with enhanced P-uptake was due to a greater decrease in the free energy of P by mycorrhizal roots due to their rapid utilization of PO_4 *i.e.* that mycorrhizal roots form a greater P-sink than non-mycorrhizal roots and so increase the rate of solubilization.

Phosphatase production by fungi

It is expected that large numbers of soil microorganisms, particularly those involved in decomposition, might be expected to produce phosphatases. Beever and Burns[3] suggest that three types of enzyme systems may be present *i.e.* acid phosphatases, alkaline phosphatases and highly substrate-specific enzymes. All

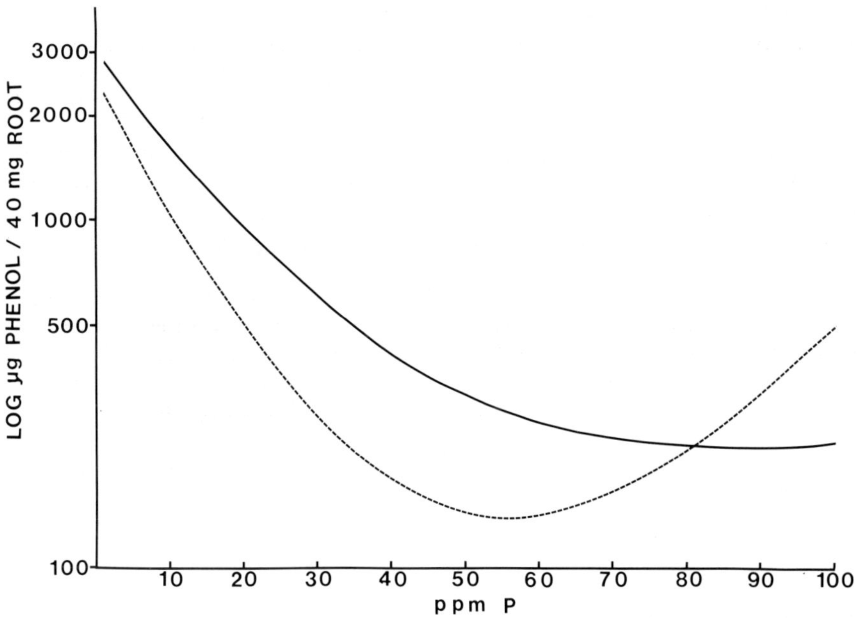

Fig. 1. Phosphatase production by 40 mg birch root system with (*pecked* line) or without (*solid* line) mycorrhizas when grown in different concentrations of orthophosphate in perlite.
Non mycorrhizal $\log y = 0.029x + 0.0002x^2 + 3.484$
Mycorrhizal $\log y = 0.043x + 0.0004x^2 + 3.405$

Table 1. Acid phosphatase production by mycorrhizal *Pinus contorta* seedling root systems grown at various levels of orthophosphate-P in Hewitts solution in perlite culture

P concentration (ppm)	Phosphatase production (µg phenol/40 mg root) of roots infected with different mycorrhizas (mean ± SE)			
	Control	*Hebeloma crustuliniforme*	*Suillus luteus*	*Amanita muscaria*
1	2271 ± 1417	1647 ± 354	2524 ± 65	2587 ± 506
2	3898 ± 1635	1399 ± 562	3374 ± 1149	1951 ± 373
4	2210 ± 457	2189 ± 393	3452 ± 676	2261 ± 399
8	3264 ± 449	1835 ± 231	2136 ± 355	2318 ± 340
12	3080 ± 1410	1950 ± 438	2382 ± 188	2507 ± 1025
25	2227 ± 563	1455 ± 301	1049 ± 518	2241 ± 145
50	2167 ± 171	961 ± 470	1858 ± 446	1737 ± 226
100	2061 ± 405	1910 ± 366	2079 ± 224	2193 ± 326
	$F = 0.48$	$F = 0.94$	$F = 2.0$	$F = 0.32$
	$P = \text{NS}$	$P = \text{NS}$	$P = \text{NS}$	$P = \text{NS}$

Comparison of control *vs.* Hebeloma $F = 8.0$ $P < 0.01$

are found in a variety of fungi. Aspergilli, yeasts and bacteria produce acid phosphatases[5,21]. A number of rhizosphere organisms produce phosphatases[6,16,17,20,24], although their role in P-solubilization and uptake by plants is poorly understood and the evidence conflicting. Martin[17] found that the presence of rhizosphere organisms with low IHP-P concentration did not increase the rate of P-uptake by wheat over sterile plants. Agrostis roots in water culture produced no more phosphatases than roots in soil[24]. In contrast Saxena[20] demonstrated a 15–40% increase in phytase activity in pea, gram, barley and wheat roots in the presence of a rhizosphere flora. The number of P-solubilizing fungi on roots increased in the absence of PO_4–P[6] and in the presence of mycorrhizas on yellow birch[16]. A number of mycorrhizal fungi exhibit phosphatase or phytase activity[4,13,22,23]. The evolution of basidiomycete mycorrhizal symbioses may have led to both obligate and facultative forms of association[9] in which the ability of mycorrhizal fungi to produce phosphatases may be significantly different from decomposer basidiomycetes. Experiments comparing phosphatase/phytase activity in mycorrhizal and decomposer basidiomycete fungi in liquid culture are described.

Acid phosphatase production by a range of sheathing mycorrhizal and two commonly occurring decomposer basidiomycete fungi *Mycena galopus* and *Marasmius androsaceus* in liquid culture containing 10 ppm P or IHP-P is shown in Table 2. In all fungi a large proportion of the phosphatase produced is retained in the fungus. Especially with organic P, phosphatase production, per gram of mycorrhizal fungus, seemed greater than for decomposer fungi.

The inverse correlation between phosphatase production by birch roots and P-content of growth medium suggests the possibility of a negative feedback mechanism controlling phosphatase production by roots and microorganisms[1,6]. Some of the fungi used in experiment 1 were grown in 10, 50 or 100 ppm PO_4–P or IHP–P. In the presence of either PO_4–P or IHP–P the mycorrhizal fungi produced more acid phosphatases than the decomposer fungi (Table 3). Phosphatase production and P-concentration were not consistently related. This suggests that phosphatase production in mycorrhizal roots is root initiated, although it is difficult to extrapolate from behaviour in liquid culture to that in a symbiotic association in nature.

Mycorrhizas are generally assumed to enhance nutrient uptake by the root system, relative to non-mycorrhizal roots. It would seem therefore advantageous for the mycorrhizal fungus, with its larger amounts of phosphatase, to release the hydrolysed PO_4 in a plant available form to the root. Since sheathing mycorrhizas have no intracellular hyphal penetration, PO_4 released by phosphatase activity would only be available if released into the soil solution adjacent to the root surface or into the intercellular spaces of the Hartig net.

Three mycorrhizal and two decomposer fungi were suplied with IHP–P at 10 ppm in liquid culture. The amount of P released from IHP and the % incorporated into the fungal biomass are shown in Table 4. *Lactarius rufus* and

Table 2. Acid phosphatase production by fungal mycelia in liquid Hagem's medium with 10 ppm orthophosphate-P or inositol hexaphosphate-P after 42 days growth at 20°C (single determinations)

Fungus (isolate code)	Orthophosphate				Inositol hexaphosphate			
	μg phenol mg^{-1} mycelium	mean ± S.E.	% phenol in medium	mean ± S.E.	μg phenol mg^{-1} mycelium	mean ± S.E.	% phenol in medium	mean ± S.E.
Lactarius rufus (MH)	26.3		1.1		22.2		4.8	
Paxillus involutus (MH)	17.1		14.4		10.2		7.2	
Paxillus involutus (3)	9.7		14.6		10.1		15.2	
Lactarius pubescens (6)	12.8	13.8	0	17.2	7.6	12.3	4.6	7.2
Hebeloma crustuliniforme (3)	29.8	± 3.5	62.2	± 7.5	23.9	± 2.4	0	± 2.9
Suillus luteus (4)	5.0		10		9.2		23.6	
Amanita muscaria (147)	5.8		1.3		8.8		0	
Amanita muscaria (156)	3.9		33.6		6.6		1.8	
Mycena galopus	0.9	10.4	17.8	48.1	0.2	0.9	25.5	21.0
Marasmius androsaceus	20.0	± 9.5	78.4	±30.3	1.7	± 0.8	16.4	± 4.6
		z* = −0.5		z = 1.6		z = −2.1		z = 1.83
		NS		NS		$P < 0.05$		$P < 0.05$

* Comparison between mean data for mycorrhizal and saprophytic fungi made by Mann Whitney U test.

Table 3. Acid phosphatase production by fungal mycelia in liquid Hagem's medium at different concentrations of orthophosphate and sodium inositol hexaphosphate (2 replicate determinations)

Fungus	μg phenol mg^{-1} fungus							
	PO_4–P concentration (ppm)				IHP–P concentration (ppm)			
	10	50	100	Mean*	10	50	100	Mean*
[b] *Paxillus involutus*	33.0	18.3	20.0		25.0	18.7	61.8	
[a] *Lactarius rufus*	9.3	5.9	30.3	36.9	22.9	68.9	31.4	44.5
[b] *Suillus luteus*	11.9	11.1	10.0	±10.6	9.6	9.1	27.8	±14.3
[a] *Amanita muscaria*	54.5	0	182.1		177.4	63.3	26.1	
[c] *Mycena galopus*	2.4	2.7	1.5	1.4	0.6	2.1	0.4	0.9
[c] *Marasmius androsaceus*	0.4	0.6	0.8	± 0.3	0.3	0	1.6	± 0.2
				$F = 5.5$				$F = 4.8$
				$P < 0.05$				$P < 0.05$

[a] 51 days growth; [b] 40 days growth; [c] 24 days growth
* Mean of all P levels
Comparisons between P- levels for any fungus in either PO_4- or IHP-solution were not significant by analysis of variance.

Table 4. Release of PO_4–P from 10 ppm inositol hexaphosphate in Hagem's medium by fungi grown in liquid culture at 20°C

Fungus	Total P released μg P mg^{-1} fungus	% P released from IHP retained in fungus
Paxillus involutus	62.4 ± 6.9	16.7
Lactarius rufus	86.1 ± 18.4	9.2
Suillus luteus	7.1 ± 0.7	92.5
Mycena galopus	13.4 ± 4.3	76.7
Marasmius androsaceus	2.7 ± 1.4	34.0
	$F = 17.29$	
	$P < 0.001$	

Comparison between mean data for mycorrhizal and saprophytic fungi for total P released by Mann Whitney U test $z = -2.12$ $P < 0.05$.

Paxillus involutus hydrolysed more IHP–P than *Suillus luteus*, or the two decomposer fungi Mycena and Marasmius. The proportion of released P incorporated into fungal biomass was much lower in Lactarius and Paxillus. These two commonly occurring mycorrhizal fungi could in the decomposition processes hydrolyse organic forms of phosphorus and make it available to host roots. Saprophytic basidiomycetes after hydrolysing phosphate usually incorporate it into their own biomass. *Suillus luteus* behaved here more like the decomposer fungi than the other mycorrhizal fungi.

Although vesicular-arbuscular mycorrhizas produce phosphatases not involved in P-solubilization sheathing mycorrhizas have the capacity to solubilize P from inorganic and organic complexes. The P released is not always incorporated into fungal biomass but may be supplied to the plant host. This may help explain competitive effects between mycorrhizal fungi and decomposers in soil[7,8], where litter decomposition is suppressed in the presence of mycorrhizal roots. If mycorrhizal fungi act as decomposers and regulate P-supply to hosts decomposition and mineralisation by saprophytic organisms may be circumvented.

Phosphatase activity has only been studied in mycorrhizal fungi in culture and on the surface of mycorrhizal roots. Fungal hyphae and rhizomorphs of sheathing mycorrhizas extend considerable distances from the root surface. The contribution of these hyphae to nutrient transport and P hydrolysis has yet to be assessed.

Acknowledgements I wish to thank Mr. M. R. Smith for his technical assistance.

References

1 Alexander I J and Hardy K 1981 Surface phosphatase activity of Sitka spruce mycorrhizas from a Serpentine site. Soil Biol. Biochem. 13, 301–305.

2 Bartlett E M and Lewis D H 1973 Surface phosphatase activity of mycorrhizal roots of beech. Soil Biol. Biochem. 5, 249–257.

3 Beever R E and Burns D J W 1980 Phosphorus uptake, storage and utilization by fungi. Adv. Bot. Res. 8, 127–219.

4 Calleja M, Moussin D, Lecouvreur B and D'Auzac J 1980 Influence de la carence phosphatée sur les activités phosphatases acides de trois champignon mycorrhiziens: *Hebeloma edurum* Metrod., *Suillus granulatus* (L. ex Fr.) O Kuntze et *Pisolithus tinctorius* (Pers.) Coker et Couch. Physiol. Vég. 18, 489–504.

5 Casida L E Jr. 1959 Phosphatase activity of some common soil fungi. Soil Sci. 87, 305–310.

6 Chhonkar P K and Subba-Rao N S 1967 Phosphate solubilization by fungi associated with legume root nodules. Can. J. Microbiol. 13, 749–753.

7 Gadgil R L and Gadgil P D 1971 Mycorrhiza and litter decomposition. Nature London 233, 133.

8 Gadgil R L and Gadgil P D 1975 Suppression of litter decomposition by mycorrhizal roots of *Pinus radiata*. N. Z. Jl. For. Sci. 5, 33–41.

9 Giltrap N J 1982 Production of polyphenol oxidases by ectomycorrhizal fungi with special reference to *Lactarius* spp. Trans. Br. Mycol. Soc. 78, 75–81.

10 Harrison A F and Pearce T 1979 Seasonal variation of phosphatase activity in woodland soils. Soil Biol. Biochem. 11, 405–410.

11 Hayman D J and Mosse B 1972 The role of vesicular-arbuscular mycorrhiza in the removal of phosphorus from soil by plant roots. Rev. Ecol. Biol. Sol. 11, 463–470.

12 Hewitt E J 1952 Sand and water culture methods used in the study of plant nutrition. Tech. Commun. Commonw. Bur. Hort. Plantn. Crops No. 22, 241 p.

13 Ho I and Zak B 1979 Acid phosphatase activity of six ectomycorrhizal fungi. Can. J. Bot. 57, 1203–1205.

14 Hoffman G 1968 Eine photometrische Methode zur Bestimmung der Phosphatase – Aktivität in Böden. Z. Pflanzenernaehr Dueng. Bodenkd. 118, 161–172.

15 John M K 1970 Colorimetric determination of phosphorus in soil and plant materials with ascorbic acid. Soil Sci. 109, 214–220.

16 Katznelson H, Peterson E A and Rouatt J W 1962 Phosphate-disolving microorganisms on seed and in the root zone of plants. Can. J. Bot. 40, 1041–1180.

17 Martin N K 1973 The influence of rhizosphere microflora on the availability of ^{32}p-myoinositol hexaphosphate phosphorus to wheat. Soil Biol. Biochem. 5, 473–483.

18 Mason P A, Dighton J, Last F T and Wilson J 1983 Procedure for establishing sheathing mycorrhizas on tree seedlings. For. Ecol. Manage. 5, 47–53.

19 Morrison T M 1962 Absorption of phosphorus from soils by mycorrhizal plants. New Phytol. 61, 10–20.

20 Saxena S N 1964 Phytase activity of plant roots. J. Exp. Bot. 15, 654–658.

21 Shieh T R and Ware J H 1968 Survey of microorganisms for the production of extracellular phytase. Appl. Microbiol. 16, 1348–1351.

22 Theodorou C 1967 Inositol phosphates in needles of *Pinus radiata* D. Don and the phytase activity of mycorrhizal fungi. Trans. 9th Int. Cong. Soil Sci. 3, 483–490.

23 Theodorou C 1971 The phytase activity of the mycorrhizal fungus *Rhizopogon luteolus*. Soil Biol. Biochem. 3, 89–90.

24 Woolhouse H W 1969 Differences in the properties of the acid phosphatases of plant roots and their significance in the evolution of edaphic ecotypes. *In* Ecological Aspects of the Mineral Nutrition of Plants. Ed. T H Rorison. Blackwell, Oxford. pp 357–380.

Plant and Soil 71, 463–467 (1983). A-52

Increased uptake of anions by plants with vesicular-arbuscular mycorrhizas

J. G. BUWALDA, D. P. STRIBLEY and P. B. TINKER
Rothamsted Experimental Station, Harpenden, Herts, AL5 2JQ, UK

Key words Anion uptake Barley *Hordeum vulgare* Mycorrhiza Phosphorus pH Rhizosphere *Triticum aestivum* Wheat

Summary In two field experiments sown in 1982 to test the effect of vesicular-arbuscular mycorrhizas (VAM) on growth and phosphorus nutrition of (i) spring wheat and spring barley, (ii) winter wheat and winter barley, we measured the concentrations of the major cations (K^+, Ca^{2+}, Mg^{2+} and Na^+) and anions (Cl^-, SO_4^{2-}, $H_2PO_4^-$ and NO_3^-) in shoot tissue. In all cases the sum of the anion concentrations (ΣA) was increased strongly by mycorrhizal infection but not by P additions, confirming earlier observations[2] on spring wheat. The concentration of total cations (ΣC) was generally reduced by P additions, hence P and VAM both reduced the cation excess ($\Sigma C - \Sigma A$) but by different mechanisms.

These results suggest that increased uptake of anions by plants with VAM may be a general phenomenom which would have important implications for the elemental composition of crops. The effect may also be manifested by other types of mycorrhizal association.

Introduction

Many published papers have shown the major effects on plant growth and nutrition which result from infection with vesicular-arbuscular mycorrhizas[4,5,6]. In a few cases these effects have been shown to be due to an increased supply of trace metals, and there have been claims that infection causes an increased supply of water, or the production of phytohormones. However, almost all these changes have been associated with an increased supply of phosphorus to the plant via the mycorrhizal hypae and it has been very rare to find an effect which could certainly not be attributed to a change in the phosphorus status of the plant. An effect of mycorrhizal infection which cannot have been caused by an improved phosphorus supply was reported by Buwalda *et al.*[2] When spring wheat was grown in the field or in pots, it was found that an increased level of soil phosphorus had little or no effect on the anion content of the plants, except for the increased phosphorus concentration. However when plants were infected by mycorrhizal fungi (*Glomus mosseae*) there was a consistent increase in the content of chloride, bromide, sulphate and phosphorus. The increase of bromide was originally detected in field experiments where methyl bromide had been used to eliminate indigenous mycorrhizal fungi, but the effect was repeated in pot experiments under controlled conditions. In this paper we report further experiments on the same topic.

Methods

The methods used were reported by Buwalda *et al.*[2] Two field experiments were done with either spring wheat (*Triticum aestivum* var. Timmo) and spring barley (*Hordeum vulgare* var. Triumph) or winter wheat (*Triticum aestivum* var. Avalon) and winter barley (*Hordeum vulgare* var. Igri) in the growing season 1982. The sites had previously been sterilized by methyl bromide, so indigenous mycorrhizal fungi were at a low level. The designs were factorial, with two levels of phosphorus (0 and 60 kg P ha^{-1}, as superphosphate), with or without methyl bromide fumigation (98 g m^{-2}, released under polythene sheets), and with or without inoculation with *Glomus mosseae* (350 g m^{-2} of inoculum grown on leek plants in the greenhouse). Samples of the shoots were taken from all plots at about anthesis, weighed fresh, dried, reweighed, ground and analysed for phosphorus, chloride, nitrate, sulphate, potassium, calcium, magnesium and sodium[2]. In both experiments a significant response in total yield was obtained to inoculation with the mycorrhizal fungus on the plots where the indigenous fungi had been eliminated by methyl bromide, both with and without phosphorus additions.

Table 1. Effects of mycorrhizal inoculation on the levels of cations and anions (meq. 100 g^{-1}) in spring wheat grown at two levels of applied phosphorus

Ions	P applied (kg.ha^{-1})	Inoculation −	Inoculation +	$\bar{x}$	Significance	Phosphorus	Inoculation
Cations							
K^+	0	46.73	46.04	46.38	*P*	<0.05	NS
	60	43.20	40.85	42.02	$LSD_{(0.05)}$	5.04	4.99
	$\bar{x}$	44.97	43.45				
Ca^{2+}	0	18.94	17.28	18.11	*P*	NS	NS
	60	18.47	18.10	18.28	$LSD_{(0.05)}$	3.08	1.66
	$\bar{x}$	18.71	17.69				
Mg^{2+}	0	8.62	7.60	8.11	*P*	NS	NS
	60	7.33	6.79	7.06	$LSD_{(0.05)}$	0.85	0.73
	$\bar{x}$	7.97	7.19				
Na^+	0	0.71	0.59	0.65	*P*	NS	NS
	60	0.64	0.71	0.67	$LSD_{(0.05)}$	0.42	0.33
	$\bar{x}$	0.68	0.65				
Anions							
Cl^-	0	11.03	12.15	11.59	*P*	NS	<0.01
	60	11.67	14.83	13.24	$LSD_{(0.05)}$	3.85	1.16
	$\bar{x}$	11.35	13.49				
SO_4^{2-}	0	4.25	5.31	4.78	*P*	NS	<0.05
	60	5.05	6.71	5.87	$LSD_{(0.05)}$	1.14	0.66
	$\bar{x}$	4.65	6.01				
$H_2PO_4^-$	0	4.23	5.00	4.61	*P*	<0.01	<0.001
	60	5.40	6.18	5.78	$LSD_{(0.05)}$	0.65	0.34
	$\bar{x}$	4.91	5.59				
NO_3^-	0	2.73	2.89	2.81	*P*	NS	NS
	60	2.63	2.78	2.70	$LSD_{(0.05)}$	0.62	0.38
	$\bar{x}$	2.68	2.83				

Results and discussion

Cation and anion concentrations in spring wheat are given in Table 1. General trends were similar for spring barley, winter wheat and winter barley. The sums of cations and anions and cations minus anions for all four crops, and for a spring wheat experiment in 1980[2], are given in Tables 2, 3 and 4.

All the results with cereals are consistent. In every case the sum of anions is increased strongly by mycorrhizal infection, but not by phosphorus additions. The change in the phosphorus concentration, which is normally expected with mycorrhizal infection, is much smaller on a molar basis than the enhancement of the concentrations of the anions chloride and sulphate. In both of the experiments reported here, phosphorus decreased the total sum of cations. The effect was most marked for potassium, the major cation, and may result from dilution due to increased growth. The only cation consistently affected by inoculation was sodium, with mycorrhizal plants generally having a markedly higher concentration. This differs from previous reports[3] where sodium concentration was usually reduced by infection. In our experiments both phosphorus and mycorrhizal infection had the same final effect, a decreased cation excess. However, this effect occurred by quite different mechanisms for the two treatments; phosphorus reduced the sum of cations, while mycorrhizal infection increased the sum of anions. In the earlier experiment[2] applied phosphorus did not depress cation concentrations in this way.

Table 2. Effect of mycorrhizal inoculation on the cation-anion balance (meq. $100\,g^{-1}$) in spring wheat and spring barley at two levels of applied phosphorus

	Crop	P applied ($kg.ha^{-1}$)	Inoculation –	Inoculation +	Inoculation $\bar{x}$	Significance	Phosphorus	Inoculation
ΣC	Spring wheat	0	75.00	71.51	73.25	*P*	<0.05	NS
		60	69.64	66.44	68.04	$LSD_{(0.05)}$	5.79	7.62
	Spring barley	0	95.69	31.74	88.71			
		60	76.36	73.65	75.00			
		$\bar{x}$	79.17	60.84				
ΣA	Spring wheat	0	22.25	25.35	23.80	*P*	NS	<0.05
		60	24.73	30.47	27.60	$LSD_{(0.05)}$	5.73	2.57
	Spring barley	0	40.40	44.80	42.60			
		60	35.67	37.66	38.87			
		$\bar{x}$	30.76	34.57				
C–A	Spring wheat	0	52.76	46.18	49.46	*P*	<0.05	<0.01
		60	44.90	35.97	40.43	$LSD_{(0.05)}$	6.46	5.58
	Spring barley	0	55.29	36.45	45.87			
		60	50.54	40.25	36.63			
		$\bar{x}$	50.88	39.71				

Table 3. Effect of mycorrhizal inoculation on the cation-anion balance (meq. 100 g^{-1}) in winter wheat and winter barley at two levels of applied phosphorus

	Crop	P applied (kg.ha^{-1})	Inoculation –	Inoculation +	Inoculation $\bar{x}$	Significance	Phosphorus	Inoculation
ΣC	Winter wheat	0	93.54	80.18	86.86	*P*	<0.001	NS
		60	83.30	87.87	85.58	$LSD_{(0.05)}$	2.97	3.33
	Winter barley	0	95.84	97.52	96.68			
		60	80.54	79.06	79.80			
		$\bar{x}$	88.34	86.12				
ΣA	Winter wheat	0	34.35	39.51	36.94	*P*	<0.01	<0.001
		60	38.91	49.86	44.38	$LSD_{(0.05)}$	4.30	3.52
	Winter barley	0	44.89	51.76	48.32			
		60	45.85	63.20	54.50			
		$\bar{x}$	40.99	51.09				
C–A	Winter wheat	0	59.19	40.66	49.92	*P*	<0.001	<0.001
		60	44.40	38.01	41.20	$LSD_{(0.05)}$	4.37	5.12
	Winter barley	0	50.95	45.77	48.36			
		60	34.89	15.71	25.30			
		$\bar{x}$	47.36	35.04				

Table 4. Effect of mycorrhizal inoculation on the cation-anion balance (meq. 100 g^{-1}) in spring wheat (1980) at two levels of applied phosphorus

	P applied (kg.ha^{-1})	Inoculation –	Inoculation +	Inoculation $\bar{x}$	Significance	Phosphorus	Inoculation
ΣC	0	67.72	69.07	68.39	*P*	NS	NS
	60	68.45	70.40	69.42	$LSD_{(0.05)}$	2.70	1.81
	$\bar{x}$	68.09	69.74				
ΣA	0	28.34	35.82	32.08	*P*	<0.05	<0.001
	60	30.51	39.25	34.88	$LSD_{(0.05)}$	1.90	0.90
	$\bar{x}$	29.43	37.54				
C–A	0	39.39	33.25	36.32	*P*	NS	<0.001
	60	38.99	30.18	34.45	$LSD_{(0.05)}$	3.21	1.90
	$\bar{x}$	41.74	32.05				

Analyses given here refer only to the shoots, and the nutrient composition of roots may differ greatly. It is however difficult to extract the roots of field grown crops quantitatively and analyse them for all nutrients. In pot work [2], the general pattern of an increased anion content was the same for roots and shoots, although effects among different anions varied.

Buwalda *et al.*[2] concluded from the size of changes in chloride, bromide and sulphate concentrations that it was unlikely that these increases resulted from the direct uptake, translocation and transfer of the ions by mycorrhizal hyphae, as occurs for phosphate. They suggested that the changes might be caused by a shift in the plant metabolism, perhaps due to a change in the way the plant regulates its pH. This can occur either by the formation of malate and other organic acids which, in effect, carry the deficit of negative charges or by efflux of hydroxyl or bicarbonate ions from the cell. The latter process requires the uptake of a counter-balancing anion to maintain electrical neutrality. It is suggested that a local diversion of sugars within the root cortical cells towards the mycorrhizal fungus, perhaps especially in cells containing arbuscules, might lead to a preference for the second mechanism of pH control, consequently leading to a larger uptake of chloride and other anions.

The results in this paper are consistent with others reported [2] and occur irrespective of times of sowing and type of cereal. In addition the pH in the rhizosphere of Douglas Fir seedlings inoculated with *Hebeloma crustuliniforme* was more alkaline, by about 0.3 pH units, than that of non-inoculated seedlings, irrespective of whether ammonium or nitrate was the nitrogen source [1]. This is consistent with our finding, as it implies that hydroxyl or bicarbonate is being exchanged for acid anions. Further investigations of this effect on a wider range of crops, in different soils and with other root-fungi infections, could have interesting implications for the elemental composition and possibly the quality of economic crops.

References

1 Bledsloe C S and Zasoski R J 1983 Effects of ammonium and nitrate on growth and nitrogen uptake by mycorrhizal Douglas-fir seedlings. Plant and Soil 71, 445–454.

2 Buwalda J G, Stribley D P and Tinker P B 1983 Increased uptake of bromide and chloride by plants infected with vesicular-arbuscular mycorrhizas. New Phytol. *In press.*

3 Jarrell W M and Beverly R B 1981 The dilution effect in plant nutrition studies. Adv. Agron. 34, 197–224.

4 Mosse B 1973 Advances in the study of vesicular-arbuscular mycorrhiza. Annu. Rev. Phytopathol. 11, 171–176.

5 Mosse B, Stribley D P and Le Tacon F 1981 Ecology of mycorrhizae and mycorrhizal fungi. Adv. Microb. Ecol. 5, 137–210.

6 Tinker P B 1980 Role of rhizosphere micro-organisms in phosphorus uptake by plants. *In* The Role of Phosphorus in Agriculture. Eds. S E Kwasahneh, E C Sample and E J Kamprath. pp 617–654, American Society of Agronomy, Madison.

Plant and Soil 71, 469–476 (1983). A-53

Phosphorus-31 nuclear magnetic resonance study of polyphosphate metabolism in intact ectomycorrhizal fungi

F. MARTIN, D. CANET*, D. ROLIN**, J. P. MARCHAL* and F. LARHER**
Unité de Microbiologie des Sols, INRA-CNRF, Champenoux, F-54 280 Seichamps, France

Key words *Cenococcum graniforme* Ectomycorrhizal fungi *Hebeloma crustuliniforme* *Hebeloma cylindrosporum* Nuclear magnetic resonance Phosphorus-31 Phosphorus metabolism Polyphosphates

Summary Phosphorus-31 nuclear magnetic resonance spectra at 36.4 MHz are presented for intact ectomycorrhizal fungi grown in pure culture. Resonances from polyphosphates and intracellular orthophosphate are identified in *Cenococcum graniforme, Hebeloma cylindrosporum,* and *H. crustuliniforme.* Comparison of the NMR spectra with phosphorus fractionation of the fungi extracts leads to the statement that the NMR-observed polyphosphates is a good part of the accumulated polyphosphates. In actively growing mycelia, this fraction account for up to 17% of total P.

Introduction

Mycorrhiza can greatly increase the utilization of phosphate present in soil, especially when phosphate availability is limited[13]. Recent experiments[2,4] have suggested that phosphate accumulation in mycorrhizas was correlated to the synthesis of phosphorylated compounds presumed to be polyphosphates. The determination of these metabolites is based on the histochemical staining of the cells by basic dyes (metachromatic granules)[3,20], electron-probe microanalysis (phosphorus rich granules)[4,21], and by polyacrylamide gel electrophoresis[2]. The use of such methods must nevertheless be considered with caution. This is due to the fact that the basic dyes used to identify the granules may also react with other polymeric metabolites and that the chemical methods used can lead to partial hydrolysis of polyphosphates[12]. Thus, there are no unambiguous techniques for polyphosphates identification. Knowledge of the identity, concentration, and distribution of the compounds in purc cultures of fungi and in the association is essential if an understanding of the cellular metabolism of mycorrhizal associations is to be obtained.

Phosphorus-31 (^{31}P) nuclear magnetic resonance (NMR) spectrometry can provide detailed information about the concentrations and environments of phosphorus-containing metabolites in intact cells[7,17]. Methods have recently been developed[8,10,11,15,19] for identification and quantitation of linear

* Laboratoire de Chimie Théorique, Université de Nancy I, ERA au C.N.R.S. n° 22, BP 239, 54 506 Vandoeuvre-Nancy Cédex, France.
** Laboratoire de Biologie Végétale, Université de Nancy I, BP 239, 54 506 Vandoeuvre-Nancy Cédex, France

polyphosphates by NMR. The main advantage of NMR is that is noninvasive and nondestructive.

In this paper, we report the results of a ^{31}P NMR study of intact ectomycorrhizal fungi grown in pure culture. Spectra of suspensions of these organisms reveal the occurrence of low level of polyphosphates. The data obtained were compared to values obtained by chemical method.

Material and methods

Growth and harvesting of cells

Cenococcum graniforme, strain Kiffer 1973, *Hebeloma crustuliniforme*, strain Kiffer 1973, and *Hebeloma cylindrosporum*, strain Bruchet 1970, were grown in medium containing 2.5 m*M* diammonium tartrate, 7 m*M* KH_2PO_4, 2.7 m*M* $MgSO_4$, 28 m*M* maltose, and 110 m*M* glucose (1 liter Erlenmeyer flasks with 500 ml of medium at 25°C; shaker, 120 oscillations/min). The pH was adjusted to 5 at the beginning of growth.

Mycelia, from the exponential phase of growth (4–5 weeks), were harvested by filtration and washed 8–10 times with distilled water and stored at 4°C until needed. The NMR measurements were started within 1 h following the harvest. At this time, the mycelia were resuspended in 10% DHO and placed in 10 mm o.d. NMR sample tube to obtain the ^{31}P NMR spectrum. The pH of the fungal suspension was 5.5.

^{31}P-NMR measurements

The ^{31}P NMR spectra were obtained on a modified Bruker HX 90 spectrometer[1] operating at 36.4 MHz in the pulsed Fourier transform mode; 8 μsec pulses corresponding to a flip angle of 45° were generated every 0.5 sec and each spectrum represents the time average of 10^5 transients. In all spectra, broad-band proton irradiation was applied. The deuterium signal of the DHO allowed field frequency stabilization. Spectra were measured at 4°C. The partial saturation of resonances stemming from rapid pulsing was corrected by using estimated values of longitudinal relaxation times. The latter were deduced from series of spectra obtained under various pulsing rates; under our standard conditions, polyphosphates were fully relaxed whereas the observed intensity of the orthophosphate peak must be multiplied by a factor of about 7. Chemical shifts are reported in ppm with respect to 85% phosphoric acid. A detail account of NMR monitoring of cellular metabolism is given by Roberts and Jardetsky[17].

Obtention of phosphorus fractions

Phosphorus fractions were separated by a modification of Mudd *et al.*'s[14] procedure[18]. The following section is dealing with the obtention of acid-soluble polyphosphates fraction used in the NMR measurements.

All operations were carried out at 4°C. Freshly harvested mycelia were extracted by 0°C (melting ice) TCA (7.5% w/v) for 15 min with constant shaking, then centrifuged at 3,500 *g* for 5 min. This step was repeated 4 times and the supernatants were pooled. Orthophosphate P was immediately determined on this fraction according to Fiske and Subbarow[6]. The pH was adjusted to 4.0 with solid Ba acetate and acid-soluble polyphosphates were precipitated overnight, centrifuged, washed with cold water and redissolved in water. Ba was removed by cation-exchange resin (Biorad AG 50 W, X 8, H^+) batch chromatography for 4 h with constant shaking[5]. The aqueous solution (5 ml, 25 m*M* P) was then used for NMR. The deviation between two measurements never exceeded 10%.

To check that chemical hydrolysis of polyphosphates during the procedure was not a potential cause of artifacts, some tissue samples, devoid of polyphosphates, were extracted in TCA-containing standard polyphosphate (n = 39). The Ba acetate precipitated fraction contained most of the polyphosphate in undegraded form.

Materials

The tri-, tetra-, penta-, seventeen-, and thirtynine-polyphosphates used for assignements of the chemical shifts in these experiments were purchased from Sigma Co (St Louis, MO, USA). All other reagents were of analytical grade.

Results

Fig. 1 shows the 36.4 MHz ^{31}P NMR spectra of dense suspensions of *C. graniforme* (Fig. 1A), *H. crustuliniforme* (Fig. 1B) and *H. cylindrosporum* (Fig. 1C) mycelia grown in 7 m*M* phosphate medium and harvested during the exponential phase of growth. An interesting feature displayed in these spectra is the presence of two strong resonances. Assignments of these peaks to particular metabolites was made by comparison with known standards and with the data obtained in yeast cells[8,19]. Peak PP resonates at 21 ppm upfield from 85% phosphoric acid. This position is consistent with that obtained for the middle phosphate of long chain ($n \geqslant 5$) standard polyphosphates. Peak P is identified as intracellular orthophosphate.

The concentration of polyphosphates relative to orthophosphate is estimated after the correction for peak saturation. In our experimental conditions, the amount of detected polyphosphates represents, on a P basis, 67, 61 and 17% of the orthophosphate from *C. graniforme*, *H. cylindrosporum* and *H. crustuliniforme*, respectively.

In the extracts of *C. graniforme*, *H. cylindrosporum* and *H. crustuliniforme* the total P content was about 110, 223 and 219 µmol per gram of dry weight, of which orthophosphate comprised 19, 32 and 35 µmol of P (Table 1). Based on these data, the NMR-observed polyphosphates content is estimated to be about 13, 19 and 7 µmol P per gram of dry weight of *C. graniforme*, *H. cylindrosporum* and *H. crustuliniforme*, respectively. NMR-observed polyphosphates account thus for 11,8 and 3% of total P of actively growing ectomycorrhizal fungi mycelia.

However, it must be borne in mind that liquid phase NMR only detects resonances of P molecules with fast motional correlation times in the more homogeneous regions of the sample. Moreover, there is evidence that in many fungi, a sizeable proportion of the polyphosphates is either bound to the wall and membrane and interacting with paramagnetic components of the cells, as Mn^{++}, which leads to broad and undetectable resonances[16].

To evaluate the size of the NMR-observed polyphosphates pool in the total polyphosphates and total P contents of the fungi, chemical distribution of P have been performed on the samples used in NMR (Table 1). The orthophosphate, soluble organic P and P-lipids fractions are present at similar percentages of total

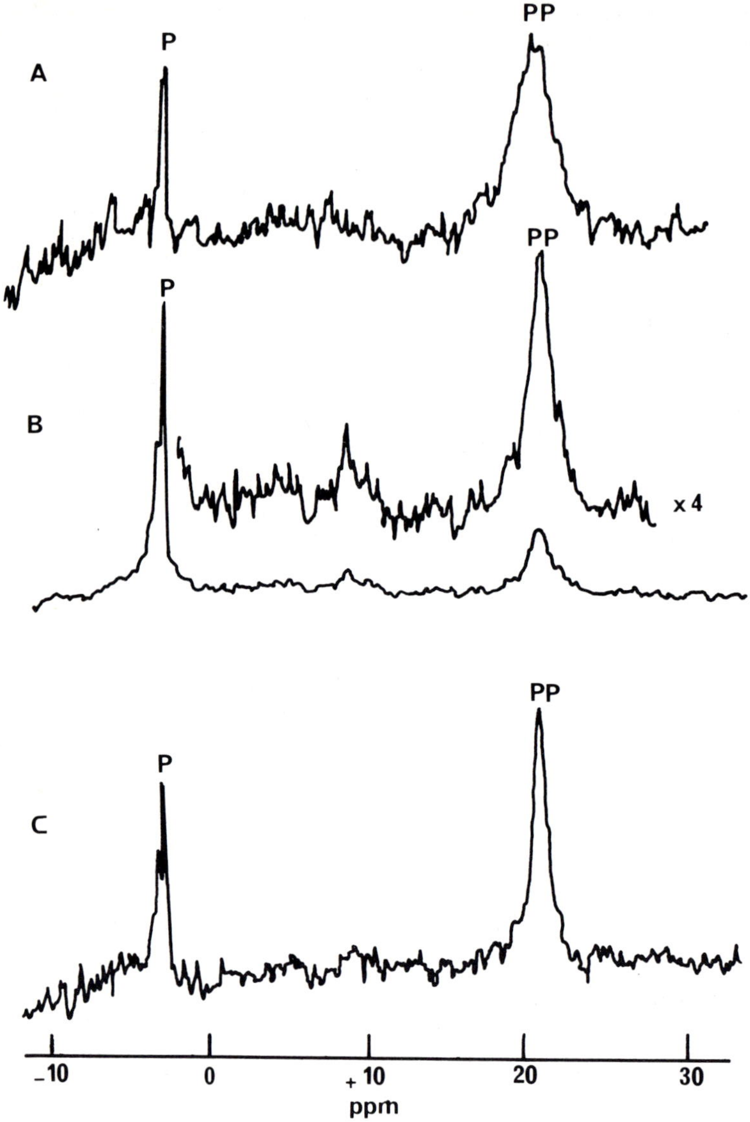

Fig. 1. 36.4 MHz ^{31}P NMR spectra of ectomycorrhizal fungi cells. Fungi were grown for 5 weeks on 7 m*M* phosphate medium and harvested as in section 2. Each spectrum represents the time average of 100 000 scans of 45° flip angle pulses with a 0.5 sec repetition rate. Spectra were measured at 4°C. Spectra: (A) *Cenococcum graniforme*; (B) *Hebeloma crustuliniforme*; (C) *Hebeloma cylindrosporum*. Peaks were assigned on the basis of chemical shifts. P, internal orthophosphate; PP, inner phosphate of polyphosphates chains.

Table 1. Chemical distribution of phosphorus in actively growing ectomycorrhizal fungi mycelia. For experimental details see Material and Methods and Rolin *et al.*[18]

Phosphate-containing fraction	*Cenococcum graniforme*		*Hebeloma cylindrosporum*		*Hebeloma crustuliniforme*	
	μmol P (g d.w.)$^{-1}$	% of total P	μmol P (g d.w.)$^{-1}$	% of total P	μmol P (g d.w.)$^{-1}$	% of total P
Orthophosphate	19	17	32	14	35	16
Acid-soluble Polyphosphates	19	17	22	10	7	3
Acid-insoluble Polyphosphates	2	2	1	0.5	4	2
Soluble organic P	40	37	34	15	52	24
Lipid P	18	16	29	13	44	20
RNA	3	3	15	8	17	8
DNA	8	7	90	40	60	27
Total P	109	100	223	100	219	100

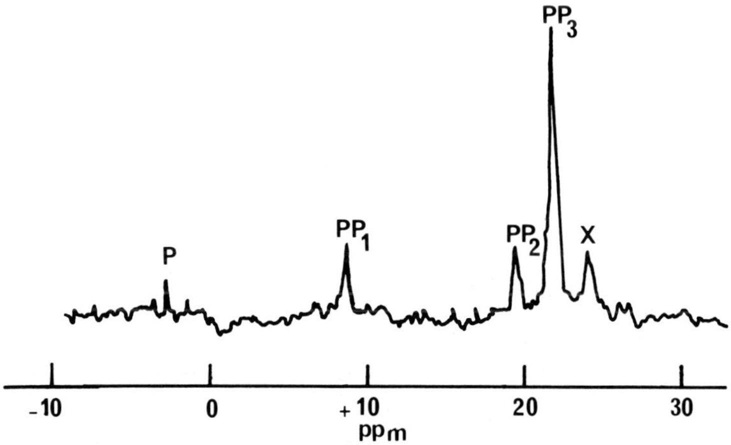

Fig. 2. ^{31}P NMR spectrum of acido-soluble polyphosphates fraction from *Cenococcum graniforme* extract. For further details see section 2. The spectrum represents the time average of 20000 scans of 45° flip angle pulses with a 0.2 sec repetition rate. Spectrum was measured at 4°C. Peaks assignments were made by comparison with known standards. P, internal orthophosphate; PP_1, PP_2, PP_3, terminal phosphates, penultimate phosphates and inner phosphates of polyphosphate chains, respectively.

P in the three species, whereas the nucleic acid content of the Basidiomycetous (Hebeloma) greatly exceed those of *C. graniforme*. The acid-soluble polyphosphates concentrations obtained by chemical fractionation are in agreement with those measured by NMR and this fraction account for 60% (*H. crustuliniforme*) to 95% (*H. cylindrosporum*, *C. graniforme*) of the total polyphosphates content. Thus, NMR detected the main part of the polyphosphates.

Fig. 2 shows the spectrum of the purified acid-soluble polyphosphates fraction from the extract of *C. graniforme*. As expected, this fraction is rich in polyphosphates. Peak PP_1 is the sum of several terminal phosphate resonances from polyphosphate sequences. Peak PP_2 corresponds to the penultimate phosphates from polyphosphates chains. Finally, PP_3 comes from the inner phosphates of polyphosphates. The peak labeled X and occurring at 24 ppm has not been identified.

According to Shulman and coworkers[15], the relative intensities of the peaks can be used to estimate the average chain length of the polyphosphate molecules; a ratio of about 4:1 for the inner phosphates relative to the penultimate phosphates corresponds to an average chain length of about 12 residues. The ratio of PP_3/PP_1 is about 4 corresponding again to an average chain length of 12 phosphates. This value falls within the range of values of average chain length found in the acid-soluble fraction[12] and high-resolution spectra from yeasts[15].

By using thin-layer chromatography we have been unable to detect short length polyphosphate chains, as pyrophosphate, tri- and tetra-polyphosphates (data not shown). Thus it appears that the purified acid-soluble polyphosphates fraction was compound mainly of long chain polyphosphates with an average length of 12 phosphate residues.

Discussion

The presence of condensed inorganic phosphates has been demonstrated and the metabolism of polyphosphates investigated in many mould fungi, mucor fungi, yeasts, mushrooms and other saprophytic Basidiomycetes[12]. Recently, ^{31}P high-resolution NMR have been used to study the metabolism of polyphosphates in yeasts and its connection with ATP[8,19]. Two purposes of the present investigation were to characterize polyphosphates in ectomycorrhizal fungi, and to compare NMR with chemical method.

Resonances from polyphosphates are identified in *C. graniforme*, *H. cylindrosporum*, *H. crustuliniforme* grown in batch culture on 7 m*M* phosphate medium. The NMR-observed polyphosphate content is species-dependent and account for 3 to 11% of total P. These values are in agreement with the concentrations of acid-soluble polyphosphates determined by the chemical method (Table 1) (3 to 18%). Moreover, no significant acid-insoluble polyphosphates were measured. Thus, there is evidence to support the statement that the polyphosphate pool observed in living material is similar to the acid-soluble polyphosphate fraction. Salhany *et al.*[19] have shown, that in yeast, NMR-observed polyphosphates are also a good part of the accumulated polyphosphates.

The low amount of polyphosphates found in the ectomycorrhizal fungi grown in pure culture was an unexpected observation in view of the high polyphosphate concentrations found in ectomycorrhizas by several authors (see Strullu *et al.*[20]). Therefore, in another set of experiments, *C. graniforme* was grown in batch culture with higher concentrations of phosphate (14 and 30 mM), in fermentor which allows short time of culture, or harvested at early stage of growth. These investigations lead to similar results (Martin *et al.*, in preparation). This is consistent, however, with the numerous observations that polyphosphates are not normal metabolites of microorganisms, but are rapidly accumulated by them under conditions of nutritional imbalance or in aged cultures[9,11].

The fact that compounds presumed to be polyphosphates are present in high amount in ectomycorrhizas leads to ask the question: to what extent, and under what circumstances, does polyphosphates are accumulated?

In conclusion, it is clear from these preliminary investigations that ^{31}P NMR can be a very useful method for the study of phosphorus metabolism in mycorrhizal associations.

Acknowledgements The authors are indebted to F. Le Tacon for useful discussions.

References

1 Brondeau J, Diter B and Canet D 1981 Flexible Fourier multi-nuclear magnetic resonance spectrometer. Rev. Scientific Instrum. 52, 542–547.

2 Callow J A, Capaccio L C M, Parish G and Tinker P B 1978 Detection and estimation of polyphosphate in vesicular-arbuscular mycorrhizas. New Phytol. 80, 125–134.

3 Cox G, Sanders F E, Tinker P B and Wild J A 1975 Ultrastructural evidence relating to host-endophyte transfer in a vesicular-arbuscular mycorrhiza. *In* Endomycorrhizas. Eds. F E Sanders, B Mosse and P B Tinker. pp 297–312. Academic Press, New York and London.

4 Cox G, Moran K J, Sanders F, Nockolds C and Tinker P B 1980 Translocation and transfer of nutrients in vesicular-arbuscular mycorrhizas. III. Polyphosphate granules and phosphorus translocation. New Phytol. 84, 649–659.

5 Ebel J P 1952 Recherches sur les polyphosphates contenus dans diverses cellules vivantes. Bull. Soc. Chim. Biol. 34, 321–335.

6 Fiske C and Subbarow Y 1926 The colorimetric determination of phosphorus. J. Biol. Chem. 66, 375–400.

7 Gadian D G and Radda G K 1981 NMR studies of tissue metabolism. Annu. Rev. Biochem. 50, 69–83.

8 Gillies R J, Ugerbil K, Hollander J A den and Shulman R G 1981 ^{31}P NMR studies of intracellular pH and phosphate metabolism during cell division cycle of *Saccharomyces cerevisiae*. Proc. Natl. Acad. Sci. USA 78, 2125–2129.

9 Harold F M 1966 Inorganic polyphosphates in biology: structure, metabolism and function. Bacteriol. Rev. 30, 772–794.

10 Hollander den J A, Ugurbil K, Brown T R and Shulman R G 1981 Phosphorus-31 nuclear magnetic resonance studies of the effect of oxygen upon glycolysis in yeast. Biochemistry USA 20, 5871–5880.

11 Jacobson L and Cohen J S 1981 Improved technique for investigation of cell metabolism by ^{31}P NMR spectroscopy. Bioscience Rep. 1, 141–150.

12 Kulaev I S 1979 The Biochemistry of Inorganic Polyphosphates. John Wiley and Sons, New York.

13 Moore P D 1979 Mycorrhizal associations. Nature London 282, 780.

14 Mudd S, Yashida A and Koike M 1958 Polyphosphate as accumulator of phosphorus and energy. J. Bacteriol. 75, 224–235.

15 Navon G, Shulman R G, Yamane T, Eccleshall T R, Lam K B, Baronofski J J and Marmur J 1979 Phosphorus-31 nuclear magnetic resonance studies of wild-type and glycolytic pathway mutants of *Saccharomyces cerevisiae.* Biochemistry USA 18, 4487–4499.

16 Ostrovskii D N, Sepetov N F, Reshetnyak V A and Siberl'dina L A 1980 Investigation of the localization of polyphosphate in cells of microorganisms by the method of high resolution ^{31}P-NMR-145.78 MHz. Biochemistry USSR 45, 517–525.

17 Roberts J K M and Jardetsky O 1981 Monitoring of cellular metabolism by NMR. Biochem. Biophys. Acta 639, 53–76.

18 Rolin D, Le Tacon F and Larher F 1982 Caractérisation des différentes fractions du phosphore chez le champignon mycorrhizogène *Hebeloma cylindrosporum. In* Les mycorhizes, partie intégrante de la plante: biologie et perspectives d'utilisation. Ed. S Gianinazzi. Colloque de l'Institut National de la Recherche Agronomique. 13, 101–106.

19 Salhany J M, Yamane T, Shulman R G and Ogawa S 1975 High resolution ^{31}P nuclear magnetic resonance studies of intact yeast cells. Proc. Natl. Acad. Sci. USA 72, 4966–4970.

20 Strullu D G, Gourret J P and Garrec J P 1981 Microanalyse des granules vacuolaires des ectomycorhizes, endomycorhizes et endomycothalles. Physiol. Vég. 19, 367–378.

21 Strullu D G, Gourret J P, Garrec J P and Fourray A 1981 Ultrastructure and electron-probe microanalysis of the metachromatic vacuolar granules occurring in Taxus mycorrhizas. New Phytol. 87, 537–545.

Plant and Soil 71, 477–486 (1983). A-54

Root growth and litter decomposition in a coffee plantation under shade trees

G. CUENCA, J. ARANGUREN* and R. HERRERA
Centro de Ecologia, Instituto Venezolano de Investigaciones Cientificas (IVIC), Apartado 1827, Caracas 1010 A, Venezuela

Key words Coffee Decomposition rate Nutrient cycle Root distribution Root productivity VA Mycorrhiza

Abstract In a non-fertilized coffee plantation under shade trees the root biomass was excavated to estimate its distribution in the soil profile. One third of total fine (less than 1 mm) roots was found in the first 10 cm of soil; the cumulative total to 30 cm reached 73%. A highly variable and transient amount of fine roots colonized the litter layer. Root production both in the litter and in the first 7.5 cm of mineral soil was estimated from sequential samplings and was 10 g m^{-2} yr^{-1} and 660 g m^{-2} yr^{-1} respectively. The decomposition rate of weighed averages of litter fractions in the coffee plantation, calculated as the ratio of litter fall rate to the amount found in the soil was $k = 4.8$. Shade tree leaves, the major component of litter descomposed slower than coffee leaves and these slower than flowers and fruits. Litter bag experiments showed considerable slower rates when mesh was 0.03 mm than 0.5 mm. Nitrogen and phosphorous showed increases in concentrations as decomposition progressed while potassium, calcium and magnesium followed a decrease in concentration that paralleled that of dry weight loss. In comparing the decomposition rate for litter with or without coffee roots growing in the bags, a tendency to show faster decomposition rates was found for the treatment with roots. These differences were however, only significant for one month for shade tree leaves litter. Nitrogen amounts remaining in shade tree leaves litter was lower in the treatment with roots than without roots. Potassium concentration in roots was positively correlated with potassium concentration in decomposing leaf litter where roots were growing. These results suggest that while roots growing attached to decomposing litter had little or no effect in speeding the decomposition process, the superficial roots seem to play an important role in absorbing very efficiently the mineralized nutrients from litter. The anatomical study of roots showed that the plantation is intensely infected with V-A mycorrhiza. External mycorrhizal hyphae did not to play a role in attachment of roots to decomposing litter while root hairs were found to grow in profusion on root surfaces oriented toward litter.

Introduction

Litter decomposition and the concomitant liberation of mineral nutrients is important to nutrient cycling in natural vegetation and crops. Soil litter acts as a transition phase between living biomass and the soil. In tropical forests the amounts of nutrients returned to the soil via litter are *ca.* three times higher than that for temperate forests[20].

In coffee plantations under shade-trees litterfall and decomposition rates are similar to tropical rainforests[3,22]. Nutrient return to the soil follows the

* Present address: Departamento de Biología y Química, Instituto Universitario Pedagógico de Caracas, Avenida Paéz, El Paraiso, Caracas.

decomposition of litter with K, Ca and Mg being released at approximately the same rate as organic matter is lost while N and P are retained longer and sometimes accumulated[1,22].

The coffee root system is superficial. Additionally, fine roots occur in the litter layer. The role of the root system in recycling nitrogen from both shade trees and coffee has been assessed by Aranguren *et al.*[2] Coffee roots have vesicular arbuscular mycorrhizas (VAM)[13] and there has been interest in assessing their role in nutrient cycling[7]. Gadgil and Gadgil[8] found that ectomycorrhiza in Pinus roots inhibit litter decomposition, an effect due to either antibiotic effects on soil bacteria or to enhanced ability to obtain mineral nutrients, by the symbiotic fungi, which do not depend directly on litter as an energy source. In contrast, in tropical rain forests it has been suggested that mycorrhizal hyphae may help decompose litter[23]. New results both from field[12] and laboratory[16] seem to support these findings.

In the present work we have tried to estimate the relative growth rate of fine coffee roots, in the litter layer and in the mineral soil of an unfertilised coffee plantation under shade trees. The objectives of this study were, 1) to estimate root biomass distribution, 2) the seasonal growth of fine roots as related to physical environmental parameters, 3) to assess the rate of litter decomposition, the release of mineral nutrients from litter and 4) to try to establish the role of VAM mycorrhiza in these processes.

Materials and methods

Experimental area

The area was situated at 1400 m altitude in Miranda State, Venezuela. Annual precipitation is 1200 m with a dry period in February–March. Mean annual temperature is 19°C with little seasonal variation. The soils are acid, derived from well weathered mica-schists, and with pronounced slopes[11].

Coffee arabica var. Mundo Nuevo was planted 25 years ago and the plantation has been managed with minimal inputs and no fertilizers. The shade trees *Erythrina* sp. and *Inga* spp. form a canopy at *ca.* 20 m; a second shade-tree stratum at 15 m is formed by *Heliocarpus americanus*, *Clethra* sp. and *Ficus* sp. Towards the edges of the plantation banana plants are abundant.

Coffee root distribution and growth

The root biomass of 4 coffee plants was excavated at increasing distances from the trunks[2], roots separated from soil and divided into two diameter classes using 1 mm as the limit between them. Samples were taken for analyses. Roots present in 50 0.25 m^2 quadrats pushed into the litter were separated and subsampled for analyses. The seasonal variation in fine root biomass was followed both in the litter layer and in the mineral soil; in the former by extracting the roots that penetrated 50 plastic mesh bags which contained litter as explained below. The changes in root biomass in the upper 7.5 cm of mineral soil were estimated from monthly samplings for a year with 13 cutting cylinders 7.56 cm in diameter pushed into the soil and subsequently extracted. The roots were then separated by sieving and flotation. Subsamples were taken for analyses.

Litter decomposition

Litter decomposition was estimated with two sets of plastic mesh bags containing litter as described by Aranguren *et al.*[2] Decomposition rates estimated as dry weight loss from 4 bags collected monthly from each treatment, were compared for coffee leaves, shade-tree leaves, twigs and flower and fruit fractions. Samples were taken from each of these for analyses.

Root growth and litter decomposition

The effect of coffee roots infected with vesicular-arbuscular mycorrhiza (VAM) was assessed by placing on the plantation floor 100 2 mm mesh bags each containing 40 g shade-tree leaves and 8 g coffee leaves. Every week the bags were checked and 50 were lifted and any roots found growing into them severed. The remaining bags were left in place. Once a month 4 bags from each treatment were carefully taken to the laboratory and their contents sorted into leaf litter fractions and roots. Samples were taken from each fraction for analyses.

Chemical analyses

All samples were dried at 80°C, weighed and ground in a steel mill prior to chemical analyses. Appropriate aliquots were digested in a mixture of $HClO_4$ and H_2SO_4 in the presence of vanadium pentoxide. N and P were analyzed colorimetrically following the method of a Technicon II autoanalizer[10] and K, Ca and Mg by flame atomic absorption spectrometry.

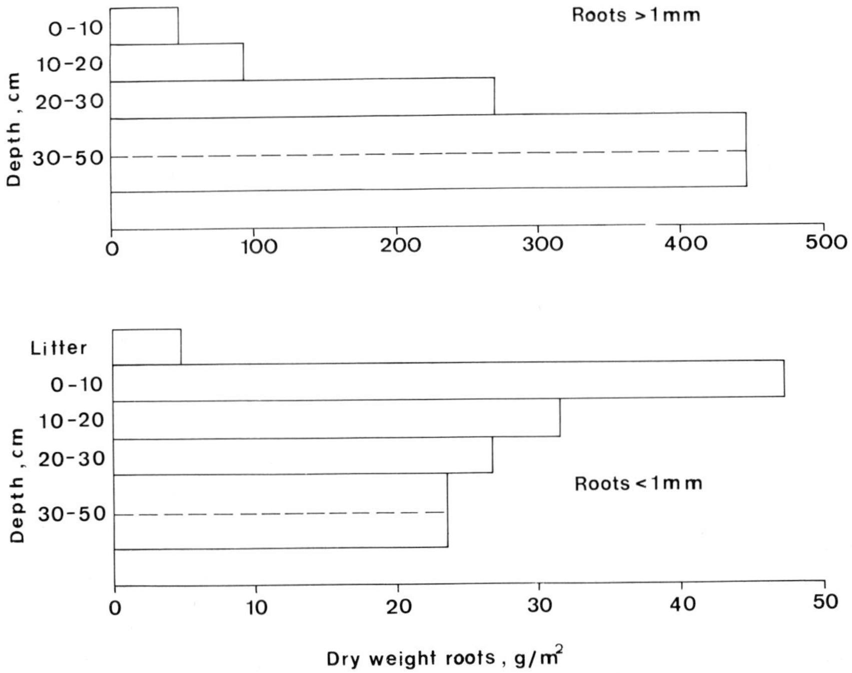

Fig. 1. Root biomass (dry weight) distribution in the soil profile and in the litter. Upper diagram refers to roots > 1 mm in diameter while lower diagram refers to those < 1 mm in diameter. The roots in the 30–40 cm and 40–50 cm intervals were pooled.

Anatomical studies

The interface between growing fine coffee roots and the leaf litter in close contact with them was studied on samples from the plantation floor fixed on collection in FAA, dehydrated in alcohol and embedded in parafin. 12 μm sections were stained with either safranin fast green or anilin blue.

Results and discussion

Fig. 1 shows coffee root distribution from the litter layer to 50 cm depth, for the fractions above and below 1 mm diameter. While the larger fraction showed a

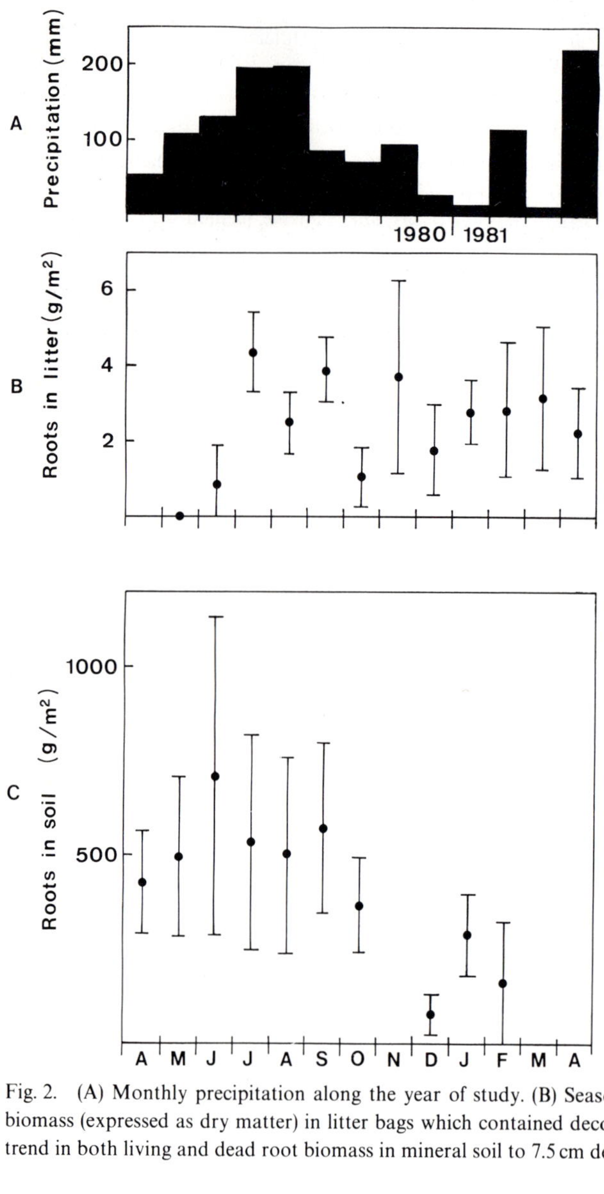

Fig. 2. (A) Monthly precipitation along the year of study. (B) Seasonal variation in living root biomass (expressed as dry matter) in litter bags which contained decomposing litter. (C) Seasonal trend in both living and dead root biomass in mineral soil to 7.5 cm depth.

steady increase in biomass with depth, the finer fraction showed a curve with maxima at 0–10 cm.

Seasonal variation in the amount of root in the litter bags on the plantation floor (Fig. 2B) and the first 7.5 cm of mineral soil (Fig. 2C) showed that the fine roots, which penetrated the litter bags were affected by water regime. When the litter bags were placed on the plantation floor there was a flush of root growth. As decomposition progressed the growth peaks which coincided with periods of high pluviosty, became progressively smaller. The effect of the presence of litter on root growth outside the mineral soil has been documented for tropical rainforests in Amazonia[14]. In order to calculate the increase in root biomass in the litter and mineral soil compartment we used the method of adding the positive changes in root biomass sampled according to Persson[18]. These calculations yielded $10\,g\,m^{-2}\,yr^{-1}$ for the roots in the litter layer and $661\,g\,m^{-2}\,yr^{-1}$ for the mineral soil fraction, although the inherent high variability of root growth does not permit a finer resolution. In the study of root biomass in the mineral soil living and dead roots were not differentiated so calculated values may overestimate production. In the litter layer it represents only living roots. Values for this compartment are lower than that reported for a boreal forest in Sweden[19] and an order of magnitude lower those reported for a tropical rainforest in Amazonas[14].

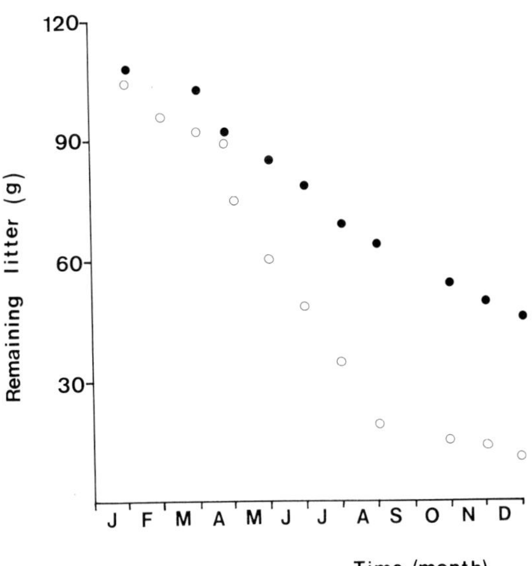

Fig. 3. Amount of remaining litter (dry weight) as a function of time in a coffee plantation. At the beginning of January 112 g of dry mixed litter was placed in each litter bag. Solid circles represent averages of 4 bags 30 microns mesh; open circles represent averages of 4 bags 0.5 mm mesh.

Litter decomposition rate

The decomposition rate for the weighed averages of litter fractions in the coffee plantation calculated as the ratio of litterfall to the amount of litter on the soil[17] is k = 4.8. This is comparable to the highest estimates for Colombian coffee plantations[22]. The fraction which decomposed fastest was flowers and fruits (k = 20), coffee leaves decomposed with k = 10 while the major component of litter in the plantation, shade-tree leaves decomposed with k = 4. With the litterbag method rates of decomposition were lower for bags of 0.5 mm mesh and much slower with 30 μm mesh bags. In both cases slower decomposition rates occurred between January and March, the drier period. The process then accelerated but slowed again towards the end of the year. Although the litter bag method introduces changes in the decomposition process which thus gives a dubious estimate of the decomposition[24], it gives a relative estimate of effects of micro and macroorganisms. The soil fauna have an important role in dismembering the litter fractions prior to decomposition by microorganisms (Fig. 3). In the 0.5 cm mesh litter bags *Trachelifus* sp., *Parajulus* sp., ants, slugs and snails and abundant earthworms were present.

Nutrient concentrations, measured in different litter fractions over a period of 11 months, showed N and P increasing in concentration as decomposition

Table 1. Mineral nutrient concentration (% dry weight) in litter fractions contained in 0.5 mm litter bags at different intervals after placing them in a coffee plantation

Litter fraction	N		P		K		Ca		Mg	
	t_1	t_2	t_1	t_2	t_1	t_2	t_1	t_2	t_1	t_2
Shade-tree leaves*	1.08	2.53	0.08	0.17	1.10	0.22	1.56	1.25	0.19	0.17
Twigs*	0.81	1.11	0.07	0.11	0.31	0.07	1.13	0.38	0.16	0.10
Coffee leaves**	1.35	2.55	0.10	0.17	0.99	0.26	1.28	1.08	0.33	0.42
Flowers and fruits**	0.99	1.56	0.10	0.16	2.01	1.00	0.97	0.33	0.32	0.19

* Fractions t_1 = 1 month, t_2 = 11 months.
** Fractions t_1 = 1 month, t_2 = 7 months.

Table 2. Effect of VAM-coffee roots on the decomposition rate of litter in comparison with litter which decomposed in the absence of roots (control) expressed as percentage of initial dry weight decomposed

Time (months)	Total litter		Shade-tree litter		Coffee litter	
	Roots	Control	Roots	Control	Roots	Control
6	67.7	64.5	62.4	59.3	93.7	90.2
12	71.0	69.2	67.9*	63.9*	100.0	100.0

* Differences statistically significant at 2.5%.

Table 3. Amounts of mineral nutrients in remaining litter in litter bags experiments both in the presence of roots (Roots) and when roots were removed from litter bags (Control)

Nutrient (mg)	Shade-tree leaves				Coffee leaves	
	6 months		12 months		6 months	
	Roots	Control	Roots	Control	Roots	Control
Phosphorus	16.9	18.2	12.7	13.1	0.58	0.88
Nitrogen	453.1	471.3	250.7*	375.8*	15.2	32.3
Calcium	170.1	240.3	162.5	261.7	10.6	17.2
Potassium	36.6	29.4	29.6	31.6	1.13	1.99
Magnesium	30.4	33.7	28.4	28.9	1.31	1.92

* Differences statistically significant at 5%.

progresses (Table 1). K, Ca and Mg decreased in concentration parallel to the loss of dry weight. In coffee leaves, litter Mg showed a small increase in concentration; a fact previously reported but without explanation[4,6,15].

Effect of VAM roots

In the treatment designed to exclude the superficial coffee roots from the litter bags, it was found that the frequency of lifting the litter bags and cutting of all roots that had grown towards them, was very effective. On the average 10.07 g dry weight roots m^{-2} yr^{-1} grew into the litter bags not lifted while in treated bags only 0.32 g m^{-2} yr^{-1}. When comparing the rate of decomposition of both litter fractions with or without roots variances were inhomogeneous and so non-parametric statistics were used to compare the weight losses month to month. The Mann-Whitney U test applied[21] has a power-efficiency of 95.5% (Table 2). Although there was a tendency for faster decomposition rates in the presence of roots, these differences were generally not statistically significant. For shade-tree leaves the values were statistically different in April 1981. Gadgil and Gadgil[9] found differences of 50% or more despite decomposition of pine needles in New Zealand proceeding much slower. VAM coffee roots appear to have no effect of the decomposition rate of litter.

Table 3 shows the results of the same experiments expressed in terms of nutrients remaining in litter after given intervals. The only significant differences found were those for the amount of nitrogen in shade-tree leaves after a year in the litter bags. When the roots that had penetrated the litter bags were analyzed for nutrients it was found however that their concentration of potassium was a significantly correlation with the potassium concentration of the litter in which they were growing (Fig. 4). These results suggest that while roots in decomposing litter have no effect on the rate of the process, they very efficiently absorb the mineralized nutrients from that source in an otherwise nutrient poor soil.

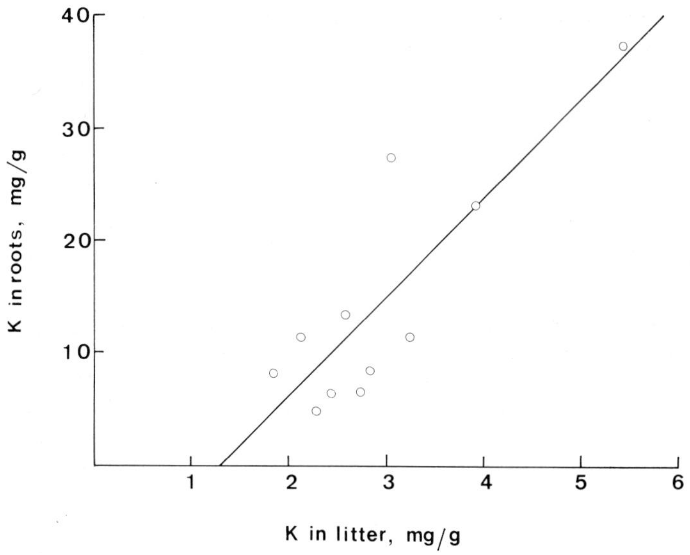

Fig. 4. Correlation between K concentration in litter and in roots that had penetrated the litterbags (correlation significant to 5%).

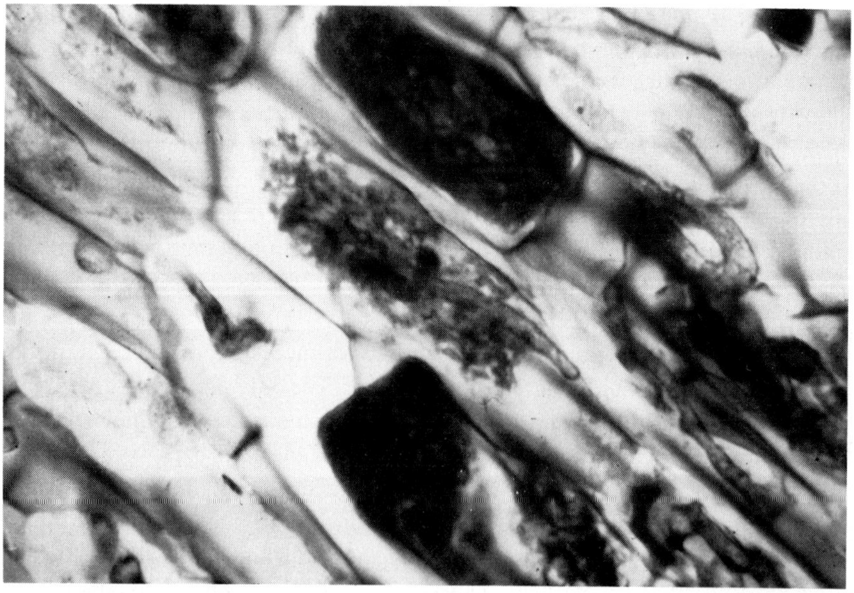

Fig. 5. VA-mycorriza arbuscles in section of a coffee root.

The anatomical study of the coffee roots collected from the field showed the plantation was intensely infected with VA mycorrhiza [5]. Fig. 5 shows the VAM of a coffee root. The adhesion of roots to the surface of decomposing leaves is quite strong and resisted the fixation and sectioning treatments. We observed that the attachment occurs through root hairs that were found to grow in more profusion on the root side towards the leaf surface. No external hyphae were observed to intervene in the contact between the two structures as is the case in Amazonian rain forest [12], where both ectomycorrhiza and VA mycorrhiza have been found.

Acknowledgements The authors wish to thank Dr V. García for his collaboration and suggestions in relation to the anatomical study of roots, to Dr T. St. John for fruitful discussion during the experimental period and to E. Orellana, G. Escalante L. Martin and S. Flores for laboratory assistance. One of us (J.A.) received partial funding from the Venezuelan CONICIT. The junior author wishes to express his gratitude to the Department of Ecology and Environmental Sciences at the Swedish Agricultural University, Uppsala for the use of their facilities during a sabbatical leave.

References

1 Aranguren J 1980 Contribución de la caída de hojarasca al ciclo de nutrientes en cultivos bajo árboles de sombra (café y cacao), M.Sc. Thesis, Caracas, Venezuela, 285 p.

2 Aranguren J, Escalante G and Herrera R 1982 Nitrogen cycle of perennial crops under shade trees. I. Coffee. Plant and Soil 67, 247–258.

3 Bernhard R F 1970 Etude de la litière et de sa contribution au cycle des eleménts mineraux en fóret ombrophile de cote-D'Ivoire. Oecol. Plant. 5, 247–266.

4 Bonner J and Galston A 1959 Principios de fisiología vegetal, Aguilar, 48 p.

5 Cuenca G 1982 Papel de las raíces micorrícicas del café (*Coffea arabica*) en la descomposición de la hojarasca. M.Sc. Thesis. Caracas, Venezuela. 136 p.

6 Epstein E 1972 Mineral nutrition of plants: principles and perspectives. John Wiley and Sons. Canada. 300 p.

7 Fogel R 1980 Mycorrhiza and nutrient cycling in natural forest ecosystems. New Phytol. 86, 199–212.

8 Gadgil R and Gadgil P D 1971 Mycorrhiza and litter decomposition. Nature London 233, 133.

9 Gadgil R and Gadgil P D 1975 Suppression of litter decomposition by mycorrhizal roots of *Pinus radiata*. N.Z.J. For. Sci. 5, 33–41.

10 Gales M E and Booth H 1974 Simultaneous and automated determination of phosphorous and total Kjeldahl nitrogen. Environmental Protection Agency, EPA-6701 4-74-002.

11 García J and Herrera R 1971 Propiedades fisicas, químicas y mineralógicas de una clinosecuencia de suelos ácidos. Agron. Trop. 21, 411–420.

12 Herrera R, Mérida T, Stark N and Jordan C F 1978 Direct phosphorous transfer from leaf litter to roots. Naturwissenschaften 65, 208–209.

13 Janse J M 1897 Les endophytes radicaux de quelques plants javanaises. Ann. Jard. Botan. Buitenz. 14, 52–212.

14 Jordan C F and Escalante G 1980 Root productivity in an Amazonian Rain Forest. Ecology 6, 14–18.

15 Larcher 1977 Ecofisiología Vegetal. Omega, Barcelona. 66 p.

16 Linkins A E and Antibus R K 1979 Growth and metabolism of cellulose and crude oil by selected mycorrhizal fungi which have extracellular cellulases and aryl hydrocarbon indroxylases 4th North American Conf. on Mycorrhiza. Colorado State University, Fort Collins.

17 Olson J S 1963 Energy storage and the balance of procedure and decomposers in ecological systems. Ecology, 44, 322–331.

18 Persson H 1978 Root dynamics in a young Scots pine stand in Central Sweden. Oikos 30, 508–519.

19 Persson H 1980 Spatial distribution of fine-root, growth, mortality and decomposition in a young Scots pine stand in Central Sweden. Oikos 34, 77–87.

20 Rodin L E and Bazilevich N I 1967 Production and mineral cycling in the terrestrial vegetation. Oliver and Boyd London, pp 208–240.

21 Siegel S 1978 Estadística no paramétrica aplicada a las ciencias de la conducta. Edit. Trillas, México. 346 p.

22 Suárez de Castro F and Rodríguez A 1955 Equilibrio de materia organica en plantaciones de café. Boletín de la Federación de Cafeteros de Colombia 5, 5–28.

23 Went F W and Stark N 1969 The biological and mechanical role of soil fungi. Proc. N.A. Sc. 60, 497–504.

24 Witkamp M and Olson J S 1963 Breakdown of confined and non confined oak litter. Oikos 14, 138–147.

Plant and Soil 71, 487–493 (1983). A-55

Growth and spatial distribution of nutrient-absorbing organs: selective exploitation of soil heterogeneity

T. V. ST. JOHN, D. C. COLEMAN and C. P. P. REID*
Natural Resource Ecology Laboratory, Department of Forest and Wood Sciences, Colorado State University Fort Collins, CO 80523, USA*

Key words Extramatrical mycelium Hyphal growth Negative binomial distribution Root growth Soil heterogeneity Tropical forest Vesicular-arbuscular mycorrhizae

Summary The soil is a heterogeneous medium with respect to decomposition activity and consequently with respect to nutrient availability. Roots of forest trees and hyphae of vesicular-arbuscular mycorrhizal (VAM) fungi, principal nutrient absorbing organs, have previously been observed to associate with decomposing organic matter and other localized sites of nutrient availability in the soil.

In two experiments in forests on contrasting soil types in Brazilian Amazonia, the association of tree roots with decomposing organic matter was shown to be a response of new root growth in the organic matter, not an avoidance of unfavorable conditions elsewhere in the soil volume. In laboratory experiments, the growth of VAM fungal hyphae was shown to respond to particles of decomposing organic matter in a way analogous to the response of forest tree roots. In both cases localized nutrient-rich sites (organic matter) were encountered by random growth of individual roots (hyphae) followed by increased branching after the encounter. Length of roots (and probably of hyphae) followed the negative binomial distribution, which is also found in other spatially aggregated organisms in heterogeneous environments.

Although roots and hyphae operate in different size scales, the selective exploitation of localized nutrient-rich sites and the mechanism by which the advantageous physical placement comes about is strikingly similar. The highly branched roots and hyphae in association with decomposing organic matter should be advantageous when ion uptake is limited by either diffusion rate or competition with other organisms.

Introduction

The heterogeneous nature of soil is well known, and a wide range of chemical and biological conditions may be present at localized points in the root zone of an individual tree. Some portions of the root system are located in decomposing organic matter, while other portions of the root system are surrounded by mineral particles. In one case, the humus or roots and leaves in various stages of decomposition have high ion exchange and water holding capacities, often contain active microbial populations, and may be sites of nutrient mineralization. In the other case the mineral particles, depending on their composition and size distribution, may contribute little to the plant's mineral nutrition.

On a microscopic scale, the hyphae of the plant's mycorrhizal fungi are divided between such contrasting microsites as fragments of sloughed root cortex, arthropod exoskeletons, mineral particles coated with humic materials, and relatively inert sand particles and soil voids.

While root systems and mycorrhizal mycelia are very different in size, they are both nutrient absorbing organs in heterogeneous environments. Within the soil volumes exploited by each kind of absorbing system, some microsites are capable of yielding proportionately large amounts of nutrients and other microsites are of essentially no value in supplying nutrients. It is thus of interest to ask whether these nutrient absorbing organs are distributed at random through the soil volume or are in some way concentrated at nutrient-rich microsites. Any selective spatial distribution of the absorbing system may have considerable significance for nutrient acquisition by the whole plant.

Recent experiments by the authors with root systems in two Amazonian rain forests[19] and with vesicular-arbuscular mycorrhizal (VAM) mycelium in pot cultures[20] are compared to examine the hypotheses that nutrient-absorbing organs associate physically with decomposing organic matter. The mechanisms by which the association comes about are inferred from indirect evidence and are shown to be similar in roots and VAM hyphae. Finally, the ecological significance of selective physical placement of nutrient absorbing organs will be discussed.

Methods

Details of the methods for the two experiments, reported elsewhere[19,20], are briefly outlined here. In both cases chambers containing either inert sand (control) or sand plus organic matter were introduced into a soil rich in tree roots (Amazonian forests) or extramatrical mycelium (VAM pot cultures). After a suitable incubation period, the chambers were withdrawn and length of absorbing organ in each chamber was determined.

Experiments with forest tree roots

Experiments were carried out at two sites in an old-growth lowland evergreen rain forest near Manaus, Brazil. Site 1, 30 km east of Manaus, was a high-diversity forest on a non-flooded, very heavy yellow oxisol and had little accumulated surface litter over most of the plot. The vegetation was described by Prance *et al.*[17] Site 2, 60 km northeast of Manaus, was a low-diversity, low-biomass forest on nonflooded white sand. Humus and litter had accumulated over most of the site. The forest floor layer ranged from 4 to 10 cm depth, and most of the fine roots of the forest were associated with this layer. The vegetation was described by Anderson *et al.*[1]

Loose-weave, plastic mesh bags were filled with a mixture of sand and *Theobroma cacao* leaf litter. Each bag received 1 kg of sand mixed with an equal volume of leaf material. Controls consisted of 2 kg of sand, an inert physical support for roots.

At both sites, the two treatments were interspersed throughout a rectangular area of approximately 20 by 30 meters. The bags were inserted into holes made in the forest soil with a 10-cm-diameter soil-sampling tube. Fifteen bags of each treatment were placed at each site and left in place for 30 d. At harvest they were withdrawn and the length of roots in each bag was determined by and intersect method[14].

Experiments with VAM hyphae were carried out using two clay containers (27 cm across by 7 cm deep) drilled and fitted with glass wool wicks. The dishes contained a very sandy, nutrient-poor soil amended with 17 µg/g P as hydroxyapatite. Host plants were *Fragaria* sp. (Applewood Seed Co., Golden, Colorado) grown from surface-sterilized seed. The host plants were inoculated and placed in the containers 92 d before the beginning of the experiment. The fungus was a local isolate of *Glomus fasciculatum* (Thaxter sensu Gerdemann) Gerdemann and Trappe, and infection was established gnotobiotically by inoculating sterile seedlings with pregerminated spores[18]. The cultures were maintained in a room kept free of dust with HEPA absolute filter systems (0.3 µm porosity) Bacterial contamination occurred during or before the experiment but no fungal contamination could be detected.

The method for determining the response of hyphae to the test materials was similar in concept to the experiments with forest roots. Test organic materials (300- to 1000-µm organic particles wet-sieved from a local prairie soil) were contained in rigid chambers with a 45 µm mesh screen separating the contents from the soil. Silica sand alone, sieved to the same size range as the organic particles, was used in the control treatment. The chambers, flattened boxes with the top open, were constructed of polycarbonate. One side consisted of a rectangular piece of stainless steel sieve cloth (45-µm openings) through which hyphae could enter the chamber.

At harvest, six d after planting, mycelium in the chambers was stained and hyphal length was determined by a line intercept method[13].

A second experiment was carried out in one of the pots used for the previous experiment to determine whether VAM hyphae are more likely to arrive at an OM chamber than at a control chamber. This experiment was only 48 h in duration.

Table 1. Root growth experiments in two tropical terra firme forests near Manaus, Brazil

	Control	OM
Site 1		
Mean root lengths ± SE	9.28 ± 2.36	135 ± 44.9
Sum of ranks	138.5	239.5
Number of observations	14	13
Site 2		
Mean root lengths ± SE	14.3 ± 3.70	125 ± 45.8
Sum of ranks	179.5	285.5
Number of observations	15	15

Table 2. Hyphal lengths in decimeters (± SE) in the VAM hyphal growth experiment

Pot 1	
Organic matter	Control
17.9 ± 4.82	7.45 ± 3.08
Pot 2	
Organic matter	Control
10.7 ± 2.21	3.53 ± 0.783

Results and discussion

The Wilcoxen Rank Sum Test[8] showed much greater total root length in organic treatments than in controls at forest site 1 ($P < 0.01$) and site 2 ($p < 0.05$) (Table 1). Likewise, the hyphal growth experiment (Table 2) shows a greater hyphal length in organic treatments than controls. An analysis of variance, performed on log-transformed data (including some treatments not reported here) showed the difference between organic and control treatments to be significant ($p < 0.01$).

The experiments show that associations between roots and organic matter[5,9,10,23] and between VAM hyphae and organic particles[6,11,15] can come about through a dynamic process. They need not result from coincidental or passive mechanisms such as negative geotropism[12] or avoidance of aluminum toxicity[7] or other soil property. Hyphal association with organic particles could have been hypothesized to originate from mycorrhizal inoculum in the particles[11] but in these experiments dit not because the material was autoclaved before the experiment.

In both sets of experiments, variability was high: Coefficients of variation (c.v.) in the root experiments were 120% and 142% in organic treatments, and 95.2% and 100% in controls at sites 1 and 2, respectively. In the hyphal growth experiments, c.v's were 76.2% and 58.4% in organic treatments, and 117% and 62.7% in control treatments, in pots 1 and 2, respectively. In all cases the frequency distributions appeared to be highly skewed: Most observations fell below the mean but a few fell well above the mean. Because of these non-normal distributions, statistical analyses were carried out by nonparametric methods or with transformed data.

The hypothesis that nutrient-absorbing organs associate preferentially with organic material was supported in both cases. To test the hypothesis that the association comes about through a locational mechanism (*i.e.*, that growing root or hyphal apices respond to an environmental cue and grow toward OM sites) a further analysis was carried out on the root data and a separate experiment was performed with VAM mycelium. Table 3a shows that the number of chambers without roots was not significantly different in organic and control treatments. The proportion lacking roots is 1 minus the proportion with roots, and can be considered as an indicator of the probability that each type of treatment will be 'found' by roots during a given time period. Fisher's Exact Probability Test (used when expected values in a contingency table are too small to apply chi square) shows that this contingency could occur by chance 23% of the time. In the hyphal growth experiments, no chambers were free of mycelium so the original data could not be reanalyzed by this method. A separate, shorter experiment showed no significant difference in number of unentered chambers in the two treatments (Table 3b).

Since the hypothesis of a locational mechanism was rejected in both cases,

Table 3. Comparison for OM and control treatments of the number of chambers containing no roots or hyphae

a. Two forest sites combined; p of this or a more extreme table = 0.226, NS, by Fisher's exact probability test[21]

	Control	Organic matter
No roots	5	2
Roots present	24	26

b. The number of chambers with no hyphae ($\chi^2 = 0.54$, NS).

	Control	Organic matter
No hyphae	7	9
Hyphae present	8	6

other mechanisms by which greater length could appear in organic chambers must be considered. The roots or hyphae could have encountered each chamber after undirected growth through the soil, then instead of growing through the chamber and back into the soil, turned repeatedly and grown back through the organic material. This would have resulted in a coiled, unbranched appearance which was not apparent on inspection of either experiment. Instead, both roots and hyphae appeared more highly branched in organic treatments, suggesting that the greater absorbing length came about by increased branching in the chamber rather than increased numbers or length of unbranched roots.

Accumulation of length in a branching root or hypha should lead to a rapid increase in absorbing length at the site of branching, provided carbon import from the shoot is at least as great as it would have been had the apex continued to grow in an unbranched mode. The combination of a random process (encounter of an organic site by a root or hypha) and a rapid increase in length at the site (increased branching following the change encounter) are analogous to conditions that lead to a negative binomial distribution[22]. The negative binomial distribution, usually applied to discrete data, is very commonly found in organisms that are spatially aggregated[22]. If root length per chamber are divided into classes and tested with the U statistic[4] the distribution is shown to fit the negative binomial (U $\pm$ standard error = 0.043 ± 0.94). The hyphal data cannot be formally tested for fit to negative binomial because of the small number of observations in each cell. However, the skewed distribution, resulting from a random followed by an exponential process, provides good reason to expect that hyphal length can also be described by the negative binomial distribution. Knowledge of the frequency distribution is useful in that it facilitates statistical analysis of future experimental data by allowing a specific normalizing transformation[2] and emphasizes the similarity of the mechanisms by which two kinds of nutrient absorbing mechanisms become physically associated with rich microsites.

Significance of selective exploitation

The association between roots and rich microsites means that the plant is rooted in a medium much richer than the bulk soil. Diffusion often limits uptake of nutrients in soil, and additional roots in a rich site lead to increased uptake in many circumstances[3].

Examined in a different way, roots that pass through the relatively poor soil matrix essentially connect roots that have encountered rich microsites. The fraction of connecting roots is minimized and the fraction of actively absorbing roots is maximized when most absorbing length is spatially allocated to rich sites. Thus, not only is total uptake greater but uptake per unit of carbon allocated to the root system is greater. The arguments are similar in principle for VAM hyphae.

Most theoretical work on ion uptake from soil assumes that the soil is uniform except for gradients near the root[16], even though this condition is never satisfied in nature. We have shown that soil heterogeneity is exploited in strikingly similar ways by two different kinds of nutrient absorbing systems.

Acknowledgements We thank J. A. Trofymow, J. W. B. Stewart, K. A. Vogt, R. E. Koske, H. W. Hunt, P. B. Tinker and J. A. Logan for discussions and for critically reading an earlier draft of the manuscript.

Portions of this work were supported by a grant by the Brazilian National Research Council to Dr. Paulo de Tarso Alvim and by National Science Foundation grants DEB 78–11201 and 80–04193 to Colorado State University.

References

1 Anderson A B, Prance G T and De Albuquerque B W P 1975 Estudos sobre a vegetacão das campinas Amazonicas. III. A vegetacão lenhosa da campina da reserva biológica INPA-Suframa (Manaus-Caracarai, km 62). Acta Amazonica 5, 225–246.

2 Anscombe F J 1950 Sampling theory of the negative binomial and logarithmic series distributions. Biometrika 37, 358–382.

3 Barley K P 1970 The configuration of the root system in relation to nutrient uptake. Adv. Agron. 22, 159–201.

4 Bliss C I and Fisher R A 1953 Fitting the negative binomial distribution to biological data, and a note on the efficient fitting of the negative binomial. Biometrics 9, 176–200.

5 Damman A W H 1971 Effect of vegetation changes on the fertility of a Newfoundland forest site. Ecol. Monogr. 41, 258–268.

6 Dowding E S 1959 Ecology of Endogone. Trans. Br. Mycol. Soc. 42, 449–457.

7 Foy C D 1974 Effects of aluminium on plant growth. pp 601–642 *In* The Plant Root and its Environment. Ed. E W Carson. Univ. of Virginia Press, Charlottesville, Virginia.

8 Hollander M and Wolfe D A 1973 Nonparametric Statistical Methods. John Wiley and Sons, New York.

9 Kimmins J P and Hawkes B C 1978 Distribution and chemistry of fine roots in a white spruce-subalpine fir stand in Britisch Columbia: Implications for management. Can. J. For. Res. 8, 265–279.

10 Klinge H 1973 Root mass estimation in lowland tropical rain forests of central Amazonia, Brazil. I. Fine root masses of a pale yellow latosol and a giant humus podzol. Trop. Ecol. 14, 29–38.

11 Kokse R E, Sutton J C and Sheppard B R 1975 Ecology of Endogone in Lake Huron sand dunes. Can. J. Bot. 53, 87–93.

12 Lyford W H 1975 Rhyzography of non-woody roots in trees in the forest floor. pp. 179–196 *In* The Development and Function of Roots. Eds. J. G. Torrey and D T Clarkson. Academic Press, London.

13 Marsh B a'B 1971 Measurement of length in a random arrangement of lines. J. Appl. Ecol. 8, 265–272.

14 Newman E I 1966 A method of estimating the total root length in a sample. J. Appl. Ecol. 3, 139–145.

15 Nicolson T H 1959 Mycorrhiza in the Gramineae. I. Vesicular-arbuscular endophytes with special reference to the external phase. Trans. Br. Mycol. Soc. 42, 421–438.

16 Nye P H and Tinker P B 1977 Solute movement in the soil-root system. Univ. of Calif. Press, Berkeley and Los Angeles.

17 Prance G T, Rodrigues W A and Da Silva M F 1976 Inventario florestal de um hectare de mata de terra firme km 30 da estrada Manaus-Itacoatiara. Acta Amazonica 6, 9–35.

18 St John T V, Hays R I and Reid C P P 1981 A new method for producing pure vesicular-arbuscular mycorrhiza-host cultures without specialized media. New Phytol. 89, 81–86.

19 St John T V 1983 Growth and spatial distribution of tree roots in two lowland Amazonian rain forests. Can. J. For. Res. (*In press*)

20 St John T V, Coleman D C and Reid C P P 1983 Association of vesicular-arbuscular mycorrhizal hyphae with soil organic particles. Ecology (*In press*).

21 Siegel S 1956 Nonparametric Statistics for the Behavioral Sciences. McGraw-Hill Book Company, New York.

22 Southwood T R E 1966 Ecological Methods with particular reference to the Study of Insect Populations. Methuen and Company, Ltd., London.

23 Woods F W 1957 Factors limiting root penetration in deep sands of the Southeastern Coastal Plain. Ecology 38, 357–359.

Plant and Soil 71, 495–499 (1983). A-56

Rooting of Sitka spruce cuttings from hedges, and after chilling

R. VAN DEN DRIESSCHE
Research Branch, British Columbia Ministery of Forests, Victoria, B.C., Canada V8W 3E7

Key words Chilling Cuttings Hedging *Pincea sitchensis* Rooting Sitka spruce

Summary Four clones of Sitka spruce (*Picea sitchensis* (Bong.) Carr.) were established from cuttings of two and four-year old material in 1968. Within each clone half the trees were randomly hedged at 1 m in 1977. Cuttings from hedges rooted more freely than cuttings from the lower crown which, in turn, rooted more readily than upper crown cuttings. Rooting occurred most readily during January and early February. Concentrations of sugars in stems and foliage showed little correlation with rooting.

Chilling must be completed for most rapid rooting of dormant Sitka spruce cuttings and this requirement can be satisfied by 10 weeks at 2°C.

Introduction

Vegetative propagation offers the opportunity for rapid genetic improvement[2], particularly in a species such as Sitka spruce (*Picea sitchensis* (Bong.) Carr.) where seed production does not normally occur abundantly in trees less than 20 years old. Sitka spruce cuttings from young trees also root relatively easily[10]. Consequently reforestation might advantageously be carried out with rooted cuttings of this species. Experiments were therefore carried out to examine the effects of hedging, time of cutting and chilling on rooting of Sitka spruce cuttings so as to develop a procedure for obtaining maximum rooting success. There was evidence from other studies[1,7] that auxin treatment and bottom heat would favour rooting and so these techniques were used throughout.

Materials

In March, 1968, cuttings were removed from four Sitka spruce seedlings for rooting. Clones 1 and 2 were from Cowichan Lake, Vancouver Island. Clones 3 and 4 were from Pallant Creek on Moresby Island, in the Queen Charlotte Islands. Second generation cuttings were planted in the field in spring 1972. In February, 1977, when the material was 13 (clones 1 and 2) and 11 (clones 3 and 4) years old half the trees in each clone, selected at random, were cut at 1 m height and then clipped each spring to hedge them.

Rooting

Cuttings consisted of current growth of third order branches, 15–18 cm long, with stems 3–5 mm diameter at the base. Needles were removed from the basal 3 cm, and cuttings were then soaked for 24 h in 50 ppm IBA containing 75 ppm Benlate fungicide. Rooting took place in outdoor frames where bottom heat maintained the rooting medium (equal volumes of peat, sand and perlite) at

18°C ± 2°C[1]. In all experiments each cutting was removed from the rooting medium and inspected every two weeks. A cutting was considered rooted when it possessed at least one root 1 cm long.

Experiment 1

Cuttings were obtained from 9 hedged plants and 9 trees in each of clones 3 and 4 in November, 1979, when the material was 13 years old. Cuttings were removed from the upper one third of the tree crowns. The cuttings were randomized in six replicates of 420 in two rooting boxes.

Experiment 2

On 1 December, 1981, and again on 22 February, 1982, cuttings were removed from hedges and the upper one third and lower on third of the crowns of trees in clone 3. Each replicate consisted of 102 cuttings in 1981 and 135 cuttings in 1982. Replicates were randomized in two rooting boxes and the data from the two years were analyzed together.

Experiment 3

Between October 1980 and March 1981 cuttings were removed from hedges and upper crowns of clones 1, 2 and 4 on six dates. Six hedges and six trees in each clone were used and 9 cuttings were removed from each plant on each date. Two replicates of 144 cuttings were used for rooting and two replicates of 18 cuttings for carbohydrate analysis each date. Replicates were randomized in three rooting boxes.

The cuttings for carbohydrate analysis were separated into needles and stems before freeze drying. The concentration of starch, estimated in terms of potato starch, and the concentration of 80% ethanol soluble sugar, estimated in terms of glucose, were determined on milled needles and stems by replicates. Concentrations of fructose, glucose and sucrose were determined by gas chromatography after silylation.

Experiment 4

On 2 November, 1981, 28 cuttings were removed from each of 24 hedges and placed into bundles of seven, labelled to show clone and hedge. Each of the four bundles was either prepared for immediate pricking out, or else cold stored at 2°C ± 1°C. for 4, 10 or 15 weeks before pricking out. In February, 1982 when cuttings had been stored for 15 weeks a further 14 cuttings were removed from each of the 24 hedges for immediate pricking out.

Results and discussion

Hedging increased the percentage of cuttings rooting (Table 1). In experiment 1 there was a significant ($p = 0.01$) difference between clones with the increase

Table 1. Percentage of cuttings rooting by form of ortet and crown position for experiments 1 and 2.

Source of cuttings	Hedge	Tree lower crown	Tree upper crown
Experiment 1			
Clone 3	60.3 a*		7.8 b
Clone 4	80.1 a		37.9 b
Experiment 2			
Means for two years	40.0 a	23.6 b	10.1 c

* Means on a line differ significantly at $p = 0.05$.

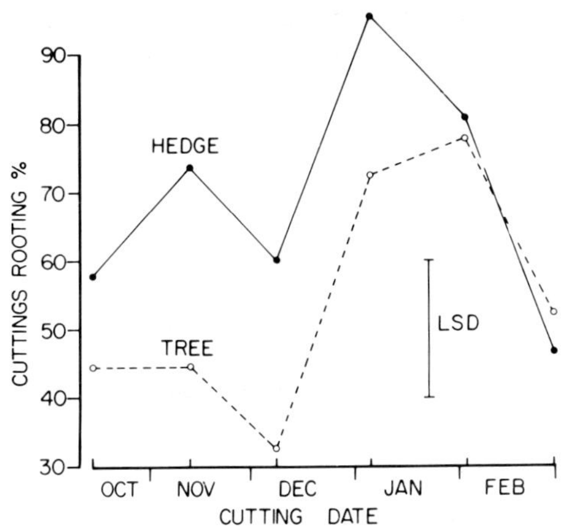

Fig. 1. Percent of cuttings rooting obtained from hedges (continuous line) and trees (broken lines) according to cutting date. Each point based on 144 cuttings.

due to hedging being greater for the less readily rooting clone 4. In experiment 3 it appeared that, as the ability to root decreased during February, there was no difference between rooting of cuttings from hedges and trees (Fig. 1) but otherwise cuttings from hedges rooted more readily. In Monterey pine (*Pinus radiata*) hedging increased the percentage of cuttings rooting compared with cuttings from the upper crowns of trees.[9].

Cuttings from the lower part of the crown normally root more easily than from the top of the crown[5]. This was true for all cuttings in experiment 2, but in 1982 alone there was no difference between rooting of cuttings from upper and lower crown. Differences between 1981 and 1982 could have been related to different

Table 2. Percentage of cuttings rooted and stem concentrations of starch and sugar shown by tree form and clone

Measurement	Tree form		Clone		
	Hedge	Tree	1	2	4
Cutting rooted (%)	69.0	53.9*	43.1 a+	60.2 b	81.1 c
Stem starch (mg g^{-1})	3.2	4.8*	2.8 a	3.5 b	5.6
Stem total sugar (mg g^{-1})	67.3	71.5*	67.6 a	68.1 a	72.6 b
Stem sucrose (mg g^{-1})	28.2	30.5	31.5 a	28.3 a	28.3 a
Stem glucose (mg g^{-1})	11.1	12.0	10.1 a	11.6 b	13.0 b
Stem fructose (mg g^{-1})	14.1	14.6	11.6 a	13.7 b	17.7 c
Needle N (%)	1.62	1.40*	1.59a	1.44b	1.49c

* Significant difference between forms ($p = 0.01$)
+ Values on a line followed by the same letter do not differ at $p = 0.05$.

seasons of cutting. Cuttings from hedges rooted more readily than cuttings from lower crowns in both years. Shaffalitzky de Muckadell[11] suggested that there is a maturation gradient from the juvenile central base of a tree to mature distant branches. Such a gradient might explain the increased rooting achieved in the order upper crown, lower crown, hedge.

It is generally considered that high levels of carbohydrates in pine cuttings favour rooting[3] and there is some evidence that sucrose applications are effective in increasing rooting[6]. In experiment 3 the most freely rooting clone had the highest carbohydrate concentrations and the most difficult to root clone had the lowest at the time of cutting (Table 2). Otherwise, however, the relationship between carbohydrate concentrations at cutting and percent rooting was weak. The strongest correlation ($r = 0.44$, $p = 0.01$ with 34 degrees of freedom) was between stem fructose concentration and number of cuttings rooting.

When dormant Sitka spruce cuttings from trees were rooted in a greenhouse, heated to 18–28°C, higher percentages of rooting occurred when cuttings were taken during December to March than in the autumn or after March[7]. This agrees well with results for cuttings from trees in experiment 3 (Fig. 1). However 80% rooting of non-dormant, 8-year old material was obtained by cutting in July, although rooting took 5 months[10].

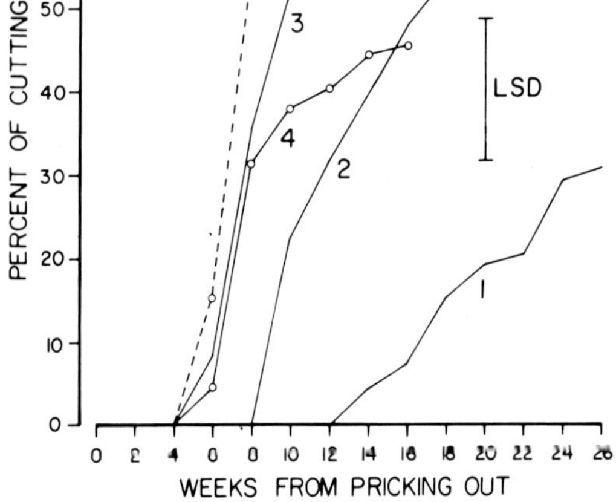

Fig. 2. Percent of cuttings rooting with time from pricking out. Curves 1–4 are for cuttings taken on 2 November, 1981 and pricked out after 0 weeks (curve 1), 5 weeks (curve 2), 10 weeks (curve 3), and 15 weeks (curve 4) cold storage at 2°C. Curve 5 is for cuttings taken and pricked out on 15 February.

Cold storage can improve rooting of dormant conifer cuttings[1], and a chilling requirement of about 8 weeks at 4°C must be satisfied in Fraser fir (*Abies fraseri* (Pursh) Poir.) to obtain greatest rooting[8]. The finding that dormant Sitka spruce cuttings require about 10 weeks chilling at 2°C to obtain most rapid and complete rooting agrees with these observations (Fig. 2). Work with hardwood cuttings suggests that chilling may both remove inhibitors and result in production of root promoting cofactors[4]. In the present experiments rooting medium temperature was constant, but air temperature fluctuated with temperature outside, although it was normally 5°C above ambient air temperatures. Thus cuttings taken during autumn received chilling, although perhaps not as effectively as material remaining in the open. This may largely explain why cuttings taken in October and November did not start rooting for about 3 months, and then rooted slowly. However, even if chilling were not a factor, cuttings pricked out in the fall would be subject to lower light intensities and air temperatures during the winter than those pricked out in late January or February.

Dormant cuttings taken from hedges in the autumn and provided 10 weeks chilling at 2°C, or cuttings taken in late January and February, which are naturally chilled, root most easily using the rooting procedures used in these experiments.

Acknowledgements Technical assistance of C Hollefreud and R Planden, both of British Columbia Ministry of Forests Research Branch, is gratefully acknowledged.

References

1 Brix H and Barker H 1973 Rooting studies of Douglas fir cuttings. Can. For. Serv. Pac. For. Res. Cent. Inf. Rep. BA-X-87, 1–45.

2 Brix H and van den Driessche R 1977 Use of rooted cuttings in reforestation. British Columbia Forest Service/Canadian Forestry Service Joint Report No 6, 1–16.

3 Cameron R J and Rook D A 1974 Rooting stem cuttings of radiata pine: environmental and physiological aspects. N.Z.J. For. Sci. 4, 291–298.

4 Gesto M D V, Vazques A and Vietez E 1981 Changes in the rooting inhibitory effect of chestnut extracts during cold storage of cuttings. Physiol. Plant. 51, 365–367.

5 Girouard R M 1971 Rooting plain and heel cuttings of spruce. Common. For. Rev. 50, 28–31.

6 Greenwood M S and Berlyn G P 1973 Sucrose-indole-3-acetic acid interactions on root regeneration by *Pinus lambertiana* embryo cuttings. Am. J. Bot. 60, 42–47.

7 Griffith B F 1940 Effect of indolebutyric acid, indoleacetic acid, and alpha napthalene-acetic acid on rooting of cuttings of Douglas fir and Sitka spruce. J. For. 38, 496–501.

8 Hinesley L E and Blazich F A 1981 Influence of post-severance treatments on the rooting capacity of Fraser fir stem cutting. Can. J. For. Res. 11, 316–323.

9 Libby W J and Hood J V 1976 Juvenility in hedged radiata pine. Acta Hortic. 56, 91–98.

10 Roulund H 1971 Experiments with cuttings of *Picea abies*, *Picea sitchensis* and the hybrid *Picea omorka* × *Picea sitchensis*. Forest Tree Improvement, Arboretet Hørsholm 3, 25–57.

11 Schaffalitzky de Muckadell J 1954 Juvenile stages in woody plants. Physiol. Plant. 7, 782–796.

Plant and Soil 71, 501–505 (1983). A-57

Influence of auxin and mycorrhizal fungi on the *in vitro* formation and growth of *Pinus pinaster* roots

A. DAVID, M. FAYE
Laboratoire de Biologie et Physiologie Végétales, Université de Bordeaux I, F-33405 Talence cedex, France

and M. RANCILLAC
Station de Physiologie Végétale, INRA, Centre de Recherches Agronomiques de Bordeaux, F-33140 Pont de la Maye, France

Key words Growth of roots *Hebeloma cylindrosporum* *Pinus pinaster* *Pisolithus tinctorius* Rooting

Summary Experiments, performed with *Pinus pinaster* cloned shoots submitted to an auxin treatment (NAA 10^{-6} *M*, 18 days), demonstrated that rooting ability *in vitro* persists over 5 successive induction cycles (through out a 9-month period). Rooting ability needs a permanent synthesis of auxin synergists which activate the metabolism of cell dedifferentiation and root primordium initiation. Agar culture permitted intense meristem initiation, but prevented active root elongation. In the presence of a mycorrhizal fungus, *Pisolithus tinctorius* or *Hebeloma cylindrosporum*, roots resumed growth and short lateral root formation was stimulated. These two phenomena induced by fungal association improve the quality of the root systems required to facilitate successful transplantation from test-tubes to field conditions.

Introduction

The *in vitro* vegetative propagation of Gymnosperms in forestry has been discussed previously[1]. Its advantages include mass-cloning of selected genotypes, production of desease-free clones, establishment of gene banks and studies of mycorrhizal symbioses under controlled conditions.

The conditions for root initiation in cloned shoots of *Pinus pinaster* have been determined[10] and the formation of mycorrhizal short roots without fungal colonization studied[2].

This paper presents results that answer the following questions:

(a) is a shoot capable of synthesizing auxin synergists required for root initiation on a continual basis?

(b) what is the influence of mycorrhizal fungi on the growth of newly formed root systems?

Materials and methods

Cloned shoots of *Pinus pinaster* (Sol.) were obtained from seedlings by axillary budding. They had been multiplied for a 3-year period by an *in vitro* vegetative propagation procedure[10].

Shoots were cultured on a modified Risser and White[11] medium (RW) supplemented with naphtalenacetic acid (NAA, 10^{-6} M) and agar (Bactoagar, Difco, 0,8%) for an 18-day period. Shoots transferred to an auxin-free medium showed slight root elongation ($\leqslant$ 5 mm) within the first two weeks after which growth stopped. These shoots, with their root systems in arrested growth, were used to investigate the influence of mycorrhizal fungi.

To study the resumption of growth of roots, either plantlets and inoculum of a fungus (*Pisolithus tinctorius* or *Hebeloma cylindrosporum*) were introduced simultaneously, or plantlets were transferred to a substrate inoculated for 18 days. Substrates (agar or perlite) were supplemented with either the mineral solution RW or an organic solution MA (malt extract 1%, L-asparagine 0.1%, sucrose 1%) which supported the best growth of fungal mycelium.

Experiments were carried out under sterile conditions in growth chambers[10].

Results

Successive inductions of rooting

To determine if rooting ability is maintained on a continual basis, successive cycles of rooting of the same shoots were conducted at 2-month intervals (5 treatments over a 9-month period, from November to August). Before each new hormonal induction treatment, the stem base was severed and the root system discarded.

The data indicated that during successive cycles neither rooting ability (percentages of rooted shoots), nor rooting intensity (mean number of root meristems per shoot) decreased significantly (Table 1). Standard errors were high.

During the experimental period no teratological characteristics were observed in root system.

Parallel experiments indicated that agar was better than perlite as a substrate for expression of rooting potentialities. The mean number of roots per shoot was 8.9 ± 2.2 in agar and only 3.3 ± 2.3 in perlite. Optimal moisture was difficult to maintain in perlite although in agar newly formed roots soon stopped growing.

Table 1. Induction of successive rootings by repeated auxinic treatments on sets of cloned shoots

	Rooting per cent		Intensity of rooting*	
Clone number	24	40	24	40
1st induction	99	98.3	6.7 ± 3.5	6.2 ± 3
2nd induction	92.6	94	3.7 ± 2.2	5.1 ± 2.7
3rd induction	89	87.5	4.7 ± 3.3	4.5 ± 3.5
4th induction	100	100	7.5 ± 3.5	7.1 ± 2.7
5th induction	100	100	9.2 ± 1.9	10.5 ± 2.9

Basal part of the stem with the root system was removed before each induction, observations 30 days after the end of the hormonal treatment.

* Mean number of root meristems per shoot.

Table 2. Comparison of the influence of two mycorrhizal fungi on root system growth

	Control	Pisolithus	Hebeloma
Number of explants	15	12	12
Mean number of elongated roots (> 10 mm) per shoot	3.7 ± 2.1	4.8 ± 1.2	5.6 ± 2
Mean length of a root (mm)	8.1 ± 5.2	21.3 ± 11.1	15.2 ± 10.7
Mean length of a root sytem (mm)	30.1 ± 17.5	103.2 ± 39.1	84.8 ± 43.9

Shoots (clone 24) with adventitious roots in arrested growth were subcultured on a medium (perlite + MA) without preconditioning (shoots and fungal inoculum were transplanted at the same time). Observations 100 days after transplantation.

Influence of mycorrhizal fungi on the development of root systems

The influence of *Pisolithus tinctorius* on the arrested root systems was studied on plantlets cultivated in agar medium supplemented with either mineral RW or organic MA elements. In both media, control (fungus-free) explant roots did not resume growth. Mean root length was the same after 2 months as it was initially (6.3 ± 3 mm). However, transferring explants to a medium preconditioned for 18 days with *P. tinctorius* induced root growth, the mean length being twice (12 ± 5 mm) that of the control.

Fig. 1. Effect of mycorrhizal fungi on root growth.
T fungus-free control, H explants cultivated with Hebeloma, P explants cultivated with Pisolithus. Bar represents 5 cm. There is no callus at the base of the stem (dark zone is rest of mycelium and perlite).

Two fungi, *P. tinctorius* and *H. cylindrosporum* were compared for their ability to stimulate root regrowth (Table 2 and Fig. 1). Sets of cloned shoots, distributed uniformly between different treatments according to the number of root meristems per shoot, were planted in perlite supplemented with MA nutrient medium. Fungal inoculum was added at transplanting.

Both strains of mycorrhizal fungi stimulated root development compared with the control. This stimulation was expressed by an increase in the number of elongated roots per shoot and the length of each root. Both parameters contribute to root system length which increased in the presence of the tested fungi.

The number of short lateral roots with mycorrhizal fungi was twice (9 ± 1 roots per cm of mother root) that of the control.

Discussion

Rooting in cloned shoots of *Pinus pinaster*, originating from seedlings by axillary budding, was obtained with a 15- to 18-day hormonal treatment period (NAA 10^{-6} *M*). This produced a high proportion rooting (80–100%) in various clones[10].

Auxin alone does not predispose cells to induce root primordium initiation[6]. It interacts with other factors including phenolics; oxidizing enzymes, enzyme activators to synthesize 'root-inducing auxin-phenol conjugates' *e.g.* ref.[7]. Results indicated that shoots' are able to synthesize permanently, either spontaneously or as a consequence of the auxin treatments, the auxin synergists required for root primordium initiation. Endogenous factors are still present after each root initiation phase.

Agar medium is used routinely for in vitro vegetative propagation. The current study indicates this substrate is suitable for root initiation but prevents further root growth. Inoculation with mycorrhizal fungi can partly overcome this disadvantage. Perlite, an aerated substrate, supports the best root and fungus development, as long as moisture is properly controlled. In the presence of fungi, perlite induces more roots to grow than agar. Consequently a two-step method of root system formation: primordium initiation in agar including hormonal treatment and root system development in perlite (or any aerated substrate) inoculated with a mycorrhizal fungus is suggested. Various substrates have been tested for mycorrhizal symbioses: perlite[9], vermiculite[5], a mixture of perlite and vermiculite[8] and filter paper[3]. Moisture remains a problem over a long period.

Homogeneous results within a clone were expected for cloned shoots. In root systems, however, the number of roots per explant were very variable. Heterogeneity could be reduced if shoots were carefully standardized (*e.g.* length, stem diameter, equal number of cytokinin treatments for clonal propagation, ...). Fungal mycelium did not reduce heterogeneity.

Mycorrhizal fungi can promote growth resumption in arrested root system.

Symbionts also stimulate rhizogenesis in *Pinus halepensis* hypocotyles[4]. Under our experimental conditions, Pisolithus was not more efficient than Hebeloma. Both fungi increased (a) the number of elongating roots (b) individual root length and (c) the intensity of short root branching. These parameters determine the quality of a root system and influence the ability of plantlets to overcome the critical period of transplantation from in vitro to greenhouse or nursery conditions.

More studies are required to permit early development of optimal mycorrhizal infection which can act competition with environmental stresses (pedological, nutritional and pathological) which are detrimental in tree establishment.

References

1 David A 1982 *In vitro* propagation of gymnosperms. *In* Tissue Culture in Forestry. Eds. J M Bonga and D J Durzan. pp 72–108. Martinus Nijhoff/Dr W Junk Publishers, The Hague, The Netherlands.

2 Faye M, Rancillac M and David A 1981 Determinism of the mycorrhizogenic root formation in *Pinus pinaster* Sol. New Phytol. 87, 557–565.

3 Fortin J and Piche Y 1979 Cultivation of *Pinus strobus* root-hypocotyl explants for synthesis of ectomycorrhizae. New Phytol. 83, 109–119.

4 Gay G 1982 Influence de champignons ectomycorhiziens sur l'enracinement d'hypocotyles de Pin d'Alep. Application à la micropropagation. Bull. Soc. Bot. Fr., Actual. Bot. (*In press*).

5 Hacskaylo E 1953 Pure culture syntheses of pine mycorrhizae in terra-lite. Mycologia 45, 971–975.

6 Haissing B E 1970 Influence of indole-3-acetic acid on adventitious root primordia of brittle willow. Planta Berlin 95, 27–35.

7 Haissig B E 1974 Influence of auxins and auxin synergists on adventitious root primordium initiation and development. N. Z. J. For. Sci. 4, 311–323.

8 McKay H M 1982 Multiple-plant module for aseptic nutritional studies and the synthesis of mycorrhizas. Plant and Soil 66, 257–262.

9 Poitou N 1979 Cultures mycéliennes de champignons mycorhiziens et techniques d'incorporation aux plantes-hotes. Mush. Sci. X (Part I), 891–902.

10 Rancillac M, Faye M and David A 1982 In vitro rooting of cloned shoots in *Pinus pinaster*. Physiol. Plant. 56, 97–101.

11 Sommer H E, Brown C L and Kormanik P P 1975 Differentiation of plantlets in longleaf pine (*Pinus palustris* Mill.) tissue cultured *in vitro*. Bot. Gaz. 136, 196–200.

Plant and Soil 71, 507–512 (1983). A-58

A note on modelling methods for studies of ectomycorrhizal systems

F. E. SANDERS,
Department of Plant Sciences, University of Leeds, Leeds LS2 9JT, UK

J. G. BUWALDA and P. B. TINKER
Rothamsted Experimental Station, Harpenden, Herts AL5 2JQ, UK

Key words Dynamics Ectomycorrhizas Host response Infection Modelling

Summary Possibilities for modelling developing ectomycorrhizal root systems are discussed in the light of recent experience with vesicular-arbuscular mycorrhizal systems. Emphasis is given to experimental techniques and measurements which can provide suitable data on which to base models.

Introduction

Mycorrhizas have been studied for many years, including their taxonomy, anatomy, physiology and ecology, and their role in the nutrition of crops in agriculture and forestry. However, most workers have tended to approach the topic from the particular viewpoint of their own discipline, so that the study of the mycorrhizal plant as a whole system has been neglected.

The main aim must be to explain the response of the host to mycorrhizal infection in a quantitative way. The first requirement for this is to understand the dynamics of mycorrhizal systems, where two or more distinct organisms interact with each other in space and time under the influence of the environment. This interaction may be very complex. It is certain that we will make little progress towards quantitative explanation if we continue to carry out experiments in which we make measurements only once in the life of the system.

The role of modelling

Deterministic mathematical models provide an ideal tool for establishing and testing hypotheses concerned with the dynamics of mycorrhizal systems. Much progress has been made with such models in other areas of plant science and biology in general[7,13,16]. The principles evolved have been applied to the study of vesicular-arbuscular mycorrhizal systems[2,3,10,12,14] but there have been no corresponding developments in research on ectomycorrhizas, perhaps because ectomycorrhizal systems are less tractable experimentally. The aim of the subsequent discussion is to set out very briefly the possibilities for modelling the response of the host to ectomycorrhizal infection, and to indicate some of the experimental measurements that appear to be essential.

Processes involved in the host response

The growth of mycorrhizal plants involves at least one feedback system. The growth of the host provides space and substrate for the fungus and hence affects the latter's growth. The growth of the fungus influences the rate at which the host receives mineral nutrients, and may also influence root system morphology and the partitioning of assimilates within the host. The changed nutrient supply may in turn influence the overall rate of growth of the host.

A complete model for this feedback system will involve the description of many processes. The following seem most important at this stage. 1) The initial infection process and subsequent changes leading to the development of a single mycorrhizal short root. 2) The spread of mycorrhizal infection in the developing plant root system. 3) The transfer of host assimilates to the fungus, the growth of fungal biomass and its partition between mycorrhiza and mycelium in the soil. 4) The uptake of mineral nutrients and water from the soil and their transfer to the host through the mycorrhizas and associated external mycelium. 5) The influence of altered nutrition on host growth rate and the partitioning of assimilates between root and shoot. Later, we may wish to examine the effects on all these processes of changes in environmental variables and of biological interactions, such as above- and below-ground competition between host plants and competition between different mycorrhizal fungi in colonisation of the roots and in nutrient uptake from the soil.

These processes may be examined and modelled separately, to be integrated in a whole-system model later. However, it may be more efficient to obtain data for the validation of simple whole-system models simultaneously during single experiments.

What measurements should be made?

It is clearly impossible to predict accurately the measurements on ectomycorrhizal systems that will be required for future practical and theoretical developments. All we can do at present is to list those measurements that seem to be essential by analogy with current works on endomycorrhizal systems.

Sequential harvesting

This is essential for appreciation of the dynamic situation. The same considerations apply as in plant growth analysis, where frequent, small harvests may be more useful than fewer and larger ones[6]. The number of harvests required is clearly determined by the type of process being modelled, its complexity and the number of model parameters to be estimated. Increasing the number of parameters in a model may improve its fit to the data, but more harvests and/or replicates will be required to compensate for the loss of degrees of freedom. The spacing of harvests will ideally relate to increments in the principal state variables, and not necessarily to time. However, without advance knowledge of the dynamics of the system, ideal spacing of harvests may be impossible to achieve. As a general rule, it seems best to make as many harvests, equally spaced in time, as possible to ensure coverage of all phases of development. As this may be incompatible with other experimental objectives, experiments intended to provide data for modelling are best designed exclusively for this purpose.

Description of the growth of root systems

In endomycorrhizal research it has been convenient to describe the growth of mycorrhizal root systems in terms of changes in total and infected root lengths, since the formation of mycorrhizas involves no gross alteration of root morphology. In ectomycorrhizal systems there are considerable changes in morphology and length may therefore not be the most appropriate variable. Chilvers and Gust have recently published an important paper[4] on the development of ectomycorrhizal root systems of eucalypts. This paper describes the measurement of variables more suited to the description of ectomycorrhizal systems, for example, the numbers of mycorrhizal and non-mycorrhizal root apices. Until these questions are settled, experiments should aim to measure length, number of apices and biomass of both infected and uninfected roots.

The theory of uptake of nutrients by plant root systems from soil lays stress on root length as an important variable[1,9]. In the case of woody perennials with ectomycorrhizal root systems, where root density is very low in comparison to that of annual plants, the soil mycelium may be of chief importance in nutrient uptake. The mycorrhizas themselves may function less as organs of direct nutrient uptake than as the interface between mycelium in the soil and host, in which case their length may be of less importance.

The soil mycelium

If this is as important as we suspect, its measurement will be vital. In endomycorrhizal systems, if the hyphae can be separated from the soil, estimates of quantities of external mycelium can be obtained easily, if tediously, by weighing and/or measurement of hyphal length[11,15]. In sterile artificial growth media, weights of mycelium can be estimated by techniques involving chitin assay[5]. In the case of ectomycorrhizal fungi, structural differentiation of the external mycelium into hyphal strands and rhizomorphs and high levels of soil organic matter in most growth media, may make the quantitative determination of external mycelium a formidable, though necessary, undertaking. By analogy with roots, length of mycelium per unit volume is probably most important for nutrient uptake. In suitable circumstances, lengths might be measured by an adaptation of Newman's method[8].

Host growth

Fresh and dry weights of tops and roots should be measured as frequently as possible. We lay great stress on this because relative growth rate will be central to any model of plant response to mycorrhizal infection. Accurate estimation of relative growth rate requires frequent and accurate determination of plant mass. If modelling is to include aspects of photosynthesis, measurements such as leaf area should also be considered.

Nutrient content

Accurate measurement of plant nutrient content, again at frequent intervals, is also essential. Such measurements permit the calculation of nutrient uptake rates, which can be expressed per unit length or mass of absorbing organ, be it root, mycorrhiza, rhizomorph or hypha. In endomycorrhizal plants, the concept of inflow (rate of uptake per unit length of root)[1] has proved an essential key to the understanding of mycorrhizal enhancement of phosphorus uptake by root systems. In modelling, inflow provides a means by which the activity of the fungus in the soil may be linked to host nutrition[10]. By analogy, in ectomycorrhizal systems, calculated rates of nutrient uptake per unit weight of mycorrhiza, or per unit weight or length of soil mycelium, may prove to be equally useful. There is a pressing need for this question to be investigated experimentally.

It is certain that the use of nutrient concentration in the host tissue as a sole index of mycorrhizal effects on host nutrition has very little value, though it is often found in the literature.

Soil measurements

It is normal practice to report basic information such as pH, extractable nutrient levels in soils, etc. but this information may be insufficient in modelling. For example, phosphorus uptake from soil may often be limited by diffusional supply to root or fungal surface. The description of this process demands, as a minimum, determination of the concentration of phosphorus in the soil solution, the buffering power of the solid phase and the volumetric moisture content[9].

Other measurements

Additional measurements for special purposes may also be considered. For example, the concentration of soluble sugars in roots may need to be determined if it is hypothesised that this may control the rate of growth of the fungus. In any growth experiment, morphology and growth stage of the host and fungus should also be monitored and recorded. At present, it seems most unlikely that a successful model can take into account changes such as a shift from vegetative to reproductive growth in host or fungus. However, information on such changes, while not obviously important for the immediate purposes of a particular experiment, may prove to be of great value to another worker in the future.

Mathematical analysis and reporting of data

The way in which data from sequential harvests are analysed will depend to a large extent on the enthusiasms of the experimenter and the resources available to him. It must be emphasised that the use of a mainframe computer installation and the aid of professional mathematicians and statisticians are not essential prerequisites for success. Indeed, the minutiae of mathematical analysis may serve to confuse rather than to clarify the biological implications of models. Analysis of simple models, sufficient for the development of ideas, can easily be carried out on a microcomputer sited on the experimenter's desk. Data should always be published as fully as possible, even if there is little attempt at mathematical analysis, so that it can be of use to others.

Conclusions

Recent progress in understanding host response in endomycorrhizal systems indicates that this may be an appropriate time to extend models to ectomycorrhizal systems. Models of ectomycorrhizal systems will necessarily be more complex those for endomycorrhizal plants but it seems probable, that the same general principles will apply. Experiments to provide data for modelling exercises should ideally be specifically designed for the purpose. Frequent sequential harvests are essential, the measurements made should be as comprehensive as resources allow and data should be reported fully for the benefit of other workers.

References

1 Brewster J L and Tinker P B 1972 Nutrient flow rates into roots. Soils Ferts. 35, 355–359.

2 Buwalda J G, Ross G J S, Stribley D P and Tinker P B 1982 The development of endomycorrhizal root systems. III. The mathematical representation of the spread of vesicular-arbuscular mycorrhizal infection in root systems. New Phytol. 91, 669–682.

3 Buwalda J G, Ross G J S, Stribley D P and Tinker P B 1982 The development of endomycorrhizal root systems. IV. The mathematical analysis of the effects of phosphorus on the spread of vesicular-arbuscular mycorrhizal infection in root systems. New Phytol. 92, 391–399.

4 Chilvers G A and Gust L W 1982 The development of mycorrhizal populations on pot-grown seedlings of *Eucalyptus St-Johnii* R. T. Bak. New Phytol. 90, 677–699.

5 Hepper C M 1977 A colorimetric method for estimating vesicular-arbuscular mycorrhizal infection in roots. Soil Biol. Biochem. 9, 15–18.

6 Hunt R 1978 Plant Growth Analysis. Studies in Biology no. 96. Edward Arnold. 67 p.

7 Maynard Smith J 1974 Models in Ecology. Cambridge University Press. 146 p.

8 Newman E I 1966 A method for estimating the total length of root in a sample. J. Appl. Ecol. 3, 139–145.

9 Nye P H and Tinker P B 1977 Solute Transport in the Soil-Root System. Blackwell. 342 p.

10 Sanders F E and Sheikh N A 1983 The development of vesicular-arbuscular mycorrhizal infection in plant root systems. Plant and Soil 223–246.

11 Sanders F E, Tinker P B, Black R L B and Palmerley S M 1977 The development of endomycorrhizal root systems. I. Spread of infection and growth-promoting effects with four species of vesicular-arbuscular endophyte. New Phytol. 78, 257–268.

12 Smith S E and Walker N A 1981 A quantitative study of mycorrhizal infection in Trifolium. Separate determination of the rates of infection and mycelial growth. New Phytol. 89, 225–240.

13 Thornley J H M 1976 Mathematical Models in Plant Physiology. Academic Press, London. 318 p.

14 Tinker P B 1975 Effects of vesicular-arbuscular mycorrhizas on higher plants. Symp. Soc. Exp. Biol. 29, 325–349.

15 Tisdall J M and Oades J M 1979 Stabilisation of soil aggregates by the root systems of ryegrass. Aust. J. Soil Res. 17, 429–441.

16 Van der Plank J E 1975 Principles of Plant Infection. Academic Press, London. 216 p.

Plant and Soil 71, 513–515 (1983). A-59

An overview of the meeting

Synthesis – progress and challenges in work on tree root systems and their mycorrhizas

J. E. JACKSON
East Malling Research Station, Maidstone, Kent ME19 6BJ, UK

Key words Anchorage Establishment Mathematical models Mycorrhizas Nutrition Tree roots

Summary There has been considerable progress in many areas of study on tree root systems and their mycorrhizas but some important subjects and approaches with respect to both orchard and forest tree roots are very inadequately researched. These include anchorage, root growth and emergence of new roots after transplanting, factors controlling nutrient uptake by different parts of mature tree root systems and seasonal variations in nutrient uptake and the use of mathematical models in relation to mycorrhizal infection and its effects on plant growth.

Overview

The title of the conference, 'Tree Root Systems and their Mycorrhizas', has been interpreted in very wide sense with papers ranging from ultracytological studies of processes involved in phosphate and carbohydrate exchange in mycorrhizal roots to acoustic studies relating to the anchorage of mature forest trees. This breadth of approach reflects a diversity of objectives, from an understanding of root function to the improvement of forest, nursery and orchard management.

In some areas of tree root system research there has clearly been great activity and progress, other have received much less attention.

This uneven emphasis can be seen firstly by considering the function of roots. They take up water and mineral nutrients, they anchor the trees and they are an important site of plant-growth-substance synthesis. By far the greatest progress has been made in work on water and nutrient uptake, including mycorrhizal aspects. Very much less has been done about anchorage in spite of the cost of the forest of stakes which fruit growers use to support their poorly-anchored trees and the damage done by windthrow in real forests. Studies on the factors controlling tree root development and on the non-nutritional effects of roots on the above-ground portions are also in their infancy, the fairly recently developed capability to induce tree root formation *in vitro* offering exciting prospects for rapid advance.

We also see an uneven emphasis if we categorise research on roots in relation to stages of tree development. Work on micropropagules is developing rapidly because of the interest in 'micropropagation' as a technique for producing planting material and because of the special advantages which material propagated in this way offers for some types of research. There have been great strides in our knowledge of seedling root systems (including their infection by mycorrhizal fungi) and of adventitious root formation on vegetatively-propagated cuttings. Propagation, either *in vitro* or in the nursery is, however, followed by transplanting and the scientific problems of tree establishment are only just beginning to receive attention commensurate with their importance. There is clear evidence that in some situations in the United Kingdom the greater part of the transplanted root system dies, to be replaced, after a period of very slow growth, by adventitious roots. Similarly, the mycorrhizal fungi which were inoculated, or infected the roots naturally, in the nursery may fail to survive for any substantial period in their new environment. Work centred on the problems of establishing new root systems after transplanting, and on the dynamics of mycorrhizal development and new infection in the orchard and forest situation might well prove more rewarding from the practical viewpoint than efforts to improve still further the growth and mycorrhizal status of root systems in the test-tube or nursery. Information on the functioning of mature tree root systems, with their attendant problems of access, size and complexity, is still fragmentary with little underlying theory to explain the observed data. Some of the older concepts relating function to structure seem increasingly untenable with the accumulating evidence that brown, secondarily thickened roots can function as effectively as white ones in water and nutrient uptake. Newer results indicate that roots at different depths can have very different uptake efficiencies, or capacities, and that shoot factors control the seasonal pattern of uptake by roots. It may well be that concepts relating to the plant as a whole, and the relations between its parts, will, in future, play as dominant a role in our thinking on nutrient uptake as they do with respect to water uptake at present.

Consideration of shoot-root interactions brings with it questions of optimal allocation of materials to root and shoot including, as a case of particular interest to this meeting, the question as to whether mycorrhizal infection involves a potentially excessive cost in terms of use of carbon resources. It is by no means certain that this is the case. There is evidence that plants with mycorrhizal roots have higher rates of photosynthetic C-fixation than those with non-mycorrhizal roots as well as having higher rates of root respiration. Whether they have higher rates of photosynthesis than non-mycorrhizal plants with similar levels of leaf phosphorus is, however, still unresolved and is the key question with regard to whether supply of P by fertilizer application (including foliar treatment) or by mycorrhizal inoculation can lead to the highest growth rates.

The formulation of precise questions to single problems leads naturally to the use of mathematical modelling of more complex systems. Indeed a major advantage of the modelling approach is that it involves definition of the parameters of the system under study and specification of the relevant quantitative information. It was salutary to hear that the vast literature on mycorrhizas had not yielded information which could be used to test current models of the development of V.A. mycorrhizal infection. Following our discussions Sanders *et al*[1] have outlined, in a paper written subsequently but included in this volume, the data which they consider most relevant to the formulation and testing of general laws relating to mycorrhizal infection. I hope that this will encourage other to record their observations in a compatible way. What has been achieved with respect to annual plants represents only a start in relation to the complexities of the perennial tree which, as we have seen elegantly demonstrated, is subject to successional infection by different mycorrhizal fungi as it ages. Our understanding is, however, most likely to grow in a satisfactory way only if we attempt to develop theory applicable to simple situations to the level where it can cope with complex ones.

In this brief overview I have emphasised those parts of our conference which seemed to me to focus attention on challenges and opportunities as yet largely un-met and unexploited. It is not really possible to summarise the vast amount of research on long-established topics which has been reported, the papers in themselves so often being either detailed reviews or condensations of much experimental work. The volume of such work does, however, reflect the unique importance of research on root systems to the production and performance of orchard and forest trees. This arises from the need to induce rooting of cuttings on the one hand and to the requirement for effective root performance in cold and frequently waterlogged soils of low nutrient status on the other. In the past this has led to some polarisation of research with a concentration on rooting of vegetative cuttings of orchard tree rootstocks and an emphasis on mycorrhizal infection of forest trees. With increasing interest in vegetative propagation of forest trees and evidence of problems of phosphorus nutrition in fruit trees grown in fumigated and herbicide-treated nursery and orchard soils there is a rapidly developing community of purpose. This conference has exemplified such a bringing together of objectives and approaches by root research scientists from the forest and fruit tree spheres in a way which can only be beneficial for our joint futures.

Reference

1 Sanders F E, Buwalda J G and Tinker P B 1983 A note on modelling methods for studies of ectomycorrhizal systems. Plant and soil 71, 507–512.

Subject index

Abies amabilis 79
A. balsamea 55, 57, 277
A. grandis 167, 169
abscisic acid 203
Acacia 334
Acaulospora laevis 333
Acer pseudoplatanus 211
A. rubrum 182
acetylene reduction 5
acid phosphatase 456, 457, 458
acid rain 15
acoustical analysis 180
actinomycetes 5
actinorhizal plants 5
aeration 275, 381
Agaricus 16
Agropyron repens 333
Agrostis tenuis 333, 458
alkaline phosphatase 200, 201, 212, 456
Allium cepa 211, 230, 232, 239, 335
Alnus 275, 279, 309
A. crispa 277, 334
A. glutinosa 277, 334
A. rugosa 277, 281
aluminium 15, 16, 490
Amanita fulva 18
A. muscaria 10, 16, 17, 18, 20, 247, 249, 251, 254, 257, 261, 281, 320, 321, 455, 459
A. rubescens 320, 321
amenity trees, see also shade trees 171
amide 404, 407
Ammonium, see nitrogen
anaerobic conditions 5, 11, 12, 384, 387, 389, 392, 515
see also waterlogging effect on nutrition 392
metabolism 391
anchorage, see also uprooting 103, 155, 171, 175, 185, 186, 189, 513
anion, see also sulphate *etc.* 463
annual ring 164, 167, 168
antibiosis 299, 301, 478
apical meristem 285
apoplast 441
apple 23, 25, 147, 371, 381, 409, 410
apricot 404
arbuscule, see endomycorrhizas
arbutin 404
arginine 404, 407, 408, 409, 410
Armillaria mellea 266
Arrhenius plot 379
Artemisia herba alba 363
aseptic conditions, see axenic
asparagine 404, 408, 410
aspartic acid 408
Aspergillus 458
assimilate, see also carbohydrate partitioning 426–429, 508
transport 440–441
ATPase 202, 211, 212, 214, 474
autofluorescence 301
auxin 4, 276, 495, 501, 502, 504
axenic 3, 276, 277, 285, 290, 353, 355, 501, 504, 513
Azotobacter vinelandii 5

β-glycero phosphatase 212
bacteria 88
bark 10
basidiospore 269, 271
biomass, see root *etc.*
beech, see *Fagus*
Betula 9, 16, 247, 248, 249, 257, 259, 263, 264, 269, 309, 455
B. glandulosa 277
B. nana 395, 397, 398
B. papyrifera 160
B. pendula 17, 257, 263, 269, 271
B. pubescens 17, 257, 263, 269, 455
bicarbonate 467
biological control 299, 345
black locust 5
Boletaceae 266
boreal forest 89
bräcke scarifier 111
bromide 463

calcium 15, 200, 371, 375, 392, 401, 404, 410, 463, 477, 482, 483
storage 405
adsorption 405
Calluna vulgaris 91, 97, 285, 289
cambium 159
Cantharellus 16
carbohydrate 2, 51, 88, 97, 123, 161, 197, 198, 369, 401, 415, 441, 495, 510, 513
glucose 404
partitioning 428–429, 441
reserves 403, 406

seasonal periodicity 406
soluble 403
sorbitol 404
starch 125, 127, 403, 408, 409, 497
sucrose 498
carbon 3, 166, 415, 441, 492
carbon dioxide 276
casparian strip 149
Castanea sativa 266
Casuarina 334, 363
C. equisetifolia 363
cation : anion balance 448, 464, 465, 466
C.E.C. 445, 450, 452, 487
Cedrus 363
C. atlantica 363
cell wall 147, 149, 151
Cenococcum geophilum 49, 53, 314, 321
C. graniforme 469, 471, 474
Centaurea incana 366
Chamaenerion angustifolium 333
Cherry 217, 404
Colt 29
F12/1 29
chilling 495, 499
Chlamydospores 301, 310
chloride 463, 464
chlorophyll 428
clay 445, 447, 448
Clitocybe, see also red clover 16
clover 5
Coffea arabica 477, 482–483
Collybia 16
compaction, see mechanical impedance
Complexipes moniliformis 313, 316, 319
Copper 202, 371, 375
paint 107
coppice 266
Coprinus 16
cortex, see root
Cortinarius 17, 248, 266, 281
Coronilla varia 333
Cox orange pippin, see also apple 24, 25, 28
critical rooting density 176
cross sectional area 159, 162, 166, 173, 176, 182
cucumber 347, 349
cultivation 63, 171
cuttings 131, 137, 495, 498–499, 514, 515
cyclosis 200
cytokinin 4, 203, 404
cytoplasm 151

day length 61, 409
decomposition 75, 88, 455, 456, 461, 477, 478, 481, 487
periodicity 482–483
defoliation 409
dendrochronology 167
dermatocalyptrogen 286, 289
desert species 363
disease resistance 197, 203, 204
dithane m45 353, 355–361
DNA 5
drainage 171
drought resistance 76

E. Strain 293, 297, 314, 315, 316, 319, 320, 321
ecosystem 433, 440
ectomycorrhizae, see ectomycorrhizas 9, 303
ectomycorrhizas 1, 9, 15, 16, 44, 49, 50, 55, 75, 153, 247, 257, 263, 269, 275, 278, 300, 303, 309, 313, 319, 339, 345, 353, 363, 367, 415, 419, 423, 433, 440, 445, 455, 469, 478, 501, 507, 513
biomass 77, 97, 458, 460
carbohydrate 2, 415, 428–429, 440, 508
competitivity 9, 303
development 248, 254, 260, 311, 339, 340, 342, 355, 356–359, 428–429, 504, 508
distribution 18, 250, 265, 368
dynamics, see periodicity
ecology 272, 433, 440–441
effectiveness 248
nitrogen uptake 445
food base 263
fruiting bodies 247, 248, 257, 258, 259, 272, 320, 321, 369
germination 269, 272, 275
Hartig net 2, 4, 76, 296, 310, 363, 364, 435, 458
infection 19, 263, 266, 272, 325, 340, 419, 426, 428, 433, 435, 448, 505, 507, 509, 514
effect light 424, 425, 429
effect nitrogen 424, 425, 429
inoculation 247, 250, 252, 259, 271, 275, 278, 279, 303, 313, 314, 315, 319, 322, 325, 329, 343, 353, 355, 419, 424, 425, 426, 428, 433, 435
inoculum 259, 261, 278, 315, 424, 428
interplant connections 433, 435, 436, 441
mantle 2, 60, 76, 310, 364, 441
mycelial strands 263, 266, 275, 276, 278, 281, 433, 435, 441
periodicity 367, 507
pouch method 278

pure culture 247, 353
rhizomorph 263, 433, 436, 440–441, 509
sclerotium bodies 261, 275, 278, 281
sheathing species, see separate entries 9, 10, 20, 52, 247, 251, 257, 458
spores 272, 433, 436
sporocarp, see fruiting bodies
structure 433, 436, 438–439, 441
early stage 261, 263, 266, 269, 272
late stage 261, 263, 266, 269, 272
succession 20, 247, 248, 254, 255, 256, 263, 269, 319, 322, 514, 515
synthesis 5, 16, 19, 51, 52, 275, 276, 433
translocation 1, 223, 508
turnover 81, 82, 507
uptake 1, 276, 448
vessel hyphae 433, 436, 441
ectendomycorrhizas 293, 297
chlamydospores 293
endophyte 293
growth 294, 295
infection 296
races 293
species, see separate entries
synthesis 293, 294
Eleagnus umbellata 334
endodermis, see root
endoplasmic reticulum 151
endomycorrhizàe, see endomycorrhizas
endomycorrhizas 1, 5, 9, 23, 32, 33, 197, 198, 202, 211, 217, 223, 224, 309, 331, 363, 433, 442, 461, 463, 477, 483, 487, 508, 513
activity
litter decomposition 483
phosphate inflow 223, 238
phosphate release 461
photosynthesis 416
arbuscule 201, 202, 212, 214, 230, 309
biomass 198, 488
carbohydrate 198, 224, 231, 416
competitivity 331, 335, 336
development 223, 224, 488
ecology 488
entry points 234, 235, 238
exclusion mechanisms 332
growth depression 198
host, see also species 331
hydraulic conductivity 203
hyphae 199, 238, 263, 487, 489, 490–491
infection 33, 198, 220, 223, 224, 227, 230, 232, 233, 234, 237, 238, 241, 243, 465, 514
inocula 331, 343
inoculation 220, 331, 335, 465
interface 214
internal hyphae 201, 212, 230, 311
relative efficiency 199, 223, 331, 488
spores 32, 225, 226, 227, 230, 310, 312, 331, 332, 335, 343
spores: germination 219, 225
strategies 331
uptake 5, 32, 202, 203, 334
vesicles 311
Entoloma 16
enzymes, see also separate enzymes 2
ericoid mycorrhizas 285, 289, 290, 398, 433
endophyte 290
fungal sheath 290
infection 286
Eryobotria japonica 363
establishment 513, 515
ethanol 391
external hyphae 2, 199, 201, 223, 230, 233, 278, 279, 310, 487, 508, 509
Eucalyptus cupressus 363
E. globosa 363
exudation 3

Fagus 364–348
F. sylvatica 303
fertilizers 44, 49, 53, 309, 313, 329, 374, 403, 407, 409, 410, 416, 429, 447, 453, 514
phosphate 387, 453, 465, 466
field capacity 126
field root growth, see root systems
fine root, see root branching
Flammulina 16
Frankia 5
fumigation 309, 325, 327, 463, 515
fungal hyphae, see also ectomycorrhizas, endomycorrhizas 2, 76, 230, 289, 301, 487
fungicide 353
Fusarium oxysporum 203, 299, 301, 351

general yield class (GYC) 15
genetic manipulation 4, 5
geotropism 131, 135, 137, 490
giberellins 203
Gigaspora calospora 333
G. gigantea 333
G. margarita 217
Globularia glypum 363
Glomus 9, 309, 331
G. caledonium 228, 232, 235, 238, 239, 331, 333, 335

G. calosporum 333
G. clarum 225, 230, 331, 335
G. fasciculatum 333
G. geosporum 332, 333, 335
G. mosseae 32, 201, 211, 232, 331, 333, 335
G. tenuis 332, 333
glucose, see carbohydrate 404
glucosides 404
glutamic acid 408
glutamine 408
glycogen 198
gnotobiotic 301
golden delicious, see also apple 24
gramoxone 353, 355–361
grass 26, 28, 30
grazing 81
growth 223, 252, 331, 509
growth analysis 508, 509
growth regulator 161, 165, 198, 309, 513
growth ring 155, 159, 166, 167

Hagen – Poiseuille equation 441
Hartig net, see ectomycorrhizas
Hebeloma 16, 249, 250, 251, 257, 258, 259, 260, 264, 265, 270
H. cylindrosporum 303, 469, 471, 474, 501, 503
H. crustuliniforme 17, 248, 258, 261, 281, 305, 306, 307, 345, 346, 347, 348, 349, 445, 455, 459, 467, 469, 471, 474
H. fragilipes 17
H. leucosarx 17, 269, 270
H. mesophaeum 17
H. populinum 260
H. sacchariolens 17, 20, 247, 260, 261, 266, 270, 455
H. vaccinum 17
H. velutipes 247
hedges – cutting, see also cuttings 495, 496
Helianthus annuus 154
Helianthemum lippii 363
hemicellulose 403
herbicide 26, 28, 44, 353, 515
Herniaria fontanesii 363
Hippophaea rhamnoides 334
Hordeum vulgare 463, 466
hormones, see also individual hormones 4, 197, 290, 416, 429, 463
host response 507
Hyacinthoides nonscripta 332, 333
hydroxyl 467
Hygrophorus 16
Hymenogaster tener 16, 17
hyphae, see fungal hyphae

induced resistance 299, 301
infection, see ectomycorrhizas, endomycorrhizas
inflow, see nutrient
Inocybe 16, 249, 257, 258, 260, 269
I. lacera 269, 270
I. lanuginella 17, 248, 269, 270
I. petiginosa 17
instability 171
interfacial matrix 211, 212, 214
ion, see nutrient
iron 371, 375, 388, 390
irradiance 371, 376, 409, 424, 425, 426
irrigation 23, 165, 381

Juniperus phoenica 363
juvenility 131

Koch's postulate 275

Laccaria amethystina 313, 314, 320, 321
L. laccata 10, 16, 17, 247, 248, 249, 250, 252, 253, 257, 258, 259, 260, 270, 299, 301, 313, 314, 316, 320, 321
Lactarius 16, 265, 455
L. glyciosmus 17
L. hepaticus 304, 305
L. pubescens 17, 248, 250, 257, 258, 260, 261, 264, 266, 269, 270, 281, 455, 459
L. rufus 320, 321, 459
L. tabidus 320, 321
L. turpis 320, 321
leaf area 27, 31, 395, 397, 407, 409, 422, 423, 510
Leccinum 248, 257, 258, 260, 261, 269, 270
L. roseofracta 17
L. scabrum 17
L. subleucophaeum 17
L. versipelle 17
Ledum palustre 395, 397, 398
L. hepiota 16
L. hepista 16
light 413, 414
lipid 198
litter 478, 481

M factor 3
magnesium 15, 371, 375, 392, 463, 477, 482, 483
malate 391
Malus, see apple

manganese 15, 371, 375, 388, 390, 392
mankozeb 353
mantle, see ectomycorrhizas
Marasmius 16
  M. androsareus 458, 459
mechanical impedance, see also soil 41, 72
membrane permeability 4, 153, 202
meristem 289, 290, 296
microbial antagonism 272
microconidia 301
microflora, see also ectomycorrhizas, endomycorrhizas 299
micronutrient, see also elements 197, 202
micropropagation, see also axenic 217, 250, 257
microsite 116
mineral deficiency, see nutrient
mineral nutrition, see nutrient
mitochondria 150, 289
mixed inocula 331
modelling 172, 179, 185, 186, 189, 190, 223, 231, 234, 238, 381, 507, 511, 513, 514
mucigel layer 289
Mycena 16
  M. galopus 458, 480

Narcissus pseudonarcissus 333
negative binomial distribution 487
negative feedback 456, 458
neutral phosphatase 211, 214
NIF gene 5
nitrogen 1, 3, 4, 15, 40, 371, 415, 424, 482, 483
  ammonium 3, 45, 50, 375, 445, 448, 450, 452, 469
  content 477

open rooted 123
organic matter 477, 490–491, 509
outplanting 55, 111, 313
oxalate 404

paraquat 353
passage cells 147, 153
Paxillus involutus 250, 251, 252, 253, 269, 270, 272, 281, 314, 320, 321, 345, 346, 347, 459
peanut 5
pear 404
Peganum harmala 363
pesticide 354
petunia 9
Peziza badia 16, 17
Pezizella ericae 285, 290
pH 15, 212, 220, 375, 388, 445, 447, 463, 469

phellogen 11
phenolic 147, 149, 299, 301
    root 406
    shoot 420, 427, 428, 452, 453
  deficiency 375
  diurnal variation 378
  fertilizers, see fertilizers
  fixation 5, 197, 277, 309, 331, 333
  in solution 373
  isotopes 3
  nitrate 3, 15, 202, 375, 388, 445, 448, 450, 452, 464, 469
    inflow 375, 379
    root surface 374, 425
    soil solution 374, 424
  N/P ratio 429
  reserves 401, 409, 410
  seasonal veriation 379, 401, 403, 405, 407
  uptake 325, 328, 372, 402, 445
Noaea mucronata 363
'N Serve' 448
nuclear magnetic resonance 469–475
nursery practice 55, 59, 113, 124, 247, 297, 309, 313, 316, 319, 322, 330, 339, 343, 345, 346, 347, 353, 355
nutrient
  content 510
  cycling 2, 15, 75, 88, 331, 447, 478
  deficiency 63, 103, 153, 285, 363, 515
  exchange 211, 487
  ions 202, 371
    imbalance 387, 393
  supply 88, 160, 371, 402, 442, 507, 508
  transport 3, 149, 223, 442, 461, 508, 510
  uptake 5, 76, 97, 223, 328, 371, 372, 393, 402, 404, 442, 487, 508, 513
    per unit root 372, 510
Nyssa sylvatica 387

oak, see Quercus
onion, see Allium

Phleum nodosum 25, 30
phloem 159, 161, 165, 166, 401, 403
phloridzin 404
phosphatase, see also acid *etc.* 455, 456, 458, 461
phosphate 2, 15, 197, 199, 218, 371, 415, 455, 464
  application, see also fertilizer 335
  partitioning 199
  solution

concentration 378, 457, 510
depletion 372, 378
uptake 9, 31, 199, 223, 238, 325, 328, 372, 375, 378, 381, 383, 445, 458, 463
per unit root 375, 510
phosphorus
availability 469
in soil 202, 224, 387, 445, 456, 510
in solution 458
release by mycorrhizas 460
solubilization 456, 458, 461
content 234, 240, 242, 331, 375, 391, 452, 463, 471, 477, 482, 483, 513
leaf 33, 329, 392, 426, 428, 453
root 202
shoot 420, 453, 467
deficiency 5, 199, 200, 224, 429
fractionation 473
isotopes 381, 382, 469
metabolism 469
photosynthesis 127, 161, 166, 198, 276, 401, 415, 416, 419, 420, 425, 428–429, 440–441, 510, 514
Phylloporus rhodoxanthus 340, 342
phytase 199, 455, 458
Phytophthora 203

Picea glauca 55, 57, 75, 103, 277
P. mariana 55, 57, 111, 275, 277
P. omorika 167, 169
P. sitchensis 11, 15, 49, 147, 155, 167, 169, 172, 177, 185, 189, 247, 252, 253, 313, 319, 495
Pinus 275, 455
P. banksiana 55, 57, 111, 277
P. caribaea 123, 339
P. contorta 103, 415, 424, 426, 429, 455, 457
P. densiflora 325, 327
P. elliottii 297
P. halepensis 363, 364, 504
P. roraiensis 325
P. maritima 363, 364
P. patula 17, 249
P. pinaster 501
P. radiata 2, 5, 11, 37, 79, 123, 248, 314, 367
P. resinosa 55, 57, 79, 161, 293, 297
P. rigida 79, 325, 327
P. strobus 80, 277, 290, 297
P. sylvestris 3, 15, 79, 91, 97, 277, 356, 357, 358, 359, 426
P. taeda 14, 44, 387, 415, 416, 419, 428
P. thunbergii 325, 327
Pisolithus tinctorius 280, 315, 325, 340, 343, 415, 419, 425, 426, 501, 503
Pistacia lentiscus 363
planting deficiencies 169
planting density 24
plasmalemma 149, 153, 202, 212, 214
plasmodesmata 151, 289
Pleurotus 16
plum, pixy 29
st. Julien 29
Pluteus 16
polyphosphate 200, 469–475
degradation 200, 469
identification 469
NMR spectra 472–473
rinase 200
synthesis 200, 469
polysaccharides, see also hemicellulose 211, 214
fibrils 212
Populus 173, 279, 309
P. tremuloides 277, 281
potassium 15, 45, 202, 371, 375, 392, 401, 404, 409, 463, 477, 482, 483, 484
protoplasmic streaming 201
protoxylem 296
principal components analysis 381
proplastid 289
provenances 11, 55, 111, 114, 339
prunasin 404
pruning, see also root 162, 409
Prunus, see cherry, plum *etc.*
Psathyrella 16
Pseudoclitocybe 16
Pseudotsuga menziesii 78, 79, 80, 167, 169, 248, 249, 299, 445
pulse labelling 421, 423
Pyrus, see pear
Pythium 345
P. ultimum 345, 346, 348, 350
infectivity 345

Quercus 76, 266, 346, 347, 363
Q. ilex 363
Q. imbricaria 309
Q. petraea 15
Q. robur 131

rainfall 382
Ramaria 17
red clover, see also clover 230, 232, 234, 235
redox potential 388

relative growth rate 239, 341
relative multiplication rate 235
relative root growth 58
respiration 198, 387, 406, 415, 421, 422, 423, 428, 514
rhizatron, see root laboratory
Rhizoctonia solani 351
Rhizobia 198
rhizoplane 14, 20
Rhizopogon luteolus 340, 343
  R. nigrescens 340, 343, 344
  R. roseolus 2, 5
  R. rubescens 343
rhizosphere 445, 458, 463
  pH 448
Rhodotorula glutinis 269
ridge planting 12
Robinia 309
  R. hispida 334
  R. pseudoacacia 334
root
  absorption, see also nutrients and individual elements 23, 491
  anatomy
    bark 403
    cap 290
    cortex 3, 10, 52, 147, 230, 234, 286, 289, 296, 301, 310, 335, 426
    endodermis 147, 149, 364
    epidermis 230, 286, 289, 290
    membranes 393
    pericycle 149
    stele 149
  architecture 3, 158, 171, 186, 189, 190
  ball 179
  biomass 13, 14, 38, 49, 50, 56, 61, 75, 77, 78, 79, 87, 93, 95, 166, 167, 241, 397, 404, 407, 424, 425, 426, 428–429, 450, 477, 481
  branching 169, 176, 190, 491
    lateral roots 9, 11, 26, 46, 49, 60, 75, 87, 89, 93, 97, 98, 106, 131, 160, 176, 177, 181, 182, 190, 191, 220, 278, 293, 296, 314, 315, 316, 322, 325, 363, 395, 422, 424, 428, 477, 478, 501, 505, 507
  breakage 177, 179, 186
    carbohydrate, see also carbohydrate 415
    colonization 261, 266, 301, 345, 435
    density 383
      la 25, 26, 30, 41, 76, 442
      lv 12, 44, 234
  diameter 25, 60, 161, 165, 166, 173, 295, 296, 375, 396, 397
distribution 13, 14, 23, 37, 41, 63, 65, 67, 90, 94, 109, 155, 182, 477, 479–480, 487, 488
exudation 422
growth 11, 14, 23, 24, 25, 37, 41, 81, 89, 111, 159, 162, 167, 230, 232, 234, 253, 264, 269, 332, 397, 406, 407, 410, 477, 478, 487, 489, 501, 507, 509, 513
  extension 76, 77, 103, 167, 168, 182, 230, 504, 505
geometry, see root architecture
hairs 375, 442, 485
hydraulic conductance 44, 203
hypocotyl 276
initiation 4, 157, 495, 497, 499, 501, 502, 504
leakiness 387, 392
length 13, 25, 26, 30, 38, 50, 88, 125, 190, 234, 235, 242, 285, 286, 373, 389, 390, 489, 503, 505, 509
morphogenesis 167, 185, 203, 297, 364, 513
morphology 293, 309, 319, 320, 364, 435, 508, 509
necromass 4, 80, 87, 91, 93, 403, 481
number 50, 190, 234, 258, 264, 271, 295, 502
pathogens, see also pythium *etc.* 20
  protection, see biological control 301, 345
periodicity 26, 49, 57, 60, 75, 77, 88, 92, 97, 332, 396, 403, 406, 410, 477, 478
primary 131, 158, 169, 299, 363, 433, 514
pruning 39, 105, 123, 124, 127
regeneration 131, 502, 513, 514
resistance to uprooting, see also anchorage 155, 165, 171, 189, 194
seedling 261, 264, 266, 269, 271, 343, 345, 367
sinker 168, 182, 183
storage 402
strength 171, 174, 176, 182
  plastic deformation 173, 186
  plastic limit 173, 179, 186, 193
structural 13, 24, 39, 155, 175
  suberized 23, 24, 44, 81, 147
  woody 23, 24, 157, 161, 175, 178, 422, 423, 514
surface area 63, 69, 395, 397, 429
system 23, 38, 39, 111, 118, 139, 155, 163, 167, 230, 289, 293, 355, 361, 371, 477, 478, 508, 513
radial 167
tap 131, 135, 141, 168
tip 285, 289, 290, 326
turning moment 195
turgidity 126
volume 63, 167

weight, see biomass
weight/length relations 29, 43, 66
root growth capacity 103, 106, 107, 111, 120, 123, 124, 127
root laboratory 23, 135, 141, 276, 371
root promoting cofactor, see also root initiation 504
root : shoot relations 9, 10, 11, 60, 61, 162, 165, 224, 316, 328, 355, 390, 392, 395, 397, 398, 424, 426, 428–429, 508, 513, 514
root soil interface 445
root soil plate 175, 177, 178, 183, 185
rootstock 32, 33, 70, 217, 373, 409
Rosmarinus Tournefortii 365
Russula 16, 248
R. betularum 17, 266
R. emetica 320, 321
R. grisea 17, 266
R. ochroleuca 320, 321
R. versicolor 17, 266

Salix 279
S. planifolia 277
saprophytes 455
Scabiosa stellata 366
scions 371, 374, 409
Scleroderma bovista 340
S. granulatus 340, 342
S. texense 340, 343
seed germination 326, 390
seedlings 313, 322, 328, 356, 358, 359, 393, 419, 420, 428, 448, 513
death 347, 349, 448
Serpula lacrymans 439
shade trees 477, 482, 483
sheathing mycorrhizas, see ectomycorrhizas
shoot growth 59, 87, 105, 107, 115, 161, 166, 234, 295, 316, 324, 325, 335, 340, 342, 355, 388, 396, 397, 407, 419, 420, 422, 423, 426, 448, 450, 463, 509
shoot: root relationships, see root: shoot
sieve tube 166, 402
silvopastoral systems 367
Sitka spruce see *Picea sitchensis*
sodium 390, 463
soil
fauna 81, 88
infectivity 303, 345, 347, 348, 351
litter 477
moisture 14, 15, 25, 28, 30, 31, 71, 97, 126, 160, 165, 167, 275, 367, 388, 441
flow 31
oxygen 388
physical properties 381
bulk density 381
porosity 384
tensile strength 161, 171, 173, 175, 179, 185, 194
profile 13, 88, 91, 97,
heterogeneity 487, 492
type 313
soil management 26
sorbitol, see carbohydrate
sorghum 309
source-sink relations 415, 416
specific absorption rate 223
stability 171, 185
starch, see carbohydrate
stem
mechanical properties 171, 178, 182, 193
structure 9, 10
sterilization 360
Stipa tenacissima 363
stomatal conductance 420
stress, see transplanting, water *etc.*
sucrose, see carbohydrate
Suillus bovinus 433, 439
S. granulatus 340, 415, 425, 426
S. luteus 367, 455, 459
sulphate 15, 202, 203, 463, 464
summer pruning 409
survival 40, 105, 106, 115, 342, 343, 345, 448, 498
swecon 89, 91, 97
sycamore, see acer
symbiosis 197
symplast 392

Tamarix 363
tap root, see root
temperature 38, 294, 295, 297, 368, 371, 379, 402, 408, 495
Thelephora terrestris 3, 16, 17, 248, 250, 313, 314, 316, 319, 320, 321, 325, 328, 340, 342, 346
Thielaviopsis basicola 203
thinning 171
Tilia cordata 247
tissue nutrients, see elements
tomato 9
trace elements, see also elements 463
transfer cell 147, 152, 153
translocation, see also elements *etc.* 415, 436
transpiration, see also water 30, 125, 392
transplanting 38, 55, 250

Tricholoma 16
Triticum aestivum 463, 466
Tsuga heterophylla 167, 169
tundra 395
turnover, see also root 75, 80, 87
tussock 395

uprooting, see also anchorage 171, 179, 183, 185, 189

VA mycorrhizas, see endomycorrhizas
Vaccinium uliginosum 395, 397, 398
  v. vitis-idaea 91, 97
vacuolar polyphosphate, see also polyphosphate 200
vacuole 147, 149, 200, 289, 290
VAM, see endomycorrhizas
vegetative propagation 495, 514
vital stains 83
Vitis champini 63
  V. labruscana 63
  V. rotundifolia 63
  V. vinifera 63

water
  inflow 30, 31, 97, 197, 363, 441, 463, 508, 513
  potential 38, 46, 123, 126, 389, 390, 436, 441
  stress 123, 126, 161, 363
    mycorrhizas 15, 16, 203
    root 15
  transport, see also transpiration 149, 433, 436, 441
Waterlogging, see also anaerobic 131, 135, 141, 392
water table 131
weed competition 12
winching 171, 177, 179, 184
wind 172, 179, 186, 189, 192
wind throw 171, 175, 177, 184, 185, 189
wood, see root *etc.*
woodlands 331, 440

xylem 10, 124, 147, 159, 161, 403, 410, 442

yeast 458, 474

zinc 202, 371, 375, 387, 390